SAP S/4HANA® Sales Certification Guide

SAP PRESS is a joint initiative of SAP and Rheinwerk Publishing. The know-how offered by SAP specialists combined with the expertise of Rheinwerk Publishing offers the reader expert books in the field. SAP PRESS features first-hand information and expert advice, and provides useful skills for professional decision-making.

SAP PRESS offers a variety of books on technical and business-related topics for the SAP user. For further information, please visit our website: *www.sap-press.com*.

Aditya Lal
SAP Activate Project Management Certification Guide:
Certified Associate Exam
2021, 530 pages, paperback and e-book
www.sap-press.com/5194

Murat Adivar
SAP S/4HANA Business Process Integration Certification Guide:
Application Associate Exam
2021, 520 pages, paperback and e-book
www.sap-press.com/5238

Christian van Helfteren
Configuring Sales in SAP S/4HANA
2020, 766 pages, hardcover and e-book
www.sap-press.com/4907

James Olcott, Jon Simmonds
Sales and Distribution with SAP S/4HANA: Business User Guide
2021, 434 pages, hardcover and e-book
www.sap-press.com/5263

Lauterbach, Sauer, Gottlieb, Sürie, Benz
Transportation Management with SAP: Embedded and
Standalone TM (3rd Edition)
2019, 1054 pages, hardcover and e-book
www.sap-press.com/4768

Darío Franco, Jon Simmonds

SAP S/4HANA® Sales Certification Guide

Application Associate Exam

Editor Meagan White
Acquisitions Editor Emily Nicholls
Copyeditor Yvette Chin
Cover Design Graham Geary
Photo Credit Shutterstock.com: 95096560/© Sofiaworld
Layout Design Vera Brauner
Production Graham Geary
Typesetting SatzPro, Krefeld (Germany)
Printed and bound in Canada, on paper from sustainable sources

ISBN 978-1-4932-2124-0

1st edition 2021

Library of Congress Cataloging-in-Publication Data:
Names: Franco, Dari´o, author. | Simmonds, Jon, author.
Title: SAP S/4HANA sales certification guide : application associate exam / by Dari´o Franco and Jon Simmonds.
Description: 1st edition. | Bonn ; Boston : Rheinwerk Publishing, [2021] | Includes index.
Identifiers: LCCN 2021029322 | ISBN 9781493221240 (hardcover) | ISBN 9781493221257 (ebook)
Subjects: LCSH: SAP HANA (Electronic resource)--Examinations--Study guides. | Selling--Data processing--Examinations--Study guides. | Customer relations--Data processing--Examinations--Study guides.
Classification: LCC HF5438.35 .F73 2021 | DDC 658.850285/53--dc23
LC record available at https://lccn.loc.gov/2021029322

Contents at a Glance

Dear Reader,

Standardized testing has changed since the last time I had to sit down to take one. Gone are the days of waking up at 6:00 am to drive 30 minutes or more to the closest testing center. Gone are the days of shambling sleepily into a large auditorium with 100+ of your fellow test-takers. Gone are the days of sharpening your favorite brand of no. 2 pencil as you wait for the test to begin. Instead, these days you can sit down at home—hopefully a little better rested—turn on your webcam, show the room to your proctor, and then get started. What a luxury!

What hasn't changed are the steps you should take to prepare yourself. First, get a good night's sleep. Next, eat a hearty breakfast. Finally, in the weeks and months leading up to your exam: review, review, review. While this book won't make you breakfast in the morning or tuck you into bed, it does contain all the information you need to pass your exam. From key concept reviews and terminology definitions to practice questions with detailed answers, this book will lead you through every exam topic.

What did you think about *SAP S/4HANA Sales Certification Guide: Application Associate Exam?* Your comments and suggestions are the most useful tools to help us make our books the best they can be. Please feel free to contact me and share any praise or criticism you may have.

Thank you for purchasing a book from SAP PRESS!

Meagan White
Editor, SAP PRESS

meaganw@rheinwerk-publishing.com
www.sap-press.com
Rheinwerk Publishing · Boston, MA

Contents

Preface

The SAP PRESS Certification Series is designed to prepare readers to take an SAP certification exam with all of the review, insight, and practice they need to pass the exam. The series is written in practical, easy-to-follow language that provides targeted content focused on what you need to know to successfully pass the exam.

Target Audience

This book is specifically written for readers preparing for one of the following SAP Certified Application Associate exams:

- C_TS462_2020: SAP Certified Application Associate – SAP S/4HANA Sales 2020
- C_TS460_2020: SAP Certified Application Associate – SAP S/4HANA Sales 2020 Upskilling

If you've purchased this book, you're obviously interested in learning how to successfully pass one or both certification exams. We'll focus on core knowledge of the SAP S/4HANA Sales profile to help you become an SAP Certified Application Associate – SAP S/4HANA Sales.

By reading this book, you'll walk away with a thorough understanding of the structure of both exams and what to expect when taking them. You'll receive a refresher on key concepts covered in the exams, and you'll be able to test your skills via practice questions and answers. This book is closely aligned with the course syllabus and the exams' structure, so all the information provided is relevant and applicable to test preparation. In the next section, we'll explain how the book is structured to correspond to each exam.

We'll also explain SAP products and features through practical examples and straightforward language, so you can prepare for the exams and improve your skills for your day-to-day work as an SAP S/4HANA Sales consultant. Each book in the series has been structured and designed to highlight what you really need to know.

Structure of This Book

Each chapter begins with a clear list of learning objectives, for example, the following section.

Techniques You'll Master

- How to prepare for the exam
- Understand the general structure of the exams
- Practice questions and preparation

From this list of objectives, you'll dive into the chapter and get right into the test coverage.

Throughout the book, we've also provided several elements that will help you access useful information:

- *Tips* call out useful information about related ideas and provide practical suggestions for how to use a particular function.
- *Notes* provide other resources to explore or special tools or services from SAP that will help you with the topic under discussion.
- *Examples* contain concrete illustrations of concepts in action.
- *Warnings* draw your attention to points that often cause confusion or are ignored.

The following boxes are examples of these elements.

Tip

This book contains screenshots and diagrams to help your understanding of the numerous information modeling concepts.

Note

This certification guide covers all the topics you need to successfully pass the exam. It provides sample questions similar to those found on the actual exam.

Example

This book contains real-word examples that show the concepts in action.

Warning

This book is meant to review and supplement the knowledge gained from standard SAP Training and Adoption training.

Each chapter covers an exam topic and is organized in a similar fashion, helping you become familiar with the structure of the exams and to easily find the information you need. Here's an example of a typical chapter structure:

- **Techniques You'll Master**
  The beginning of each chapter discusses the techniques you must master to be considered proficient in the topic for the certification examination.

- **Introduction**
  This section provides a general idea of the topic at hand to frame future sections. It also includes objectives for the exam topic covered.
- **Real-World Scenario**
  This part illustrates a scenario in which these skills would be beneficial to you or your company.
- **Objectives**
  This section provides the information necessary to successfully pass this portion of the test.
- **Key Concept Refreshers**
  These sections outline the major concepts of the chapter. They identify the tasks you'll need to understand or perform properly to answer the questions on the certification examination.
- **Important Terminology**
  Just before the practice examination questions, we'll review important terminology, including definitions of various terms from the chapter.
- **Practice Questions**
  The chapter then provides a series of practice questions related to the topic of the chapter. These questions are structured in a similar way to the actual questions on the certification examination.
- **Practice Question Answers and Explanations**
  Following the practice questions section, we'll discuss their solutions. As part of the answer, we'll explore why an answer option might be correct or incorrect. Each question is explained, and you'll get detailed justifications about the right and wrong answers.
- **Takeaway**
  This section provides a takeaway or reviews the areas you should now understand.
- **Summary**
  Finally, we'll conclude each chapter with a summary.

Now that you have an idea of how the book is structured, the following list will dive into the individual topics covered in each chapter:

- **Chapter 1: SAP S/4HANA Innovations and Analytics**
  This chapter introduces you to key principles in SAP S/4HANA, including important user interface (UI) changes and the concepts of embedded analytics using standardized SAP Fiori apps, SAP Smart Business, and core data services (CDS) within the virtual data model (VDM). This chapter also covers the use of extended warehouse management as a fully functioning tool within SAP S/4HANA. Additionally, this chapter will help test takers understand the roles and responsibilities of internal sales representatives within SAP S/4HANA, the functions they carry out, and how SAP S/4HANA is designed to assist them in their tasks.

- **Chapter 2: Organizational Structures**
  This chapter covers the meaning of and customizing settings for the various organizational units relevant for sales processes.
- **Chapter 3: Master Data**
  This chapter reviews some of the most important master data records for the sales process, such as material master data, customer-material information records, and master data for pricing.
- **Chapter 4: Sales Document Customizing**
  This chapter reviews customizing settings for sales document types, item categories, and schedule line categories.
- **Chapter 5: Basic Functions Customizing**
  This chapter reviews some of the basic functions within the customizing of sales and distribution, such as incompletion procedures, partner determination, free goods, and material determination.
- **Chapter 6: Cross-Functional Customizing**
  This chapter covers customizing settings that are relevant for the whole sales process chain. Among other topics, the settings for text control, copy control, and output determination are included in this chapter.
- **Chapter 7: Availability Check and Advanced ATP**
  This chapter reviews the central approach for performing material availability checks in sales documents as well as reviews key features in advanced available-to-promise (ATP).
- **Chapter 8: Pricing**
  This chapter reviews the relevant customizing and information records for using the condition technique for pricing.
- **Chapter 9: Sales Process**
  This chapter reviews the sequence of typical and special business processes in sales as well as describes some helpful tools within the application (e.g., document flows).
- **Chapter 10: Shipping Process and Customizing**
  This chapter reviews the main features of the shipping process (e.g., creation of the delivery document, picking, and posting of the goods issue) and the corresponding customizing settings.
- **Chapter 11: Billing Process and Customizing**
  This chapter reviews the meaning of the major billing document types and the corresponding customizing settings.
- **Chapter 12: Simplifications**
  Every SAP S/4HANA release comes with a list of simplifications. This chapter reviews how the simplifications can be put to good use, focusing on the key areas of business partners, analytics, business rule frameworks, and amendments to the sales data structure.

- **Chapter 13: Next-Generation Technologies**
  This chapter reviews SAP Business Technology Platform (SAP BTP), which encompasses artificial intelligence (AI), robotic process automation, machine learning, big data, and the native SAP chat bot, SAP CoPilot.

Most chapters review topics required for both the C_TS462_2020 and the C_TS460_2020 exams, whereas a minority of chapters cover areas that are only relevant to the C_TS460_2020 exam. Table 1 provides a roadmap for each exam, so you know what chapters and/or sections of this book can be skipped if you're only preparing to take the C_TS462_2020 exam. Should you wish to take the C_TS460_2020 exam, then you should read all the chapters and understand all the topics reviewed in this book.

Thus, your first decision should be determining which certification exam is the right one for you. The SAP S/4HANA Sales 2020 exam is an excellent certification for consultants who will be working on SAP S/4HANA support and ongoing maintenance. We would recommend that, for consultants who are embarking upon SAP S/4HANA greenfield implementations, the SAP S/4HANA Sales 2020 Upskilling certification is probably a better fit, as this certification will give you a more rounded knowledge of the innovations and leading-edge vision of SAP S/4HANA.

As shown in Table 1, if a chapter has a checkmark ✓, the whole chapter is relevant for that certification exam. On the other hand, the × indicates the whole chapter can be skipped since the content reviewed in that chapter is not needed to pass the exam. If a chapter shows the titles of specific sections within the chapter, then only those sections are relevant and the other sections can be ignored when preparing for the specific exam.

Chapter Number	Chapter Title	C_TS462_2020	C_TS460_2020
1	SAP S/4HANA Innovations and Analytics	SAP Fiori User Experience (Section 1.2) Sales Analytics (Section 1.6)	✓
2	Organizational Structures	✓	✓
3	Master Data	✓	✓
4	Sales Document Customizing	✓	✓
5	Basic Functions Customizing	✓	✓
6	Cross-Functional Customizing	✓	✓
7	Availability Check and Advanced ATP	✓	✓

Table 1 Chapters and Sections Relevant to Each Certification Exam

Chapter Number	Chapter Title	C_TS462_2020	C_TS460_2020
8	Pricing	✓	✓
9	Sales Process	✓	✓
10	Shipping Process and Customizing	✓	✓
11	Billing Process and Customizing	✓	✓
12	Simplifications	Business Partners (Section 12.3) Output Using BRFplus (Section 12.5)	✓
13	Next-Generation Technologies	×	✓

Table 1 Chapters and Sections Relevant to Each Certification Exam (Cont.)

Practice Questions

Before we really get started, we want to give you some background on the test questions before you encounter the first few sample question in the chapters themselves. Just like in the exam, each question conforms to the following basic structure:

- **The actual question**
  Read the question carefully and be sure to consider all the words used in the question because they can impact the answer.
- **Question hint**
  Not a formal term, we call it a *hint* because we'll tell you how many answers are correct. In the exam, you'll always see a *Note* after the question indicating the number of correct answers. In this book, if only one answer is correct, normally you'll be asked to choose the correct answer. If more than one is correct, then, as in the actual certification examination, the hint will indicate the correct number of answers.
- **Answers**
  The answers to select from depend on the question type. The following question types are possible:
  - *Multiple response*: More than one correct answer is possible.
  - *Multiple choice*: Only a single answer is correct.
  - *True/false*: Only a single answer is correct. These types of questions are not used in the exam but are used in this book to test your understanding.

Summary

With this certification guide, you'll learn how to approach the content and key concepts highlighted for each exam topic. In addition, you'll have the opportunity to practice with sample test questions in each chapter. After answering the practice questions, you can review the explanations of the answers, which dissect the questions by explaining why certain answers are correct or incorrect. The practice questions provide insight into the types of questions you can expect, what the questions look like, and how the answers relate to the question. Understanding the composition of the questions and seeing how the questions and answers work together is just as important as understanding the content. This book gives you the tools and understanding you need to be successful. Armed with these skills, you'll be well on your way to becoming an SAP Certified Application Associate in SAP S/4HANA Sales.

Acknowledgments

The authors would like to collectively thank the tireless editorial team of Emily Nicholls and Meagan White for all their hard work on this project. This final product would not have been possible without your professionalism, attention to detail, flexibility, and understanding in the face of all the challenges of a disruptive year battling our way through a global pandemic.

Darío would first like to sincerely thank James Olcott and his coauthor Jon for their generosity in allowing him to access a generic and already configured SAP S/4HANA box for this book. A kind gesture that shows their human quality.

Darío's special thanks also go to his coauthor and friend Jon, for introducing me to Emily Nicholls and making this dream come true. Darío will always be grateful for his support and invaluable contributions. Jon has demonstrated a relentless commitment to this project while writing another SAP PRESS book and working full time. Darío could have not asked for a better partner in crime.

Last but not the least: Darío would like to thank his darling wife Laura for all the support he needed to keep himself going in difficult times. And to his two little angels Oscar and Lucas, who always put a smile on his face after long days of work and writing.

Jon would like to thank, first and foremost, Darío, for his exceptional sales and distribution knowledge and for magically juggling full-time work, writing a book, and parenting two very young children in a pandemic, while also remaining patient and calm throughout the process.

And finally, Jon must thank his loving support crew at home: Susan, Ella, and Charlie—for never complaining when Jon took up evenings, weekends, and "days off" to write—thank you!

Introduction: The Path to Certification

Techniques You'll Master

- Understand the relevant certification offerings for sales and distribution in SAP S/4HANA 2020
- Find the courses required for certification
- Learn techniques for taking the certification exams
- Identify further relevant SAP Training and Adoption offerings in SAP S/4HANA
- Explore additional resources for sales and distribution in SAP S/4HANA
- Expand your knowledge and keep your skills up to date

To begin your journey into SAP S/4HANA Sales certification, you should understand the different levels and offerings available for certification. To build on this certification information, we'll identify additional key resources that should be helpful for expanding your overall understanding. Initially, we'll uncover whether the SAP S/4HANA Sales 2020 Certification exam or the SAP S/4HANA Sales 2020 Upskilling Certification exam is the right certification for you. As a part of this exploration, we'll go over the certifications in detail and describe the SAP Training and Adoption materials that are recommended to achieve the certification.

Finally, we'll explore some additional resources that are available to you when preparing for the exam.

Audience for This Book

Sales processes are fundamental to almost every SAP S/4HANA implementation project, just as they were for SAP ERP implementations. As a result, a certified SAP S/4HANA Sales consultant with a solid background in SAP ERP and SAP S/4HANA is a much sought-after resource for both SAP partner firms and SAP customers.

You'll find that this book covers many in-depth sales configuration and business processes found within the SAP S/4HANA system. For this reason, this book can be a good starting point for those of you just getting introduced to sales processes in SAP. For example, you might be a new hire at an SAP implementation partner, an IT support consultant whose company is upgrading to SAP S/4HANA, or even a recent graduate looking to begin a career in SAP.

This book is also a good guide for more experienced consultants to achieve a few things. First, this book can refresh an experienced consultant's knowledge in the wider sales-specific scope of an SAP S/4HANA system. Perhaps, after working many years in one organization or industry sector, some aspects of the sales scope of SAP S/4HANA are not regularly used, and the consultant wants to refresh his/her memory and knowledge in these areas.

Second, the book is an excellent jumping-off point for SAP ERP 6.0 consultants who want to upskill themselves in the next-generation SAP ERP system, SAP S/4HANA.

Third, the book can be useful for experienced consultants and developers who wish to broaden their existing knowledge starting from their own functional and technical areas of expertise, into the sales and distribution area of SAP S/4HANA.

Finally, and most clearly, this book is designed with the certifications for SAP S/4HANA Sales in mind and, therefore, are ideal companions for guiding you through the required knowledge and expertise for passing those certification exams on your journey towards becoming a fully fledged and certified SAP S/4HANA Sales consultant.

The content contained in this book is specific to the most recent SAP S/4HANA release and the most recently available sales certifications. Importantly, while SAP S/4HANA's functionality evolves as each new version is released, general concepts and principles tend to remain static, and therefore, the knowledge found in this book should support future versions of the product too.

Certifications for SAP S/4HANA Sales

SAP has a wide variety of certifications, covering every functional and technical area of the SAP S/4HANA system. All certifications mentioned in this book have the same format:

- Length of exam: Up to 3 hours
- Exam questions: 80 questions, either multiple choice or multiple selection
- Exam location: Can be taken either at a certification center or online

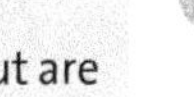

Note

Certification centers have been closed for the duration of the COVID-19 pandemic but are expected to reopen as restrictions are eased.

In this section, we'll go into detail about the online certifications covered within the scope of this book. We'll also discuss other related certifications, which, although not covered in this book, may be of interest to you as a student of SAP S/4HANA Sales.

As you'll see in the following sections, two certification exams are fully covered by the content of this book, namely, the following exams:

- C_TS462_2020: SAP Certified Application Associate – SAP S/4HANA Sales 2020, available at *https://training.sap.com/certification/C_TS462_2020/*
- C_TS460_2020: SAP Certified Application Associate – SAP S/4HANA Sales 2020 Upskilling, available at *https://training.sap.com/certification/C_TS460_2020/*

Within this book, we'll refer to these separate certifications as the SAP S/4HANA Sales 2020 Certification exam and the SAP S/4HANA Sales 2020 Upskilling Certification exam. Some content overlaps in these two certification exams, but we'll look at each in turn.

Online Certification

Online certification is a popular way to take these exams, and currently, two methods exist for taking these exams online:

1. One attempt: This option is a single, one-off fee for one attempt at the exam.
2. Certification Hub subscription: This option is a subscription that allows six attempts for any certification exam within a 12-month period. Your subscription is valid for all available exams in the Certification Hub, and you can take the same exam up to a maximum of three times in that period.

The format of the online exam is extremely flexible—times can be chosen from a 24/7 calendar on extremely short notice. You'll need a quiet, dedicated work area where you can take the exam, your own laptop with an integrated webcam, and a clear desk. To ensure the integrity of the certification, the exam will be fully proctored, and the proctor will need to see you via webcam for the duration. You'll receive instructions in advance of the exam, which will include downloading a secure browser called Questionmark. You'll need a photo ID to show the proctor, and the proctor will also expect you to show them your entire working area, so be sure to clear away any notebooks.

Once the certification exam is completed, either online or at a certification center, you'll receive the result immediately once your finalized answers are submitted. Successful examinees receive an SAP Global Certification digital badge for their accomplishments.

SAP S/4HANA Sales 2020 Certification Exam

This certification is the most common starting point for SAP sales professionals. Much of the content in this exam (but crucially, not all) is based on the same content as the SAP ERP 6.0 certification tracks. Therefore, if you're already an experienced SAP ERP consultant, you should find little difficulty using this book to add some of the missing SAP S/4HANA-specific details required to complete your successful certification.

This certification has an annual release cycle, reflecting the release of each new version of the SAP S/4HANA system. Note, however, that the certification exam is often published long after the system version is released. For example, the SAP S/4HANA 2020 system was released in September 2020, but the 2020 version of the SAP S/4HANA Sales 2020 Certification exam was not released until May 2021. As mentioned earlier, because core customizing and functionality are covered in each of the different version exams, your certification will not be obsolete as soon as a new version is released. The same goes for the validity of this book as a reference and for future releases of the certification.

As shown in Figure 1, which is a screenshot from the SAP Training web shop, the certification exam is an associate-level exam. In other words, the assumption is that the examinee already has some hands-on experience in a role relevant to sales, using SAP ERP 6.0 or SAP S/4HANA, and has good knowledge of the prerequisite education courses.

Figure 1 SAP S/4HANA Sales 2020 Exam Description and Notes

The certification exam consists of 80 questions of the following types:

- Multiple choice, for which you must select one correct answer from the available options
- Multiple selection (type 1), for which you must select the 2 correct answers out of the 4 available options
- Multiple selection (type 2), for which you must select the 3 correct answers out of the 5 available options

For the multiple selection questions, you'll be told how many answers are correct.

Only these question types exist on the exam. Note that, for multiple selection questions, you must get all the answers correct for the question to be considered correctly answered. Questions will be presented to you in a randomized order, so topics are mixed up throughout the exam.

To pass the SAP S/4HANA Sales 2020 Certification exam, you must answer 63% of the questions correctly, meaning that you'll need a total score of at least 50 out of 80. Typically, the percentage pass rate may change from year to year for the certification, depending upon the perceived difficulty rating of the exam. Once you pass, you'll be issued with a digital badge, shown in Figure 2, which can be viewed at *www.credly.com*.

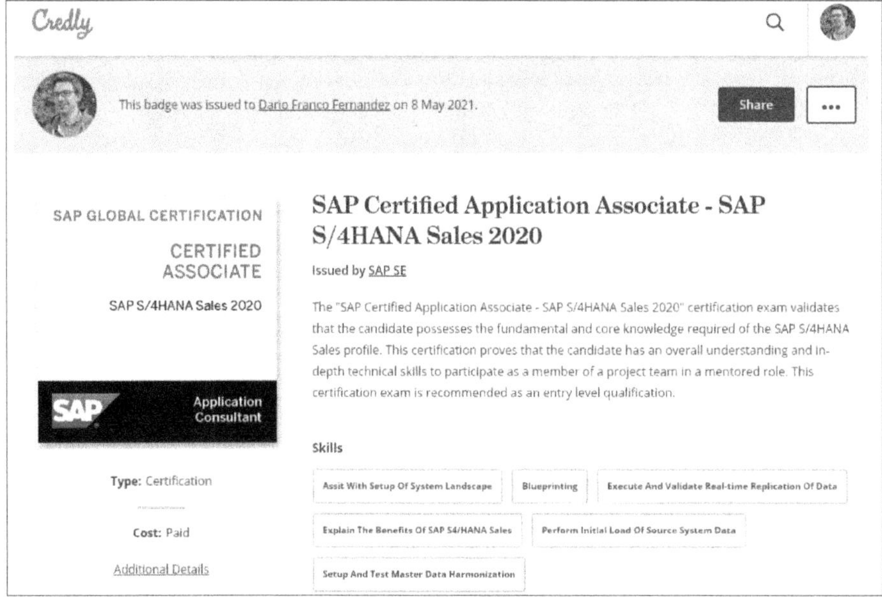

Figure 2 SAP S/4HANA Sales 2020 Certification Digital Badge

As mentioned earlier, you can retake an exam twice, for a total of three attempts per certification exam. If you fail an exam, don't panic! As soon as you can, jot down what you can remember about the exam for future reference. This step is important because the questions that you'll encounter when you retake the exam will be identical (albeit in a randomized order).

The scope of the exam is shown in Figure 3. This book replicates these scope areas for ease of use, with one small variation. Since this book covers both the SAP S/4HANA Sales 2020 Certification exam as well as the SAP S/4HANA Sales 2020 Upskilling Certification exam, the analytics portion is covered in Chapter 1, Section 1.6. This organizational choice is because the topic of analytics is more pertinent to the SAP S/4HANA Sales Upskilling exam, so this topic is covered with that content. However, SAP S/4HANA Sales 2020 Certification examinees should be aware that analytics is covered in their certification exam too.

These topics relate to a number of questions in the exam and the content in the book, as shown in Table 1. Notice that the chapters in the book are not in the same order as detailed in the certification description shown in Figure 3. First, as mentioned earlier, the questions on the exam are presented in randomized order and therefore will not fit any given sequence. Second, the sequence of topics as presented in this book fits a natural flow—conceptual issues are covered in Chapter 1, followed by the organizational structures and data setup, then a chronological sequence of SAP documents.

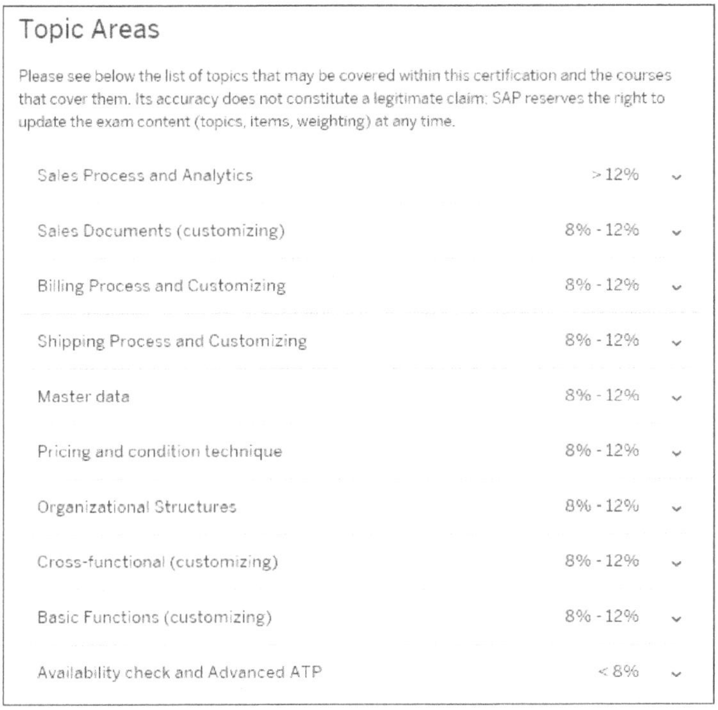

Topic Areas

Please see below the list of topics that may be covered within this certification and the courses that cover them. Its accuracy does not constitute a legitimate claim; SAP reserves the right to update the exam content (topics, items, weighting) at any time.

Sales Process and Analytics	> 12%
Sales Documents (customizing)	8% - 12%
Billing Process and Customizing	8% - 12%
Shipping Process and Customizing	8% - 12%
Master data	8% - 12%
Pricing and condition technique	8% - 12%
Organizational Structures	8% - 12%
Cross-functional (customizing)	8% - 12%
Basic Functions (customizing)	8% - 12%
Availability check and Advanced ATP	< 8%

Figure 3 SAP S/4HANA Sales 2020 Exam Topics

Topic	Percentage (%)	Approximate Number of Questions	Chapter
Sales Process and Analytics	>12	10	1 and 9
Sales Documents Customizing	8–12	8	4
Billing Process and Customizing	8–12	8	11
Shipping Process and Customizing	8–12	8	10
Master Data	8–12	8	3
Pricing and Condition Technique	8–12	8	8
Organizational Structures	8–12	8	2
Cross-Functional Customizing	8–12	8	6
Basic Functions Customizing	8–12	8	5
Availability Check and Advanced ATP	<8	6	7

Table 1 SAP S/4HANA Sales 2020 Certification Topic Areas Weighting

SAP S/4HANA Sales 2020 Upskilling Certification Exam

The SAP S/4HANA Sales 2020 Upskilling exam description, as shown in Figure 4, is suitable for existing sales consultants who already have previous experience in sales and distribution in SAP ERP 6.0. The difference, therefore, between this certification and the SAP S/4HANA Sales 2020 certification is that the Upskilling exam is designed for existing sales and distribution consultants who wish to upskill by becoming familiar with the new sales functions of SAP S/4HANA 2020.

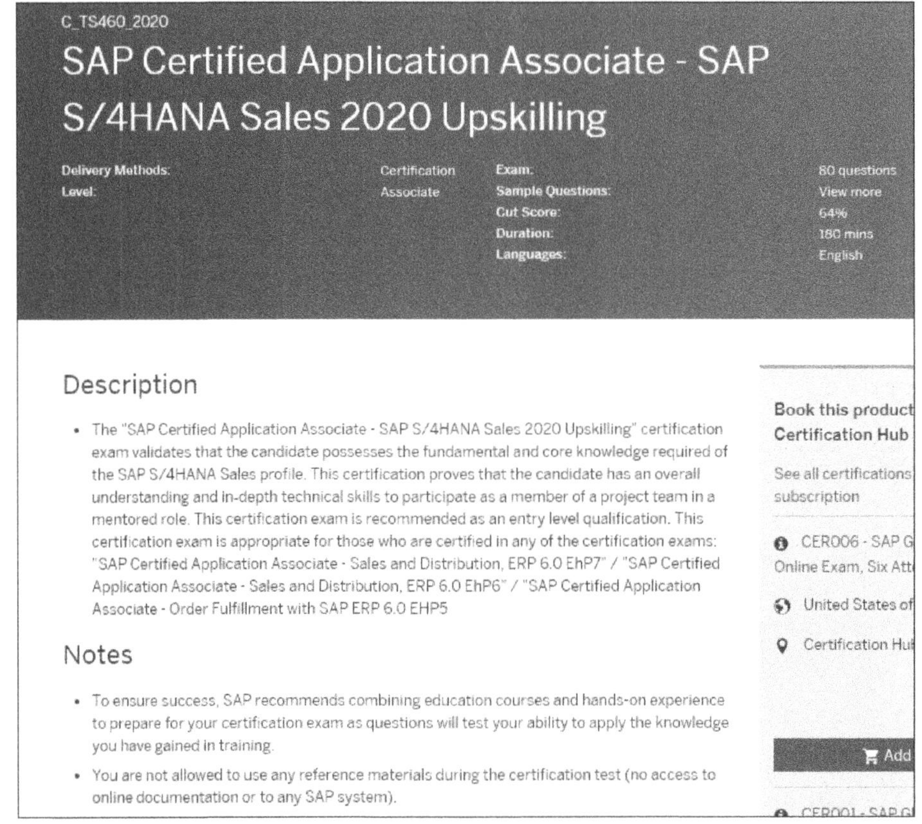

Figure 4 SAP S/4HANA Sales 2020 Upskilling Exam Description and Notes

Warning

The SAP S/4HANA Sales 2020 Upskilling exam contains much the same content as the SAP S/4HANA Sales 2020 exam, so don't be fooled into thinking that the exam only contains SAP S/4HANA-specific content. You'll need a full appreciation of the entire SAP S/4HANA Sales suite to tackle the SAP S/4HANA Sales Upskilling exam.

The format of the exam is quite similar to the SAP S/4HANA Sales 2020 certification exam, with a slightly higher passing threshold—64% instead of 63%, meaning you must answer 51 out of 80 questions correctly.

The topic areas covered in the SAP S/4HANA Sales Upskilling Certification exam are shown in Figure 5.

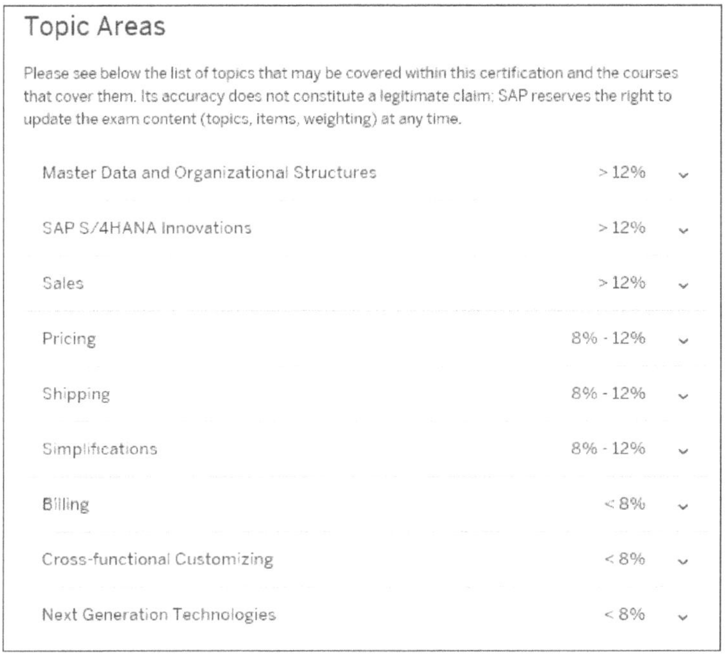

Topic Areas

Please see below the list of topics that may be covered within this certification and the courses that cover them. Its accuracy does not constitute a legitimate claim; SAP reserves the right to update the exam content (topics, items, weighting) at any time.

Master Data and Organizational Structures	> 12%
SAP S/4HANA Innovations	> 12%
Sales	> 12%
Pricing	8% - 12%
Shipping	8% - 12%
Simplifications	8% - 12%
Billing	< 8%
Cross-functional Customizing	< 8%
Next Generation Technologies	< 8%

Figure 5 SAP S/4HANA Sales 2020 Upskilling Topic Areas

The weighting for the questions in the exam and the related chapters in the book by topic are shown in Table 2.

Topic	Percentage (%)	Approximate Number of Questions	Chapter
Master Data and Organizational Structures	>12	12+	2 and 3
SAP S/4HANA Innovations	>12	12+	1
Sales	>12	12+	4 and 9
Pricing	8–12	8	8
Shipping	8–12	8	10
Simplifications	8–12	8	12
Billing	<8	6	11
Cross-Functional Customizing	<8	6	6
Next-Generation Technologies	<8	6	13

Table 2 SAP S/4HANA Sales 2020 Upskilling Topic Areas Weighting

When comparing the weighting of topic areas for the SAP S/4HANA Sales 2020 Upskilling exam with the SAP S/4HANA Sales 2020 exam, you should be aware of a few things.

First, while the topic headings of the two certifications might be quite similar, you'll notice some differences. For instance, the topic "Availability Check and Advanced ATP" is not included in the SAP S/4HANA Sales 2020 Upskilling Certification exam topic areas. Don't be fooled by this apparent omission! The content related to this topic has been rolled into the "Shipping" topic area. As a result, while we've made our best efforts to align this book as closely as possible with the topic areas of both certification exams, mirroring them both precisely has not been possible.

Second, notice that fewer topic areas exist in the SAP S/4HANA Sales 2020 Upskilling certification exam. Again, this difference is simply because some topics have been combined.

Finally, be aware that the first three topics—Master Data and Organizational Structures, SAP S/4HANA Innovations, and Sales—all have weightings greater than 12%. Thus, some simple math shows us that the weighting of these topics is greater than the weighting of the "less than 12%" topics in the SAP S/4HANA Sales 2020 exam. As a result, expect more questions in these areas.

As with the SAP S/4HANA Sales 2020 exam, passing the certification will provide you with a digital badge from *www.credly.com*, which you can add to your social media accounts. An example of the SAP S/4HANA Sales 2020 Upskilling digital badge is shown in Figure 6.

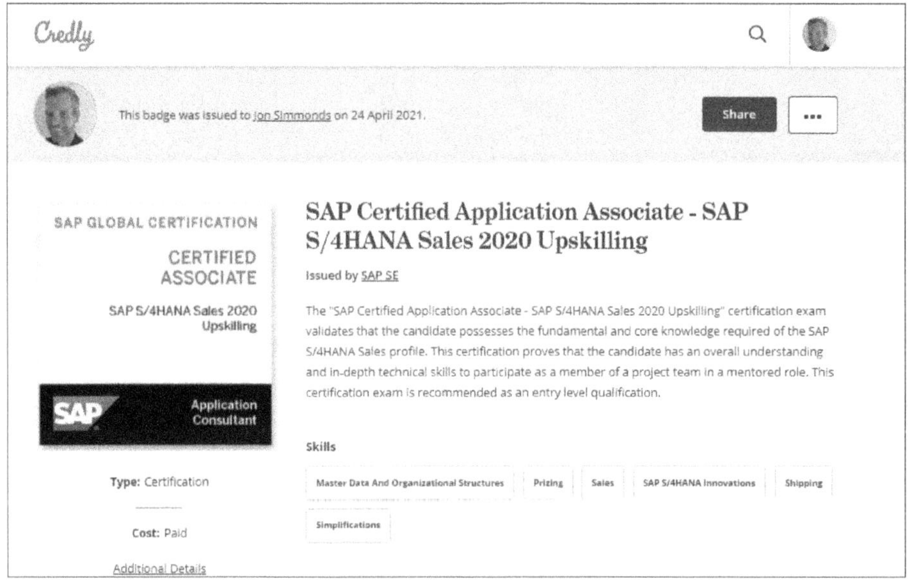

Figure 6 SAP S/4HANA Sales 2020 Upskilling Digital Badge

Related Certifications

Many other certifications are interrelated and may be of some interest to aspiring SAP S/4HANA Sales consultants. Four of the more useful certifications related to the SAP S/4HANA Sales area include the following:

- C_ACTIVATE13: SAP Certified Associate – SAP Activate Project Manager (*https://training.sap.com/certification/C_ACTIVATE13/*). This certification builds upon basic consultancy skills in any functional area and extends these skills into the project management space. This certification exam tests communication and presentation skills, as well as the candidate's knowledge of the SAP agile methodology, known as SAP Activate (more on this topic in Chapter 1, Section 1.5).
- C_S4EWM_1909: SAP Certified Application Associate – Extended Warehouse Management with SAP S/4HANA (*https://training.sap.com/certification/C_S4EWM_1909/*). As you'll learn in Chapter 1, Section 1.8, some knowledge of the strategic logistics solution SAP Extended Warehouse Management (SAP EWM) is crucial to becoming an SAP S/4HANA Sales consultant. Candidates interested in this area would do well to follow this certification track as a useful add-on to their sales experience.
- C_TM_95: SAP Certified Application Associate – SAP Transportation Management 9.5 (*https://training.sap.com/certification/C_TM_95/*). Consultants who find themselves applying their knowledge predominantly in the logistics area may find they have exposure to the related SAP functionality in SAP Transportation Management (SAP TM). This certification track is recommended as an add-on to an SAP S/4HANA Sales certification rather than a standalone certification.
- C_TS410_2020: SAP Certified Application Associate – Business Process Integration with SAP S/4HANA 2020 (*https://training.sap.com/certification/C_TS410_2020/*). This certification is a really useful starting point for SAP S/4HANA consultants and provides a sound platform of knowledge around key business process fundamentals in the system. Overall, this certification is a good place from which to begin your certification journey.

SAP Training and Adoption Courses

Several courses can feed into your learning and assist you with passing the certification exams. Typically, these courses would normally be offered as classroom-based training. However, since the global COVID-19 pandemic in 2020 and 2021, these courses have been listed as virtual courses only. We do anticipate that classroom training will return once pandemic restrictions are lifted.

SAP S/4HANA Sales 2020 Certification Courses

The list of courses offered in this section contain all the knowledge you need to sit and pass the SAP S/4HANA Sales 2020 Certification exam. These courses are enough for a beginner consultant to gain the knowledge required to sit the exam. Nevertheless, we recommend that certification candidates also have some hands-on experience in the SAP S/4HANA 2020 system before attempting the exam.

We recommend the following courses to help you prepare for this certification exam:

- S4600 Business Processes in SAP S/4HANA Sales: *http://s-prs.co/v530802*
- S4605 Sales Processing in SAP S/4HANA Sales: *http://s-prs.co/v530803*
- S4610 Delivery Processing SAP S/4HANA: *http://s-prs.co/v530804*
- S4615 Billing in SAP S/4HANA Sales: *http://s-prs.co/v530805*
- S4620 Pricing in SAP S/4HANA Sales: *http://s-prs.co/v530806*
- S4650 Cross-Functional Topics in SAP S/4HANA Sales: *http://s-prs.co/v530807*
- S4270 Advanced Available-to-Promise in SAP S/4HANA: *http://s-prs.co/v530808*

SAP S/4HANA Sales 2020 Upskilling Certification Courses

Remember that the SAP S/4HANA Sales 2020 Upskilling Certification assumes consultant-level knowledge in at least SAP ERP 6.0 as a minimum. Therefore, the list of courses in this section covers the additional material required to add to your existing knowledge. Your existing knowledge should be represented by the SAP S/4HANA Sales 2020 Certification courses listed in the previous section, by earlier SAP ERP 6.0 courses, or by relevant experience. The link for the SAP S/4HANA Sales 2020 Upskilling Certification will detail these SAP ERP courses, if required.

We recommend the following courses to help you prepare for this certification exam:

- S4H01 SAP Business Suite to SAP S/4HANA Delta: *http://s-prs.co/v530809*
- S4SD1 SAP S/4HANA Sales – Functions and Innovations: *http://s-prs.co/v530810*
- S4LG1 Innovative Logistics Processes in SAP S/4HANA Enterprise Management: *http://s-prs.co/v530811*

SAP Learning Hub

The SAP Learning Hub (*https://learninghub.sap.com/*) is a cloud-based offering, with an annual per-user subscription. This digital learning solution can help you build and upgrade your SAP skills in all applications in the SAP suite. All content required for the certifications can be found on SAP Learning Hub, in addition to a wealth of useful content. SAP Learning Hub has various editions available:

- SAP Learning Hub, professional edition—full access to the SAP Learning Hub
- SAP Learning Hub, business edition—licensed for everyone in your organization
- SAP Learning Hub, enhanced student edition—licensed for higher education institutions or as individual student subscriptions
- SAP Learning Hub, partner edition—licensed for SAP partners only

Additionally, SAP Learning Hub has editions that are tailored to a given solution. You can choose from eight areas: SAP S/4HANA Cloud, digital platforms, procurement, analytics, customer experience, Internet of Things (IoT)/digital supply chain, people engagement, and finance.

SAP Learning Hub, discovery edition, is a free edition that provides 14 days of free access to preview the content of selected courses.

To identify the services that SAP offers by edition, you can download the SAP Learning Hub catalog, which shows you this detail, at the following link: *http://s-prs.co/v530812*.

Once you're logged on to SAP Learning Hub, you can search for courses and add courses to your individual learning plan, as shown in Figure 7.

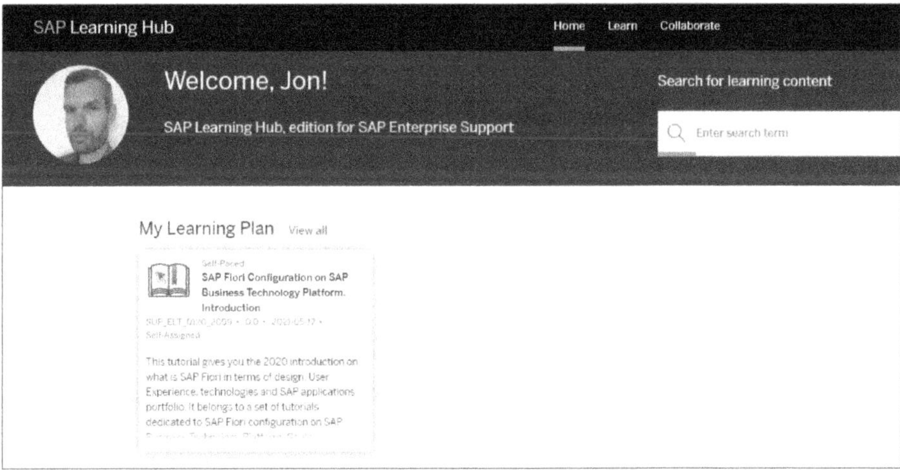

Figure 7 SAP Learning Hub

Additional functionality is available by clicking the **Collaborate** button at the top of the screen, which gives you access to SAP Learning Rooms (see Figure 8), where you can interact with SAP subject matter experts. Note that you must have a valid subscription to access this content.

Another useful function is SAP Live Access, again only available (for an additional fee) to subscribers to SAP Learning Hub. SAP Live Access gives you the opportunity to practice your skills and experience in a prebuilt SAP environment. These environments are all fully built to support the certification courses and thus can be quite useful. Normally, when you purchase an SAP Training and Adoption course, you'll get access to this kind of system for up to a week after the completion of the course.

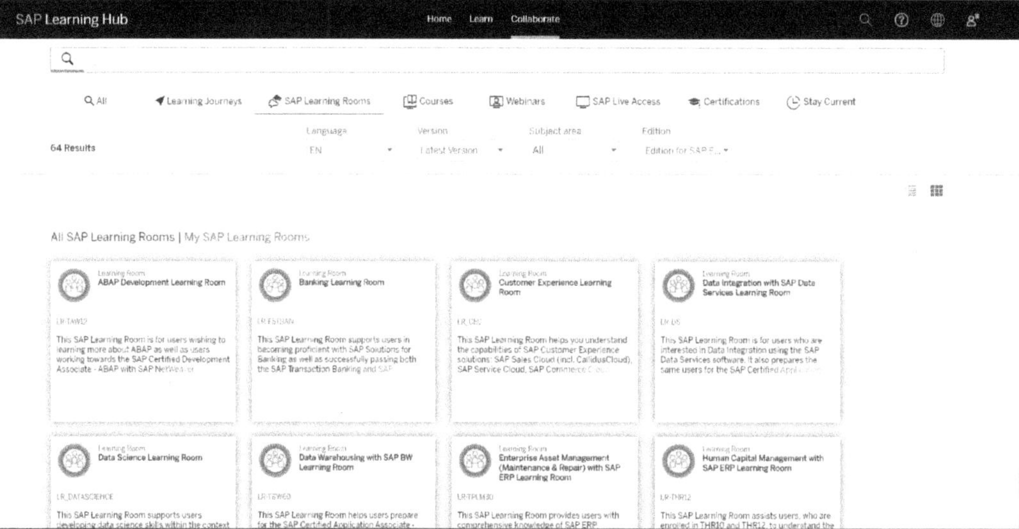

Figure 8 SAP Learning Hub: Learning Rooms

Tips for Taking the Certification Exams

Let's briefly review a few tips now to help you better prepare for the exam. The initial and most obvious thing to do is to study any of the education materials you have—the content therein will cover everything in the certification exams. However, to help you further, let's look at the types of questions you'll encounter and cover a few tips and tricks that should help.

Another point to press home is that the certification exams are designed purely to test your knowledge in the SAP S/4HANA Sales area. As any SAP consultant will gladly tell you, real life is quite different from education. The certification exams will try to mimic real life by offering practical examples and asking you to apply your knowledge to solve a problem. However well intended, this approach is certainly not the same as a real project challenge. As a result, a recent graduate, who perhaps has the test-taking methodology all figured out and the experience of taking exams fresh in his/her memory, will have an equal chance of passing the exam as a seasoned and experienced professional SAP consultant. While this fact may seem unfair, this is the reality!

What to Expect When Taking the Exam

As mentioned earlier, the exam is based on multiple choice and multiple selection questions. All the questions are based entirely on the course material that support the certification. This book, if absorbed completely, will provide you with all the answers you require.

Let's look at an example of the type of questions covered.

Multiple Choice Questions

An example multiple choice question may look something like the following question:

1. Your organization would like to block the creation of deliveries for all sales orders until they are reviewed manually and released. Where would you configure this block in SAP S/4HANA?

☐ A. The sales order type

☐ B. The sales organization

☐ C. The item category

☐ D. The sold-to party

This type of question is the most straightforward and is normally a clear question, with a single specific answer. In this example, the intent is to test your knowledge that a delivery block can be added to the configuration of the sales order type—thus, the only correct answer is answer A.

Note

You may feel that some certification questions are designed to trick you, which is not the case. In our example, sales orders could be blocked using a delivery block assigned to the sold-to party, but the question explicitly says that ALL sales orders should be blocked for delivery. This question is not a trick question—rather, the goal is to test your powers of analysis to clearly understand the requirement and design the solution to match it. In other words, read every element of the question clearly and carefully. The 3 hours you're given to take the exam will be plenty, and you should take your time.

Multiple Selection Questions

An example multiple selection question may look something like the following question:

1. The client you're working for is embarking on a conversion from SAP ERP 6.0 to SAP S/4HANA. One of the queries they have is why they must use the business partner instead of the customer master. For which of the following

reasons is the business partner approach is preferable? (There are three correct answers.)

☐ A. All products in the SAP ecosystem use the business partner approach, so it offers a unified solution.

☐ B. Customers are redundant completely in SAP S/4HANA, meaning the eradication of an entire master data object.

☐ C. One business partner can play multiple roles (customer, supplier, etc.).

☐ D. One business partner can have multiple addresses associated with it.

☐ E. Customers are now exclusively housed in the SAP CRM system.

Again, a reasonably straightforward question, but easy to slip up on. What's important is that you rely on your knowledge. Many multiple selection questions will offer up answers that may at first seem correct but upon closer inspection just don't look right. For this example question, the correct answers are A, C, and D. You may be tempted to think that answer B is also correct, but you should rely on your knowledge here—you know that business partners can be created as customers and the data is written back to the customer master data tables. Therefore, as a result, customers do still exist in SAP S/4HANA; thus, this answer is incorrect. The best approach to take in this case is to read through the answers and mark any answers that you're 100% sure of. This approach may eradicate any "tricky" answers by a process of elimination. Concentrate on the real dilemma—don't keep going over the answers you know to be correct.

True or False Questions

In this book, we will use true or false questions to test your understanding of key principles. The questions will be phrased in a similar way to the certification exam questions, but be aware that true or false questions do not appear on the exam itself. An example true or false question may look something like the following question:

1. True or false: Standard available-to-promise (ATP) in SAP S/4HANA 2020 is now longer available and has been replaced by advanced available-to-promise (sometimes called aATP).

☐ A. True

☐ B. False

Another straightforward question, but in typical certification exam style, you must read the question carefully. The correct answer is B (False). Advanced ATP is an option in SAP S/4HANA 2020 but has not replaced standard ATP. The difference is subtle but vital, and with questions like this, you can use your time wisely by reading and re-reading the question to ensure you understand the true nature of what is being asked.

Additional Tips and Tricks

You'll have 3 hours to go through and answer all the exam questions, which gives you a little over 2 minutes per question. While this time may not seem like much, the beauty of multiple choice and multiple selection questions are that they are relatively quick to answer—they are designed to take no more than 30 seconds to read through. The exam system keeps track of the questions that you have not yet answered in a grid on the bottom right of the screen. From this grid, you can also flag questions as queries to come back to later. Before submitting your exam, make sure that you've answered all the questions. You won't be penalized for incorrect answers, so submitting no answer is a wasted chance at an educated guess.

Some other useful tips that may help you pass the exam include the following:

- Scan through your training materials for bulleted lists. These lists are a great resource for test takers answering multiple selection questions, so keep your eyes open for them.
- Use the practice questions in this book wisely. If you find you cannot answer all of them correctly, revisit the chapter area where they are covered and study until you're sure of your knowledge.
- Remove stress factors about the exam: Get a good night's sleep (more important than last-minute cramming). If you're doing the exam in person, then get to the testing location early. If you're taking the online exam, make sure that your desk is clear and you've already downloaded the secure browser Questionmark well in advance. (You'll be sent instructions about downloading and installing Questionmark as soon as you purchase the exam.)
- Many practice exam questions are available on various sites throughout the internet. Treat these with great caution because many of them are poorly written and have incorrect answers flagged as correct. This book, on top of any experience and courses, will give you plenty to work with.
- Don't let the time worry you. In most cases, you'll finish way before the time limit and have plenty of time to go over your answers.
- Read and re-read the question. We cannot stress this point enough. Every part of the question is important, as is every part of each answer option.
- Eliminate the answers that you immediately know are wrong. In many cases, this method will give you the correct answer(s). By the same token, immediately flag the answers that you know are definitely correct.
- In some cases, answer options might be paired. If one answer is correct, then the other must be wrong. This approach allows you to eliminate incorrect answers effectively.

Additional Learning Resources

While the experience you have in addition to course materials and this book will be sufficient to pass the certification exams with flying colors, you must be aware that nothing in the SAP world stands still for long. The certification exam is no different, and the SAP S/4HANA Sales 2020 exam has some additional material that was not covered in the SAP S/4HANA Sales 1909 exam. Therefore, a good idea is to have some additional material at your disposal to keep your knowledge of the latest SAP innovations up to date. Many blogs, articles, and forums are available for gaining SAP S/4HANA knowledge, and while these resources can also be extremely useful, caution should be exercised here. Since these kinds of sites are not official SAP material, there is no guarantee that the content is 100% accurate. This section will run through a list of available resources, focusing on official SAP content, to maintain the integrity of the knowledge contained therein.

SAP Best Practices Explorer

SAP S/4HANA system deployment and the SAP Activate methodology is closely integrated with SAP Best Practices. A new system will generally use a best practices package as the baseline upon which you can build your own processes (or in some cases you might change your processes to match best practices). The homepage for SAP Best Practices Explorer, shown in Figure 9, is available at *https://rapid.sap.com/bp/*. SAP Best Practices are explored in detail in Chapter 1, Section 1.2.

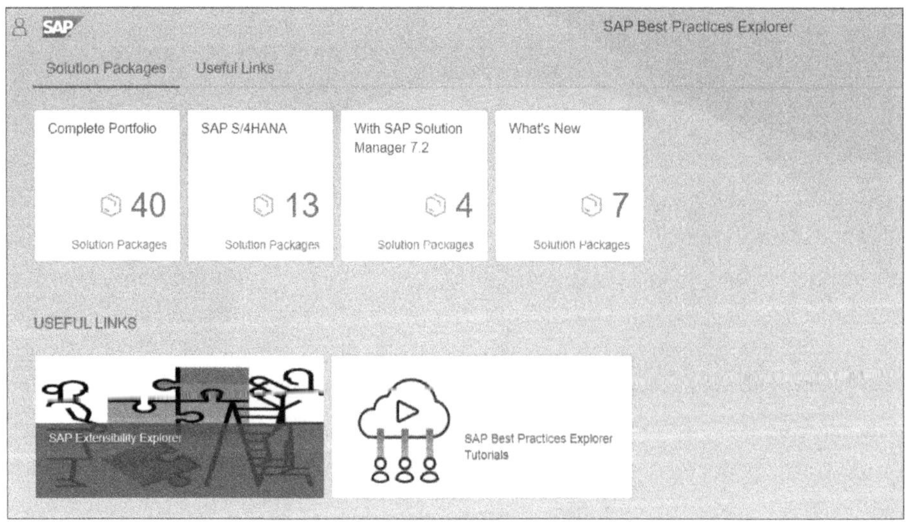

Figure 9 SAP Best Practices Explorer

SAP Help Portal

SAP Help Portal is an invaluable weapon in the arsenal of any SAP consultant. Every conceivable section of the SAP portfolio of products has the extent of its documentation detailed on SAP Help Portal, including links to other sites that contain additional information, such as the SAP Fiori apps reference library. The homepage for the SAP Help Portal, shown in Figure 10, is available at *https://help.sap.com/viewer/index*.

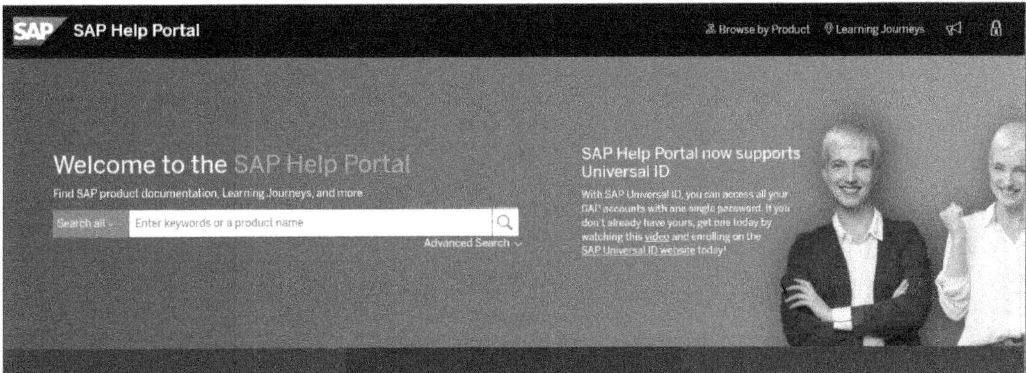

Figure 10 SAP Help Portal

This important information resource is updated every time a new release of SAP S/4HANA is made available. This website includes the following sections:

- **What's New Viewer** offers a complete overview of the new features and functions in every new SAP S/4HANA release.
- **Feature Scope Description** provides an overview of the available features.
- **Installation Guide** helps you install SAP S/4HANA.
- **Upgrade Guide** describes the upgrade process for different software versions.
- **Conversion Guide** supports conversions to SAP S/4HANA.
- **UI Technology Guide** gives information on user interface (UI) implementation.
- **Security Guide** enables the implementation of a secure system landscape.
- **Operations Guide** helps you operate your system.
- **Simplification List** describes features that aren't in scope for SAP S/4HANA in comparison to SAP ERP 6.0 and provides alternatives.

SAP Fiori Apps Reference Library

SAP Fiori apps make up some of the key innovations within SAP S/4HANA and are explained in more detail in Chapter 1, Section 1.2. In the SAP Fiori apps reference library, you can explore every SAP Fiori app and its functionality and implementation information, as well as easily browse through new and updated apps. Each app has its own page in the library, similar to the page shown in Figure 11.

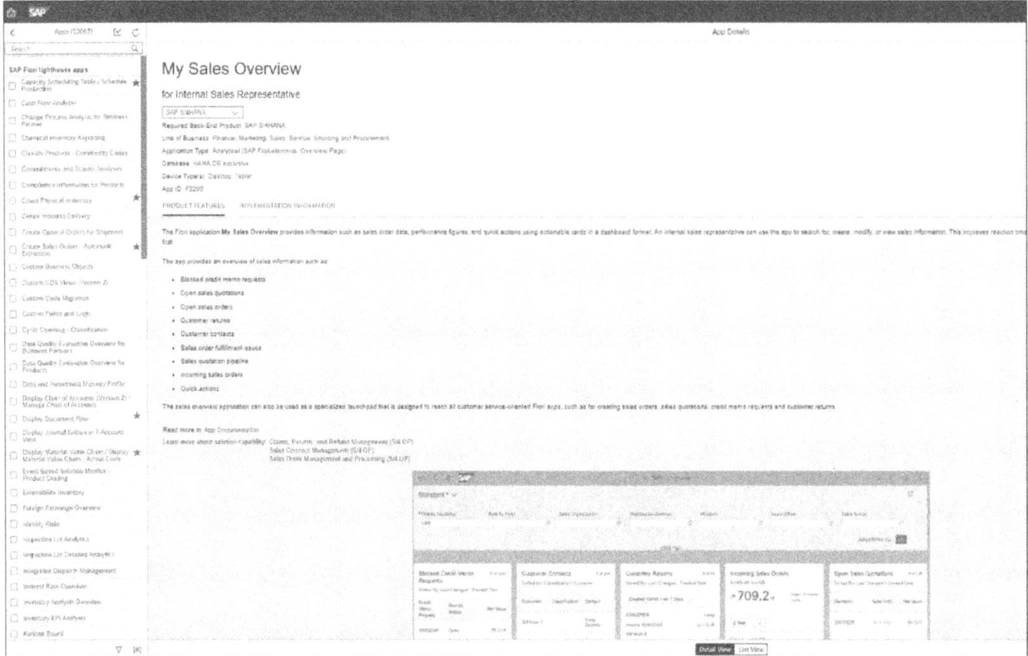

Figure 11 SAP Fiori Apps Reference Library: My Sales Overview App

The search box in the SAP Fiori apps reference library is useful, and you can use many different parameters, such as line of business, product, release, and so on, to filter the list of available apps and find the one you need.

The homepage for the SAP Fiori apps reference library can be found at *https://fioriappslibrary.hana.ondemand.com/sap/fix/externalViewer/*.

SAP Community

SAP Community is another great way to gain valuable knowledge and get assistance from other expert community members. Many blog posts are available at this resource, including how-to guides and plenty of questions answered. SAP Community, shown in Figure 12, can be found at *https://community.sap.com/*.

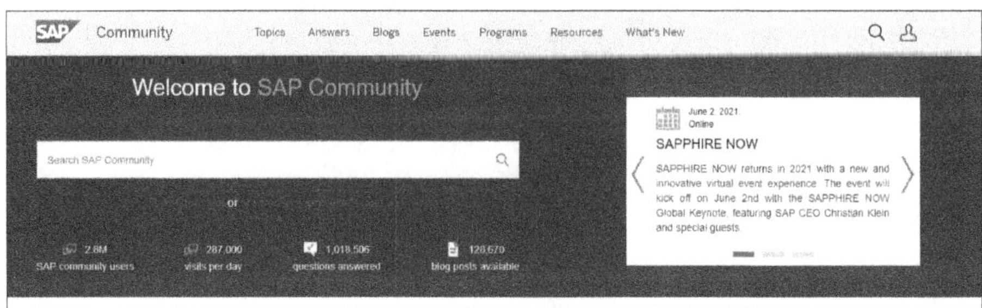

Figure 12 SAP Community

openSAP

A beneficial tool for your learning journey, openSAP (*https://open.sap.com*), shown in Figure 13, is a free, enterprise-level massive open online course (MOOC) offering.

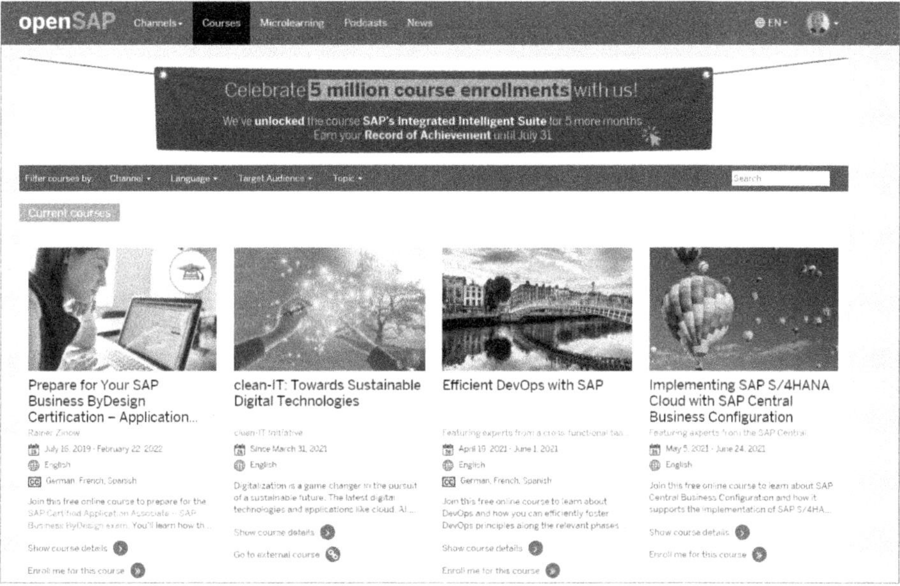

Figure 13 openSAP

The best way to use openSAP is to give you an introduction to new innovations. The detail in the courses is not particularly deep, but for an overview, it is a great resource. openSAP has a wealth of ongoing and previously run courses for several topics including machine learning, big data, IoT, and more.

Summary

You should now understand the certification options for SAP S/4HANA Sales that this book covers and that are relevant for you, as the target audience. With this knowledge, you can identify which certification exam, either for the SAP S/4HANA Sales 2020 Certification or SAP S/4HANA Sales 2020 Upskilling Certification, is most relevant for you.

You now know about the exams' structure, topic weighting, and required scores to pass. You also know which SAP Training and Adoption training courses you can review or attend for your certification exam and which related SAP courses and resources can complement and enhance your knowledge and skills.

In the next chapter, we'll launch straight into the certification exam content by exploring the innovations available in SAP S/4HANA.

Chapter 1
SAP S/4HANA Innovations and Analytics

Techniques You'll Master

- Make the best use of the SAP Fiori user experience (UX)
- Understand SAP Best Practices and how they are used
- Describe in detail the concept of embedded analytics within SAP S/4HANA
- Understand SAP Smart Business
- Gain an overview of the functions available to an internal sales representative when using the SAP Fiori UX
- Understand the concept and uses for SAP Extended Warehouse Management (SAP EWM)

This long chapter is key to understanding some of the fundamentals of SAP S/4HANA—many questions in the certification exams are centered around the topics discussed here. In this chapter, we'll provide an overview of some of the concepts behind SAP S/4HANA, with which you must be familiar in order to complete the SAP S/4HANA Sales Upskilling Certification exam successfully. These innovations in SAP S/4HANA will provide a good foundation for your knowledge journey from an SAP ERP 6.0 sales consultant into an SAP S/4HANA Sales consultant. Alternatively, perhaps you've never used SAP ERP 6.0, and your learning journey will begin with SAP S/4HANA, in which case these innovations will form the core of your sales knowledge.

Real-World Scenario

As an SAP S/4HANA Sales consultant on an SAP S/4HANA implementation project, you're part of a team that operates using the SAP Activate implementation methodology. You must use the approach defined in the SAP Activate methodology to design and deliver your solutions. Additionally, many requirements you'll face will relate to reporting capabilities. Some of these capabilities are freely available as standard within the embedded analytics of SAP S/4HANA and are accessible via the SAP Fiori UX.

1.1 Objectives of This Portion of the Test

The goal of this portion of the SAP S/4HANA Sales Upskilling Certification exam is to test your understanding of the following topics:

- SAP Fiori UX
- SAP HANA database
- SAP Best Practices
- SAP Activate methodology
- SAP S/4HANA embedded analytics
- SAP Smart Business and key performance indicators (KPIs)
- SAP EWM

Note

The SAP S/4HANA Innovations and Analytics topic is included in the SAP S/4HANA Sales Upskilling Certification exam and will comprise more than 12% of that test. However, the topics included in this chapter will also feature in the SAP S/4HANA Sales Certification exam as part of the Sales Process and Analytics topic.

1.2 SAP Fiori User Experience for SAP S/4HANA

In this section, we'll introduce the concept of the "user experience," including how the system uses SAP Fiori apps to provide an effective UX and also how you can personalize your SAP Fiori launchpad. To bring these lessons together, we'll examine a use case for the internal sales representative role.

1.2.1 User Interface versus User Experience

Let's begin this section by identifying the difference between a *user interface (UI)* and the *user experience (UX)*. This key difference is central to the SAP Fiori approach.

From a software perspective, a UI is the relationship between a person and the device. The overriding factor with a UI is efficiency of interaction—a person should be shepherded to the end goal as quickly and as efficiently as possible in order to maximize productivity.

UX approaches the interaction between user and device quite differently. The driving factor for UX is not to lose the productivity approach that the UI uses but, rather, to combine it with viewing the whole experience from the user's perspective. In this way, UX can enhance motivation and encourage a positive attitude towards the experience. In other words, the aim of UX is to "delight" the user.

Let's look at an example: displaying a sales order. Normally, displaying a sales order would be achieved in a UI environment, such as SAP ERP 6.0, with Transaction VA03, as shown in Figure 1.1.

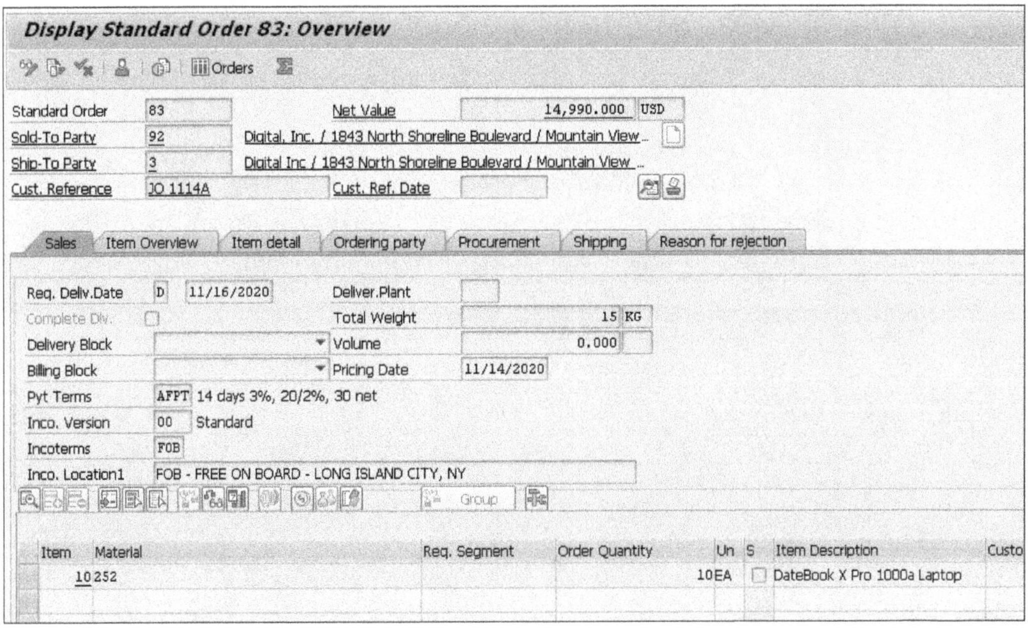

Figure 1.1 Displaying a Sales Order in SAP GUI

A vast amount of data must be presented to the user, which can be problematic for a UI to achieve efficiently. With a UX strategy, such as the strategy employed by SAP Fiori, you can present only the required data to the user, in a format that is both efficient and delightful, for example, the screenshot from the Sales Order factsheet app shown in Figure 1.2. We'll discuss different types of SAP Fiori apps later in this section.

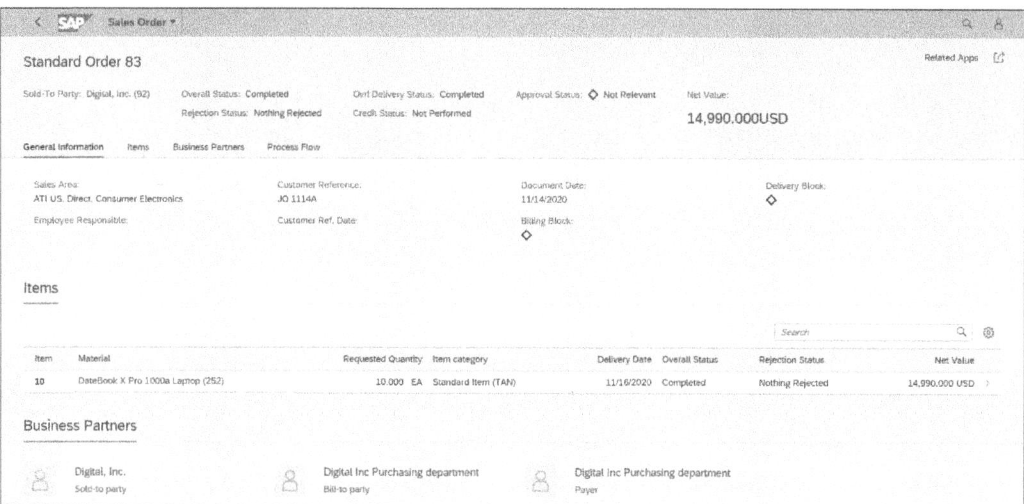

Figure 1.2 Displaying a Sales Order through the Sales Order Factsheet App in SAP Fiori

Note

SAP Fiori is device agnostic, so SAP Fiori apps can be used as standard on PCs, Macs, tablets, or smartphones. All delivered SAP Fiori apps resize automatically to fit the screens of these devices.

The SAP Fiori launchpad is designed to work in a web browser (any HTML5-ready browser) or an SAP Business Client (as of SAP ERP 6.0) or an SAP Fiori Client (on all mobile devices from Apple, Google, and Microsoft).

All available SAP Fiori apps can be viewed from the SAP Fiori apps reference library, as shown in Figure 1.3, located at *www.sap.com/fiori-apps-library*.

Figure 1.3 SAP Fiori Apps Reference Library

1.2.2 SAP Fiori Apps

Let's look now at how the SAP Fiori UX strategy is utilized by SAP Fiori. The strategy consists of the following three main areas of focus:

- **New applications**
  The "new" section of the strategy involves new apps that have been developed for SAP Fiori as well as additional new standard integrations with existing applications like SAP Customer Experience, SAP Concur, SAP Fieldglass, and SAP SuccessFactors.
- **Renewed applications**
  The "renew" section consists of SAP Fiori apps that are generally based upon the classic ABAP transactions you might be familiar with from SAP ERP 6.0. The difference, and why the term "renew" is employed, is that these transactions have been repurposed to fit the role-based nature of SAP Fiori—an important concept we will discuss later in Section 1.2.3. These renewed applications are also linked within other SAP Fiori apps, resulting in a seamless solution with drill-down functionality for more efficient and intuitive processing. Additionally, "renew" means that SAP Fiori is used as the common UI for existing SAP applications such as SAP Fieldglass, SAP Ariba, or SAP SuccessFactors.
- **Enable customers**
  The "enable" section of the strategy allows you to design and deliver your own apps to suit your business requirements. SAP provides several tools to help you achieve these custom apps, based upon SAP Fiori elements.

SAP Fiori elements are a key differentiator in the drive towards an effective UX strategy and can serve as a production line for the building of your own UIs. An SAP Fiori element is a template, often called a *floorplan*, which you can use to create any type of SAP Fiori app. Five floorplans enable you to create different sorts of SAP Fiori apps. The following floorplans are available:

- Overview pages
- Analytical list pages
- List report pages
- Object pages
- Worklists

Each floorplan has a specific look and feel and is specifically designed to adhere to SAP's design principles emphasizing usability, efficiency, and beauty. In this way, apps are consistent and comfortable for users.

SAP offers several tools for their UX strategy, namely, SAP Build—a free website where you can design your own prototype SAP Fiori apps based upon SAP Fiori elements and design principles—and SAP Screen Personas—an SAP GUI tool that you can use to simplify SAP GUI-based screens.

In general, three different types of SAP Fiori apps exist:

1. **Transactional apps**
   These apps resemble the classic ABAP transactions found in SAP GUI. These apps are available for SAP S/4HANA as well as for SAP ERP 6.0 on any database. An example of a sales-relevant transactional app is the Change Sales Orders app, shown in Figure 1.4, which is a web-based representation of the classic SAP GUI Transaction VA02.

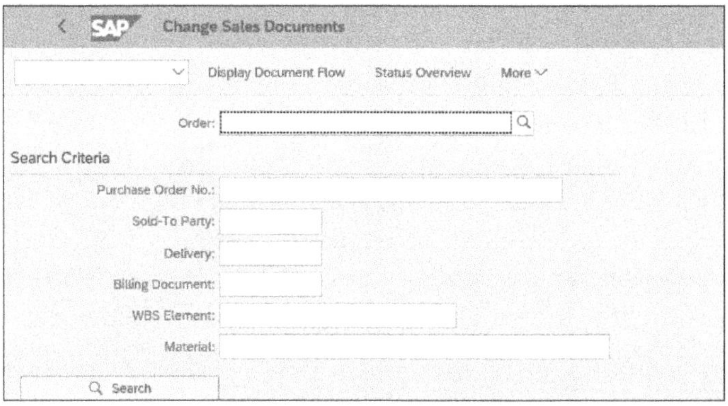

Figure 1.4 Changing a Sales Order in SAP Fiori

2. **Analytical apps**
   These apps have embedded analytical capabilities that utilize the power of the SAP HANA database, which will be discussed in more detail later in Section 1.6. The aim of analytical apps is to provide business insights at speed using complex algorithms. These apps are available for SAP S/4HANA and SAP Business Suite on SAP HANA. An example of an analytical app is the Sales Order Fulfillment app, shown in Figure 1.5.

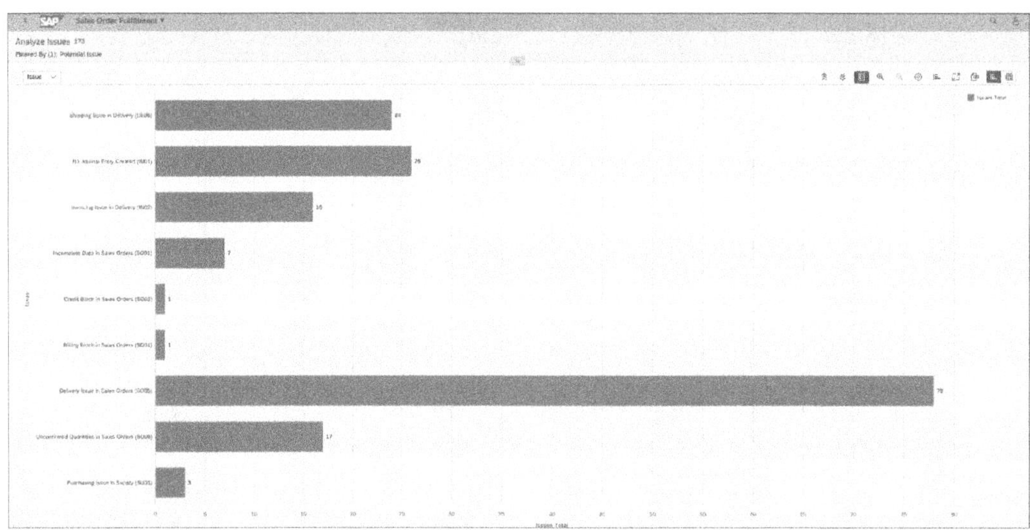

Figure 1.5 Sales Order Fulfillment App in SAP Fiori

3. **Factsheet apps**
   These apps use the enterprise-level search capabilities of the SAP HANA database to provide search results in specific contexts. These apps are available for SAP S/4HANA and SAP Business Suite on SAP HANA. We've already seen an example of a factsheet app, shown earlier in Figure 1.2.

Tip

Factsheet apps, while being apps in their own right, are normally accessed from another app—either a transactional app or an analytical app. This route of access is because their purpose is to provide targeted information based upon the activities carried out in previous apps or searches.

1.2.3 SAP Fiori Personalization

The SAP Fiori platform operates as a role-based application. In other words, whenever a user logs on to the SAP Fiori launchpad, the applications that are available to them depends on the roles that have been assigned to the user. SAP has provided standard roles such as the internal sales representative, shipping specialist, and pricing specialist roles, among others, to fit common business processes. Of course, you can define your own roles from scratch or base custom roles on standard predelivered roles from SAP. This methodology means that SAP Fiori apps are based on the user according to their role, not based upon the overall function. This definition of roles is a key part of the SAP Fiori personalization approach.

Personalization in SAP Fiori is achieved through tiles, catalogs, and groups, which can be arranged to comprise the full personalization of SAP Fiori, as shown in Figure 1.6.

Each tile is a front-facing display and launches an SAP Fiori app when clicked. Tiles are housed in catalogs and, from a catalog, are assigned to groups. The role assigned to your user defines which catalogs you have access to. Groups are then used to group together specific tiles from catalogs. These groups are displayed on your SAP Fiori launchpad along the top of the screen. Personalized groups, known as *user groups*, can be defined by users by editing their homepage. Thus, several ways to personalize your users' SAP Fiori experience are available, such as the following:

- Create a user's own group, which is only visible to them and is not available centrally. This task can be performed by the user.
- Add specific catalogs to the groups that are assigned to users. From these catalogs, a user can personalize their groups by adding new tiles from new catalogs.
- Personalize existing groups by adding tiles from catalogs that are assigned to the group.

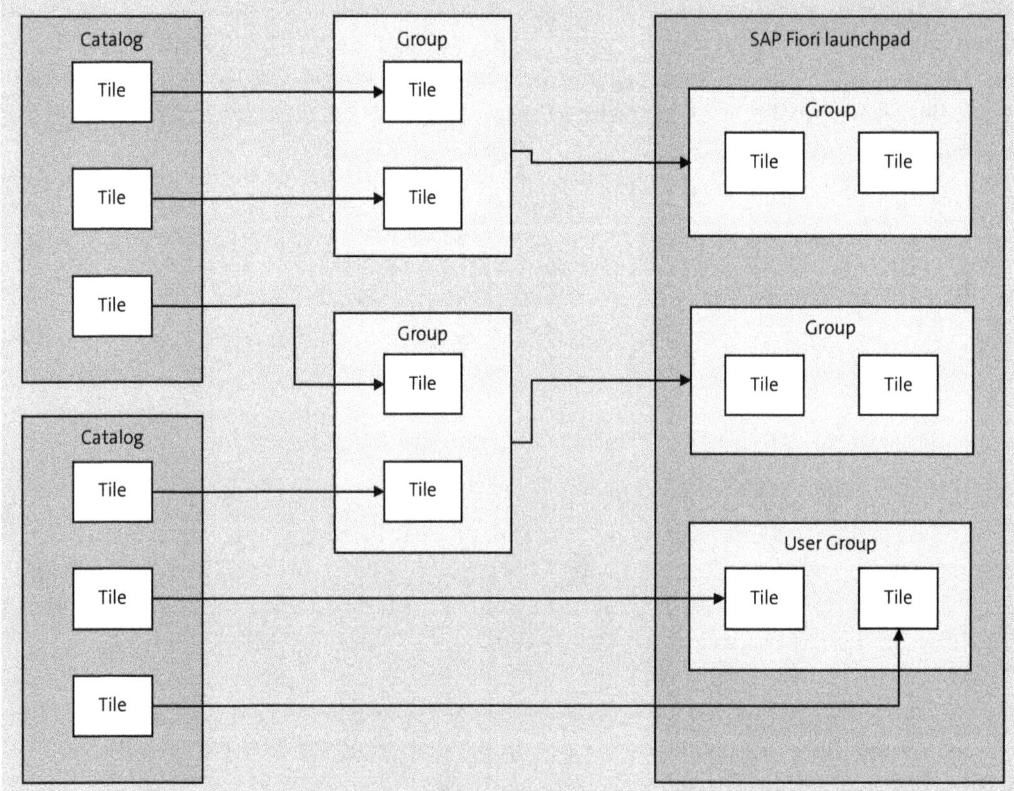

Figure 1.6 SAP Fiori Personalization

Warning

SAP Fiori apps and their associated tiles can be assigned to more than one catalog.

Note that one user's view of the apps in a specific catalog may be different from another user's view, which is controlled by security.

Figure 1.7 shows groups at the top of the page. The first group, **My SD Favorites**, is a user group that has been freely defined by the user. All other groups are standard groups. Nevertheless, as long as sufficient authorization is granted, any group—user or standard—may be personalized.

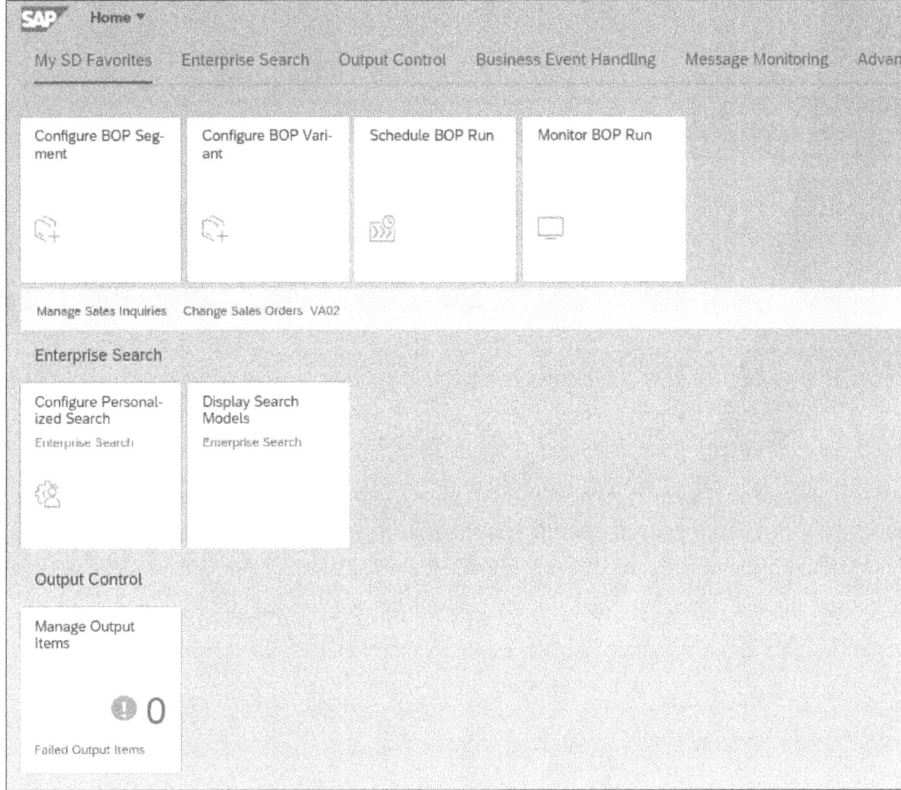

Figure 1.7 SAP Fiori Groups

1.2.4 Use Case: Apps for the Internal Sales Representative Role

The internal sales representative role is a standard SAP S/4HANA role defined for users who carry out creation, amendment, and monitoring tasks within SAP S/4HANA to support sales teams. For the SAP S/4HANA Sales Upskilling Certification exam, you'll won't need to know all the tasks, SAP S/4HANA transactions, and SAP Fiori apps used by this role, but you should know the core functions carried out, especially the functions using SAP Fiori. Table 1.1 details some of the more commonly used SAP Fiori apps for internal sales representatives.

Area	SAP S/4HANA Function and Associated SAP Fiori Apps
Sales	Creation and amendment of sales documents using SAP Fiori apps such as the following: ▪ Credit Memo Request ▪ Debit Memo Request ▪ Sales Contract

Table 1.1 SAP Fiori Apps for Internal Sales Representatives

Area	SAP S/4HANA Function and Associated SAP Fiori Apps
Sales (Cont.)	▪ Sales Quotations ▪ Manage Credit Memo Requests ▪ Manage Debit Memo Requests ▪ Manage Sales Contracts ▪ Manage Sales Inquiries ▪ Manage Sales Orders ▪ Manage Sales Quotations ▪ Condition Contract ▪ Customer Return ▪ Manage Customer Materials
Customers	Creation and amendment of customers using SAP Fiori apps such as the following: ▪ Customer – 360° View ▪ Assign Sold-to Parties ▪ Customers Overview
Products	Amendments to utilization of product portfolio using SAP Fiori apps such as the following: ▪ Product ▪ Compliance Information for Products ▪ Batch Overview/Manage Batches ▪ Configure Product Allocation ▪ Manage Characteristic Catalogs
Logistics	Creation and amendment of deliveries using SAP Fiori apps such as the following: ▪ Outbound Delivery ▪ Returns Delivery ▪ Manage Delivery Scheduling Processing
Order Monitoring and Analytics	Full monitoring of sales process using SAP Fiori apps such as the following: ▪ My Sales Overview ▪ List Sales Documents by Object Status ▪ Sales Order Fulfillment – Resolve Issues/Sales Order Fulfillment Issues ▪ Sales Order Fulfillment Issues (Version 2) ▪ Track Sales Order Details ▪ Track Sales Orders

Table 1.1 SAP Fiori Apps for Internal Sales Representatives (Cont.)

Area	SAP S/4HANA Function and Associated SAP Fiori Apps
Order Monitoring and Analytics (Cont.)	■ List Incomplete Sales Documents ■ Manage Duplicate Sales Documents ■ Predicted Delivery Delay
Billing	Creation and amendment of billing documents using SAP Fiori apps such as the following: ■ Resolve Payment Card Issues ■ Resolve Payment Card Issues – Reauthorizations ■ Billing Document ■ Billing Document Request

Table 1.1 SAP Fiori Apps for Internal Sales Representatives (Cont.)

1.3 SAP HANA Database

The SAP HANA database is central to SAP S/4HANA's ability to offer innovation at speed. As a result of SAP HANA's unique abilities in this area, the SAP S/4HANA core only runs on an SAP HANA database, unlike previous versions of SAP ERP, which could operate on any database. The SAP HANA database is one of the first of its kind: able to handle both transactions and analytics in-memory on a single dataset. As a result, SAP HANA can be used for both online analytical processing (OLAP) and online transactional processing (OLTP).

SAP HANA's data storage capabilities are quite different from traditional databases, which store data in rows and on hard disks. SAP HANA is an *in-memory database*, meaning the data is stored in RAM memory, allowing for much faster retrieval. Additionally, the data within SAP HANA is stored in a column store, which optimizes both transactional and analytical data. Data in column stores are automatically compressed, but what is a column store?

Traditionally, database tables store data in rows. As an example, let's look at some example data entries, shown in Table 1.2.

Country	Material	Sales per Year
US	1234	400
US	9876	900
MX	1234	500
GB	9876	600

Table 1.2 Example Data Table

Using the data shown in Table 1.2, a row store of this data would look like the table shown in Table 1.3.

Row 1	US
	1234
	400
Row 2	US
	9876
	900
Row 3	MX
	1234
	500
Row 4	GB
	9876
	600

Table 1.3 Example Row Store Table

However, column store data would look like the table shown in Table 1.4.

Country	US
	US
	MX
	GB
Material	1234
	9876
	1234
	9876
Sales per year	400
	900
	500
	600

Table 1.4 Example Column Store Table

Notice how, in the column store, only three entries exist (country, material, and sales per year) as opposed to four entries in the row store. Furthermore, the column store contains repeated adjacent entries, for which the SAP HANA database can employ highly efficient compression methods.

For example, imagine that you want to generate a report from the data in your example database to answer the question "What are my total sales per year?" In a row store, every row must be read to fetch the total. In a column store, only one read operation is necessary since all the sales data is in the same store, thus yielding significant performance and sizing improvements in the database.

Warning

The SAP HANA database holds the data in *both* row stores and column stores.

An additional benefit to using SAP HANA over traditional databases is the overall database size. Due to a change in the design of SAP S/4HANA, database tables have been eliminated and simplified, and index tables have been eradicated. In addition, with the in-memory and columnar store approach, your database footprint can be reduced significantly with an SAP HANA database.

1.4 SAP Best Practices

SAP Best Practices are preconfigured business processes that can be used to kick-start your SAP S/4HANA implementation. This standardized content library has been designed by SAP in conjunction with customers and partners and is available for both cloud-based and on-premise solutions.

By installing SAP Best Practices into your SAP S/4HANA client, you can accelerate and significantly simplify your implementation through content preconfigured for core business processes.

SAP Best Practices are included in the initial license fee and therefore are ready and waiting to be activated. SAP Best Practices encompass the following areas:

- Preconfigured, ready-to-run business processes
- Country-specific business processes
- Guided configuration steps
- Operational reporting and analytics
- SAP Fiori UX content
- Integration with SAP's cloud-based solutions such as SAP SuccessFactors and SAP Ariba
- Test scripts
- Full documentation describing business processes

SAP Best Practices can be examined via the SAP Best Practices Explorer at *http://rapid.sap.com*. At this website, separate SAP Best Practices are available for cloud-based and on-premise solutions as well as further drilldown access to resources on version- and country-specific solutions. When accessing the *rapid.sap.com* website, you should select **Solution Packages** from the SAP Fiori-style screen layout, as shown in Figure 1.8.

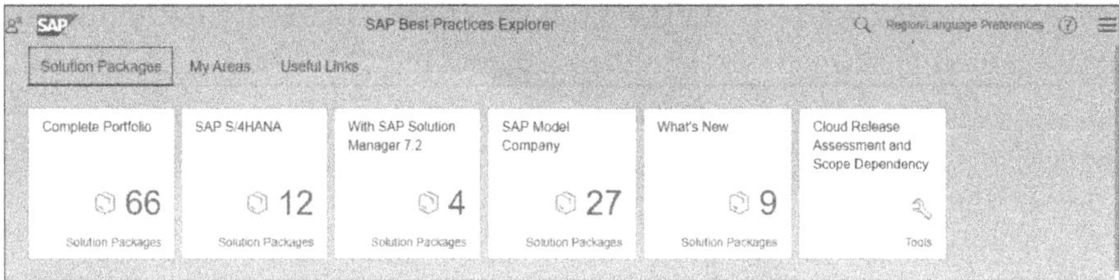

Figure 1.8 SAP Best Practices Explorer: Overview Screen

Once selected, you'll be navigated to the screen shown in Figure 1.9, where you can select your relevant solution package.

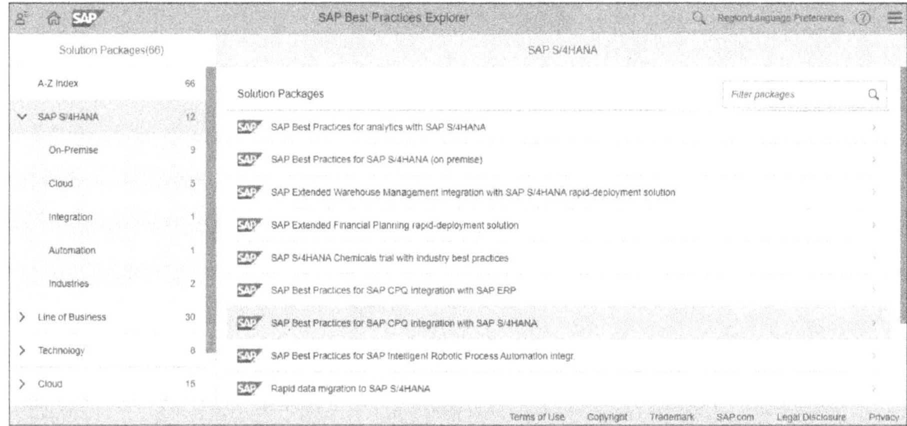

Figure 1.9 SAP Best Practices Explorer: Solution Packages

Once selected, the solution package will open and show the content library in full, as shown, for example, in Figure 1.10.

Figure 1.10 SAP Best Practices Explorer: Individual Solution Package

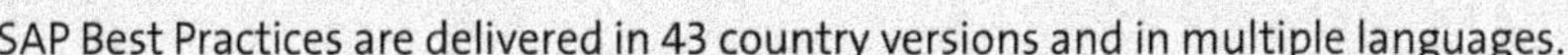

Note

SAP Best Practices are delivered in 43 country versions and in multiple languages.

The content library assigned to the solution package is made up of the following elements:

- **Solution overview**
  Description of the solution package, including software requirements.
- **Solution scope**
  Solution scope items are assigned to solution packages and explain in detail the solution area as predefined business functionality. For example, within the SAP S/4HANA on-premise solution package is the solution scope item 2TT (Sales Commissions - External Sales Representative). Within this solution item, you'll find process flows and test scripts, as shown in Figure 1.11.

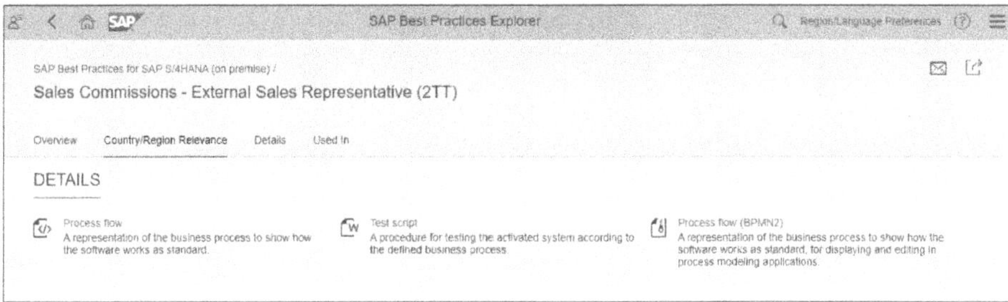

Figure 1.11 SAP Best Practices Explorer: Solution Scope Item

Another section, under the solution scope, is the **Accelerators** section. These templates and documents can be used for your own projects and can help you with key areas such as configuration and master data. Accelerators are divided into **General Documents** and **Implementation** sections for ease of access, as shown in Figure 1.12.

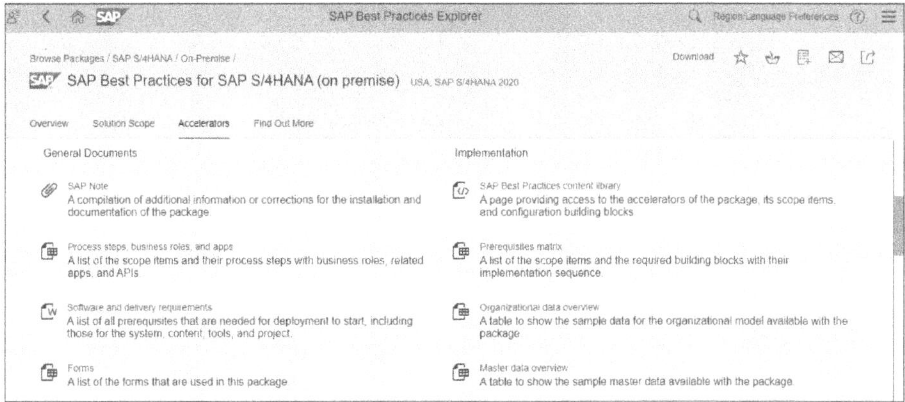

Figure 1.12 SAP Best Practices Explorer: Accelerators

The solution package can be downloaded by clicking the **Download** link in the top-right corner of the screen shown in Figure 1.12. Be sure you select the option with your country code and version suffixed (e.g., BP_OP_ENTPR_US_2020).

Once downloaded, the files appear in your **Downloads** folder, as shown in Figure 1.13.

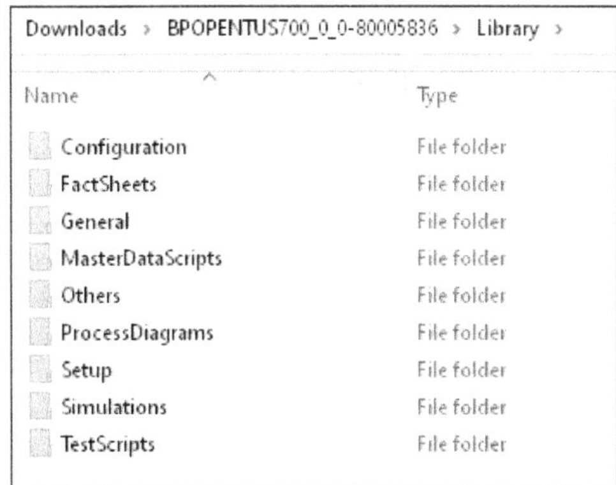

Figure 1.13 SAP Best Practices Explorer: Solution Package Downloaded

To make the best use of SAP Best Practices, we recommend importing them directly into your SAP Solution Manager client. Once in the client, the file structure is automatically generated, and you can easily access the *building blocks*, which provide the necessary configuration needed to build solution items. Building blocks, as with configuration guides, can be applied to multiple scope items.

SAP Best Practices are a key part of the SAP Activate methodology for implementing SAP S/4HANA, which we'll describe next.

1.5 SAP Activate Methodology

The SAP Activate methodology is SAP's recommended approach to implementation and conversion projects for deploying SAP S/4HANA on-premise and cloud solutions. Prior to the advent of SAP S/4HANA, SAP's recommended approach was Accelerated SAP (ASAP), but the SAP Activate methodology has superseded that and marries the use of SAP Best Practices with an agile approach to implementation.

Tip

SAP Activate methodology is not only recommended for SAP S/4HANA on-premise and cloud deployments, but also for SAP BW/4HANA, SAP SuccessFactors, SAP Customer Experience, and SAP Ariba implementations.

SAP Activate is comprised of six project phases, as shown in Figure 1.14.

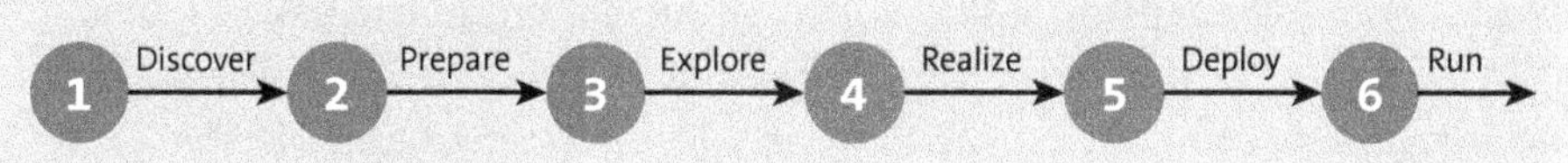

Figure 1.14 SAP Activate Phases

Let's briefly look at each phase next:

- ❶ **Discover phase**
  During this phase, the project team will research and build knowledge about the solution capabilities available with SAP S/4HANA. The goals of this phase include understanding the business benefits of the solution, determining the best strategy for implementation, and defining the target architecture. In this phase, your business prepares the business case for implementing SAP S/4HANA.
- ❷ **Prepare phase**
  Planning and preparation for the project occur during this phase. The goals of the phase include securing executive sponsorship, assigning project team members, establishing governance and tracking standards, and defining roles and responsibilities.
- ❸ **Explore phase**
  Fit-gap analysis takes place in the explore phase, when business users familiarize themselves with SAP Best Practices to identify whether they fit the business requirements. From this analysis, design activities are undertaken to bridge the identified gaps. Additional activities in this phase include master data load preparation.
- ❹ **Realize phase**
  The solution is incrementally built, tested, and validated using agile methodologies. Master data is loaded in this phase and ongoing operations are planned.
- ❺ **Deploy phase**
  End-user training is finalized, the production environment is established, and cutover activities take place during this phase. Business readiness is assessed and confirmed, and the system goes live.
- ❻ **Run phase**
  The system is fully operational, and optimization continues through continuous improvement, which is based upon the monitoring standards established in earlier phases.

1.6 SAP S/4HANA Embedded Analytics

SAP S/4HANA embedded analytics is a key differentiator for SAP S/4HANA, allowing the system to open a myriad of new possibilities in terms of functionality and insight. To understand the concept of real-time embedded analytics in SAP S/4HANA, first let's explore two key concepts: OLTP and OLAP. After gaining background knowledge about the combined analytical and transactional nature of SAP S/4HANA, we'll then explore how SAP enables embedded analytics through core data services (CDS) views.

1.6.1 Online Transactional Processing and Online Analytical Processing

OLTP systems are exclusively used for transactions to generate records of data—either master data or transactional data.

In contrast, OLAP systems are a part of a broader scope within a business intelligence suite of systems used for multidimensional report writing and data mining. Typically, the data that resides in the OLAP system is fed nightly from the OLTP system. As a result, neither real-time reporting nor any write-back functionality can occur from an OLAP system to an OLTP system.

The separation of OLTP and OLAP systems has several key disadvantages:

- Data in the OLAP system is not real-time data.
- Data in the OLAP system is limited to predefined subsets and therefore is not complete.
- An extract, transform, and load (ETL) process must sync both systems.
- Redundancy occurs because the data replicated in two systems.
- Complexity is added when combining data sources into a single data model in the OLAP system.

Figure 1.15 shows the additional process steps required when operating with separate OLTP and OLAP systems, including multiple replication steps, duplicate copies of the data, and business insights being hampered by old data.

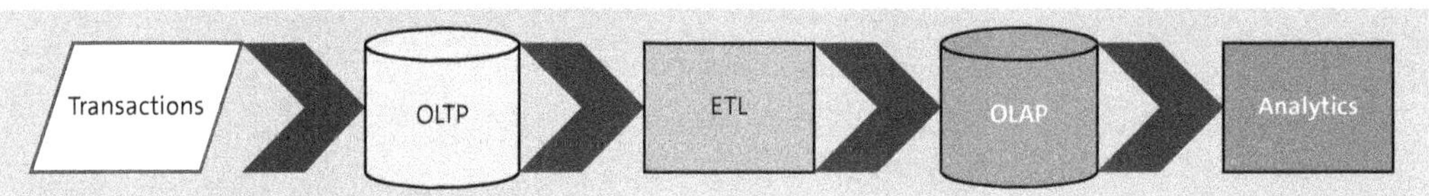

Figure 1.15 Separation of OLTP and OLAP Systems

As a result of these disadvantages, the goal for SAP S/4HANA is to create a unified OLTP and OLAP system on a state-of-the-art, high-speed database such as SAP HANA. Figure 1.16 shows a simplified view, with available insight to action on live data, a single source of the truth with no replication of data, and no ETL or batch processing required.

Figure 1.16 Combined OLTP and OLAP Systems

Note, however, that real-time embedded analytics in SAP S/4HANA does not mean the end of the road for data warehouses. The OLAP model employed by SAP S/4HANA and by SAP BW/4HANA support two different operations. The differences between the models employed are shown in Table 1.5.

SAP S/4HANA Real-Time Analytics	SAP BW/4HANA
Analytics directly on the original transactional data	Analytics using complex logic, KPIs, and benchmarking
Zero latency of data/real-time data	Integration with multiple applications including third-party data
Easily extendable to add new analytics because data is housed in the same system	Preconfigured content available, with options to amend and extend that content
Analytics consumed using open standards such as SQL, Multidimensional Expressions (MDX), and OData	Ability to use historical data, including archived data
Ability to embed analytics in OLTP transactions	Planning and modeling based on aggregated levels from multiple sources
Analytics built in a uniform model across all applications in SAP Business Suite	Better governance, including a more comprehensive view of analytics across the enterprise with mature governance processes

Table 1.5 Analytics Architecture in SAP S/4HANA versus SAP BW/4HANA

1.6.2 Core Data Services

The concept of real-time analytics is employed in SAP S/4HANA as a virtual data model (VDM) by utilizing CDS for operational reporting on live transactional data. CDS views are developed in the ABAP layer of the SAP S/4HANA system and uses the power of the SAP HANA database to extract and transform the data. In other words, CDS views are defined and consumed on the database server, thus providing an additional persistence layer, rather than carrying out these tasks at the application layer. Due to these advantages, SAP S/4HANA uses CDS views for operational reporting instead of the traditional Logistics Information System (LIS) reporting structure. LIS operates on data that is redundant, preaggregated, and presented on old UI technology. As a result, reporting with LIS does not meet the high standards expected of embedded analytics, such as access to real-time, fast, readily aggregated data.

A good way to view a CDS view is as a combination of multiple tables combined into a single overall view, as shown in Figure 1.17.

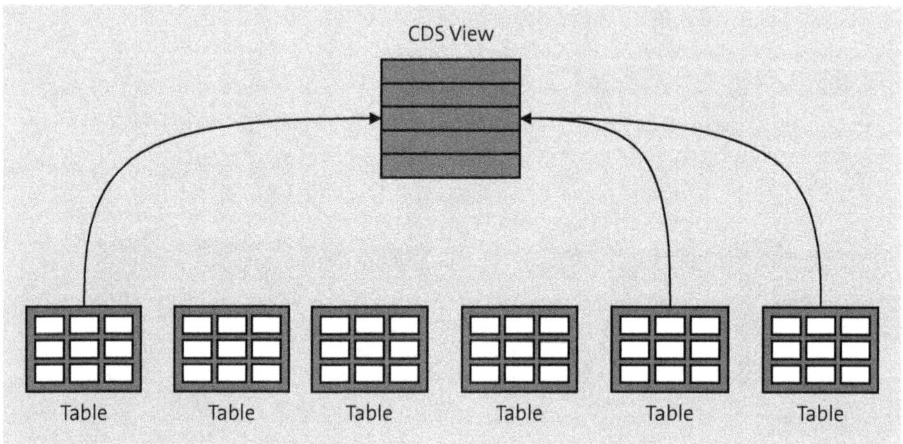

Figure 1.17 CDS Views

Besides pulling data from multiple tables into a combined view, CDS views have the power to aggregate and group fields from these tables. The CDS view is then available to be analyzed and reported against from multiple SAP Fiori apps and external applications.

The advantages of VDMs such as CDS views are numerous and include the following:

- Execution is run wherever possible, for instance, within the SAP HANA database instead of the application server.
- Operational advantages over standard ABAP tables include allowing aggregation and groupings as well as nested views.
- CDS views can be accessed by business intelligence frontends as well as other ABAP selections.
- A VDM replaces need for multiple complex ABAP queries.

CDS views can be used with SAP S/4HANA as part of the multidimensional reporting client (MDRC), which is based on Web Dynpro for ABAP. An example of multidimensional reporting is the standard SAP Fiori app Incoming Sales Orders, as shown Figure 1.18. Within these multidimensional reports, the functionality to sort, filter, pivot, and rearrange data graphically or in tabular format are all provided as standard.

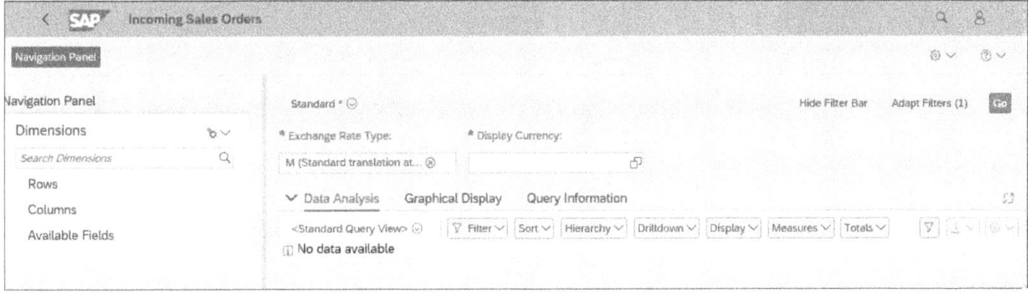

Figure 1.18 Incoming Sales Orders App in SAP Fiori

From within an MDRC view of a CDS view, you can run reports in an entirely flexible way. Freely definable options include adding your own drilldowns, sorts, and filters as key capabilities. The report can then be viewed in graphical or tabular format.

CDS views can be consumed by SAP Smart Business apps (more on this topic in Section 1.7) and by the Query Browser app in SAP Fiori.

Although CDS views are the most common approach for embedded analytics, an additional option is the use of OData services. These services are standard open interfaces that can be consumed by SAP or any external application or program and are used for querying and updating data.

1.7 SAP Smart Business

SAP Smart Business is a framework that you can use to create visually appealing and operationally valuable analytical content, in the form of charts and tiles. This content can be created without the need for coding and thus is open to anyone with the necessary skills and abilities. The intention with SAP Smart Business is to offer targeted statistics and analysis based on an exception model. To demystify SAP Smart Business, let's first look at KPIs as a method to define metrics to drive your SAP Smart Business analysis. Then, we'll move on to a case study using SAP Smart Business to demonstrate how this solution can be used effectively in a standard SAP Fiori app.

1.7.1 Key Performance Indicators

KPIs can be defined as metrics for measuring the performance of your company. SAP Smart Business relies on KPIs that you can use to generate SAP Fiori dashboards to visualize KPI data. KPIs are normally checked regularly, often many times per day, and therefore, their accuracy is vital, which includes the requirement that the data KPIs rely on is up to date in real time.

You can design SAP Smart Business apps using a standard set of SAP Fiori apps, as shown in Figure 1.19.

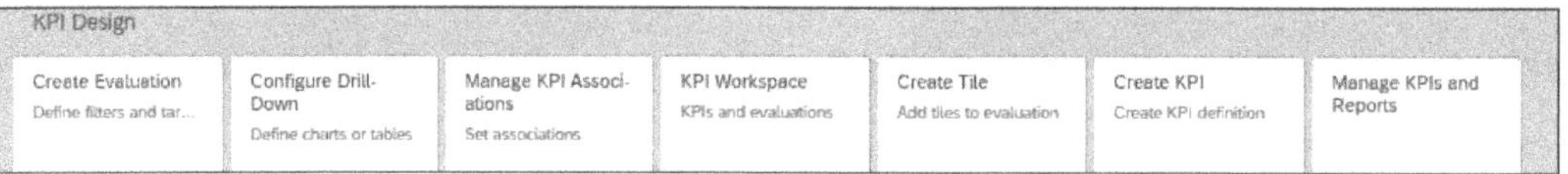

Figure 1.19 KPI Tiles in SAP Fiori

Standard out of the box apps are also delivered by SAP Fiori, which make good use of the SAP Smart Business technology. Apps like the Sales Order Fulfillment Issues app are good examples, and we'll analyze this example later in this section.

For creating a new KPI for use with SAP Smart Business, you'll need to follow a defined process, shown in Figure 1.20.

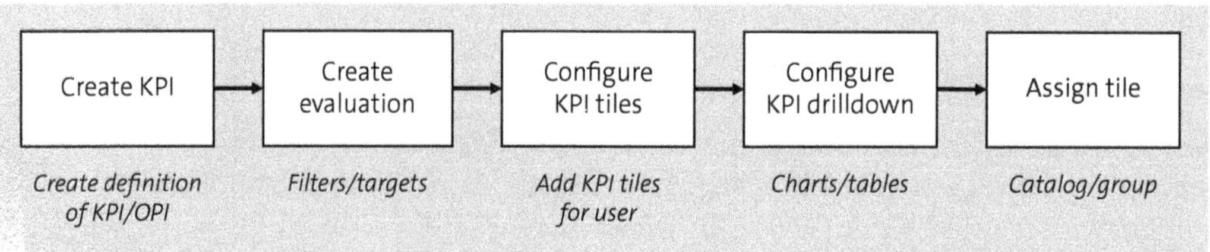

Figure 1.20 Creation Process for KPIs

Using SAP Smart Business apps comes with many advantages, such as the following:

- Analysis of KPIs in real time, at the touch of a button
- Follows standard SAP Fiori design guidelines, offering a unified look and feel
- Combination of operational key figures and strategic key figures into a single cockpit
- Simplified security model following the SAP Fiori role-based approach
- Full flexibility from within the app to filter, drill down, and navigate to other apps
- Ability to add dynamic data to the front of the tile to represent immediate values

KPIs can be created and activated using the Create KPI app, shown in Figure 1.21.

When creating a KPI, you can specify the CDS view or OData service it should read and add data, such as a name and a description for the KPI. Once the KPI is created, you can activate and add the evaluation.

The evaluation offers up data from the source with filters, input parameters, thresholds, and targets.

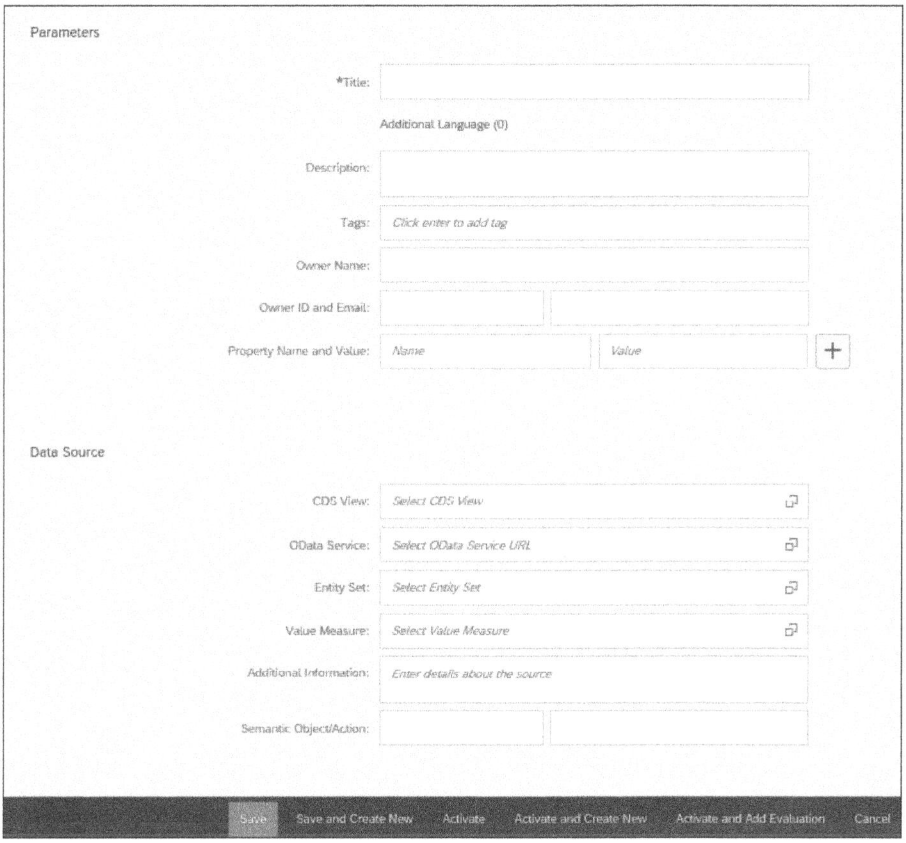

Figure 1.21 Create KPI App

Once the evaluation is created, click on **Activate and Create New** link. Now, you can select the tile format for the KPI from the following standard list:

- **Numeric Tile**
  Data on the front of the tile is presented as a number, as in the first example shown at the top of Figure 1.22.
- **Comparison Tile**
  Data on the front of the tile is presented as a list of measures with color-coded results, as in the second example shown at the top of Figure 1.22.
- **Trend Tile**
  Data on the front of the tile is presented graphically as a line graph, as in the third example shown at the top of Figure 1.22.
- **Actual vs. Target Tile**
  Data on the front of the tile is presented graphically as a bullet chart against a target value, as in the fourth example shown at the top of Figure 1.22.

- **Comparison Tile – Multiple Measures Tile**
  Data on the front of the tile is similar to the comparison tile, but in this case, you can specify multiple key figures to display, as in the fifth example shown at the top of Figure 1.22.
- **Dual Tile**
  Data on the front of the tile shows any combination of two of the other tile types, as in the sixth example shown at the top of Figure 1.22.
- **Blank Tile**
  Only the title and subtitle are displayed—no dynamic data is shown.

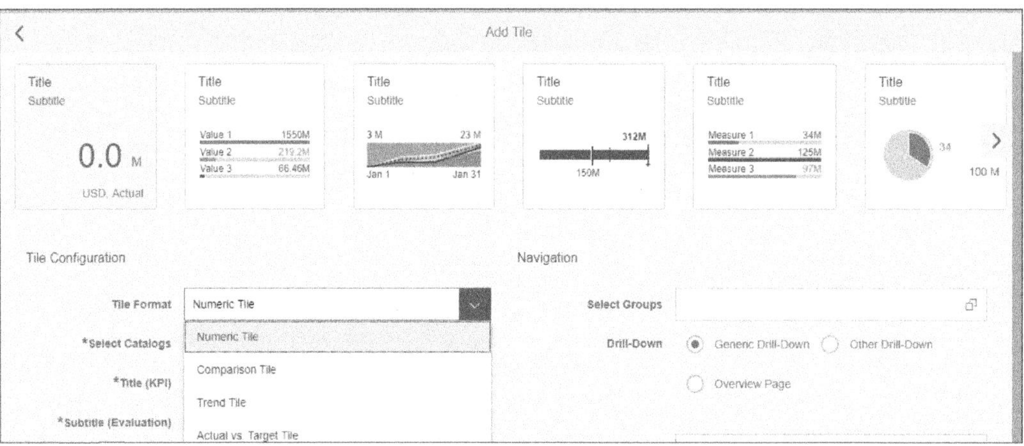

Figure 1.22 KPI Tile Formats

Once the tile format is completed, add the tile to the required catalog and group and click on **Save and Configure Drilldown**.

You can use drilldown capabilities to configure charts and different views and any associated KPIs to enrich your own KPI. Once completed, your KPI is published as a tile and should be available in your selected SAP Fiori group.

By following this process, the visualization of your KPI is presented in an SAP Fiori app by SAP Smart Business. In this way, the KPI tile can follow all the design guidelines and advantages afforded by SAP Fiori, such its responsive design for multiple devices and the ability to tailor the app to specific business roles.

The best way to demonstrate the effectiveness of SAP Smart Business is to discuss some examples in SAP S/4HANA that are provided as standard out of the box SAP Fiori apps. Most likely, reference to some of these examples from a sales and distribution perspective will be made in the SAP S/4HANA Sales Upskilling Certification exam.

1.7.2 Use Case: SAP Smart Business for Sales Order Fulfillment

SAP Fiori offers a robust suite of apps available for sales order fulfillment. All these apps are built on SAP Smart Business capabilities with underlying KPI definitions. The goal of sales order fulfillment apps is to allow you to carry out a number of tasks in terms of monitoring and resolving the following:

- Incomplete data
- Unconfirmed quantities
- Credit blocks
- Billing blocks
- Delivery blocks in sales orders and other delivery issues
- Shipping blocks in deliveries and other shipping issues
- Invoice issues and accounting issues for invoices
- Issues in supply in purchase orders (POs) and manufacturing orders

The layout of the SAP Fiori app is shown in Figure 1.23.

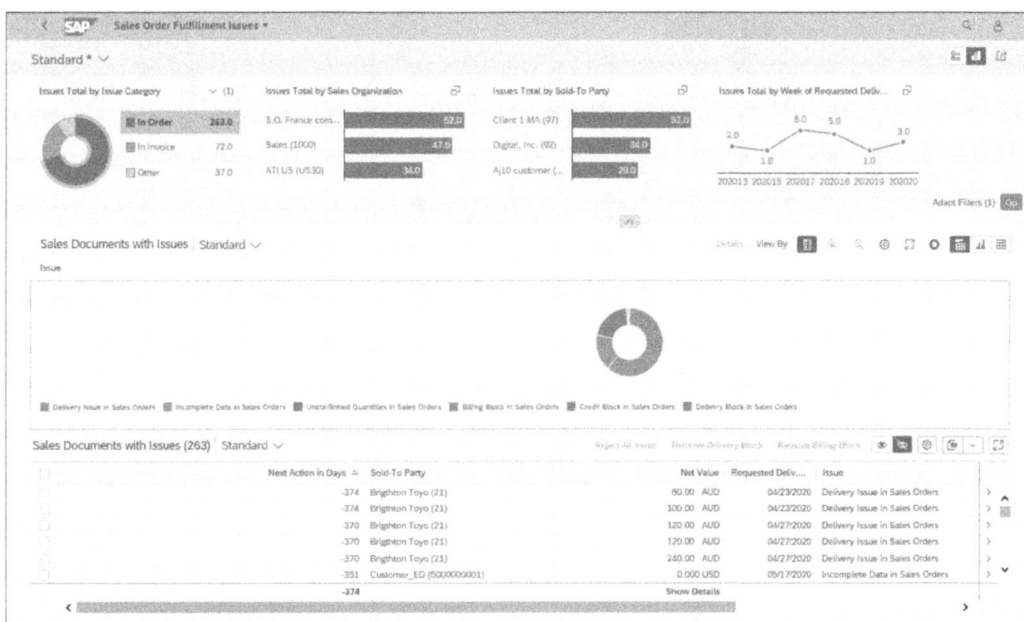

Figure 1.23 Sales Order Fulfillment App: Issues

In this app, issues are summarized by issue (order, invoice, or other); by sales organization; by sold-to party; and by week. These filters are defined when the KPI is created and therefore can be amended.

The second section shows a donut chart of all the issues, color coded. This chart type can be changed by selecting the chart type, as shown in Figure 1.24.

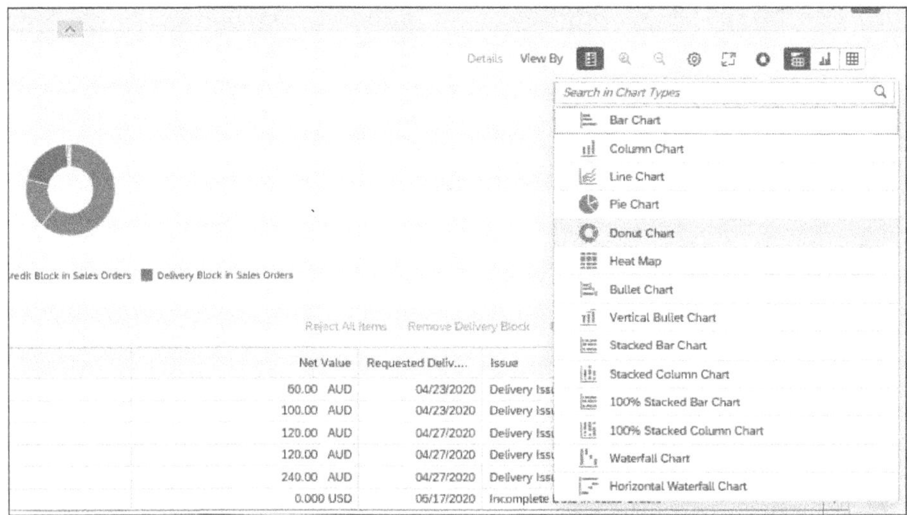

Figure 1.24 Sales Order Fulfillment App: Chart Types

Finally, the third section, shown earlier in Figure 1.23, shows the table format of the documents selected. Now, a user can make quick amendments to the selected documents by selecting a line and clicking one of the available options, as shown in Figure 1.25.

Sales Documents with Issues (263) | Standard

Reject All Items | Remove Delivery Block | Remove Billing Block

Next Action in Days	Sold-To Party	Net Value	Requested Deliv....	Issue
-374	Brigthton Toyo (21)	60.00 AUD	04/23/2020	Delivery Issue in Sales Orders
-374	Brigthton Toyo (21)	100.00 AUD	04/23/2020	Delivery Issue in Sales Orders
-370	Brigthton Toyo (21)	120.00 AUD	04/27/2020	Delivery Issue in Sales Orders
-370	Brigthton Toyo (21)	120.00 AUD	04/27/2020	Delivery Issue in Sales Orders
-370	Brigthton Toyo (21)	240.00 AUD	04/27/2020	Delivery Issue in Sales Orders
-351	Customer_ED (5000000001)	0.000 USD	06/17/2020	Incomplete Data in Sales Orders

Figure 1.25 Sales Order Fulfillment App: Fast Actions

In this third section, you'll see a column called **Next Action in Days**, which shows how many days exist between the current date and the date for the next step in the sales document's progress. The examples shown in Figure 1.25 have negative figures because the demo system we're using houses old data and the scheduled next step in the process is in the past. The calculation that the app uses is shown in Table 1.6.

Further information is available by clicking the **Show Details** option at the bottom of the screen, as shown in Figure 1.26. Now, you'll see the net value of all the items shown.

Issue Category	Details	How Date Is Calculated
In sales order	Delivery block Credit block Incomplete delivery data	Days are calculated between the current date and the material availability date on the delivery schedule line.
In sales order	Billing block Incomplete business data	If a sales order is not yet delivered, days are calculated between the current date and the next planned shipping date in the schedule line. If a sales order is delivered, days are calculated between the current date and the billing date in the delivery document.
In supply	Purchasing	Days are calculated between the current date and the delivery date of the PO item.
In supply	Manufacturing (process order or production order)	Days are calculated between the current date and the scheduled and confirmed start and finish dates for production.
In delivery	Transportation planning	Days are calculated between the current date and the transportation planning date.
In delivery	Picking	Days are calculated between the current date and the picking date.
In delivery	Shipping	Days are calculated between the current date and the shipping date.
In invoice	Delivery not yet billed	Days are calculated between the current date and the delivery billing date.
In invoice	Invoice not passed to accounting	Days are calculated between the current date and the invoice billing date.

Table 1.6 Sales Order Fulfillment App: Calculation of Next Action Days

Figure 1.26 Sales Order Fulfillment App: Show Details Option

Additional functionality is available by clicking the **>** at the right-hand side of each line. By selecting this option, users can navigate to the SAP Fiori app Track Sales Orders, shown in Figure 1.27. In this app, the process flow of sales documents can be viewed, and amendments to documents can be made.

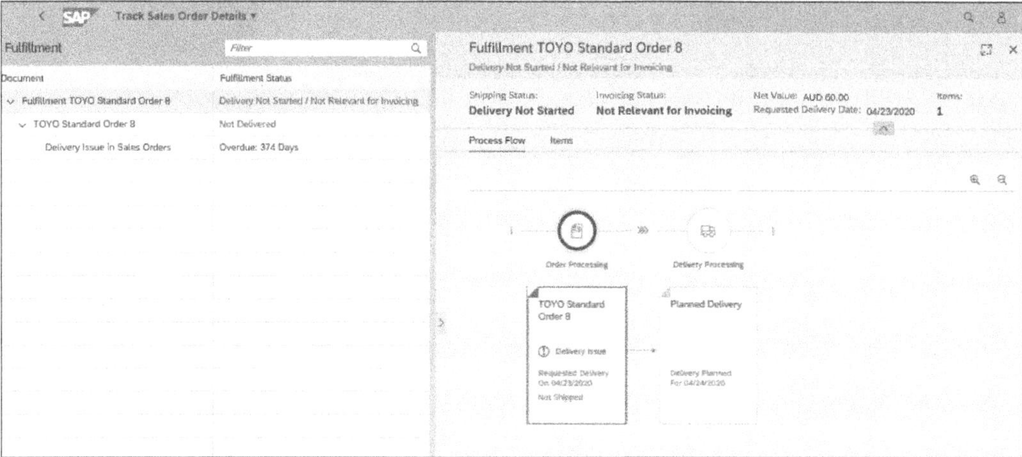

Figure 1.27 Track Sales Orders App

By clicking on a document in the Track Sales Orders app, additional options for navigation will be displayed, as shown in Figure 1.28.

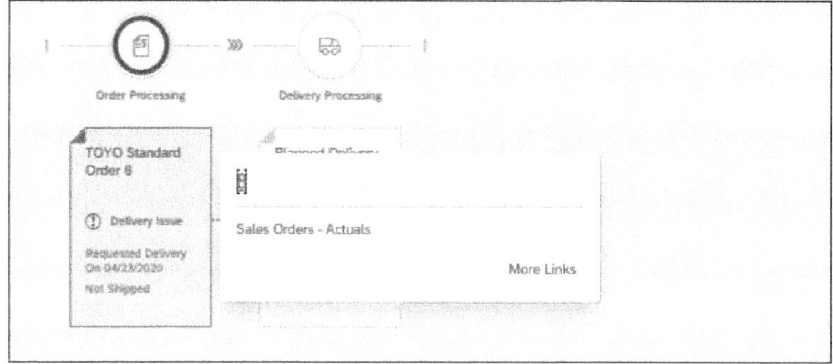

Figure 1.28 Track Sales Orders App: Additional Options

Clicking the **More Links** option on the screen, shown in Figure 1.28, will enable you to add additional links to the selection box. Numerous options are available, as shown in Figure 1.29, thus allowing you to add significantly to your standard drill-down capabilities.

Now that you have a fair understanding of SAP Smart Business and how useful this solution can be, let's take a quick look at an example of a specific role within SAP S/4HANA that uses this technology.

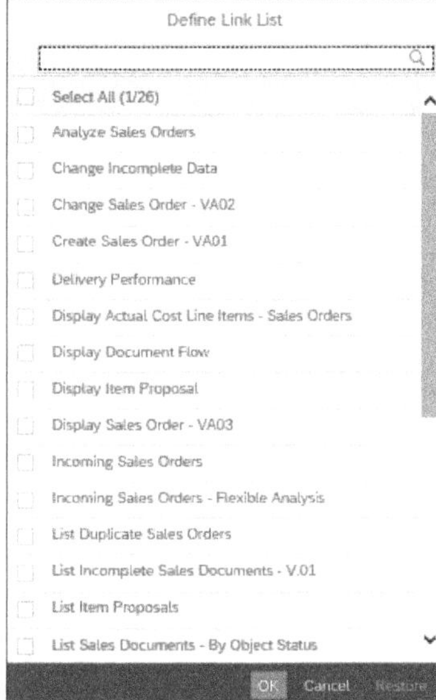

Figure 1.29 Track Sales Order App: More Links

1.8 SAP Extended Warehouse Management

As a certified SAP S/4HANA Sales consultant, you must be aware of the hand-off between sales functions and logistics functions. In SAP S/4HANA, you'll need a good grasp over the processes involved in SAP EWM. SAP EWM is the strategic platform for SAP S/4HANA—all other warehouse management solutions in SAP don't have any innovations planned for the future. The SAP S/4HANA Sales Upskilling Certification exam will test your knowledge in the SAP EWM area to establish whether you understand the different types of material flows and how they fit into the overall sales process.

Traditional management of warehousing within SAP has always been handled by Warehouse Management (WM). SAP then introduced SAP EWM as an add-on application alongside the SAP ERP 6.0 solution. SAP EWM allowed for more complex warehouse processing tasks in addition to the standard WM coverage. Let's familiarize ourselves with a definition of what we mean by a warehouse management system (WMS) first. Once you're familiar with the concept, we'll move on to warehousing options for deploying SAP EWM, before exploring the benefits and reasons why an SAP EWM implementation may fit your SAP S/4HANA landscape.

1.8.1 Warehouse Management Systems

A WMS performs a number of functions:

- Tracks the volume/amount of a given material that is stored as inventory in the warehouse

- Tracks the location of all storage bins in the warehouse, which are used to house materials
- Tracks all movement of goods in the warehouse, including determining put-away and picking locations for materials
- Allows mobile data entry through radio frequency (RF) devices
- Can interface with external applications for the purposes of transportation, automation, etc.

Additional functionality is often required in more complex warehouse environments—functionality such as serialization, batch number management, minimum shelf life, vendor-managed inventory, yard management, and value-added services (VAS).

Note

A WMS is not always essential in an SAP S/4HANA environment, and many clients run their simple warehouses using standard inventory management. However, if you need to know the location of your inventory at a more granular level than just the storage location, then you'll need a WMS.

1.8.2 SAP's Warehousing Options

Before SAP S/4HANA 1610, SAP EWM was only available as an add-on or as a decentralized version. Now, SAP EWM is fully embedded into the SAP S/4HANA ecosystem as part of the core SAP S/4HANA license, although some advanced functionality is available through an additional license fee. The advantages of a fully embedded EWM system are shown in Figure 1.30.

The other option for deploying SAP EWM is to use a decentralized SAP EWM that is integrated with SAP S/4HANA. This approach was the default approach before the embedded option became available, and SAP plans to maintain this option going forward. Some clients prefer this option since it gives them the flexibility to plan properly around downtimes, connect the system to multiple backend enterprise resource planning (ERP) systems, and maintain different release cycles. Since both embedded EWM and decentralized EWM are supported, the processes match almost entirely, and in the future, innovations will be deployed on both options.

Running a decentralized EWM system means that your master data and transactional data must be replicated in both systems, using a BAPI interface. The replication of this data is shown in Figure 1.31.

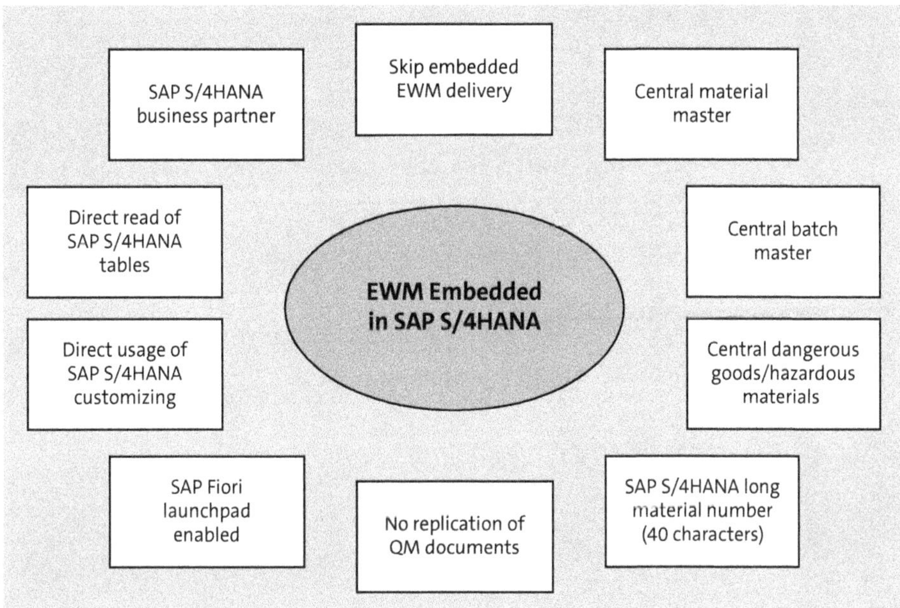

Figure 1.30 Embedded EWM in SAP S/4HANA

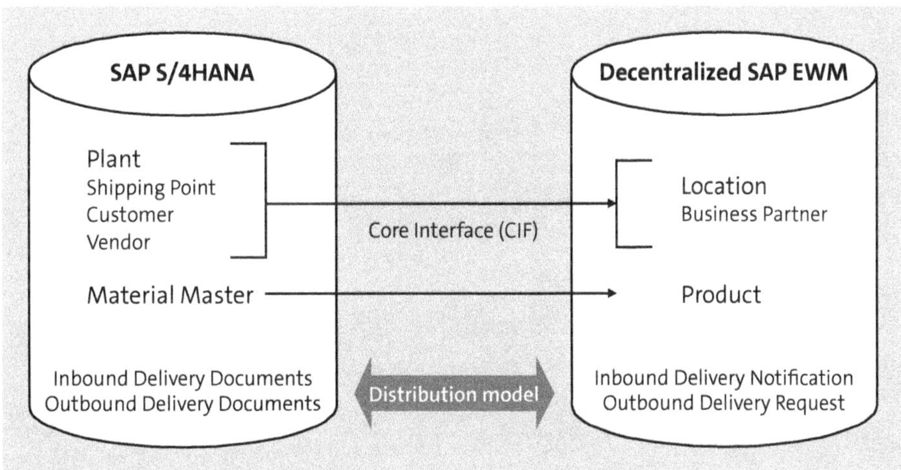

Figure 1.31 Decentralized EWM

Companies upgrading to SAP S/4HANA from an SAP ERP 6.0 system may wish to continue to use their existing WM features without making use of the additional features afforded by SAP EWM. The WM functionality is only supported by SAP until 2025, but an extension of this functionality can be realized by using stock room management. Stock room management offers broadly the same functionality as WM and is relevant for smaller, less complex warehouses. Some functions in WM that are not available as part of stock room management include the following:

- Task and resource management
- Warehouse control unit interface
- VAS
- Yard management
- Cross-docking
- Wave management
- Decentralized WM

If any of these components are active in your source SAP ERP 6.0 system, we recommend migrating to embedded EWM in SAP S/4HANA.

Warning

Stock room management will not be subject to future innovations because the strategic SAP S/4HANA solution is SAP EWM.

1.8.3 Why Choose SAP Extended Warehouse Management?

Managing a warehouse can be a daunting task and, without a fully equipped, state-of-the-art application to help you, may seem impossible. Embedded EWM in SAP S/4HANA is recommended where the warehouse in question has any of the following features:

- Fast-moving or high throughput of goods and materials, with multiple differences, such as size, weight, serial numbers, country of origin, batch numbers, and valuation types.
- Complex cross-docking processes
- Complex packaging requirements
- Complex shipping, transportation, and logistics requirements

Table 1.7 shows the capabilities of SAP EWM/embedded EWM lined up against WM.

Function	WM Coverage	SAP EWM Coverage
Putaway strategies	Good coverage	Extended coverage
Removal strategies	Good coverage	Extended coverage
Wave management	Partial coverage	Extended coverage
Replenishment	Good coverage	Extended coverage
Handling unit management	Good coverage	Extended coverage
Yard management	Good coverage	Extended coverage
RF technology	Partial coverage	Extended coverage
Resource management	Partial coverage	Extended coverage
Value-added services/kitting	Partial coverage	Extended coverage

Table 1.7 WM Capabilities versus SAP EWM Capabilities

Function	WM Coverage	SAP EWM Coverage
Cross-docking	Partial coverage	Extended coverage
Dynamic cycle counting	Partial coverage	Extended coverage
Multiclient warehousing	Light coverage	Extended coverage
Flexible process modeling	Light coverage	Extended coverage
Layout modeling	Light coverage	Extended coverage

Table 1.7 WM Capabilities versus SAP EWM Capabilities (Cont.)

Figure 1.32 shows how embedded EWM in SAP S/4HANA can be differentiated from the standard WM system at a high level.

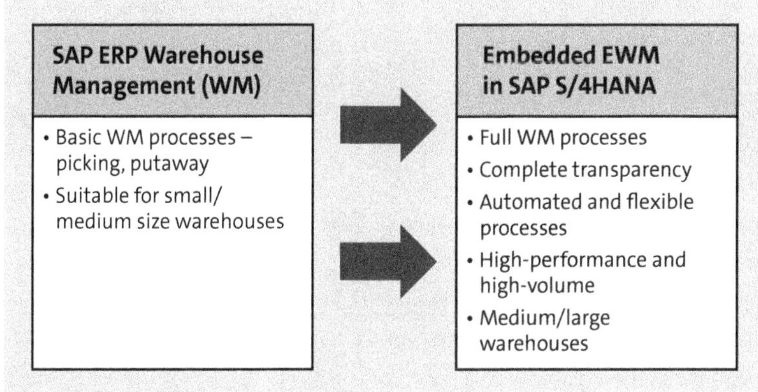

Figure 1.32 WM versus SAP EWM

Additional functions available in SAP EWM/embedded EWM but *not* available in WM include the following:

- Internal routing
- Expected goods receipts
- Unloading of transport units
- Deconsolidation
- Slotting/rearrangement
- Labor management
- Decentralized quality inspection
- Warehouse automation via material flow systems (MFS)
- Warehouse management monitor

Overall, the advantages of using embedded EWM in SAP S/4HANA over WM include better process efficiency, reduced costs, better productivity, process transparency, and highly effective monitoring.

One of the additional functions available in embedded EWM in SAP S/4HANA is the warehouse management monitor. This key advantage over WM gives full visibility into the process flows of the entire warehouse, thus enabling detailed planning of labor and resources. Through the warehouse management monitor, you can perform the following activities:

- Check stock and storage bins
- Monitor all the warehouse employees
- Check warehouse throughput and workloads in the pipeline
- Activate alerts for potential problems

1.8.4 Processes

SAP EWM has four types of processes: inbound processes, storage and operations, outbound processes, and general processes. Let's explore the processes in each grouping next.

Inbound Processes

The following inbound processes are available:

- Transportation unit processing: *Transportation units* arrive at the inbound gate, and prechecks are performed before moving the goods to the unpacking area.
- Goods receipt, management, and optimization: This key process within inbound processing controls the movement of products from the warehouse door to the final bin destination.
- Inbound quality management: When goods are received into the warehouse, they may need inspecting, which is carried out using the quality inspection engine (QIE).
- Internal routing: This process determines how products and handling units are transferred within the warehouse. The determination is based upon master data and configuration settings.

Storage and Operations

The following storage and operations processes are available:

- Physical inventory: Carrying out an inventory count of products and bins. There are three types of physical inventory procedures: periodic physical inventory, continuous physical inventory, and cycle counting.
- Replenishment or rearrangement: According to demand and usage, you can replenish stock within a picking area using replenishment control. Rearrangement is used to optimize your storage bins to cater to fast-moving items.
- Scrapping: Scrap from stock is an available internal process.
- Kit-to-stock management: Simply put, this capability is used to create kits from a list of products and then transfer them into your inventory.

Outbound Processes

The following outbound processes are available:

- Outbound planning (route, wave, bin): The route and the bin are determined based on customizing, and items from different warehouse requests can be combined or split in waves.
- Picking optimization, packing, staging, and load management: This process represents the goods issue process, comprising of the picking, packing, staging, and loading of products from the warehouse.

General Processes

The following comprise the full list of available general processes:

- Resource and labor management: Used to ensure that the correct resources are allocated to a warehouse order. Additionally, labor management is used to plan, measure, simulate, and visualize the end-to-end activities in the warehouse. Evaluation and comparison of performance can be carried out using the labor management functionality.
- Cross-docking: The movement of goods directly from goods receipt to goods issue without the use of any storage in the warehouse.
- Yard management: Used to manage transportation units that arrive to deliver or pickup goods to or from the warehouse.
- Import/export of products, compliance, hazardous handling: Integration with SAP Environment, Health, and Safety Management (SAP EHS Management), to ensure compliant transportation is carried out. Integration with SAP Global Trade Services (SAP GTS) is used to facilitate import and export control.
- Native technologies: Technologies include pick by voice and MFSs. In addition, data entry via RF devices and radio frequency identification (RFID) tags can be used for all warehouse processes. Some warehouses operate with automated storage and retrieval systems, and for this capability, the MFS can be used for the required interfaces.

1.8.5 Organizational Structure

The first thing to note about embedded EWM in SAP S/4HANA is that it uses a *warehouse number* in the same way as WM. The only difference is that the warehouse number in embedded EWM in SAP S/4HANA is a 4-character field rather than a 3-character field. This organizational unit represents the actual physical warehouse where the goods are stored. A warehouse number is linked to a plant and a storage location in that plant.

Tip

A warehouse number can be linked to multiple storage locations.

Once a warehouse is configured, setting up the remaining organizational units should follow the physical setup of the warehouse. An example of a typical setup is shown in Figure 1.33.

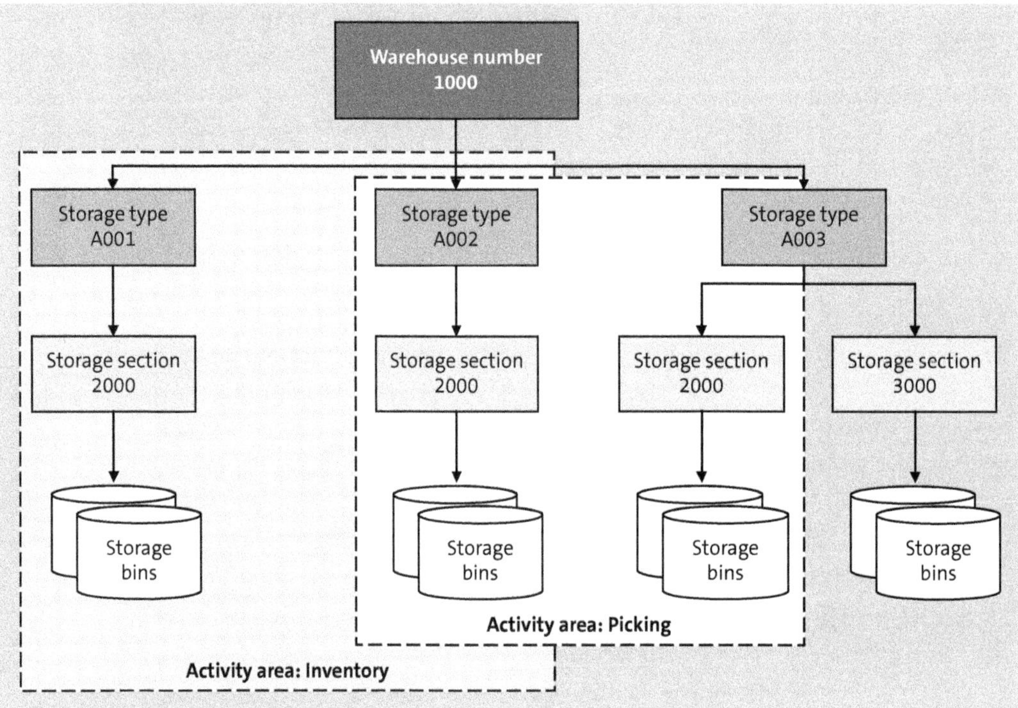

Figure 1.33 Embedded EWM in SAP S/4HANA Organizational Structure

Now that we've discussed the warehouse number as an organizational object, let's explore the use of other objects in embedded EWM in SAP S/4HANA:

- **Storage types**
  A storage type is a physical area within a warehouse, such as bulk storage, fixed-bin picking area, etc. The configuration of storage types, which is a 4-character field, controls putaway, stock removal, and goods movement control.
- **Storage section**
  A storage section is a subdivision of a storage type. For example, a specific storage type may have two sections: one for fast-moving items and one for slow-moving items. In other words, a storage section is a group of storage bins with similar characteristics. Storage sections are not mandatory.
- **Storage bin**
  Storage bins, the lowest organizational object in embedded EWM in SAP S/4HANA, are master data and assigned to storage types and storage sections. A storage bin is the precise location of a product, and the bin coordinates inform the identification number assigned to this organizational unit. A storage bin identifier is 18 characters in length and gives precise coordinates for the storage

bin—for example, coordinate 2000-01-10-03 might be storage type 2000, aisle 01, stack 10, level 03.

- **Activity area**
  Activities carried out in the warehouse, such as picking and putaway, are performed within activity areas. The beauty of activity areas is that you can use them to optimize the sequence of tasks to be completed. This optimization can be performed via the assignment of storage bins to an activity area.

1.8.6 Process Flows

Now that you understand the organizational structure of embedded EWM in SAP S/4HANA, let's look at the process flows. Each step that requires a warehouse worker to carry out a specific job is called a *warehouse task*. A warehouse task can be a movement of stock between bins, a goods issue, a goods receipt, or any other task carried out in the warehouse. Each warehouse task is made available for processing by grouping them together into a *warehouse order*. This grouping is based on customization of warehouse order creation rules (WOCRs).

Inbound Processes

When goods arrive in your warehouse from a PO or a manufacturing order (such as a process order or production order), an inbound delivery must be created. At this point, SAP S/4HANA will check whether the plant and storage location combination is assigned to a warehouse and whether that warehouse is an SAP EWM warehouse or an embedded EWM warehouse. If an SAP EWM/embedded EWM warehouse is found, the inbound delivery is created in SAP EWM/embedded EWM. All SAP EWM/embedded EWM processing—goods receipt, packing, and putaway—are performed with reference to this inbound delivery. In the case of decentralized SAP EWM, the goods receipt against the inbound delivery is posted back to the SAP S/4HANA system.

Two different methodologies are in play (decentralized SAP EWM and embedded EWM), as shown in Figure 1.34.

Outbound Processes

In an outbound delivery process in embedded EWM in SAP S/4HANA, if the outbound delivery is relevant for picking in a storage location that is managed by embedded EWM, then the outbound delivery order is created in embedded EWM.

> **Note**
> An outbound delivery order in an outbound process is subject to an outbound delivery request in the case of decentralized SAP EWM only.

An outbound delivery order serves as the worklist for embedded EWM and is the basis for the entire goods issue process. The outbound process is shown in Figure 1.35.

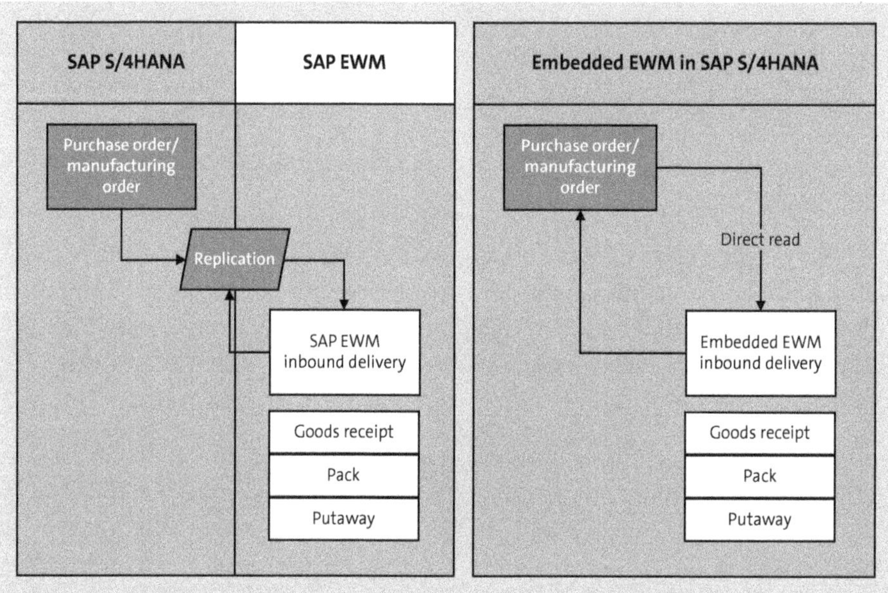

Figure 1.34 Embedded EWM Inbound Process

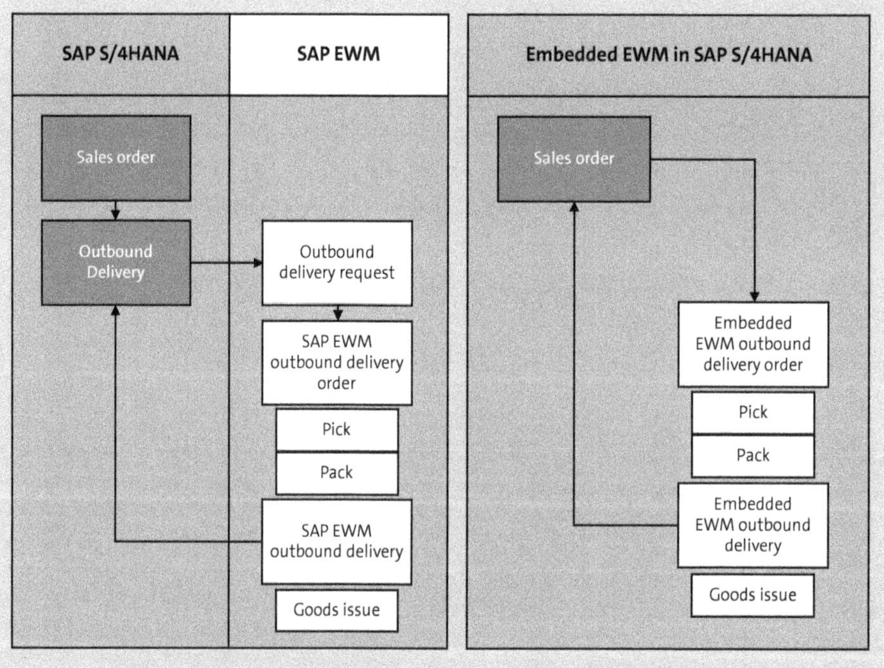

Figure 1.35 Embedded EWM Outbound Process

1.9 Important Terminology

In this chapter, the following terminology was used:

- **SAP Fiori**
  SAP Fiori is the user experience (UX) application utilized by SAP S/4HANA. A role-based application, SAP Fiori is designed for ease of use across multiple devices.
- **New, renew, and enable**
  SAP has used these concepts to define and structure SAP Fiori apps either from the ground up or based upon existing functionality or by allowing clients to define their own approach.
- **Floorplans**
  SAP Fiori screens are designed using specific templates, known as *floorplans*, to keep the UX consistent, delightful, and efficient.
- **Transactional apps**
  SAP Fiori apps that are based upon transactional capabilities, which might be familiar to users of the SAP GUI. These apps are often renewed in SAP Fiori to add additional functionality, such as drilldown capabilities and filters. These apps can be used in SAP S/4HANA or SAP ERP 6.0, in combination with any database.
- **Analytical apps**
  SAP Fiori apps that have analytical capabilities and utilize the power of the SAP HANA database. Analytical apps can be used in SAP S/4HANA or SAP ERP 6.0, but only in combination with an SAP HANA database.
- **Factsheet apps**
  SAP Fiori apps launched in response to specific contexts, presenting only relevant data aligned with the specific search that was carried out. Factsheet apps can be used in SAP S/4HANA or SAP ERP 6.0, but only in combination with an SAP HANA database.
- **Personalization**
  A user's ability to personalize their SAP S/4HANA UX through the use of SAP Fiori user groups.
- **Groups**
  An SAP Fiori group is a set of tiles grouped together and assigned to a specific business role.
- **User groups**
  An SAP Fiori user group is a group defined and personalized by a specific user.
- **Catalogs**
  An SAP Fiori catalog houses SAP Fiori apps for allocation to groups. Access to a given catalog is dependent upon the security role assigned to the user.
- **Tiles**
  An SAP Fiori tile is a container that represents an SAP Fiori app. The tile can display dynamic data in the format of text, numbers, KPIs, counters and charts.

- **SAP HANA database**
  An SAP HANA database is a state-of-the-art, next-generation database, combining in-memory data with column-oriented storage. SAP S/4HANA can only use SAP HANA as its database.
- **Online transactional processing (OLTP) and online analytical processing (OLAP)**
  SAP S/4HANA brings together both OLTP and OLAP functionality in one system, moving away from the traditional separation of the two approaches.
- **Column store**
  SAP HANA is a columnar database and, as such, stores data in columns rather than in rows. This approach makes SAP HANA especially suited for analytical query processing due to the speed and efficiency of the data retrieval.
- **SAP Best Practices**
  SAP Best Practices is a content library provided by SAP to accelerate and standardize your implementation of SAP S/4HANA. This library contains ready-to-run business processes, configuration guides, process flows, and test scripts.
- **Solution packages**
  A package contained within the SAP Best Practices content library that relates to the application for which SAP Best Practices content is applicable.
- **Solution scope items**
  Scope items are assigned to solution packages with SAP Best Practices and offer information at a more granular level around specific functions within an application.
- **Building blocks**
  Building blocks are contained within solution scope items in SAP Best Practices and offer configuration guides for setting up your SAP solution.
- **SAP Activate methodology**
  The SAP Activate methodology is the recommended approach for implementing SAP along agile guidelines. This methodology is designed to supersede the Accelerated SAP (ASAP) waterfall methodology.
- **SAP S/4HANA embedded analytics**
  This functionality is used for fast, real-time analytical model construction in SAP S/4HANA. This functionality runs exclusively on SAP HANAs database and uses core data services (CDS) views or OData services as its method of delivery. The presentation of embedded analytics uses SAP Fiori as its frontend.
- **Virtual data model (VDM)**
  A VDM is a representation of views or tables from the SAP HANA database into a single view, such as a CDS view.
- **Core data services (CDS) views**
  As a VDM, a CDS view is a structural representation, based on Open SQL, of multiple sources of data taken directly from an SAP HANA database. A CDS view may also aggregate and filter the data before presentation within the view. CDS views are normally consumed as embedded analytics within SAP Fiori apps.

- **Multidimensional reporting client (MDRC)**
  Reports built upon CDS views to offer insightful answers to specific business questions are defined as multidimensional reports. A standard SAP S/4HANA example of an MDRC is the Incoming Sales Orders app in SAP Fiori.
- **OData services**
  A standard web protocol used for querying and updating data. OData services are open standard services that can be consumed by SAP or any other application or program.
- **SAP Smart Business**
  A methodology for designing and running SAP Fiori apps based upon customizable key performance indicators (KPIs) that present exception-based analysis in a graphical and tabular format.
- **Key performance indicator (KPI)**
  A KPI is an analytical metric used to enable the analysis of performance, usually through SAP Smart Business.
- **Internal sales representative role**
  A standard SAP S/4HANA role defined assigned to users who support the sales team through the creation and amendment of sales and distribution documents as well as monitoring.
- **SAP Extended Warehouse Management (SAP EWM)**
  A next-generation supply chain management solution provided as embedded within SAP S/4HANA as standard. SAP EWM supports full warehouse and inventory management in addition to full analytical capabilities of labor and resources.
- **Stock room management**
  A warehouse management solution that supports small and manual warehouses and which is available for clients who wish to carry over the fundamental setups of their warehouse management solution (WMS) from SAP ERP 6.0 to SAP S/4HANA.
- **Storage type**
  An area of the warehouse with similar characteristics. A storage type is a subdivision of a warehouse.
- **Storage section**
  A further subdivision below storage type, used to group together storage bins with similar characteristics.
- **Storage bin**
  Master data, the lowest organizational level in SAP EWM.
- **Activity area**
  An area of the warehouse used for specific activities, such as picking or packing.
- **Warehouse task**
  Every task carried out by a warehouse worker is called a warehouse task.

- **Warehouse order**
  Warehouse tasks can be grouped together via customizing settings into warehouse orders, so they can be processed in the correct optimum sequence.
- **Outbound delivery order**
  Serves as the SAP EWM worklist and is the basis for the entire goods issue process.

1.10 Practice Questions

These practice questions will help you evaluate your understanding of the topics covered in this chapter. The questions shown are similar in nature to those found on the certification examination. Although none of these questions will be found on the exam itself, they will allow you to review your knowledge of the subject. Select the correct answers and then check the completeness of your answers in the next section. Remember that, on the exam, you must select all correct answers and only correct answers to receive credit for the question.

1. Which of the following are types of SAP Fiori apps? (There are three correct answers.)

☐ A. Analytical app

☐ B. Floorplan app

☐ C. Transactional app

☐ D. Factsheet app

☐ E. Overview app

2. Analytics can be executed from within SAP S/4HANA or from a data warehouse. Under which of these circumstances would using a data warehouse be preferable? (There are two correct answers.)

☐ A. When carrying out complex analytics, including KPIs and benchmarking, with multiple data sources

☐ B. When data drawn from multiple database tables is required

☐ C. When real-time data is essential

☐ D. When you need to add archived data into your query

3. True or false: The “renew” approach to UX with SAP Fiori repurposes SAP GUI transactions for fitting into the SAP Fiori approach.

☐ A. True

☐ B. False

4. When implementing SAP using the SAP Activate methodology, fit-gap analysis is carried out in which phase of the project?

☐ A. Discover phase

☐ B. Explore phase

☐ C. Prepare phase

5. Where is the best place to get information on SAP Best Practices from?

☐ A. Your SAP sales rep

☐ B. The SAP support forum

☐ C. The *rapid.sap.com* website

☐ D. SAP Jam

6. You're running SAP ERP 6.0 on an external database (non-SAP HANA) and would like to use some standard SAP Fiori apps to enrich your users' experience. Which types of apps are available to you?

☐ A. Analytical apps

☐ B. Transactional apps

☐ C. Factsheet apps

☐ D. All the above

7. One of your SAP Fiori users is unclear as to the terminology used in SAP Fiori and asks you what is meant by SAP Fiori launchpad. Your answer should be:

☐ A. The SAP Fiori launchpad is the home screen for all users in SAP S/4HANA and, for convenience, looks the same for all users in the system.

☐ B. The SAP Fiori launchpad is a shell that houses SAP Fiori apps, organized into groups based upon your role in the business.

☐ C. The SAP Fiori launchpad is a transaction used from within your SAP GUI to launch the SAP Fiori system.

☐ D. The SAP Fiori launchpad is the area used by developers to customize the look and feel of an SAP Fiori application.

8. True or false: SAP Best Practices, while delivered in multiple country formats, are only available in English.

☐ A. True

☐ B. False

9. How is personalization of the SAP Fiori launchpad best carried out?

☐ A. Using tiles, catalogs, and groups

☐ B. Using CDS views

☐ C. Using customized ABAP programs

10. What are the advantages of combining OLTP and OLAP? (There are three correct answers.)

☐ A. Data is available for querying in real time.

☐ B. Data can be combined with externally sourced data more easily.

☐ C. Queries can be run on the full dataset rather than predefined subsets of data.

☐ D. Security around access to data and queries is more easily monitored.

☐ E. There is no redundancy of data in multiple systems and therefore no need for complex ETL processes.

11. True or false: CDS views are a valuable tool for embedded analytics and require an additional SAP license to make best use of them.

☐ A. True

☐ B. False

12. A client you're working for is preparing a requirements specification for an ABAP report that combines several tables into one view in SAP S/4HANA. You suggest to the client that this goal would be better achieved by using CDS views, but the client is skeptical and asks for more details about CDS views. How should you explain the basics of CDS views? (There are three correct answers.)

☐ A. CDS views are tied to specific user groups.

☐ B. CDS views cannot be accessed from the SAP Fiori launchpad.

☐ C. Executing a query with a CDS view is always pushed, wherever possible, to the SAP HANA database, thus improving performance.

☐ D. CDS views can be consumed by other applications.

☐ E. Aggregating, grouping, and nesting of tables can be done within an CDS view itself, rather than dynamically by a complex ABAP report.

13. True or false: The Incoming Sales Orders app in SAP Fiori is an example of a multidimensional report.

☐ A. True

☐ B. False

14. Which of the following are true of SAP Fiori tiles? (There are two correct answers.)

- ☐ A. The only purpose of an SAP Fiori tile is to launch an SAP Fiori app.
- ☐ B. SAP Fiori tiles can show dynamic data on the front of them.
- ☐ C. An SAP Fiori tile is assigned to a single catalog in a one-to-one relationship.
- ☐ D. SAP Fiori tiles can be seen in the SAP Fiori launchpad.

15. True or false: The LIS provides a useful alternative method for running embedded analytics in SAP S/4HANA.

- ☐ A. True
- ☐ B. False

16. The storage method of the SAP HANA database allows for faster queries in which of the following ways? (There are three correct answers.)

- ☐ A. Column stores are compressed.
- ☐ B. Data is held on the hard disk, which allows for faster retrieval.
- ☐ C. Data is held within column store and row stores.
- ☐ D. Data is stored "in-memory."
- ☐ E. Data is only held in row stores.

17. True or false: SAP Best Practices building blocks are configuration guides and, as such, can be assigned to multiple scope items.

- ☐ A. True
- ☐ B. False

18. Creating your own KPIs involves a number of steps, including which of the following? (There are three correct answers.)

- ☐ A. Activating KPIs in the SAP Fiori launchpad
- ☐ B. Creating a background job to populate the CDS views used by KPIs
- ☐ C. Creating an evaluation and assigning it to the KPI
- ☐ D. Selecting an SAP Fiori tile format and assigning it to the correct catalog and group
- ☐ E. Configuring the KPI drilldown

19. Which of the following are valid tile formats for KPIs? (There are two correct answers.)

- ☐ A. Analytical tile
- ☐ B. Comparison tile

☐ C. Overview tile

☐ D. Numeric tile

20. True or false: the SAP Smart Business app Sales Order Fulfillment Issues can be used to identify and remove delivery and billing blocks.

☐ A. True

☐ B. False

21. In the SAP Smart Business Sales Order Fulfillment Issues app, how is the **Next Action in Days** value calculated for problems in sales orders? (There are three correct answers.)

☐ A. If the sales order is not delivered, then the replenishment lead time of the item material represents the number of days used.

☐ B. If the sales order is blocked for delivery, then the material availability date from the schedule line is considered.

☐ C. If the sales order is delivered, then the billing date from the delivery document is considered.

☐ D. If the sales order is delivered, then the billing date from the sales order is considered.

☐ E. If the sales order is not delivered and is blocked for billing, then the planned shipping date in the schedule line is used in the calculation.

22. Which of the following are organizational objects for embedded EWM in SAP S/4HANA? (There are three correct answers.)

☐ A. Warehouse

☐ B. Storage type

☐ C. Yard

☐ D. Storage section

☐ E. Door

23. True or false: Scrapping is an example of an outbound process in embedded EWM in SAP S/4HANA.

☐ A. True

☐ B. False

24. In SAP EWM, with outbound processing in the decentralized deployment option, what is created that triggers the outbound delivery order in the SAP EWM system?

☐ A. An interface run in a regular batch job

☐ B. A delivery Intermediate Document (IDoc) from the source ERP system

☐ C. An outbound delivery request

1.11 Practice Question Answers and Explanations

1. Correct answers: **A, C, and D**
   SAP Fiori apps can be either transactional apps, which are renewed versions of SAP GUI transactions; analytical apps, which harness the power of SAP HANA to provide built-in analytics and insights; or factsheet apps, which provide contextual information aligned with the specific search carried out by the user.
2. Correct answers: **A and D**
   Analytics can be drawn from either embedded analytics in SAP S/4HANA or a data warehouse such as SAP BW/4HANA. Using a data warehouse for your analytics is preferable when a query uses complex KPIs and the data is drawn from multiple data source applications. Additionally, querying archived data is usually best from a data warehouse.
3. Correct answer: **A**
   True. SAP Fiori apps designed using the "renew" approach take their inspiration from SAP GUI transactions. In many cases, these apps may be identical to the SAP GUI transaction, but the "renew" philosophy may mean some elements have been redesigned to fit SAP Fiori standards.
4. Correct answer: **B**
   Fit-gap analysis is a key component of the explore phase of the SAP Activate methodology. Once a project moves into the explore phase, all preparation activities are complete, including establishing the business benefits and setting up the project team. The purpose of fit-gap analysis is to identify where gaps exist from the standard solution, to plan for customization, and to anticipate the effort required.
5. Correct answer: **C**
   SAP Best Practices are housed on the *rapid.sap.com* website. From this website, you can establish your solution package and delve down into individual solution scope items to view process flows, test scripts, and configuration building blocks.
6. Correct answer: **B**
   Transactional apps are the only kind of apps available for use with a non-SAP HANA database. This limitation exists because both analytical apps and factsheet apps use the power and efficiency of the SAP HANA database.

7. Correct answer: **B**

   The SAP Fiori launchpad is the home screen for all your SAP Fiori apps, organized according to your business role into groups, catalogs, and tiles.

8. Correct answer: **B**

   False. SAP Best Practices are available in many languages, one of which is English.

9. Correct answer: **A**

   Personalization of SAP Fiori is carried out through editing your homepage to amend your groups, to add or remove groups, or add or remove tiles from groups. The tiles you see in groups are allocated to catalogs which are assigned through security authorizations.

10. Correct answers: **A, C, and E**

    In a system such as SAP S/4HANA, which combines OLTP and OLAP capabilities, the entire dataset is available for querying against in real time. As a result, as opposed to traditional approaches where OLTP and OLAP systems are separate, you no longer need ETL batch processes to load data into your OLAP system. Thus, you also no longer need to hold that same data in two systems. However, if you have a business requirement to report against the data found in SAP S/4HANA in combination with other, externally sourced data, you may be advised to utilize the capabilities of a data warehouse as a separate OLAP system. Similarly, if you have complex security and monitoring requirements around your queries, a separate OLAP data warehouse may be advisable.

11. Correct answer: **B**

    False. Although CDS views are a valuable tool for embedded analytics, you do not require an additional license to use them. CDS views are included with your core SAP S/4HANA licensing.

12. Correct answers: **C, D, and E**

    For performance reasons, the execution of a CDS view is pushed, wherever possible, to the SAP HANA database. Another advantage of CDS views is that, once defined, they can easily be consumed from other applications. When querying against a CDS view, additional functionality is available as standard, such as aggregating, grouping, and nesting of tables, meaning that these functions do not need the additional complexity of detailed ABAP code.

13. Correct answer: **A**

    True. The Incoming Sales Orders app is a true multidimensional report, meaning that this app pulls data from many different tables into a single consolidated view with aggregation, grouping, and filters available to the user to answer specific business questions.

14. Correct answers: **B and D**

    SAP Fiori tiles are more than just launchpads for SAP Fiori apps. They can display real-time analytical data in text, numerical, or graphical form. Each tile can be assigned to multiple catalogs, and SAP Fiori tiles are one of the three components that make up the SAP Fiori launchpad (the other two components being groups and catalogs).

15. Correct answer: **B**

    False. The LIS has been replaced by the embedded analytics capabilities of CDS views. Because the data in the LIS is written out to separate tables, this data is redundant and makes up a subset of rather than the entire dataset. Thus, the LIS has specific disadvantages when compared to CDS views, which are real time, do not involve data redundancy, and are presented in an up-to-date UI.

16. Correct answers: **A, C, and D**

    In the past, traditional databases held data in row stores directly on hard disks. SAP HANA has revolutionized this approach by storing data within automatically compressed column stores for ease of access, in addition to row stores. Additionally, data is held in-memory in RAM rather than on hard disks, thus allowing for significantly faster retrieval times.

17. Correct answer: **A**

    True. SAP Best Practices building blocks can serve as your guide to configuring scope items in question. However, due to the interrelated nature of SAP S/4HANA, these building blocks are relevant in different solution scopes and can therefore be assigned to multiple scope items.

18. Correct answers: **C, D, and E**

    KPIs are created in SAP Fiori, but this functionality is immediately available in SAP S/4HANA. As you learned in Section 1.6.2, embedded analytics uses live transactional data and therefore does not require background tasks to populate staging tables for the data. The creation of a KPI follows five steps: creating the KPI, creating the evaluation assigned to the KPI, configuring the SAP Fiori tile, configuring the KPI drilldown, and assigning the tile to a catalog and group.

19. Correct answers: **B and D**

    KPI tiles can take seven available tile formats: numeric tiles, comparison tiles, trend tiles, actual versus target tiles, comparison – multiple measures tiles, dual tiles, and blank tiles. Again, be careful not to fall into the trap of recognizing an SAP Fiori app format like "analytical" and assume that is the right answer—that is an app type, not a tile type. This kind of confusion is common in SAP certification exams, underlining the necessity to read the question thoroughly.

20. Correct answer: **A**

    True. The Sales Order Fulfillment Issues app is specifically designed for exception-based reporting to remove impediments that stand in the way of the progression of sales documents. These impediments can be delivery blocks or billing blocks, in addition to incomplete data, unconfirmed quantities, shipping blocks in deliveries, issues in production and procurement, and more.

21. Correct answers: **B, C, and E**

    Once the sales order is created, all dates that are used in the calculation are held within one of the sales documents themselves, so the replenishment lead time is irrelevant. Items blocked for delivery or not yet delivered will require the schedule line to be read—when delivery is blocked, the material availability date is used; when billing is blocked, the planned shipment date is used. If the sales order is already delivered, then the next step is billing, and this date is read from the delivery document.

22. Correct answers: **A, B, and D**

    The organizational objects in SAP EWM are warehouses, storage types, storage sections, storage bins, and activity areas.

23. Correct answer: **B**

    False. Scrapping is an example of a *storage and operations* process within SAP EWM.

24. Correct answer: **C**

    An outbound delivery request is created, containing all the relevant logistics data, and is then used to create the outbound delivery order in SAP EWM. This step is not required if the EWM is embedded into SAP S/4HANA.

1.12 Takeaway

In this chapter, you learned how SAP S/4HANA has moved away from the traditional UI approach to a more dynamic UX approach, where the interface is tailored specifically to the role a user plays in the business. You now also understand the guidelines by which the SAP Fiori UX has been developed, with its focus on "new," "renew," and "enable" principles. You're familiar with the different types of SAP Fiori apps available to users and how they are defined using SAP Fiori floorplans. By now, you should be familiar with the personalization options available with SAP Fiori and the use of groups, catalogs, and tiles to achieve personalization.

The use of the SAP HANA database has been explained in this chapter, and you're now familiar with the concepts behind in-memory databases, OLTP and OLAP processing methods, and data in column stores.

You can access the *rapid.sap.com* website to view the SAP Best Practices and can navigate through the hierarchies of solution packages, solution scope items, and building blocks to see such useful information as process flowcharts, configuration guides, and test scripts.

You can describe the phases of the SAP Activate methodology for implementing SAP S/4HANA and how they relate to SAP Best Practices. You're familiar with the specific tasks carried out in each phase of the SAP Activate methodology.

You also understand the importance of embedded analytics within SAP S/4HANA and can determine when a data warehouse is appropriate and when you can rely on embedded analytics. Important concepts such as VDM, CDS, and MDRC are all familiar to you.

You've extended your knowledge of SAP S/4HANA embedded analytics to incorporate SAP Smart Business and are fully aware of how to create your own KPIs. You know how SAP Smart Business is used in standard SAP Fiori apps, such as the Sales Order Fulfillment Issues app, and you're aware of how these apps are used by the internal sales representative role.

SAP's approach to warehouse management through the use of SAP EWM is now clear to you, and you have a high-level knowledge of the organizational structures and inbound and outbound process flows required in SAP EWM. Deployment options for SAP EWM, either decentralized or embedded, have been outlined, and you understand the reasons for taking each approach. You also understand the advantages of using SAP EWM over the more older WM module that was part of SAP ERP 6.0.

1.13 Summary

This chapter focused on a lot of introductory concepts in SAP S/4HANA. While this discussion may have seemed abstract, understanding these concepts is crucial to becoming a certified SAP S/4HANA Sales consultant. Without understanding the basics of embedded analytics, for example, you could easily fall into the trap of designing solutions that are only appropriate for SAP ERP 6.0 capabilities. This pitfall is especially true of reporting, where advances in SAP S/4HANA differ wildly from the solutions offered previously.

While SAP EWM is not an essential moving part in the sales and distribution landscape, some familiarity is essential. Any SAP S/4HANA implementation is likely to employ SAP EWM or stock room management and therefore cannot be ignored since SAP EWM is a key component of the logistics setup. The SAP S/4HANA Sales Upskilling Certification exam, while not requiring in-depth knowledge of SAP EWM, will certainly expect an overall understanding of the processes therein.

In the next chapter, we'll go into detail around the core foundation of the entire SAP S/4HANA Sales solution—the organizational structure.

2

Chapter 2
Organizational Structures

Techniques You'll Master

- Manage organizational units in financial accounting
- Maintain organizational units in sales and distribution
- Maintain organizational units in materials management and logistics
- Create references between organizational units

The focus of this chapter is to explain the configuration of organizational units, especially those relevant to sales and distribution processes. The definition and assignment of organizational structures are fundamental activities in any SAP S/4HANA implementation project.

Real-World Scenario

As an SAP S/4HANA Sales consultant, you'll be required to portray the specific organizational structure of the company you're working for. You'll need to analyze the existing legal entities; organizations responsible for selling products or services; distribution channels to get products to customers; manufacturing sites; warehouses; and more. With all this information, you'll define the different organizational units in accounting, sales, and logistics, as well as define some fundamental relationships between them. For example, you may realize that the same organization can sell products from two different plants in two different countries. To enable this particular scenario in the system, you'll need to assign those two plants to the same sales organization.

The organizational structures defined in customizing form the framework used in all business transactions, so setting up these structures is a matter of great importance to get it right. Once defined and configured, you'll establish how the data is structured from materials to business partners, which is not easy to change later. Creating the wrong assignments could lead to unnecessary master data records and confuse users. On the other hand, missing valid organizational links could result in system limitations when it comes to processing real-life scenarios.

The success of an SAP S/4HANA implementation project relies heavily on setting up the necessary organizational structures correctly. The expectation is that your knowledge of SAP S/4HANA organizational elements will contribute to building the right system for your customers.

2.1 Objectives of This Portion of the Test

The goal of this portion of the SAP S/4HANA Sales Certification exam is to test your understanding of the following topics:

- Definitions and assignments for organizational structures in accounting
- Definitions and assignments for organizational structures in sales and distribution
- Definitions and assignments of organizational structures in materials management and logistics

Note

The Organizational Structures topic makes up 8%–12% of the total SAP S/4HANA Sales 2020 exam.

The content of this chapter is included in the Master Data and Organizational Structures topic of the SAP S/4HANA Sales 2020 Upskilling exam, which makes up more than 12% of this certification exam.

2.2 Maintaining Organizational Units in Accounting

The organizational structures defined in SAP S/4HANA represent the system entities in which all business transactions are performed. These structures are mirror images of the actual legal and organizational units of your business and determine how the documents used in sales and logistics are classified in the system and how reporting is carried out. Master records like business partners, materials, or pricing condition records are only maintained in the specific areas of the organizational structure where they are valid. This separation helps a business, for example, ensure that certain products are only sold through valid distribution channels.

The *client* is at the highest level of the organizational structure hierarchy. Clients are identified by a 3-digit number and represent an enterprise group formed by one or more independent legal entities, known as *company codes*. The system requires at least one company code in each client, but you can set up several company codes in a client if you wish to manage the accounts of independent organizations simultaneously. Commonly, a single SAP S/4HANA system will have more than one client.

Figure 2.1 shows how clients are subdivided into company codes in an example of a two-client SAP S/4HANA instance.

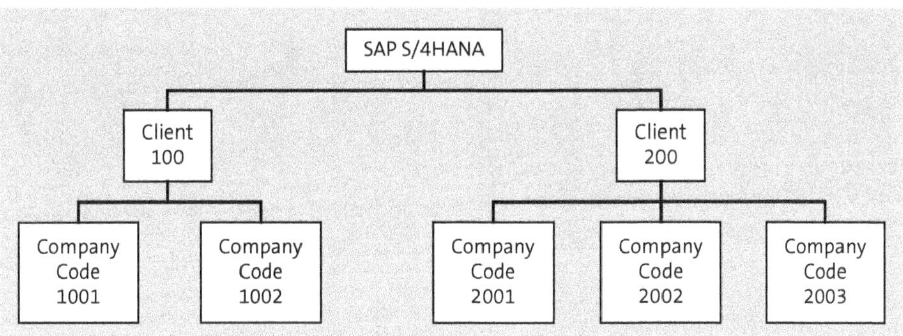

Figure 2.1 Client and Company Code Data Structure

Clients 000 and 001 are provided by default with base data and are used by administrators for maintenance work, like patching.

In technical terms, a client can be considered a subset of the database. From a business perspective, a client could be a group of corporations or subsidiaries owned by the same parent company.

Most data objects are client dependent. In other words, users can only see this data if they log on to the corresponding client. Examples of client-dependent data are materials, business partners, transactional documents, sales organizations, company codes, and so on. All these data elements are isolated from other clients. In the example shown in Figure 2.1, business data within client 100 cannot be accessed from client 200, or vice versa. However, some objects are shared across clients; these objects are known as *client-independent objects*. Output devices, ABAP programs, and directories of repository objects are examples of client-independent data since they display the same information regardless of the client a user is logged on to.

An easy way to distinguish between client-dependent and client-independent tables is by looking at field MANDT. This field represents the 3-digit client code. This field is always the first field in all client-dependent tables and is missing in client-independent tables. Figure 2.2 shows the definitions in the ABAP dictionary (Transaction SE11) of two tables: T001 and TADIR. Field MANDT is only used in client-dependent table T001.

- Client-dependent table: T001 (Company Codes)
- Client-independent table: TADIR (Directory of Repository Objects)

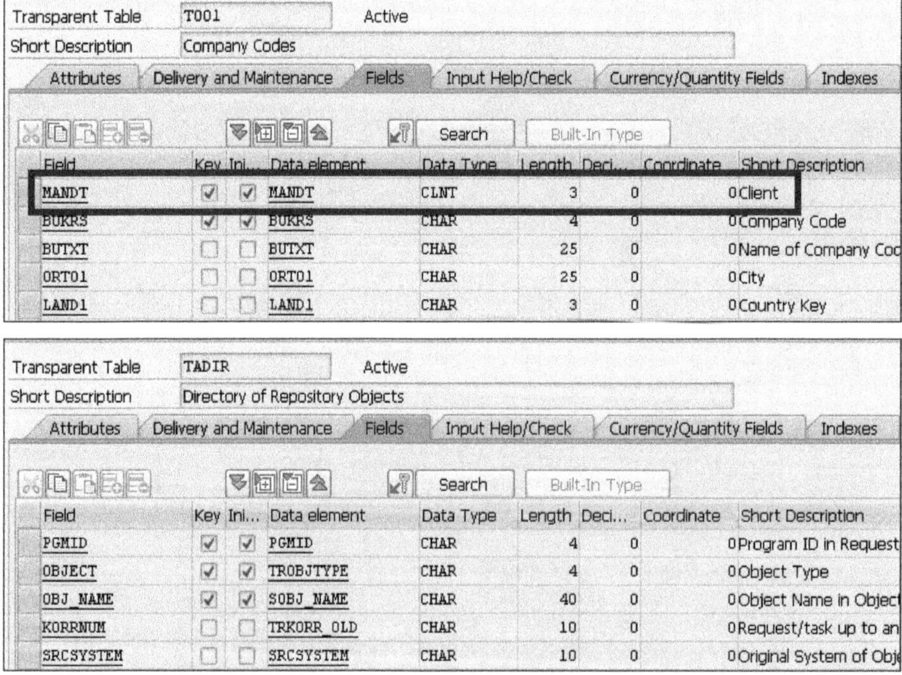

Figure 2.2 Client-Dependent versus Client-Independent Tables in SAP S/4HANA

System administrators maintain the settings for each client. We won't go into the details of all the different client attributes, which are out of scope for the certification exams, but we must mention one client attribute—the role of the client. This attribute dictates how the client is used in your SAP S/4HANA system. The following six roles are available when defining a client:

- Production
- Customizing
- Demo
- Training/education
- SAP reference
- Test

In a production client, customizing settings are protected by a client lock, meaning changes are not allowed. Changes must be made in the customizing client and then moved to the test client via client copy tools (Transaction SCC1) where testing is carried out. If the test is successful, changes are imported into the production client using transport requests.

Note

Some settings, such as posting periods or exchange rates, can be maintained in the production client directly without creating a transport request. These customizing settings are known as *current settings*.

The company code is an organizational unit used in financial accounting to represent a legal entity that has its own chart of accounts. You can define company codes in the IMG menu under **Enterprise Structure • Definition • Financial Accounting • Edit, Copy, Delete, Check Company Code.**

Figure 2.3 shows the basic information of a company code for the country USA. In the **Edit Company Code Data** customizing activity, you must also define the company code address by clicking on the icon. This address will be the company's address displayed in outputs, such as financial statements and customer invoices.

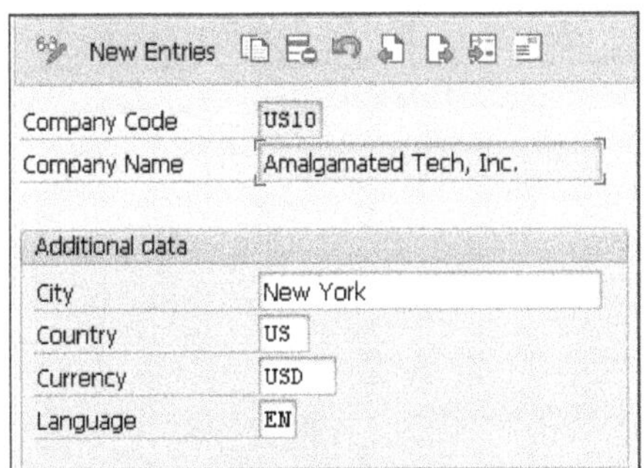

Figure 2.3 Company Code Basic Data

Another important set of data related to a company code are its global parameters. You can make various global specifications for the company code by following the menu path **SAP Customizing Implementation Guide • Financial Accounting • Financial Accounting Global Settings • Global Parameters for Company Code • Enter Global Parameters.** Figure 2.4 shows the global parameters for the sample company code US10 to give you an idea of the parameters that can be configured for a company code. However, we won't go into further details or analyze each individual field, which is beyond the scope of the exams.

Company Code US10 Amalgamated Tech, Inc. New York
Country key US Currency USD Language Key EN

Accounting organization

Chart of Accts	INT	Country Chart/Accts	
Company		FM Area	
Credit control area	0001	Fiscal Year Variant	K4
External CoCode	☐	Global CoCde	
Company Code Is Productive	☐	VAT Registration No.	
Hide Company Code in F4	☐		

Processing parameters

Document Entry Screen Variant		☐ Business Area Fin. Statements
Field status variant	0001	☐ Propose Fiscal Year
Pstng period variant	0001	☐ Define default value date
Max. exchange rate deviation	%	☐ No Exch. Rate Diff. When Clearing in LC
Sample Acct Rules Var.		☐ Tax base is net value
Workflow Variant		☐ Discount base is net value
Inflation Method		☐ Financial Assets Mgmt active
Tax Crcy Translation		☐ Purchase Acct Proc.
CoCd -> CO Area	1	☐ JV Accounting Active
Cost of Sales Accounting Actv.		☐ Hedge request active
☐ Negative Postings Permitted		☐ Enable Amount Split
		☐ Tax Reporting Date Active
☐ Manage Postg Period		

Figure 2.4 Company Code Global Parameters

If you only need one company code for your project, SAP recommends using the predefined company code 0001 (SAP A.G.) and amending the basic data and global parameters as needed. This approach will save you time by reducing the number of tables you must configure and reduces the risk of setting up something incorrectly.

Warning

Company code 0001 exists in clients 000 and 001 for the country Germany. Therefore, all country-specific parameters, such as currency, tax calculation procedures, and chart of accounts, are intended for use in Germany and may not be valid in other countries. If your project requires a company code in Canada, for example, you must first run the country installation program for Canada. This step will update the country of company code 0001 to Canada, and all its related global parameters will be set accordingly to meet Canada's legal requirements.

Since company codes have country-specific parameters, you'll need separate company codes for each country.

If more than one company code is required for your project, again, the recommendation from SAP is to use the **Copy Org. Object** function in the **Copy, Delete, Check Company Code** customizing activity, as shown in Figure 2.5. Select company code 0001—or any existing company code—as the reference company code to copy all the global parameters automatically. Then, you'll enter the 4-character key of the company code to be created. When using this functionality to copy company codes, the new company code must not have already been defined so the system can generate it.

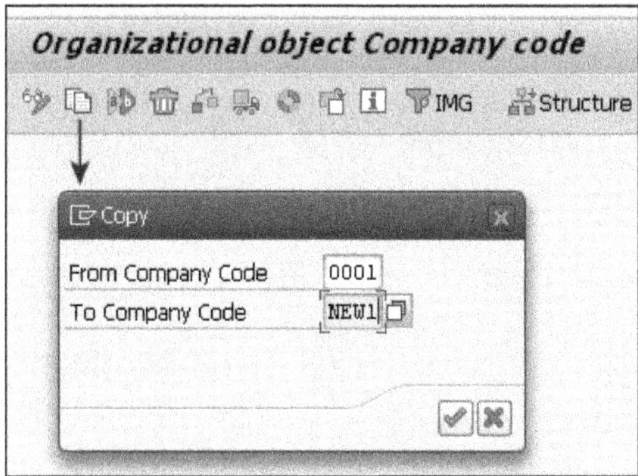

Figure 2.5 Copy Company Code

Once the company code has been created, you must maintain the basic data shown earlier in Figure 2.3, amend the company address, and adjust the global parameters that don't meet your project's requirements.

2.3 Maintaining and Assigning Organizational Units in Sales and Distribution

The sales and distribution process covers a large portion of the whole order-to-cash process in your company. In this section, we'll analyze the organizational units relevant for SAP S/4HANA Sales. SAP S/4HANA Sales is structured into the following three organizational units, which make up the framework of sales processing:

- A *sales organization* represents the legal entity responsible for the sale and distribution of products and services. This business unit defines the terms of sale and is present in all sales and distribution documents.

- A *distribution channel* represents how goods or services reach your customers. Examples of common distribution channels are retail, online sales, wholesale, direct sales, etc.
- A *division* represents a group of products or services that have similar characteristics or belong to the same line of business. The use of divisions helps salespeople specialize in certain business areas, such as data products for personal storage or storage for enterprise applications.

A sales organization, distribution channel, and division are mandatory prerequisites for processing sales transactions. The specific combination of these fields, known as a *sales area*, determines which distribution channel can be used to sell products of a certain business division in a sales organization. When creating documents such as quotations, sales contracts, or sales orders, the three organizational units that form the sales area are the first values required by the system.

Tip

The fact that sales organization, distribution channel, and division are mandatory fields to create sales and distribution documents in SAP S/4HANA doesn't mean that users must specify them every single time a sales document is created. To simplify document entry, you can store these three organizational units as user parameters in the user master record, and the system will propose these values automatically.

Besides sales and distribution documents, master data records in sales and distribution are also created at the sales area level and cannot be used in other parts of the organization. Examples of SAP S/4HANA Sales master data records are materials, business partners, and pricing condition records.

You can define the following four SAP S/4HANA Sales organizational units by following the menu path **SAP Customizing Implementation Guide • Enterprise Structure • Definition • Sales and Distribution**:

- Sales organization
- Distribution channel
- Sales office
- Sales group

Divisions, however, are maintained at the menu path **SAP Customizing Implementation Guide • Enterprise Structure • Definition • Logistics - General • Define, Copy, Delete, Check Division.**

2.3.1 Sales Organization

A *sales organization* is a 4-character alphanumeric key and the primary value of the sales area. When defining sales organizations, you should try to keep them to a minimum. The more sales organizations you create, the more records are required at the master data level. Usually, one sales organization per company code is enough. During the discovery phase of the implementation project, the following aspects must be taken into account concerning sales organizations:

- You should make sure that each sales organization is included in only one company code. In other words, you cannot assign a sales organization to multiple company codes.
- One or more plants can be assigned to the same sales organization.
- An address must be maintained for the sales organization.
- Master data records are created within a specific sales organization. Therefore, you can segregate materials and/or customers by sales organization. Sharing master data across sales organizations is not possible.
- Specific sales document types can be defined for each sales organization.
- Sales offices can be assigned to sales organizations.
- All items in a sales and distribution document belong to a sales organization.
- SAP S/4HANA Sales reports for sales documents, deliveries, and billing documents have the sales organization in the selection criteria.
- Each sales organization can have its own currency for statistics and its own calendars.
- In a system with multiple company codes, you must define a customer number for intercompany billing that represents each sales organization.

You should also know that you can hide specific values (e.g., sales organizations) in the input help, accessed by pressing F4, for sales organizations when processing sales transactions. This option is useful if a portion of your business has been acquired by another company and the corresponding sales organization must not be used going forward. Figure 2.6 shows the checkbox that, if selected, will hide the value from the input help. Note, however, the system won't return an error if a hidden value is entered by a user.

Figure 2.6 Defining Sales Organizations: Hiding a Value in Input Help

Note

The **Hide Value in Input Help** feature is available for sales organizations, distribution channels, divisions, sales offices, and sales groups.

2.3.2 Distribution Channel

The *distribution channel* is a 2-character alphanumeric key and the second field that forms a sales area. Two common distribution channels are direct sales and indirect sales. Products sold through your company's salesforce use the direct distribution channel, whereas goods sold through distributors use the indirect distribution channel. When defining new distribution channels in SAP S/4HANA, you should consider the following aspects:

- A distribution channel can be assigned to one or more sales organizations.
- One or more plants can be assigned to the same distribution channel.
- Master data records are created within a specific distribution channel. Therefore, you can segregate materials and/or customers by distribution channel.
- Specific sales document types can be defined for each distribution channel.
- Sales offices can be assigned to distribution channels.
- All items in a sales or billing document belong to the same distribution channel. However, items in the same delivery can belong to different distribution channels.
- SAP S/4HANA Sales reports for sales documents, deliveries, and billing documents have the distribution channel in the selection criteria.
- You cannot maintain an address for a distribution channel.

You can also define reference distribution channels for common conditions, customers, and materials within the same sales organization. Doing so will save you time in creating and maintaining master data since you can, for instance, create your master records in distribution channel 10 and adjust the settings so these master records are also valid in distribution channel 20. However, before this feature is enabled, you should confirm if it's in line with the business distribution strategy because, in some cases, a material must be sold through one distribution channel only.

2.3.3 Division

A division is a 2-digit alphanumeric key and the third and final field that forms a sales area. When creating a new division in SAP S/4HANA, the following points must be considered:

- A division can be assigned to one or more sales organizations.
- A division can be assigned to one or more distribution channels.
- A material master record is always assigned to just one division. Even though the division field is maintained in the **Sales: sales org. 1** view in the material master, the value is stored in table MARA (Material General Master Data) so the value is the same regardless of the sales organization.
- Customer and pricing data records are created within a specific division. Therefore, you can segregate materials and/or customers by division.

- Sales offices can be assigned to divisions.
- You can customize sales document types to ensure all items belong to the same division.
- Items in the same delivery or billing document can belong to different divisions.
- SAP S/4HANA Sales reports for sales documents, deliveries, and billing documents have the division as a selection criterion.

As with the distribution channel, you can define common divisions to share condition and customer master data within the same sales organization. As mentioned in the previous section, this approach can simplify the creation and maintenance of master data records, but you must be sure that creating common divisions is the right approach from a business standpoint.

Once your sales organizations, distribution channels, and divisions have been defined, you can maintain the relevant assignments and reproduce your company's structure in the system. This step can be carried out in customizing under **Enterprise Structure • Assignment • Sales and Distribution**.

First, to establish the interface between SAP S/4HANA Sales and SAP S/4HANA Finance, you'll assign each sales organization to a company code. Then, you'll allocate distribution channels and divisions to sales organizations, and finally, you'll set up sales areas by specifying valid combinations of sales organization, distribution channel, and division. Figure 2.7 shows an example of a basic organizational structure in SAP S/4HANA Sales consisting of four sales areas.

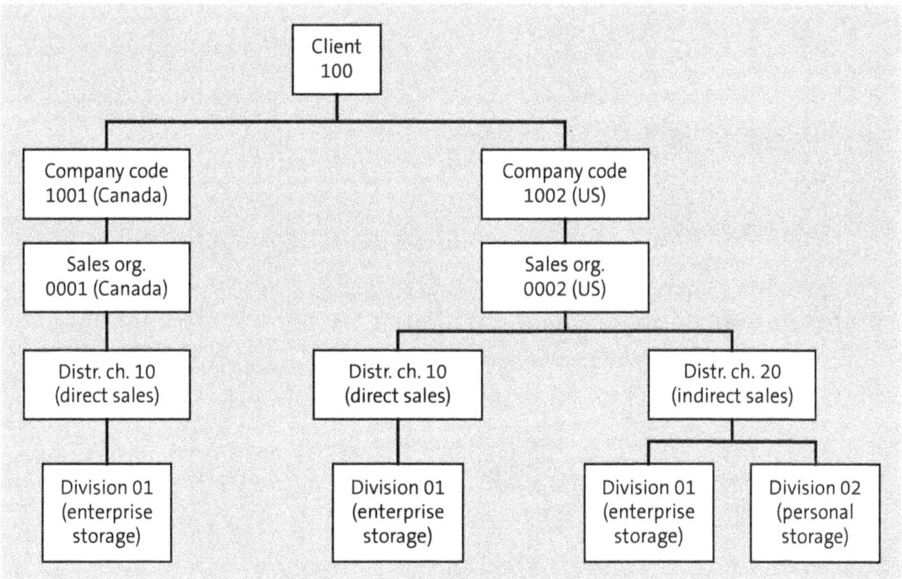

Figure 2.7 Sample Organizational Structure in SAP S/4HANA Sales

In our example shown in Figure 2.7, the following four sales areas are identified the structure (from left to right and top to bottom):

- Sales area 1: Sales organization 0001 → Distribution channel 10 → Division 01
- Sales area 2: Sales organization 0002 → Distribution channel 10 → Division 01
- Sales area 3: Sales organization 0002 → Distribution channel 20 → Division 01
- Sales area 4: Sales organization. 0002 → Distribution channel 20 → Division 02

You must also define the link between sales organizations, distribution channels, and plants. In the event you have multiple plants and multiple sales organizations within the same company code, you can specify which sales organizations can use which plants. You can also allocate a plant that belongs to a different company code to a sales organization. This scenario is shown in Figure 2.8, where sales organization 0001 can make cross-company sales for goods from plant 000B.

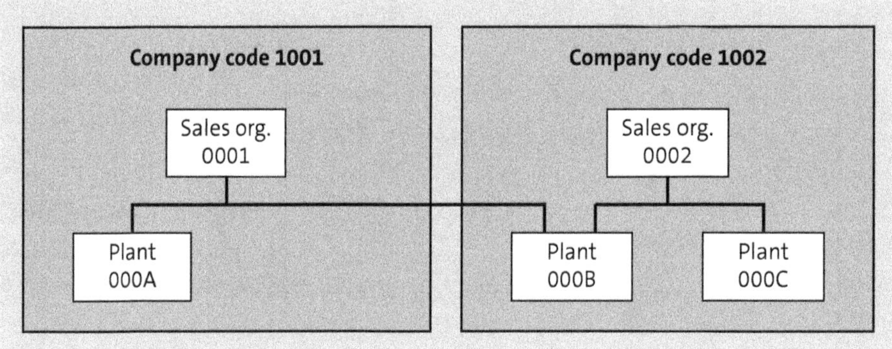

Figure 2.8 Plant Assignment to Sales Organizations

In this example, sales organization 0001 sells items from plants 000A and 000B, whereas sales organization 0002 sells items from plants 000B and 000C.

2.3.4 Sales Office

A sales office is a 4-character alphanumeric key representing a physical location in your organization that is responsible for selling products or services in a geographical area. Defining sales offices in SAP S/4HANA is not mandatory. If you decide to create sales offices, you should consider the following aspects:

- A sales office can be assigned to one or more sales areas.
- A sales office can be divided into sales groups.
- An address must be maintained for the sales office.
- You can assign business partners to a sales office.

- All items in a sales document belong to the same sales office. However, items in the same delivery or billing document can belong to different sales offices.
- The sales office field can be a selection criterion for lists of sales documents. However, this field is not a selection field for lists of deliveries and billing documents.

2.3.5 Sales Group

A sales group is a 3-character alphanumeric key representing a group of sales employees in a sales office. Like sales offices, defining sales groups is optional. If you want to create sales groups in your SAP S/4HANA system, you should consider the following aspects:

- A sales group can be assigned to one or more sales offices.
- You can assign salespeople to a sales group.
- All items in a sales document belong to the same sales group. However, items in the same delivery or billing document can belong to different sales groups.
- The sales group field can be a selection criterion for lists of sales documents. However, this field is not a selection field for lists of deliveries and billing documents.

Through various assignments, you can create an internal organization in sales and business development like the sample organization shown in Figure 2.9.

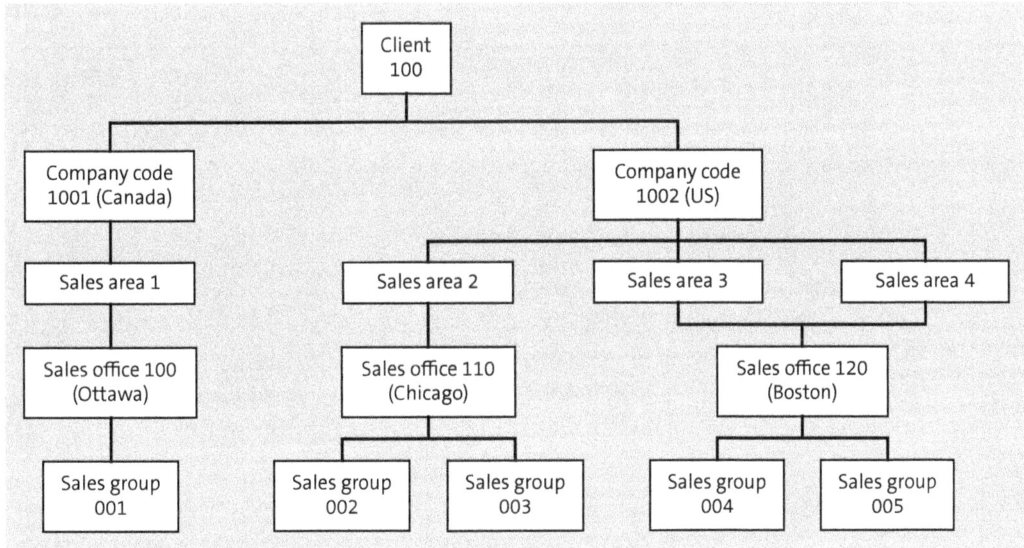

Figure 2.9 Sample Organization in Business Development and Sales

2.4 Maintaining and Assigning Organizational Units in Logistics

In this section, we'll cover the definition of the organizational structures required to store and distribute products. You must define the following four main components to enable the use of logistics applications in SAP S/4HANA:

- **Plant**
  A plant is an operational unit within a company code. Examples of plants include manufacturing facilities and branch offices.
- **Storage location**
  A physical area of the plant where goods are stored.
- **Warehouse number**
  A warehouse number in SAP S/4HANA is a complete physical warehouse complex, usually formed by several storage areas or buildings in the same geographical location. Even if you're using inventory management without warehouse management, we recommend implementing Lean WM in fixed bin warehouses to pick deliveries using the advantages of transfer orders.
- **Shipping point**
  A shipping point is a physical location from which goods are shipped. An example of a shipping point is a loading ramp.

Additionally, if required by your project, you can maintain two more optional organizational units in logistics execution:

- **Loading point**
  Basically a subdivision of a shipping point, for example, a specific loading bay.
- **Transportation planning point**
  A group of employees responsible for the creation of shipment documents in transportation processing. Each transportation planning point is assigned to a company code. If shipment documents are out of the scope of your project, then transportation planning points are not necessary.

For the certification exams, the definition and assignment of these two optional organizational units in logistics execution is out of scope. For our purposes, we'll focus on the main four organizational units (plant, storage location, warehouse number, and shipping point) next.

2.4.1 Plant

To define a plant, follow the menu path **SAP Customizing Implementation Guide • Enterprise Structure • Definition • Logistics - General • Define, Copy, Delete, Check Plant**. The plant plays a fundamental role in material valuation, material requirements planning (MRP) and production, inventory management, and costing.

A plant is defined by the following attributes:

- A plant has an address, a language, and belongs to a country.
- Material master data is created at the plant level for some of the views. (We'll elaborate further on this topic in Chapter 3, Section 3.2.)
- A plant cannot be assigned to multiple company codes, but one company code can have several plants.
- A plant can have several storage locations.
- A single plant can be assigned to different combinations of sales organization and distribution channel.
- A plant can have multiple shipping points.

2.4.2 Storage Location

Storage locations are considered organizational units in materials management and are always created for a plant. To define a storage location, follow the menu path **SAP Customizing Implementation Guide • Enterprise Structure • Materials Management • Maintain Storage Location**. You should be aware of the following attributes when defining storage locations:

- Multiple storage locations can be maintained for the same plant.
- At least one address must be created for each storage location.
- Material master data is created at the storage location level for some views.
- More than one storage location can be assigned to the same warehouse number within a plant.
- You can perform physical inventories at the storage location level.
- The valuation of stocks cannot be maintained at the storage location level.

2.4.3 Warehouse Number

The final two organizational units we'll review in this section are warehouse numbers and shipping points. These units are both considered organizational units in logistics execution and can be maintained via the SAP Implementation guide under **Enterprise Structure • Definition • Logistics Execution**.

If your company engages in high-volume warehouse operations, then you should implement embedded extended warehouse management (EWM) in SAP S/4HANA and start defining a warehouse number in the enterprise structure. But if embedded EWM in SAP S/4HANA is not required, you can decide not to define a warehouse number or to define a warehouse number in the plant where the degree of activation for a specific storage location is set to Lean WM. The Lean WM option allows you to pick deliveries using transfer orders without updating bin stock levels in warehouse management.

Consider the following characteristics when maintaining warehouse numbers:

- An address cannot be maintained for a warehouse number.
- Material master data is created at the warehouse number level for some views.
- The valuation of stocks cannot be maintained at the warehouse number level.
- You can assign multiple storage locations to the same warehouse number within a plant. The warehouse number is, therefore, linked to a combination of plant and storage location.

2.4.4 Shipping Point

Shipping points are required to create delivery documents in the system since these organizational units are responsible for managing deliveries to customers. Each delivery is processed by a single shipping point, which is defined by a 4-character alphanumeric key and has the following attributes:

- An address must be maintained for a shipping point.
- A shipping point can be assigned to multiple plants.
- In terms of shipping, the shipping point resides at the top level in the organization.
- You can divide a shipping point into loading points.
- All items in a delivery document belong to the same shipping point.
- In a sales document, the shipping point can be determined at the item level based on the plant, the loading group, and the shipping condition.
- The shipping point can be a selection criterion for lists of deliveries and when processing deliveries.

2.5 Important Terminology

In this chapter, the following terminology was used:

- **Client**
  The highest level in an SAP S/4HANA hierarchy, typically representing a corporate group. A client can have one or more company codes.
- **Company code**
  An independent legal entity in finance with its own set of accounts.
- **Sales organization**
  Represents the legal entity responsible for the sale and distribution of products and services. This organizational unit is also the business unit that defines the terms of sale and is present in all sales and distribution documents.
- **Distribution channel**
  Represents how goods or services reach your customers. Examples of common distribution channels include retail, online sales, wholesale, direct sales, etc.

- **Division**
  Represents a group of products or services that have similar characteristics or belong to the same line of business. The use of divisions helps your salespeople specialize in certain business areas, such as data products for personal storage or storage for enterprise applications.
- **Sales area**
  An organizational element formed by a combination of sales organization, distribution channel, and division.
- **Sales group**
  A group of people in the sales department who are responsible for processing sales of certain products or services.
- **Sales office**
  A physical location responsible for the sale of certain products or services within a given geographical area.
- **Plant**
  An operational unit within a company code. Examples of plants include manufacturing facilities and branch offices.
- **Storage location**
  An organizational unit within a plant where stock is physically kept. Each plant may contain one or more storage locations.
- **Warehouse number**
  Identifies a complete physical warehouse structure where the materials are stored and goods movements and physical inventory take place. Usually, warehousing facilities in the same geographical area are represented by a single warehouse number.
- **Shipping point**
  A physical location used to ship goods. An example of a shipping point is a loading ramp.
- **Lean WM**
  A simplified version of the warehouse management system (WMS) for a fixed bin warehouse that allows the use of transfer orders as pick lists while managing goods movements and stock changes at the inventory management level. You can define the degree of activation (full or Lean WM) for a warehouse number at the plant and storage location level.
- **Loading group**
  Defines a group of materials that share the same loading requirements. For example, materials that require the same amount of time for loading or that use the same loading bay.
- **Shipping condition**
  The shipping strategy for the delivery of goods to customers. Typical values include standard, immediately, or pickup.

2.6 Practice Questions

These practice questions will help you evaluate your understanding of the topics covered in this chapter. The questions shown are similar in nature to those found on the certification examination. Although none of these questions will be found on the exam itself, they will allow you to review your knowledge of the subject. Select the correct answers and then check the completeness of your answers in the next section. Remember, that on the exam, you must select all correct answers and only correct answers to receive credit for the question.

1. True or False: When creating a sales document, specifying organizational units is not always necessary since these values can be proposed automatically.

☐ A. True

☐ B. False

2. Which organizational units are absolutely necessary prerequisites for processing sales transactions? (There are three correct answers.)

☐ A. Sales group

☐ B. Division

☐ C. Sales office

☐ D. Distribution channel

☐ E. Sales organization

3. True or False: The client field (field MANDT) is only used in client-independent tables.

☐ A. True

☐ B. False

4. Which assignments are possible when configuring organizational structures for sales and distribution? (There are two correct answers.)

☐ A. Assigning a distribution channel to multiple company codes

☐ B. Assigning a plant to multiple company codes

☐ C. Assigning a shipping point to multiple plants

☐ D. Assigning a division to multiple distribution channels

5. In the organizational structure for sales, which unit represents a group of products or services that have similar characteristics or belong to the same line?

☐ A. Sales organization

☐ B. Distribution channel

☐ C. Sales group

☐ D. Division

6. True or False: A single sales organization can be assigned to multiple company codes.

☐ A. True

☐ B. False

7. Which of the following statements apply to the definition of organizational units in sales and distribution? (There are two correct answers.)

☐ A. An address must be maintained for each sales organization.

☐ B. An address must be maintained for each distribution channel.

☐ C. All the items in a sales document or delivery belong to the same sales organization.

☐ D. All the items in a sales document or delivery belong to the same distribution channel.

8. Which of the following organizational elements in SAP S/4HANA allow you to hide certain values in the input help when processing sales transactions? (There are three correct answers.)

☐ A. Plant

☐ B. Sales organization

☐ C. Company code

☐ D. Sales office

☐ E. Distribution channel

9. True or False: The division field is maintained in the material master view **Sales: sales org. 1**. Therefore, you can assign a material to a different division in each sales organization.

☐ A. True

☐ B. False

10. Which organizational elements can be defined in sales and distribution? (There are two correct answers.)

☐ A. Distribution channel

☐ B. Division

☐ C. Sales office

☐ D. Shipping point

11. To which organizational unit do you assign a shipping point?

☐ A. Plant

☐ B. Sales organization

☐ C. Storage location

☐ D. Warehouse number

12. To which objects is a warehouse number assigned in logistics execution?

☐ A. Plant and storage section

☐ B. Plant and storage location

☐ C. Plant and storage type

☐ D. Plant and company code

13. True or False. You can allocate a plant to a sales organization that belongs to a different company code.

☐ A. True

☐ B. False

14. Which of the following statements apply to distribution channels in the SAP S/4HANA system? (There are two correct answers.)

☐ A. Only one plant can be assigned to a distribution channel.

☐ B. Sales offices can be assigned to distribution channels.

☐ C. Shipping points can be assigned to distribution channels.

☐ D. Master data records are created within a specific distribution channel.

2.7 Practice Question Answers and Explanations

1. Correct answer: **A**

   True. You can store a sales organization, a distribution channel, and a division as user parameters in the user master record. The system will automatically default these values when this user creates a sales document.

2. Correct answers: **B, D, and E**

   A sales organization, distribution channel, and division are the three organizational units that form a sales area and are mandatory values for processing any sales transaction. A sales group and a sales office, however, are optional organizational units in sales documents.

3. Correct answer: **B**

   False. In fact, the opposite is true. The client field (field `MANDT`) is only used in client-dependent tables since the system must be able to segregate table entries by client. For example, when you display a client-dependent table in the data browser, the system only shows you the entries for the client to which you're logged on.

4. Correct answers: **C and D**
   You can only assign a plant to one company code, and you can only assign a sales organization to one company code. Therefore, answers A and B are incorrect. Thus, only answers C and D describe correct assignments in SAP S/4HANA Sales.

5. Correct answer: **D**
   A division is defined as a group of materials or a range of products with similar characteristics.

6. Correct answer: **B**
   False. In SAP S/4HANA you can only assign a sales organization to one company code.

7. Correct answers: **A and C**
   When defining a sales organization, you must maintain an address. Thus, answer A is correct. The sales organization field is located at the header level in both sales documents and deliveries; thus, all items in the document belong to the same sales organization. Thus, answer C is the second correct answer to this question. Answers B and D are both false statements.

8. Correct answers: **B, D, and E**
   The **Hide Value in Input Help** feature is available in the definition of sales organizations, distribution channels, divisions, sales offices, and sales groups. You cannot hide plants or company codes from the input help capability.

9. Correct answer: **B**
   False. Even though the division field is shown on the material master view **Sales: sales org. 1**, this value is stored in table MARA, which is the table for general material data. Values of all the fields in table MARA (including division) are applicable to all plants and sales organizations where the material is maintained.

10. Correct answers: **A and C**
    Divisions are defined in **Logistics - General** and shipping points are defined in Logistics Execution. Both distribution channels and sales offices are maintained in sales and distribution.

11. Correct answer: **A**
    Shipping points can be assigned to one or more plants.

12. Correct answer: **B**
    A warehouse number is assigned to a combination of plant and storage location.

13. Correct answer: **A**
    True. You can assign a plant to a sales organization that belongs to a different company code to enable cross-company sales. For example, let's say a customer places an order with sales organization 0001, which belongs to company code

1001. However, the goods are manufactured and delivered by plant 000B, which belongs to company code 1002. If a sales order is created for sales organization 0001 with delivering plant 000B, SAP S/4HANA will automatically generate an intercompany billing document from company code 1002 to company code 1001.

14. Correct answers: **B and D**
    You can allocate one or more plants to a distribution channel, so answer A is incorrect. Shipping points are assigned to plants, not to distribution channels; thus, answer C is also incorrect. You can create master data records within a specific distribution channel, for example, if you want to segregate materials by distribution channel. Also, you can assign sales offices to distribution channels, as shown earlier in Figure 2.9. (Note that the distribution channel is part of the sales area.) Therefore, answers B and D are the two correct answers for this question.

2.8 Takeaway

At this point, you should now have a solid understanding of SAP's organizational units and how to build a cohesive enterprise structure.

In addition, you learned about the basic configuration and assignment of organizational units. You now know the role of the client in SAP S/4HANA and how data is structured across clients and company codes.

Finally, we covered the main attributes of the organizational elements in financial accounting, sales and distribution, materials management, and logistics. We provided insight into the possible assignments you can make between these units and into how the organizational structure shapes both master data and transactional data.

2.9 Summary

In this chapter, you gained an understanding of the business units you must define in the system to represent the functional and legal structure of your company. As an SAP S/4HANA Sales consultant, this competence lays the foundations for master data records, sales reports, and transactions in the areas of sales and distribution, materials management, and logistics execution.

In the next chapter, we'll focus on the most important master data records in sales and distribution, such as material master data, customer-material-information records, and master data for pricing. The business partner is also an essential master data object in sales, but we'll cover this master data in Chapter 12, Section 12.3, when we discuss the main simplifications introduced by SAP S/4HANA 2020 in the sales and distribution area.

Chapter 3
Master Data

Techniques You'll Master

- Understand the material master data record
- Create and maintain customer-material info records
- Create and maintain pricing condition master records

This chapter will focus on the most influential master data objects for sales and distribution except for the business partner, which will be covered in Chapter 12, Section 12.3, since the new business partner approach is one of the main simplifications delivered by SAP S/4HANA in the area of sales. We'll explain the material master data structure with special attention to those material views and fields with higher relevance for sales transactions. Next, you'll learn how to record customer-specific information for a material in a customer-material info record. We'll end this chapter by showing you how to maintain pricing condition records in SAP S/4HANA Sales and how can they be used to fulfill your business requirements.

Real-World Scenario

Master data forms the foundation of sales and distribution processing. The accuracy of transactional data and, by extension, of reporting is based on correct master data records. Low-quality master data can lead to delays in order processing, shipping, and billing. Most companies have dedicated teams responsible for creating and updating master data to ensure the data is properly maintained.

As an SAP S/4HANA Sales consultant, you must be familiar with the key fields and functionalities available in each master data object used in sales and distribution. This knowledge will enable you to identify mandatory fields and advise your business on the values to be maintained on a specific master data object. You may be asked, for example, how to create a pricing condition record for a special price that ignores all possible discounts applicable in a sales order item. You may also be required to explain to the master data team the changes needed in the material master. For instance, perhaps the default position for a group of materials in a given sales organization is to dropship them from an external vendor, whereas other materials in the same sales organization may be shipped from a local plant by default.

3.1 Objectives of This Portion of the Test

The goal of this portion of the SAP S/4HANA Sales Certification exam is to test your understanding of the following topics:

- Managing material master records, in particular, the screens and fields most relevant for sales
- Maintaining customer-material info records
- Defining and managing pricing condition records in sales and distribution

Note

The Master Data topic makes up 8%–12% of the total SAP S/4HANA Sales 2020 exam. However, the business partner, which is a fundamental master data object, will be covered later in Chapter 12.

The content of this chapter is included in the Master Data and Organizational Structures topic of the SAP S/4HANA Sales 2020 Upskilling exam, which makes up more than 12% of the total exam.

3.2 Maintaining Material Master Data

Master data in SAP S/4HANA can be defined as the data records used to store information about the goods sold, manufactured, or purchased by the company; about bills of materials (BOMs); and about business partners. Master data also refers to data containing information about the prices of your products, customer discounts, freight charges, and customer-specific information for materials (e.g., a customer's description or code for a material).

During sales processing, a vast amount of data is transferred from master records into a sales document, which is later on passed to the shipping document and eventually to the billing document. Therefore, to obtain accurate results and accelerate business operations, you must ensure your master data is accurately maintained. For example, when a sales order is created, as soon as the customer is entered, the system reads the master data to populate fields like the shipping condition or the payment terms to mention just a few data points. Then, when the material is entered, data from the material master record is copied into the sales order item, like the unit of measurement (UoM) or delivering plant. However, at this point, the system can use both customer and material information to determine the item's price, discounts, and taxes, all of which will be passed on to the invoice. This general example shows how important making sure all the master data is correct and complete is for minimizing errors in transactional documents, which, in return, increases customer satisfaction.

The majority of the information related to the products or services your company sells is stored in the material master. This tendency is also valid for raw materials and semifinished products produced in-house.

Information in the material master can be maintained across several organizational levels. For instance, you can decide to define a product for a specific sales area only and then only in those plants where the material is manufactured. Maintaining the materials in the right organizational units with the correct values will guarantee the smooth processing of transactions.

When you create a material, you must specify the industry sector and the material type, which are defined in the following ways:

- **Industry sector**
  This field identifies the branch of industry of the material (e.g., retail, chemical industry, pharmaceuticals, or mechanical engineering). The sequence of screens and the field selection in a material master record are determined by the industry sector of the material. Once you've assigned an industry sector to a material, you cannot change the industry sector later.
- **Material type**
  This field used to group materials with similar characteristics and attributes (e.g., services, trading goods, packaging, or raw materials). The material type determines certain attributes of the material, including, for instance, whether the material number can be assigned internally or externally and whether the valuation of the material will be carried out according to the standard price (price control S) or the moving average price (price control V). As with the industry sector, when processing a material master record, the material type also determines which screens are displayed and in what sequence.

Tip
As with many other objects in SAP S/4HANA, you don't have to create a material master record from scratch. You can use the **Copy from Material** option and enter a reference material number that provides the data you need to create the new material.

The material master is divided into user department views, each of which consists of one or more tabs. The standard sequence of views is as follows:

1. Basic data
2. Sales
3. Sales text
4. Purchasing
5. Purchase order (PO) text
6. Material requirements planning (MRP)
7. Forecasting
8. Work scheduling
9. Storage
10. Warehouse management
11. Accounting

Each material type is configured with its own set of views. A *view* is linked to a specific department within the business, so you must maintain the relevant view before you can use the material for the functions carried out in that particular department. For example, you cannot create a PO for a given material if the **Purchasing** view has not been maintained. Figure 3.1 shows the material views available for standard material type HAWA (Trading Goods).

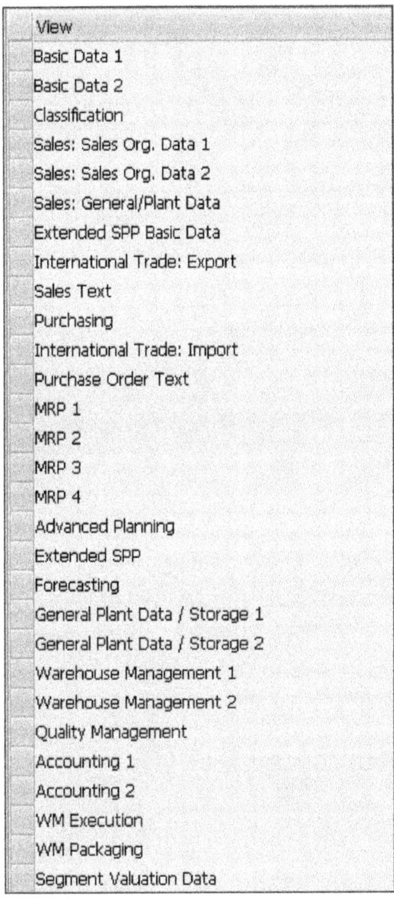

Figure 3.1 Material Master Views for Material Type HAWA

SAP S/4HANA provides over 40 standard material types to choose from, but you can create a new material type if required by your project. If you need to create a custom material type, we recommend copying and modifying an existing material type.

The following contains some of the material types delivered by SAP that can be selected when creating a material master record for a salable product:

- **DIEN – Service**
  Services provided by your company must be created as materials of type DIEN. These types of materials lack storage data and cannot be produced or supplied by a vendor. Examples of services include consulting, equipment maintenance, or security services.

- **FERT – Finished Product**
  Finished products are materials manufactured in-house. These products cannot be purchased, and so no purchasing data is maintained for these materials.
- **HAWA – Trading Goods**
  These products are purchased by your company from vendors and are later sold to customers without changing their components or properties. No internal operations are performed on these types of materials. Some examples could be a bottle of milk in a supermarket or a videogame in an electronics boutique.
- **KMAT – Configurable Materials**
  These materials can be configured via variant configuration. An automobile is a classic example of a configurable material. During the sales process, a customer will select the color of the car, the type of alloy wheels, engine size, parking sensors, etc. The **Material is Configurable** indicator is always set for this material type.
- **NLAG – Non-Stock Material**
  These materials are not held in stock at your company. Material type NLAG is typically used for nontangible materials like software downloaded from a website.

When defining a material, you don't need to maintain all the available views. You could create a material master record by only maintaining the basic data (**Basic Data 1** and **Basic Data 2** views) and then later extend the material to the areas of sales, purchasing, and quality management by maintaining the associated views, which allow your users to carry out additional activities with that material.

Note

Normally, users from different departments maintain only the views that contain data that is relevant for them. For instance, a user working in finance may be asked to maintain the **Accounting 1** and **Accounting 2** views, whereas a user in the sales department might be a good candidate to maintain all the sales views. SAP S/4HANA workflows can be used to notify users when a new material record is created and to indicate that their action is required to extend it to a specific user department.

As new views are maintained, the material master record acquires new *maintenance statuses*. The maintenance status tells you which data has already been maintained. Table 3.1 shows the list of all possible maintenance statuses in SAP S/4HANA. (The data contained in the maintenance statuses in bold cannot be updated manually.)

User Department	Maintenance Status
Accounting	B
Basic data	K
Classification	C
Costing	G
Forecasting	P
MRP	D
Plant stocks	**X**
Production resources/tools	F
Purchasing	E
Quality management	Q
Sales	V
Storage	L
Storage location stocks	**Z**
Warehouse management	S
Work scheduling	A

Table 3.1 Maintenance Statuses in the Material Master Record

Once a material master record has been created, technically, you can extend the material master in two ways:

- Extending the material to a new department area by maintaining new views. This step will update the maintenance status of the material. For example, a material record that doesn't have any quality-related information can be extended to the **Quality Management** view.
- Extend the material to a new organizational unit by populating a new segment of data without maintaining new views. For example, a material that already exists in sales organization 0001 can be extended to sales organization 0002. In this case, the maintenance status of the material remains unchanged since no new views are maintained.

When displaying a material master record, the [i] icon is available on any main screen and will provide you with the information shown in Figure 3.2 about the material record, including the material type, deletion flags, and who created the departmental data.

At this point, for a material to be sold in the system, sales views must be maintained. The same principle can be applied to other user departments, such as purchasing or finance. Analyzing each individual view in the material master is beyond the scope of this book. Instead, we'll cover the basic data views (**Basic data 1** and **Basic data 2**) and the sales views (**Sales: sales org. 1** and **Sales: sales org. 2**) since these views are the most relevant views for all sellable materials.

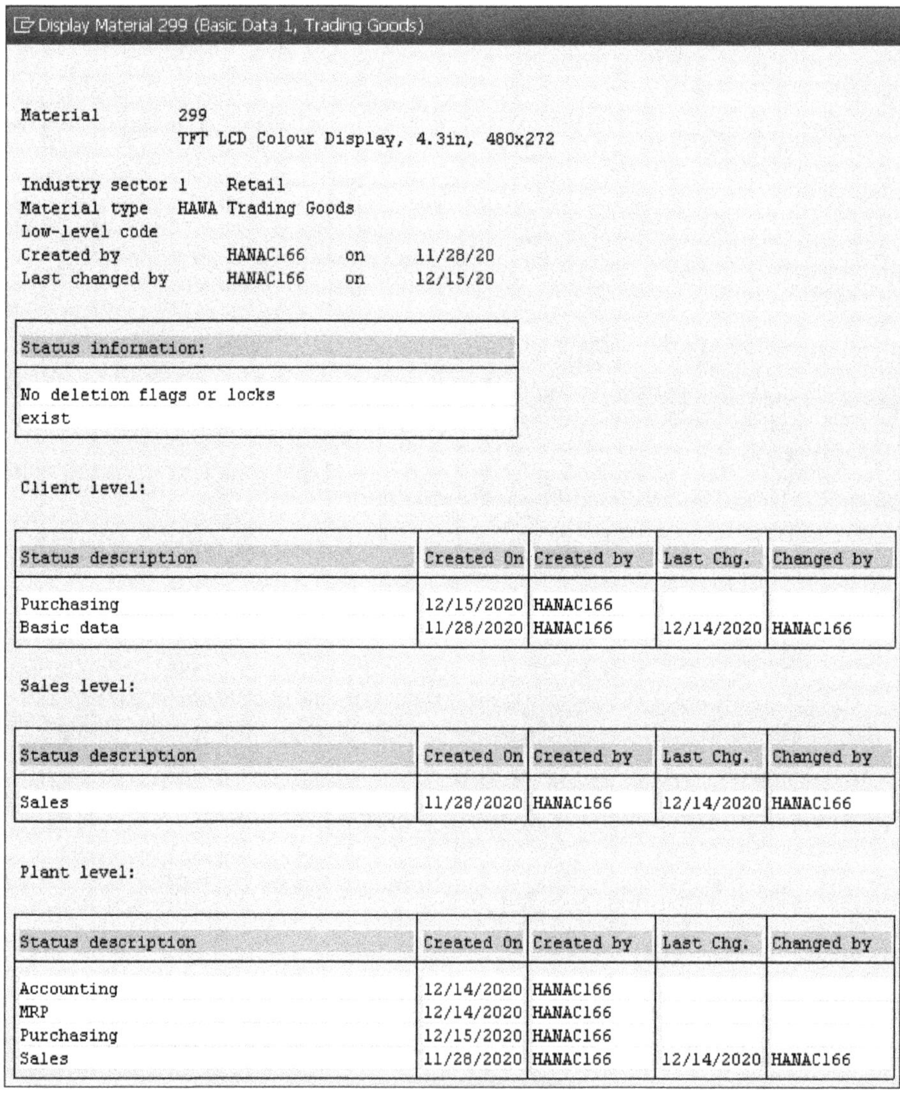

Display Material 299 (Basic Data 1, Trading Goods)

Material 299
TFT LCD Colour Display, 4.3in, 480x272

Industry sector 1 Retail
Material type HAWA Trading Goods
Low-level code
Created by HANAC166 on 11/28/20
Last changed by HANAC166 on 12/15/20

Status information:

No deletion flags or locks exist

Client level:

Status description	Created On	Created by	Last Chg.	Changed by
Purchasing	12/15/2020	HANAC166		
Basic data	11/28/2020	HANAC166	12/14/2020	HANAC166

Sales level:

Status description	Created On	Created by	Last Chg.	Changed by
Sales	11/28/2020	HANAC166	12/14/2020	HANAC166

Plant level:

Status description	Created On	Created by	Last Chg.	Changed by
Accounting	12/14/2020	HANAC166		
MRP	12/14/2020	HANAC166		
Purchasing	12/15/2020	HANAC166		
Sales	11/28/2020	HANAC166	12/14/2020	HANAC166

Figure 3.2 Information about the Material

The **Basic data 1** and **Basic data 2** views contain information at the client level that is common for all sales organizations, distribution channels, and plants. Figure 3.3 displays an example of the **Basic data 1** view of a material master data record.

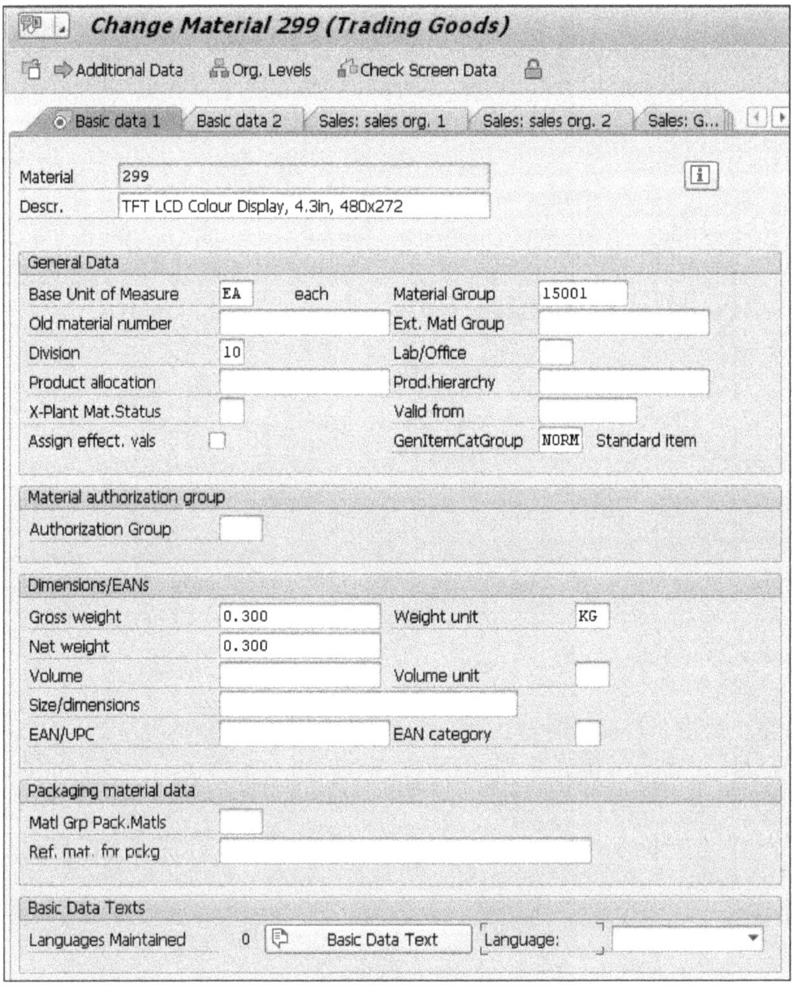

Figure 3.3 Material Master Record: Basic Data 1 View

The goal of this section is to review the most important fields in the basic data and sales screens, so we won't cover all the fields in each view. Instead, we'll focus on the following fields, which have a significant impact on sales order processing:

- **Basic data 1**
  - **Material (material number)**
    In this field, the material number is a unique number that identifies the product or service in the system. This number may be automatically assigned by the system (internal assignment) or may be a character string entered by a user when the material is created (external assignment). Previously, this field was 18 characters long, but the maximum field length of a material number was extended in SAP S/4HANA from 18 to 40 characters. By default, the extended material number functionality is not activated. To use a material number with 40 characters, you must activate this functionality by navigating to **IMG • Cross-Application Components • General Application**

Functions • Field Length Extension • Activate Extended Fields. Special characters are allowed in material numbers with external number assignments. However, certain characters—for instance, ampersands (&), asterisks (*), and commas (,)—cannot be used due to technical reasons.

- **Descr. (description)**
  This field is usually the first field maintained. A material description can be 40 characters long, so ideally, a material naming convention should be defined across your company. If needed, you can use **Basic Data Texts** button to add extra text in the **Basic data 1** view.
- **Base Unit of Measure**
  A UoM must be defined to represent how much stock exists in your warehouses. The base UoM is used to manage stocks. A material's UoM could be each, bottle, kilogram, inch, etc. Conversion factors are maintained from alternative UoMs to the base UoM (e.g., 1 box = 12 each).
- **Material Group**
  This field helps you group similar materials. A material group is 9 characters long, and only the groups already configured can be selected. For instance, companies selling electrical components may want to create material groups to segregate cables from motors, fans, and blowers.
- **X-Plant Mat.Status (cross-plant material status)**
  This 2-character field indicates the stage in the material's lifecycle and is valid across the client. You can configure the statuses available and whether the usability of the product must be restricted in certain areas for all plants. Statuses can be set up to display a warning or error when the material is used in specific process areas (e.g., materials management, plant maintenance, or production planning). Examples of cross-plant statuses are design, use, obsolete, and end of life.
- **Prod.hierarchy (product hierarchy)**
  This alphanumeric character string is used to classify your company's products. You can use product hierarchies to build a family tree. This field is 18 characters long and can have up to 3 levels. However, you can also extend the maximum number of levels to 9. This field is mainly used for price determinations and analyses. An example of a product hierarchy would be the following:
  - Level 1: Electrical components
  - Level 2: Cables and wires
  - Level 3: Single-core cables
- **Division**
  This field helps group materials that have common characteristics or belong to the same family. Each material can only be assigned to one division, but the material can be sold in multiple sales organizations. The combination of

sales organization, distribution channel, and division determines the sales area where the product can be sold.

  - **GenItemCatGroup (general item category group)**
    This field, alongside the sales document type and usage, is used by the system to determine the item category in the sales document being created. For instance, for materials that are dropshipped from a supplier, you can maintain an item category group that determines an item category in the sales order item for a third-party PO (like standard item category CB1). On the other hand, if the material is shipped from a local plant, you can maintain an item category group that determines standard item category TAN. We'll describe how to influence the item category determination in a sales order in Chapter 4, Section 4.3.
  - **Net weight and Gross weight**
    These fields are used in logistics execution, in particular during the shipping and transportation phases. If the wrong value is maintained, the system may not be able to calculate shipping costs accurately.
- **Basic data 2**
  - **Material is Configurable**
    This indicator determines if the material is configurable across all plants. If selected, you can assign a variant class to the material master record, thus allowing users to configure the relevant characteristics during sales order processing.

The data contained in the **Sales: sales org. 1** and **Sales: sales org. 2** views are used primarily in sales order processing and shipping, and they're only relevant for the sales organization and distribution channel selected. The most important fields in these views include the following:

- **Sales: sales org. 1**
  - **Sales unit**
    This field is the UoM proposed automatically in the sales order. This field must be maintained if a product is sold in a unit different than the base UoM.
  - **X-distr.chain status (cross-distribution chain material status)**
    This field serves the same purpose as the cross-plant material status from the **Basic data 1** view but only applies to the sales organization where the status is maintained for all distribution channels.
  - **DChain-spec. status (distribution chain-specific material status)**
    This field serves the same purpose as the cross-distribution channel status but only applies to the selected sales organization and distribution channel.
  - **Delivering Plant**
    If populated, this field is automatically copied as the default delivery plant for the sales order item, where it can be changed.

  - **Tax data**
    This field indicates the tax category and tax liability of the material in the countries from which the material can be shipped. A list of departure countries is proposed by the system based on the countries of the plants assigned to the sales organization. The **Tax classification** indicator is used in the access sequence for output tax determination during pricing.
  - **Min.order qty (minimum order quantity)**
    This field is the minimum quantity that can be sold to a customer in the base UoM.
  - **Min. Dely Qty (minimum delivery quantity)**
    This field is the minimum quantity that can be delivered to a customer in the base UoM.
- **Sales: sales org. 2**
  - **Acct Assmt Grp Mat. (account assignment group for material)**
    This field is used to group materials with the same accounting requirements. During billing, when creating an accounting document, this field determines the revenue or deduction account.
  - **Item Category Group**
    This field serves the same purpose as the general item category group in the **Basic data 1** view, but this value has priority in the sales organization and distribution channel concerned. For example, let's say a material has been created with NORM as the general item category group in the **Basic data 1** view. The same material can be extended to several sales areas, but only in sales organization 0001 and distribution channel 10 is the item category group changed to CBOR. If a sales order is created in sales organization 0001 and distribution channel 10, the system will use CBOR to determine the item category in the sales document item.
  - **Product hierarchy (product hierarchy sales)**
    This field serves the same purpose as the product hierarchy in the **Basic data 1** view, but this value has priority in the sales organization and distribution channel concerned.

The majority of fields in the **Sales: General/Plant** are plant specific, although you can also find basic data fields, such as for base UoM or gross weight, on this screen. The following list is a selection of the key fields in the **Sales: General/Plant** view:

- **Availability check**
  This plant-specific field determines whether the material is relevant for availability checks and, if so, how the check should be carried out and how requirements are transferred to MRP.
- **Batch management**
  This indicator specifies whether the material is stored and shipped in batches across all plants. The indicator cannot be changed if stocks exist in either the current period or in the previous period.

- **Batch management(Plant)**
  This field specifies if the material is stored and shipped in batches in the selected plant and is only activated if the **Batch management** indicator at the client level is selected. This indicator cannot be changed if stocks exist in either the current period or in the previous period in the selected plant.
- **Trans. Grp (transportation group)**
  This field is used to group products that share the same transportation requirements (i.e., refrigerated trucks, air freight, train, etc.). The value selected applies to all plants.
- **LoadingGrp (loading group)**
  This value is maintained at the plant level and is used to group materials that share similar loading requirements. The loading group, together with the shipping condition and the delivering plant, determines the shipping point in the sales document item.
- **SerialNoProfile (serial number profile)**
  This 4-digit key determines if the material is serialized in the plant concerned. Depending on the serial number profile selected and how it's been configured, the system may force the user to enter serial numbers when posting goods movements.

In addition to the main data in the different views or user departments, the material master record also holds information on a screen called **Additional Data**. You can access this area of the material master by clicking on the → **Additional Data** button. On this screen, you can maintain material descriptions in different languages, set up conversions of UoMs, specify additional International Article Numbers (EANs) or Universal Product Codes (UPCs), and maintain basic and inspection texts. Figure 3.4 shows the **Units of measure** tab for converting the UoM of a material with a base UoM of each (EA).

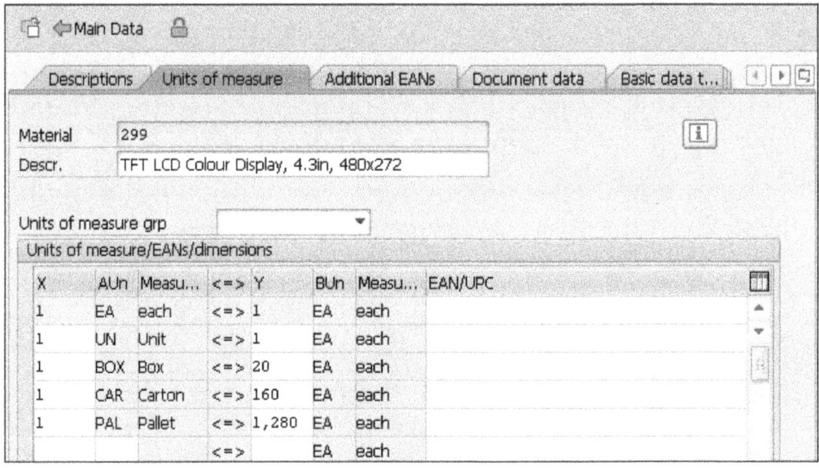

Figure 3.4 Additional Data: Units of Measure Tab

Before concluding our review of the material master record, let's not forget that you can flag materials for deletion at different organizational levels. Once the deletion flag has been set, the archive program can delete the material master record at the level selected and at all lower levels. As shown in Figure 3.5, material 299 has been flagged for deletion at the client level (the checkbox at the top), which is the highest level possible, so the system automatically selects all the checkboxes for the lower levels.

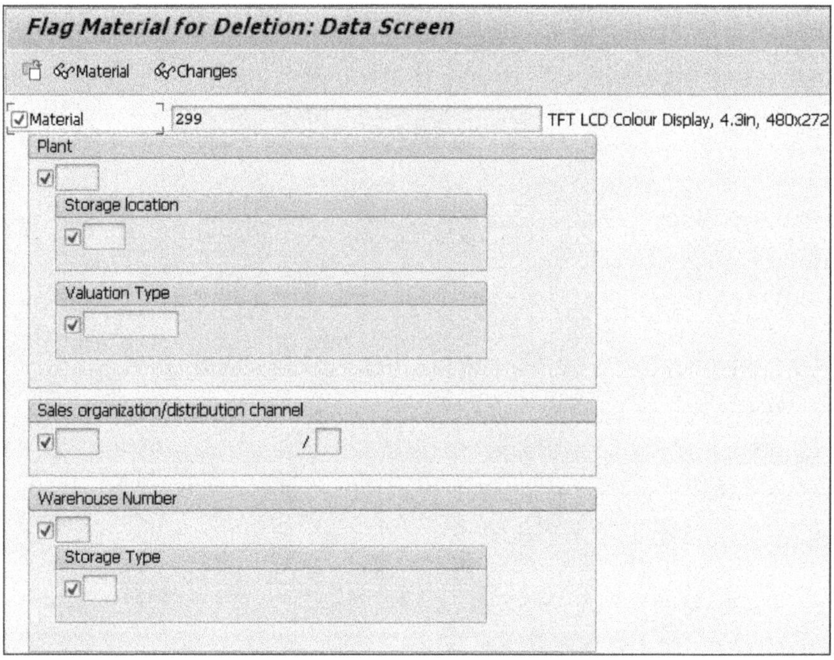

Figure 3.5 Flagging a Material for Deletion

If you decide to flag a material for deletion at the plant level, the system automatically selects a storage location and valuation type since these values both reside under the plant from a data hierarchy perspective. If a deletion flag has been set by mistake, you can cancel the deletion flag by deselecting the organizational level.

3.3 Maintaining Customer-Material Info Records

Commonly, your customers may use a material or item code different from your company's material code, so the material code on the PO sent by your customer may not correspond to a material code in your SAP S/4HANA system. A user entering the order must then find the equivalent material number in SAP S/4HANA, and often, the customer description is used for this purpose, which can lead to errors.

To solve this problem, SAP devised a master data object called the *customer-material information record*, which stores data about a material that applies only to a specific customer. A customer-material info record often includes the following information:

- A customer-specific material number
- A customer-specific material description
- Shipping-related information (such as delivering plant, delivery priority, or whether the customer allows partial deliveries)
- A customer text that can be copied to the sales order item during text determination

Once the record has been created, you can enter items in the sales order by specifying the material code used by the customer, and the system will fetch the relevant SAP S/4HANA material number. This capability makes order entry quicker and considerably reduces potential issues arising from entering incorrect material numbers.

A customer-material info record is created for a specific customer, sales organization, and distribution channel; thus, you could potentially maintain different customer information for the same material and the same customer in two or more sales organizations. Figure 3.6 shows the screen you're presented with when creating a customer-material info record in SAP S/4HANA. In this example, 299 is the material number in your company whereas PL98187 is the customer material number for this item.

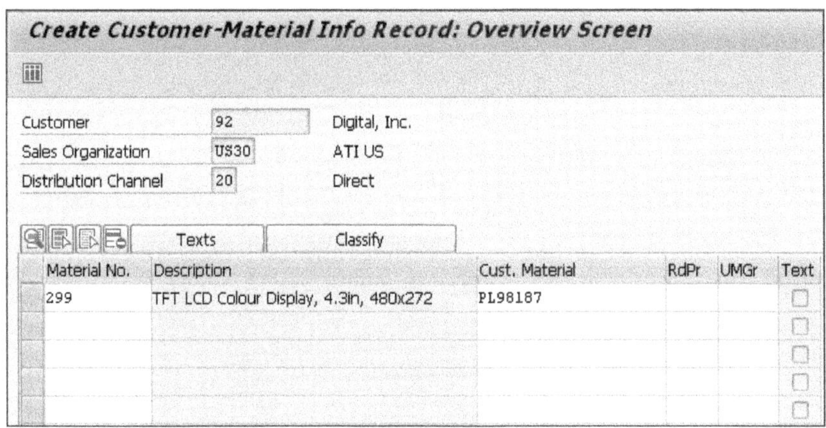

Figure 3.6 Creating a Customer-Material Info Record: Overview Screen

By selecting a line and going into the item's details, you can enter further information, including a customer description, shipping data, and partial delivery agreements. The fields available for entry on the details screen are shown in Figure 3.7.

Material	299	
	TFT LCD Colour Display, 4.3in, 480x272	
Sales Organization	US30	ATI US
Distribution Channel	20	Direct
Customer	92	Digital, Inc.
Customer material		
Customer Material	PL98187	
Customer Description		
ETag		
Shipping		
Plant	US20 Long Island City	
Delivery Priority	2 Normal item	
Minimum Delivery Qty	4 EA	
Partial delivery		
Part.dlv./item		Underdel. Tolerance %
Max.Part.Deliveries	2	Overdeliv. Tolerance %
		Unlimited Tolerance
Control data		
Item usage		
Units of Measure		
Sales unit		<-> EA
Additional Customer Materials		
Customer Materials		
Customer Material Number	Customer Description of Material	

Figure 3.7 Creating a Customer-Material Info Record: Details Screen

If known, you can enter the customer description for the product and the delivering plant. If the plant is maintained, note that, when a sales order item is created, the system determines the plant based on the following sequence:

1. Customer-material info record
2. Business partner record: **Sales and Distribution** data
3. Material master record: **Sales: sales org. 1** view

Therefore, the information maintained in the customer-material info record takes precedence over the other two master data records. However, if needed, a user can change the data copied from the info record manually in the sales document item.

From the shipping perspective, the following four additional fields are relevant:

- **Delivery Priority**
  This field is used as a selection criterion when processing deliveries collectively, so deliveries for customers with a higher priority are created first. Again, SAP S/4HANA first looks for the delivery priority in the customer-material info record and, if not found, copies the value from the business partner master record.

- **Minimum Delivery Qty (minimum delivery quantity)**
  This field specifies the minimum quantity of the material in question that can be delivered to the customer. The system first looks for this value in the customer-material info record and, if not found, fetches the minimum delivery quantity from the material master record.
- **Part.dlv./item (partial delivery at the item level)**
  This field is used to specify whether the customer allows partial deliveries or only complete deliveries are allowed, whether only one attempt should be made on the requested delivery date, and whether the customer allows unlimited partial deliveries.
- **Max.Part.Deliveries (maximum number of partial deliveries allowed per item)**
  If partial deliveries are allowed, you can use this field to specify the maximum number of partial deliveries your company can make to fulfill the item quantity requested.

Additionally, you can define underdelivery and overdelivery tolerances, as well as whether or not the customer accepts unlimited over- and underdeliveries.

On the item details screen, you can also define the sales UoM to be used and its conversion to the base UoM. If a customer has multiple material codes and descriptions for the same material in SAP S/4HANA, you can store all their codes and descriptions in the **Additional Customer Materials** section at the bottom of the item details screen.

On the item overview screen, you'll see two buttons labeled **Texts** and **Classify**, as shown in Figure 3.6. By clicking on **Texts**, you can store a free text that will be copied to the sales order item through text determination. We'll cover text determination in Chapter 6, Section 6.3. The **Classify** button can be used to define a material variant for the customer and material in the customer-material info record. This functionality will transfer the values maintained in the characteristics from the customer-material info record to the configuration of a sales document item, thus saving time during order entry since the user won't need to select the configuration chosen by the customer. However, if needed, the user can still change the characteristic values adopted in the sales document item.

Note

You can only classify a customer-material info record if the material is configurable. Only the characteristics included in the class of the configurable material (class type 300) will be passed on to sales document items.

Warning

If a user changes the customer in a sales document after the characteristic values have been copied from the customer-material info record into the sales document item, the system will not adopt the new configuration from the customer-material info record for the new customer.

Once the customer-material info record has been saved, a user creating a sales document has the option of either entering the SAP S/4HANA material number in the **Material** field or entering the customer's material number in the **Customer Material Number** field. You should be aware that values entered in the **Material** field are used to search for existing materials in the system or material determination records, but these values are never used to read customer-material info records. In the example shown in Figure 3.8, customer material code PL98187 was entered in the **Customer Material Number** field for sales order item 10, and the system determined material number 299 automatically.

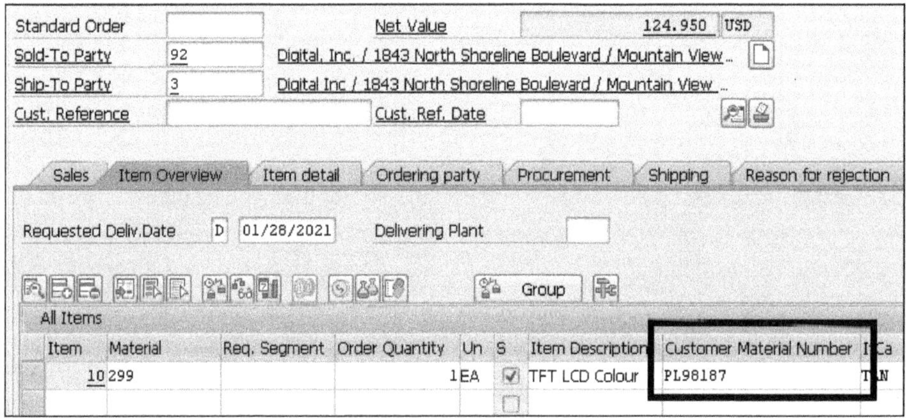

Figure 3.8 Using a Customer Material Number in a Sales Document

3.4 Maintaining Condition Master Data for Pricing

One of the priorities of any sales department is to define and maintain pricing policies for the company. This process could be quite complex since many factors must be taken into account to derive the correct expected price in every sales transaction. To achieve this goal, the system can receive price data from predecessor documents, such as contracts or quotations, or it can use pricing condition records to derive the price. In this subsection, we'll cover how to set up pricing condition records and how they are used during price determination.

Note

Since the focus of this chapter is master data, we'll review the relevant customizing activities in the area of pricing in Chapter 8, Section 8.3.

All the information required to determine the final price on a sales document item can be stored in the system in the form of *pricing condition records*. These master data objects usually contain data related to the customer, the material, the quantity, the condition type, and the organizational unit selling the goods. The system uses the information entered in the sales document and a method called the *condition technique* to find the relevant pricing condition record and determine the correct price. We'll provide an overview of the condition technique in Chapter 5, Section 5.3, since this infrastructure is used across SAP S/4HANA to define and determine master data for several other objects in addition to pricing. We'll also cover the specifics of the condition technique as used in pricing in Chapter 8, Section 8.2.4.

You must specify a condition type when creating a pricing condition record. The condition type defines whether the price stored in the condition record is a gross price, net price, discount, freight charge, or tax. No limit exists for the number of condition records you can create for a condition type. Let's start with a simple example to illustrate how the system selects the relevant pricing condition records when the same product is sold to two different customers.

Example

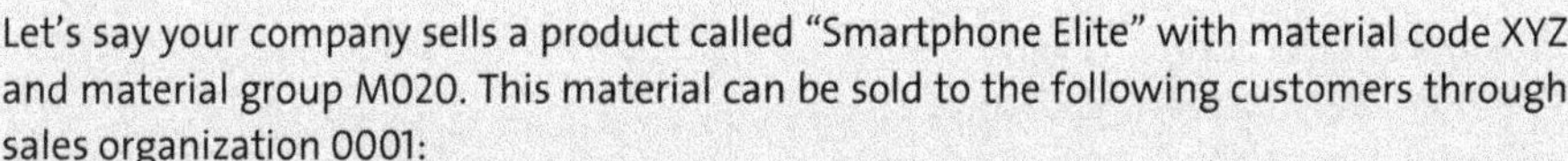

Let's say your company sells a product called "Smartphone Elite" with material code XYZ and material group M020. This material can be sold to the following customers through sales organization 0001:

- Customer A: Loyal customer that has an excellent credit score.
- Customer B: New customer that buys from the company sporadically.

For customer A, the company has agreed on a special net price of $450 for this item, whereas customer B is given a 5% discount on the gross price for all products where the material group is M020. To support this scenario, three pricing condition records can be maintained, as shown in Table 3.2, Table 3.3, and Table 3.4.

Condition Type	Sales Organization	Material	Value
PR00 (Gross Price)	0001	XYZ	$600

Table 3.2 Gross Price of Material XYZ in Sales Organization 0001

Condition Type	Sales Organization	Customer	Material Group	Value
DC00 (Customer Discount)	0001	B	M020	5%

Table 3.3 Customer Discount for Products with Material Group M020 in Sales Organization 0001

Condition Type	Sales Organization	Customer	Material	Value
PN00 (Net Price)	0001	A	XYZ	$450

Table 3.4 Customer Net Price of Material XYZ in Sales Organization 0001

Note that pricing condition PN00 has been configured to deactivate any gross prices.

For customer A, the system finds two valid pricing condition records and calculates the final price in the following way:

- Condition type PR00 = $600 (inactive because of subsequent price PN00)
- Condition type PN00 = $450
- Final price for customer A = $450

For customer B, the system also finds two valid pricing condition records, but in this case, the condition types and the final price are different and thus calculated in the following way:

- Condition type PR00 = $600
- Condition type DC00 = 5%
- Final price customer for B = $600 × (1 – 0.05) = $570

Once the system finds the relevant condition record, it transfers the amount to the corresponding condition type in the pricing schema of the sales document item and then calculates the net value (including all applicable discounts and surcharges) and the tax amount.

Tip

Although not mandatory, we recommend maintaining pricing condition records at the sales organization and distribution channel level. This approach will limit business user access to the corresponding sales organization and distribution channel. Pricing is sensitive data, and you must ensure that only authorized users can maintain or view prices.

Now, let's see how you can maintain and display pricing condition records. The following transaction codes and corresponding SAP Fiori apps are available in SAP S/4HANA:

- Condition Maintenance: Create (Transaction VK31)
- Condition Maintenance: Create with Refer. (Transaction VK34)
- Condition Maintenance: Change (Transaction VK32)
- Condition Maintenance: Display (Transaction VK33)

Transactions VK31, VK32, and VK34 allow mass maintenance. You can, for instance, maintain all material prices for a customer in one step.

After executing one of these four transactions, you'll be presented with a navigation tree, shown in Figure 3.9, where you can choose from 18 pricing reports for maintaining pricing condition records.

Each pricing report applies to specific condition tables. Pricing condition tables are defined in customizing and consist of the fields needed to find a pricing condition record. For example, if a price is valid for a customer and material number, in a given combination of sales organization and distribution channel, these four attributes form the condition table.

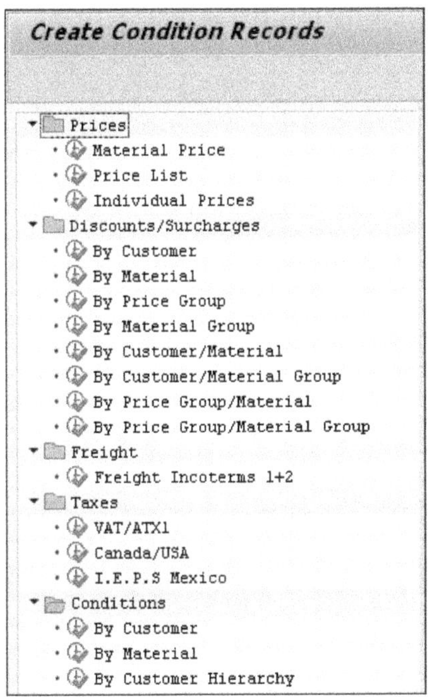

Figure 3.9 Condition Maintenance: Pricing Reports

The fields in the condition tables assigned to a pricing report and the condition type are part of the selection criteria when getting multiple condition records from the database. This feature allows you to display pricing condition records for different condition types and condition tables on the same screen. For more information about the fields available in the selection screen of a pricing report, right-click on the report and select the **Display pricing report structure** option, as shown in Figure 3.10.

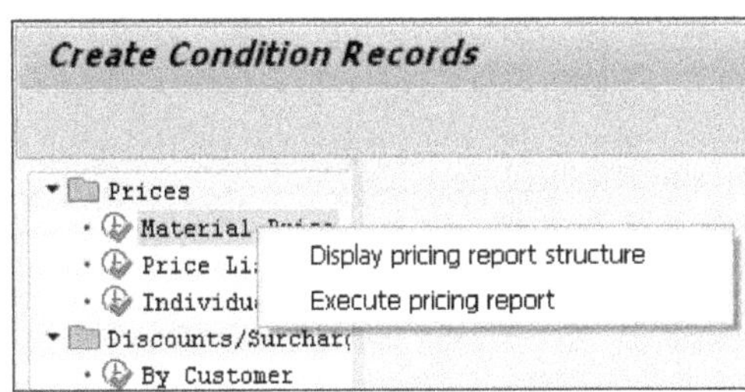

Figure 3.10 Condition Maintenance: Display Pricing Report Structure

If you select the **Display pricing report structure** option for the first pricing report (Material Price), you'll be presented with the screen shown in Figure 3.11.

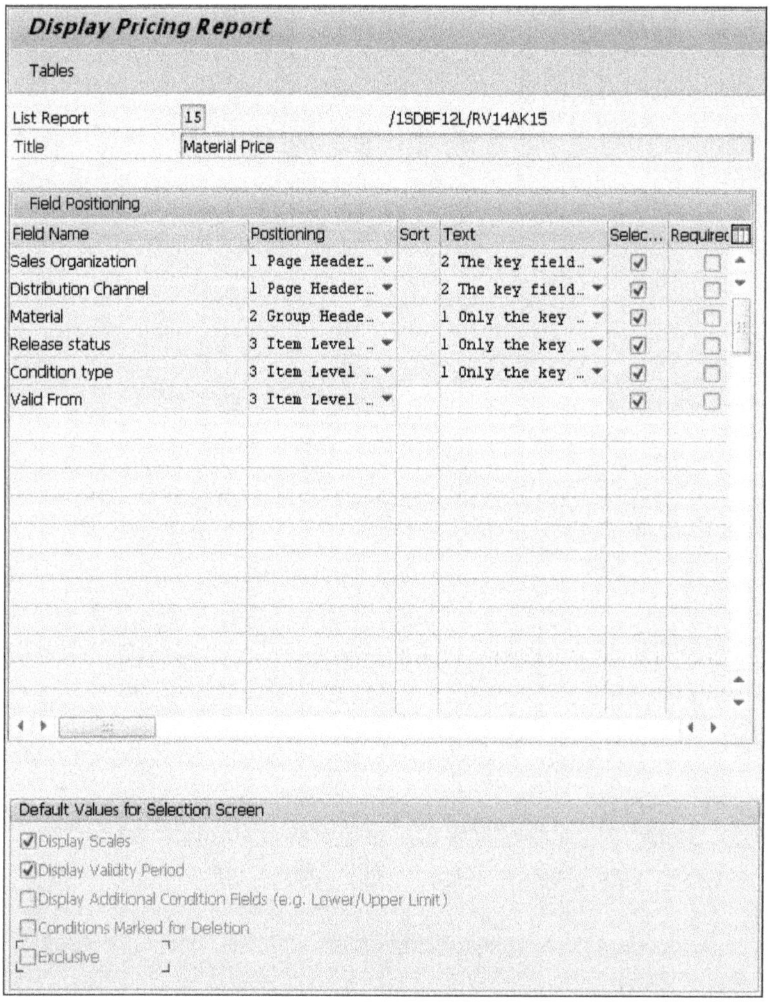

Display Pricing Report

Tables

List Report 15 /1SDBF12L/RV14AK15

Title Material Price

Field Positioning

Field Name	Positioning	Sort	Text	Selec...	Required
Sales Organization	1 Page Header...		2 The key field...	☑	☐
Distribution Channel	1 Page Header...		2 The key field...	☑	☐
Material	2 Group Heade...		1 Only the key ...	☑	☐
Release status	3 Item Level ...		1 Only the key ...	☑	☐
Condition type	3 Item Level ...		1 Only the key ...	☑	☐
Valid From	3 Item Level ...			☑	☐

Default Values for Selection Screen

☑ Display Scales

☑ Display Validity Period

☐ Display Additional Condition Fields (e.g. Lower/Upper Limit)

☐ Conditions Marked for Deletion

☐ Exclusive

Figure 3.11 Pricing Report Structure: Material Price

Click on the **Tables** button to access the condition tables preassigned to the pricing report selected. Figure 3.12 shows the two condition tables considered in the Material Price pricing report.

Display Pricing Report

Name of list 15 /1SDBF12L/RV14AK15

Title Material Price

Key combinations

Table	Short Description
004	Material
304	Material with release status

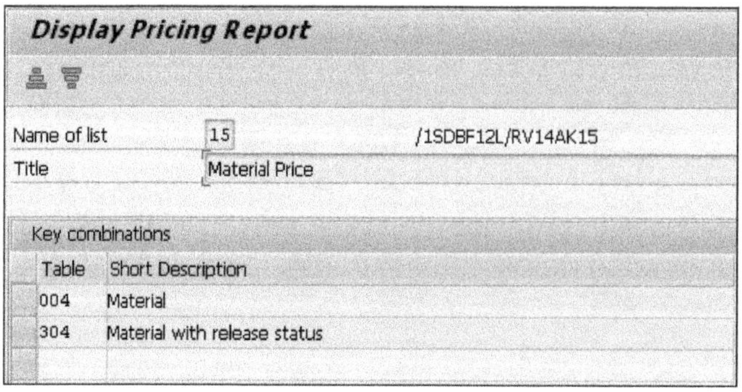

Figure 3.12 Condition Tables: Material Price Pricing Report

Now, you can display the fields contained in the table via the menu path **IMG • Sales and Distribution • Basic Functions • Pricing • Pricing Control • Condition Tables and Field Catalog • Display Condition Tables**. The fields selected in both tables shown in Figure 3.12 are the sales organization, distribution channel, and material. The difference between these two tables is the text ...**with release status,** which is only selected in condition table 304.

Tip

Alternatively, you can display condition tables in the ABAP dictionary (Transaction SE11) by adding the character "A" to the beginning of the condition table number and entering it in the **Database table** field. For instance, you would enter "A004" and "A304" based on the two tables shown in Figure 3.12.

However, if you define a new condition type that uses a custom condition table that is not considered by the 18 predefined pricing reports shown earlier in Figure 3.9, you can always maintain pricing condition records for a specific condition type through the following four classic transactions:

- Transaction VK11 (Create Condition Records)
- Transaction VK14 (Create Condition Records with Reference)
- Transaction VK12 (Change Condition Records)
- Transaction VK13 (Display Condition Records)

Alternatively, you can create your own pricing report in customizing and select the key fields and condition tables that should be taken into consideration.

Let's go back to condition maintenance via Transaction VK32. To execute a pricing report, double-click on the icon or right-click on a pricing report and select the **Execute pricing report** option. After the selection screen, you'll reach the overview screen shown in Figure 3.13 where three pricing condition records have been selected for condition type PR00 in sales organization US30 and distribution channel 20.

Change Condition Records: Overview

- Prices
  - Material Price
  - Price List
  - Individual Pric
- Discounts/Surchar(
- Freight
- Taxes
- Conditions

Sales Organization	US30	ATI US
Distribution Channel	20	Direct

SAPMV13A_3011_ALLEFELDER All...

Sel...	Variable key	Amount	Unit	per	UoM	CCon...	Num...	B...	P..	Valid From	Valid To
	CnTy Material...										
	PR00 296 ...	10.000	USD		1EA					01/01/2020	12/31/9999
	PR00 299 ...	119.000	USD		1EA					12/15/2020	12/31/9999
	PR00 341 ...	87.000	USD		1EA					12/28/2020	12/31/9999
	CnTy Material										

Figure 3.13 Change Condition Records (VK32): Overview Screen

On this screen, you'll see two icons to create condition records for the two condition tables assigned to this pricing report, which was shown earlier in Figure 3.12. If you click on the icon, condition details regarding validity dates, scales, texts, and key combinations appear at the bottom of the screen, as shown in Figure 3.14.

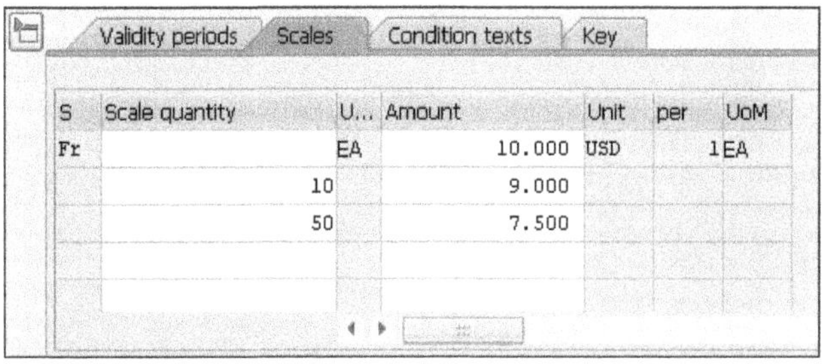

Validity periods | Scales | Condition texts | Key

S	Scale quantity	U...	Amount	Unit	per	UoM
Fr		EA	10.000	USD	1	EA
	10		9.000			
	50		7.500			

Figure 3.14 Pricing Condition Detail (Scales)

In this section of the screen, you can maintain scales (i.e., prices based on the quantity ordered) and condition texts. The other two tabs are for display purposes only. Validity periods can be maintained in the overview area shown earlier in Figure 3.13.

Analyzing all the fields in the pricing condition record are beyond the scope of this book. However, some of the most relevant fields, with a brief explanation for each, include the following:

- **Amount (condition amount or percentage)**
  This field specifies the rate used to calculate pricing and depends on the condition type. This value can be a fixed amount or a percentage. If scales have been defined in the condition record, the rate shows the first line of the scale.
- **Unit (condition unit)**
  Like the condition amount or percentage, the condition unit also depends on the condition type and determines whether the value in the condition record is based on a percentage (discount) or on an amount in a particular currency (material price).
- **UoM (condition UoM)**
  If the condition record value is based on an amount, this field specifies the UoM to which the condition amount refers.
- **Valid From and Valid To**
  These two fields define the start and end dates of the pricing condition record. The system uses the pricing date of the sales document to determine pricing elements. The default pricing date is the current date (the date when the document is created), but this date can be changed manually. A condition record is considered invalid if the pricing date of the sales document falls outside the validity period.

- **Deletion (deletion indicator for condition record)**
If this indicator is selected, the system no longer takes the condition record into account during pricing.
- **Exclusion (condition exclusion indicator)**
You can select this indicator to exclude follow-up conditions (e.g., discounts) proposed during pricing. The exclusion of preceding conditions is not possible.
- **Upper limit and Lower limit**
You can use these two fields to restrict manual changes in the price for the concerned condition type. The system will prompt you with an error message if the new price entered manually for the item exceeds or drops below the defined limits. These two fields are displayed in the **Details** view, which can be accessed by via **Goto • Details**.
- **Max.condition value (maximum condition value)**
This field specifies the maximum cumulative value permitted. For example, if a customer receives a 5% discount when ordering a particular material, you can enter a maximum cumulative amount of $10,000 in the condition record for the discount. The system keeps track of the cumulative discount value after each order and deactivates the 5% discount once it reaches $10,000. This option is only available if the **Condition Update** indicator is set for the pricing condition type in customizing. This field is displayed in the **Additional Data** view, which can be accessed via **Goto • Additional Data**.
- **Max.cond.base value (maximum condition base value)**
This field specifies the maximum cumulative quantity, weight, or volume permitted. For example, if a customer receives a 10% discount when ordering a particular material, you can enter a maximum cumulative quantity of 2,000 units in the condition record for the discount. The system keeps track of the cumulative quantity after each order and deactivates the 10% discount once it reaches 2,000 units. This option is only available if the **Condition Update** indicator is set for the pricing condition type in customizing. This field is displayed in the **Additional Data** view, which can be accessed via **Goto • Additional Data**.
- **Max.number.of.orders (maximum number of sales orders)**
Using this field, you can specify the maximum number of sales orders linked to a pricing condition record. For example, let's say a customer has been offered a special net price applicable to the first three orders only. The system keeps track of the number of orders and deactivates the special price after the maximum number is reached. This option is only available if the **Condition Update** indicator is set for the pricing condition type in customizing. This field is displayed in the **Additional Data** view, which can be accessed via **Goto • Additional Data**.

Finally, when processing pricing condition records via *condition maintenance* transactions (Transactions VK31, VK34, VK32, and VK33), you can click the icon to execute the pricing simulation tool. This feature enables you to evaluate your changes before saving. The system will simulate a billing document where you can compare the old and new net values and profit margins.

Note

Features like displaying pricing condition records for different condition types and condition tables on the same screen or the ability to simulate pricing before you save your changes are only available through condition maintenance transactions (Transactions VK31, VK34, VK32, and VK33). The traditional way of maintaining pricing condition records for a specific condition type (Transaction VK11, VK14, VK12, and VK13) does not support these features.

3.5 Important Terminology

In this chapter, the following terminology was used:

- **Material type**
  This value groups materials with similar characteristics and attributes, such as services, trading goods, packaging, or raw materials. This field determines what fields are relevant for data and whether the valuation of the material will be carried out according to the standard price or the moving average price.
- **Industry sector**
  This field identifies the branch of industry of the material (e.g., retail, chemical industry, pharmaceuticals, or mechanical engineering). Once you've assigned an industry sector to a material, it cannot be changed. The screen sequence and field selection in a material master record is determined by the industry sector of the material.
- **Screen sequence**
  The sequence of screens presented to you when processing a material master record.
- **Field selection**
  The fields displayed in the material master record screens.
- **Maintenance status**
  This field determines the departmental data that has been maintained for a material master record. This status is updated when a material is extended to a new view. The maintenance status is formed by individual maintenance status designated by a capital letter. For example, accounting is represented by the letter B, whereas basic data is K. If only these two views are maintained for a material, the maintenance status field shows BK.
- **Material number**
  This number uniquely identifies a material master record. Two types of number assignment in SAP S/4HANA exist: external assignment and internal assignment. With the external assignment approach, you must enter a character string when creating a material. With the internal assignment approach, the system can assign a consecutive number to the material. The material number was previously only 18 characters long, but the maximum field length of the material number has been extended to 40 characters.

- **Serial number**
  This string of characters is used to identify individual instances of a material for tracking purposes. The system only allows entering serial numbers for products with a serial number profile defined at the plant level. Unlike the batch number, which is assigned to several items, the serial number is unique to one item.
- **Batch number**
  This number represents a quantity of a material that has characteristics in common and, therefore, is useful from the tracking perspective. For instance, in the food industry, commonly a batch is a certain number of products produced at the same time.
- **Customer-material information record**
  This master data object is used to store data about a material that applies to a specific customer only. This record can contain the customer's material number and description, a customer text, shipping information, and even a material variant for the customer if the material is configurable. Once a customer-material info record has been created, you can enter the customer material in the sales order item in the **Customer Material Number** field.
- **Pricing condition record**
  This pricing data is stored for a specific condition type and retrieved during pricing. This value can be a percentage or a fixed amount and is always defined for a specific pricing condition type.

3.6 Practice Questions

These practice questions will help you evaluate your understanding of the topics covered in this chapter. The questions shown are similar in nature to those found on the certification examination. Although none of these questions will be found on the exam itself, they will allow you to review your knowledge of the subject. Select the correct answers and then check the completeness of your answers in the next section. Remember that, on the exam, you must select all correct answers and only correct answers to receive credit for the question.

1. Which of the following statements are true regarding the properties of the material number field in SAP S/4HANA 2020? (There are two correct answers.)

☐ A. It can be up to 18 characters long.

☐ B. It can be a consecutive number assigned by the system.

☐ C. It can contain special characters such as /, &, or *.

☐ D. It can be a character string entered by the user.

2. True or False: During sales order processing, the general item category group maintained in the **Basic Data 1** view has priority over the item category group maintained in the **Sales: sales org. 2** view.

☐ A. True

☐ B. False

3. Your company has offered a special net price to a customer for a specific material for a limited time. However, a discount already exists for this customer and material. What change can you make in the pricing condition record to ensure the special price takes preference over the discount?

☐ A. Set the condition inactivity indicator for the discount

☐ B. Select calculation type "No discount" for the special net price

☐ C. Select calculation type "Net price" for the discount

☐ D. Set the condition exclusion indicator for the special net price

4. Which of the following factors determine the screen sequence and field selection in a material master record? (There are two correct answers.)

☐ A. Division

☐ B. Material group

☐ C. Material type

☐ D. Industry sector

5. True or False: The sales organization and distribution channel fields are mandatory entries when creating pricing condition records in SAP S/4HANA Sales.

☐ A. True

☐ B. False

6. True or False: SAP S/4HANA comes with a wide range of material types to choose from when creating new materials, but defining new custom material types is not possible.

☐ A. True

☐ B. False

7. In a standard SAP S/4HANA system, which of the following search sequences is used to determine the delivering plant in a sales document item?

☐ A. Business partner master record → Customer-material info record → Material master record

☐ B. Material master record → Business partner master record → Customer-material info record

- ☐ C. Customer-material info record → Material master record → Business partner master record
- ☐ D. Customer-material info record → Business partner master record → Material master record

8. True or False: When creating a sales order, if a customer-material info record exists, the customer material number can be entered in either the **Customer Material Number** field or the **Material** field, and the system will bring the linked material into the order item.

- ☐ A. True
- ☐ B. False

9. Which features are available for condition maintenance in SAP S/4HANA through Transaction VK32? (There are two correct answers.)

- ☐ A. The system issues a warning message if the price entered for the material results in a negative profit margin.
- ☐ B. You can display pricing condition records for different conditions types and condition tables on the same screen.
- ☐ C. You can maintain pricing condition records for your own custom condition types and condition tables using predefined pricing reports.
- ☐ D. You can simulate pricing before you save your changes.

10. When maintaining pricing condition records in SAP S/4HANA Sales, which of the following options are available if the condition update indicator is not set for the condition type in customizing? (There are two correct answers.)

- ☐ A. Condition exclusion
- ☐ B. Upper and lower limits
- ☐ C. Maximum values
- ☐ D. Maximum quantities

11. True or False: The date on which a sales document is created is the date used to determine the pricing elements and cannot be changed manually.

- ☐ A. True
- ☐ B. False

12. Which of the following fields in the material master record are maintained at the plant level? (There are two correct answers.)

- ☐ A. Delivering plant
- ☐ B. Transportation group

- ☐ C. Availability check
- ☐ D. Loading group

13. Which of the following parameters can be controlled in pricing condition records? (There are two correct answers.)

- ☐ A. Scales
- ☐ B. Rounding rule
- ☐ C. Maximum number of sales orders
- ☐ D. Access sequence

3.7 Practice Question Answers and Explanations

1. Correct answers: **B and D**
   In SAP S/4HANA 2020, the field length of the material number can be extended from 18 to 40 characters; thus, answer A is not correct. You can use special characters in the material number as long as they are not the ampersand (&), asterisk (*), or comma (,) which cannot be used due to technical reasons. Ampersands and asterisks are proposed as examples of special characters in answer C, making it a false statement. Thus, only answers B and D are the two true statements. You can use either an internal or external number range on the material number field as mentioned in answers B and D, respectively.
2. Correct answer: **B**
   False. In fact, the opposite is true. To determine the item category in a sales order item, the system reads the item category group of the material maintained in the **Sales: sales org. 2** view for the relevant sales organization and distribution channel.
3. Correct answer: **D**
   Neither calculation type "No discount" nor "Net price" exist in SAP S/4HANA nor does a condition inactivity indicator exist in the pricing condition record. The closest indicator would be the deletion indicator, but even if the answer were to set the deletion indicator, this approach wouldn't be correct either. To make sure the discount is not considered when the special net price is determined, you must set the condition exclusion indicator on the condition record for the net price. The system will automatically ignore any subsequent conditions in the pricing schema.
4. Correct answers: **C and D**
   The screen sequence and field selection in a material master record is determined by the industry sector of the material and the material type.

5. Correct answer: **B**
   False. The fields you must maintain when creating a pricing condition record are defined by the condition table used. Even though the sales organization and distribution channel are used in most of the standard condition tables for pricing in SAP S/4HANA Sales, selecting these two fields is not mandatory when maintaining new pricing condition tables. You can create, for example, a condition table that contains customers and materials only, so the price stored in the condition record is valid for the combination customer/material regardless of the sales organization and distribution channel. However, having the sales organization and distribution channel on the condition table is useful because this information allows you to control the users who can display or edit pricing condition records based on the organization to which they belong.

6. Correct answer: **B**
   False. SAP S/4HANA comes with over 40 preconfigured material types that cover the most common materials used across all types of industry sectors. Nonetheless, you can create your own material type if required via customizing. If you need to create a custom material type, we recommend copying the closest standard material type and then making the required changes.

7. Correct answer: **D**
   A delivering plant can be maintained in all three objects: material master record, business partner record, and customer-material info record. However, when determining the delivering plant for a sales document, the system will always read the customer-material info record first. If no data is found, then the system will try to fetch the plant from the business partner master record. If the system still fails to find the plant, it uses the material master record as a last resort.

8. Correct answer: **B**
   False. The customer material number must be entered in the **Customer Material Number** field for the system to find the corresponding internal material number. If the customer material number is entered in the actual material number field (**Material** field), an error message will inform the user that the material has not been defined in the relevant sales organization and distribution channel.

9. Correct answers: **B and D**
   If you use Transaction VK32 or its corresponding SAP Fiori app, the Condition Maintenance: Change app, you can select multiple pricing condition records for several condition types and condition tables. Which condition types and condition tables are selected depend on the pricing report chosen and the selection criteria. This capability is particularly useful for the mass maintenance of condition records. Also, in this newer transaction, a document simulation tool has been made available to evaluate changes in prices and profit margin before saving.

10. Correct answers: **A and B**

    The feature to set a maximum value and/or a maximum quantity, weight, or volume in the pricing condition record is only available if the **Condition Update** indicator has been set in customizing for the corresponding condition type. This limitation exists because the system must update the condition record after every order to keep tracking the cumulative value, quantity, weight or volume. On the other hand, both the **Exclusion** indicator and **Upper limit** and **Lower limit** fields are always available on the condition record.

11. Correct answer: **B**

    False. The date used to determine pricing elements is the pricing date in the sales document. The system proposes the current date as the pricing date, but you can change it manually in the document.

12. Correct answers: **C and D**

    The delivering plant is contained in the **Sales: sales org. 1** view and is thus maintained at the sales organization and distribution channel level (table MVKE). The transportation group, loading group, and availability check are all fields in the **Sales: General/Plant** view. However, the transportation group is stored at the general master data level (table MARA) and thus is valid for all plants. So, answers A and B are both incorrect. On the other hand, the loading group and availability check fields are maintained at the plant level (table MARC), which makes answers C and D the only two correct answers to this question.

13. Correct answers: **A and C**

    You can maintain scales in the pricing condition record if you want to charge a customer a different amount depending on the quantity ordered. Also, assuming the **Condition Update** indicator is set for the pricing condition type in customizing, you can specify (in the pricing condition record itself) the maximum number of orders linked to the pricing condition concerned. Therefore, answers A and C are both correct. The rounding rule and the access sequence are two attributes of the pricing condition type, which are both defined in customizing, thus invalidating answers B and D.

3.8 Takeaway

You should now have a good overview of the material master record, and you understand how the data is structured in views that contain information relevant to specific user departments. You can recognize some of the key data contained in the sales views that must be maintained for a material to be sold in the system. You now know the difference between creating and extending a material master record in SAP S/4HANA.

You're familiar with the information that can be stored in a customer-material info record and the main functionalities provided by this data object. You learned about the benefits of using the customer material number in sales documents, that is, to automatically populate certain values that only apply to the customer and material selected.

In this chapter, you've seen how prices are maintained in pricing condition records and how the system uses the condition technique to determine the correct prices in a sales document item. You also understand the additional features available during condition maintenance compared to the traditional transactions to create and maintain pricing condition records for specific condition types. Finally, we listed some of the most important fields in the pricing condition record and described the functionalities provided by each field.

3.9 Summary

The material master is probably the only master data object shared across all modules. In this chapter, we provided a brief overview of the material master record with a special focus on the data used in sales and distribution.

You should now understand the customer-material information record and the fact that the system gives priority to this particular object over other master data records when retrieving information from the database during sales order processing. We also described the process of storing pricing information in the system and how you can make good use of the most important fields.

In the next chapter, we'll discuss customizing settings for sales documents.

Chapter 4
Sales Documents Customizing

4

Techniques You'll Master

- Control sales processes
- Define sales document types
- Apply item categories
- Interpret the item category determination
- Use bills of materials (BOMs) in sales documents
- Apply schedule line categories
- Interpret the schedule line category determination
- Understand sales contracts
- Set up customer-specific contract data

The focus of this chapter is to explain the role that sales document types, item categories, and schedule line categories play in controlling the processing of sales documents in an SAP S/4HANA system. We'll discuss the rules for item category and schedule line category determination and the usage of BOMs in sales documents as a typical special business process in sales. We'll end the chapter by reviewing the different categories of sales contracts and the most relevant customizing activities.

Real-World Scenario

Companies that sell products may require the creation of bundles or sales kits to group together materials that are related to or that complement each other. For example, angle grinders are usually sold alongside cutting discs, printers can be sold alongside ink cartridges, and a common practice in the retail industry is to offer an extended warranty or insurance policy when a customer buys expensive electronic equipment such as high-end tablets or mobile phones. In all these cases, a more competitive price is typically offered if the customer orders the bundle as opposed to the standalone items. These bundles can be achieved through BOMs in sales documents; thus, knowing how to set up these master data objects would allow you to meet your customers' requirements in this kind of scenario.

The sales process in SAP S/4HANA is controlled by the combination of the sales document type, item category, and schedule line category. For instance, based on the document type, you can specify whether the system should carry out a credit check or whether a delivery or billing block must be automatically applied. You can also specify whether a certain item category is relevant for delivery or billing. Finally, you can use the schedule line category to dictate how the system determines the quantity available based on the requested delivery date. As a consultant for sales and distribution, configuring these elements with confidence is essential, and you'll need to describe the available options when asked to come up with solutions for specific sales requirements.

4.1 Objectives of This Portion of the Test

The goal of this portion of the SAP S/4HANA Sales Certification exam is to test your understanding of the following topics:

- Define and assign sales document types
- Define and assign sales item categories
- Understand item category determination
- Define and assign schedule line categories

- Understand schedule line category determination
- Understand BOMs in sales documents
- Maintain sales contract data
- Understand master contracts
- Understand quantity contracts
- Understand value contracts
- Understand service contracts

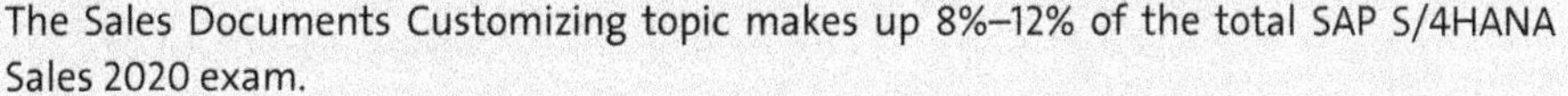

Note

The Sales Documents Customizing topic makes up 8%–12% of the total SAP S/4HANA Sales 2020 exam.

The content of this chapter is included in the Sales topic of the SAP S/4HANA Sales 2020 Upskilling exam, which makes up more than 12% of this exam.

4.2 Controlling Sales Documents

All sales transactions take place within the organizational structures discussed in Chapter 2. The system uses the information stored in master data records, together with the additional data entered by the user, to generate a sales document.

The structure of any sales document in SAP consists of data at the header level and at the item level. For each item, in some cases, you can go down one more level and store information at the schedule line level. The header of a sales document contains information that applies to the entire document, such as the sold-to party or the currency. Data that is only valid for specific items is stored at the item level. Delivering plant, material number, and net price are examples of item data. Finally, any information regarding the delivery schedule for an item is located in the schedule line. For example, if a customer orders 50 units to be delivered tomorrow and 70 units next week, you'll see 120 units as the quantity ordered for the item, but two schedule lines will be created for 50 and 70 units with the corresponding delivery dates. A sales document item can have one or more schedule lines. Unlike header and item data, schedule line data is not mandatory in all sales documents. Some types of documents like contracts or credit memo requests are not relevant for delivery; thus, these documents do not have schedule lines.

The following structure represents a hypothetical sales document with three items and two or three schedule lines per item:

- Header
  - Item 10
    - Schedule line 1
    - Schedule line 2

  - Item 20
    - Schedule line 1
    - Schedule line 2
    - Schedule line 3
  - Item 30
    - Schedule line 1
    - Schedule line 2

During the implementation of an SAP S/4HANA system, one of the critical tasks in the sales area is to define the types of data controlling how sales documents are processed. Via customizing, you can set control parameters for sales documents at the following three levels:

1. Sales document type (header level)
2. Item category (item level)
3. Schedule line category (subitem level)

In SAP S/4HANA, four different categories of sales documents exist, and each category comes with a subset of predefined sales document types. Some common standard sales document types are shown in Table 4.1.

Document Category	Sales Document Type	Document Type Description
Presales	IN	Inquiry
	QT	Quotation
	FD	Free-of-charge delivery
Sales orders	OR	Standard order
	BV	Cash sale
	SO	Rush order
Outline agreements	CQ	Quantity contract
	WV	Service and maintenance contract
	MV	Rental contract
	LP	Scheduling agreement
Complaints	CR	Credit memo request
	DR	Debit memo request
	SD	Subsequent delivery, free of charge
	RE	Returns

Table 4.1 Sample of Standard Sales Document Types for Each Sales Document Category

Tip

If the document types available in the standard version of the SAP S/4HANA system don't meet your company's needs, you can modify them or create your own sales document types. We recommend you copy a similar existing sales document type and make the required changes. This procedure will copy all the specifications of the existing sales document type, thus saving you a considerable amount of time. To easily distinguish between custom and standard sales document types, any new document type should start with the letter "Z," following SAP's guidelines, to avoid your custom objects being overwritten by a release upgrade.

Some functions controlled by the sales document type include the following:

- Availability checks
- Transfer of requirements
- Delivery dates
- Pricing
- Subsequent document types
- Texts
- Credit checks
- Messages

You can define sales document types in the IMG menu under **Sales and Distribution • Sales • Sales Documents • Sales Document Header • Define Sales Document Types**. As a reference, Figure 4.1 shows a portion of the screen where the standard sales document type OR can be configured.

Sales Document Type	OR	Standard Order		
SD Document Category	C		Sales Document Block	
Indicator				

Number systems

No. Range Int. Asst	01	Item No. Increment	10
No. Range Ext. Asst	02	Subitem Increment	10

General control

Reference mandatory		Material entry type	
Check division	2	☑ Item division	
Probability	100	☑ Read info record	
Check credit limit		Check Customer Ref	
Credit group		☐ Enter PO number	
Output application	V1	Commitment date	
		☐ Disp. Preceding Docs	

Transaction flow

Screen sequence grp.	AU	Sales Order	Display Range	UALL
Incompl.Proced.	11	Standard Order	FCode for overv.scr.	UER1
Transaction group	0	Sales Order	Quotation Messages	
Doc. Pricing Proc.	A		Outline Agrmt Mess.	
Status profile			Message: Mast.contr.	
Alt.sales doc. type1			ProdAttr.messages	A
Alt.sales doc. type2			☐ Incomplet.messages	
Variant				

Figure 4.1 Defining Sales Document Types

Details about each and every element in this customizing activity are beyond the scope of the certification exam, but you should be familiar with how the following settings influence the sales process:

- **General Data**
  - **SD Document Category**: This field is used to assign one of the predefined sales document categories to the sales document type. Inquiries (A), quotations (B), orders (C), contracts (G), and returns (H) are some of the most common document categories.
  - **Sales Document Block**: This field determines whether the sales document is blocked for use. The following three values are available:
    - <blank>: No block
    - **A**: Only automatic creation allowed (e.g., for rebate processing)
    - **X**: The sales document type is blocked
  - **Indicator**: This field is used to classify the sales document type internally. Several standard SAP S/4HANA Sales programs query this field to find out the classification of the sales document type. Some of the values available include the following:
    - <blank>: No classification
    - **B**: Delivery order
    - **C**: Scheduling agreement with delivery order
    - **D**: Invoice correction request
    - **E**: Delivery order correction
  - **Item No. Increment**: This field specifies the increment by which item numbers are increased in a sales document when the system automatically generates item numbers. A typical value for this field is 10, which allows for new items to be inserted between other items.
  - **Reference mandatory**: This field indicates whether the document can only be created with reference to a preceding document and the type of reference document required (e.g., quotation, sales order, billing document, etc.)
  - **Read info record**: If this checkbox is selected, the system considers existing customer-material info records.
  - **Propose Deliv.Date**: If this checkbox is selected, the system automatically proposes the current date as the delivery date.
  - **Incompl.Proced.** (incompletion procedure): The incompletion procedure assigned to a sales document type determines which fields must be maintained for a document to be considered complete. We'll expand on this concept in Chapter 5, Section 5.2.
  - **Incomplet.messages** (incompletion messages): If this indicator is selected, and you try to save a document where mandatory data is missing, the system will prompt you with a dialog message informing you that the document is incomplete. You can then decide whether to save the incomplete document

or to populate the missing data. If you leave the indicator blank, incomplete sales documents cannot be saved.

- **Item division**: If this checkbox is selected, the system takes the division from the material master record for every item. Otherwise, the division entered at the sales document header level is also valid for all items.
- **Check division**: This field dictates how the system reacts if the division entered at the header level differs from the division at the item level.
- **Check credit limit**: This field controls whether the system runs credit checks and how it responds if the check is unsuccessful.
- **No. Range Int. Asst** (number range, internal assignment): This field determines how documents are numbered by the system.
- **No. Range Ext. Asst** (number range, external assignment): This field determines how documents are numbered by the user.
- **Doc. Pricing Proc.** (document pricing procedure): This field specifies the key of the pricing procedure assigned to the sales document type, which determines how pricing is carried out in a sales document.

- **Shipping Data**
  - **Delivery type**: This field determines the delivery document type used by the system when a delivery document is created for the order concerned.
  - **Delivery block**: This field controls if a delivery block must be set for the entire sales document.
  - **Immediate delivery**: This field indicates if a delivery should be automatically created as soon as the sales document is saved. You can also decide whether all items or only those with a quantity confirmed on the date should be delivered.
  - **Shipping conditions**: If this field is maintained, the shipping condition from the customer master record is overwritten by the shipping condition defined in this field.
- **Billing Data**
  - **Dlv-rel.billing type** (delivery-relevant billing type): This field controls the billing document type used by the system when creating billing documents with reference to a delivery.
  - **Order-rel.bill.type** (order-relevant billing type): This field controls the billing document type used by the system when creating billing documents for sales document items that are not relevant for delivery but that require billing.
  - **Intercomp.bill.type** (intercompany billing type): This field controls the billing document type used by the system when creating billing documents for intercompany billing.

- **Billing block**: This field specifies whether a sales document is automatically blocked for billing. For example, this block could be useful in credit memo requests to ensure that documents are checked before billing.

- **Contract Data**
  - **Contract profile**: This field is the contract profile assigned to the document type. The contract profile is used by the system to calculate start and end dates and also contains information regarding the cancellation procedure, the follow-up action to be carried out before the contract expires, and the date when the follow-up action must be carried out.
  - **Contract data allwd.** (contract data allowed): This field controls whether contract data can be maintained for a particular sales document type. Contract data is stored in both the header and in the contract items. By default, the header data applies to all items. Contract data contains validity period and cancellation data.

All sales documents can be used in all sales areas by default. However, if required by your project, you can restrict the sales document type by sales area. To set up this restriction, first, you must combine sales organizations, distribution channels, and divisions by specifying a reference sales organization, distribution channel, and division.

Example

Let's imagine you've created your own document type ZQT for quotations that is a copy of the standard document type QT. Sales document type ZQT should only be allowed in the following sales area:

- Sales organization: 0001
- Distribution channel: 10
- Division: 01

First, you must combine sales organizations by assigning a reference sales organization, as shown in Table 4.2.

Sales Organization	Reference Sales Organization
0001	0001

Table 4.2 Assigning a Reference Sales Organization

Second, you'll combine distribution channels by assigning a reference distribution channel, as shown in Table 4.3.

Sales Organization	Distribution Channel	Reference Distribution Channel
0001	10	10

Table 4.3 Assigning a Reference Distribution Channel

Third, you'll combine divisions by assigning a reference division, as shown in Table 4.4.

Sales Organization	Division	Reference Division
0001	01	01

Table 4.4 Assigning a Reference Division

Finally, you can assign the sales document type allowed for this sales area, as shown in Table 4.5.

Reference Sales Organization	Reference Distribution Channel	Reference Division	Sales Document Type
0001	10	01	ZQT

Table 4.5 Assigning a Sales Document Type to a Reference Sales Area

Warning

Once you have assigned a sales document type to a sales area, you must assign all other sales document types to the sales areas where they can be used. This requirement exists because, after assigning the first sales document type, the norm "all document types can be used in all sales areas" is no longer valid, and the system expects an assignment to a sales area to consider a sales document type as valid.

SAP S/4HANA introduced the option to assign order reasons and reasons for rejection to sales document types. These options were not available in SAP ERP, but now, you can restrict both elements by sales organization and sales document type. Figure 4.2 shows the order reason restriction table and the reason for rejection restriction table.

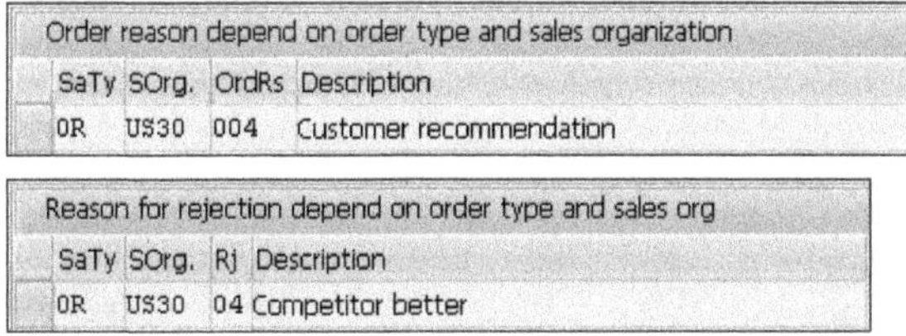

Figure 4.2 Assigning Sales Document Types and Sales Organizations to Order Reasons and Reasons for Rejection

This capability is particularly useful because reasons for rejection can be specifically defined for quotations or inquiries, removing them from the dropdown list in a standard sales order, which may help users find the relevant reason for rejection more quickly. Similar logic can be applied to order reasons.

4.3 Understanding Item Categories

Together with sales document types, item categories set the framework for the different business transactions in the system. The item category allows you to process the same material differently depending on the document type. For instance, if a material is used in a quotation, the item category determined by the system is not relevant for delivery or billing. However, if the same material is entered in a sales order, or a sales order is created with reference to the quotation, a new item category is determined, which makes the item deliverable, and the item can be billed.

Table 4.6 shows some of the item categories defined in the standard system for sales documents. If the standard item categories don't meet the requirements of your project, you can always define your own item categories in customizing to control the sales item differently.

Item Category	Description	Document Type Where Used
AGN	**Standard Item**	Quotation
AGNN	**Free of Charge Item**	Quotation
AGTX	**Text Item**	Quotation
CB1	**3rd Party with SN**	Sales order
TAN	**Standard Item**	Sales order
TANN	**Free of Charge Item**	Sales order
TATX	**Text Item**	Sales order
TAB	**Indiv.Purchase Order**	Sales order
TAD	**Service Item**	Sales order
TAE	**Explanation**	Sales order
TAP	**Pric.at Item Level**	Sales order
TAQ	**Pric.at Header Level**	Sales order

Table 4.6 Sample of Standard Sales Item Categories

You can define and assign item categories in the IMG menu under **Sales and Distribution • Sales • Sales Documents • Sales Document Item**. As a reference, Figure 4.3 shows a portion of the screen where the standard item category TAN can be configured.

As with the document type, we won't review every single function controlled by the item category, but you should be aware of the following parameters:

- **Business Data**
  - **Item Type**: This field is used to classify the item so that the system only performs relevant functions. You can specify if the item is a material, text, value item, or packaging. For instance, neither weight calculations nor pricing is carried out for text items.

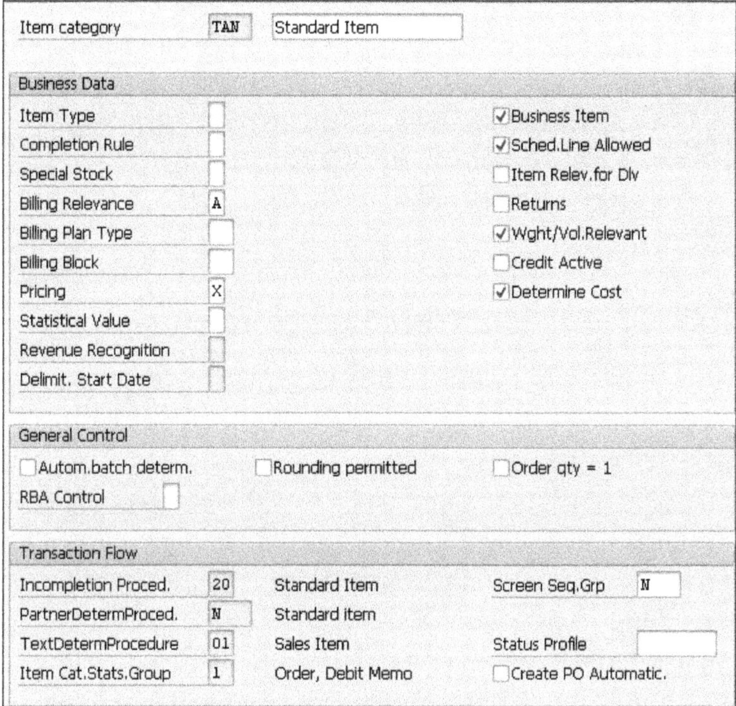

Figure 4.3 Defining Item Categories

- **Completion Rule**: This field determines the condition that must be met to mark the item as completed in a sales document. For instance, if value **B** is selected, the item is marked as completed after the full quantity has been referenced. The completion rule can only be applied if the **Update Document Flow** field is selected in the copy control (sales document to sales document) at the item level. Some of the values available include the following:
  - <blank>: Not relevant for completion.
  - **A**: Item is completed with the first reference.
  - **B**: Item is completed after the full quantity has been referenced.
  - **C**: Item is completed after the target quantity is fully referenced.
- **Pricing**: This field defines if pricing must be carried out for the item. The following four options are available:
  - <blank>: No pricing (e.g., text items)
  - **X**: Pricing standard (e.g., items with price)
  - **A**: Pricing for empties (e.g., returnable packaging such as pallets or crates)
  - **B**: Pricing for free goods (e.g., automatic 100% discount on free of charge items)
- **Sched.Line Allowed** (schedule lines allowed): If this indicator is selected, schedule lines can be created for the item. You should select this checkbox if the item category is relevant for delivery.

- **Credit Active**: If selected, the item is relevant to be processed by the credit management functions and the value of the item will be updated in the total liability.
- **Business Item**: This field specifies whether business data at the item level can deviate from the values at the header level. For example, if this checkbox is selected, you can enter payment terms *30 days* at header level and *60 days* at the item level.
- **Incompletion Proced.** (incompletion procedure): This field indicates which fields are part of the incompletion log, that is, which fields must be populated for the item to be completed.
- **PartnerDetermProced.** (partner determination procedure): The procedure maintained in this field specifies which partner functions are allowed at the item level and which are mandatory.
- **Create PO Automatic.**: If this indicator is selected, the system automatically creates both a purchase requisition and a purchase order (PO) when the sales document is saved for third-party schedule lines.
- **Structure scope**: This field indicates how a BOM is processed in a sales document. You can specify, for example, whether or not the material structure is exploded and, if so, whether only the first level must be exploded or a multilevel explosion must be carried out.

- **Shipping Data**
  - **Item Relev.for Dlv** (item relevant for delivery): If this checkbox is selected, the text or value item is relevant for delivery; thus, the system copies the item into the delivery note. As shown earlier in Figure 4.3, you don't need to select this item for standard items that are relevant for delivery (e.g., item category TAN). This field only applies to value or text items (item types A and B, respectively).
  - **Wght/Vol.Relevant** (weight/volume relevant): If selected, the system determines the weight and the volume of the item.
  - **Create Delivery Group**: This field determines if the subitems of a BOM that has been exploded in a sales document must be delivered together. The following three options are available:
    - <blank>: The system will not form delivery groups, and the BOM components will be delivered as they become available.
    - **X**: The system will use the last date of the items in the group to confirm the schedule line.
    - **A**: The system generates correlated schedule lines. Assuming the **Complete Dlv** (complete delivery) indicator is not selected, on the first date when something is available (quantity available > 0) for all items in the BOM, a confirmed quantity is calculated based on the ratio of the BOM.

- **Billing Data**
  - **Billing Relevance**: This field specifies whether the item is relevant for billing and, if so, the basis for billing. More than 20 different options are available in this dropdown list. Covering all these options is beyond the scope of this book, but we'll describe a few of the most common values:
    - <blank>: Not relevant for billing.
    - **A**: Delivery-related billing document. The billing for this item is based on the outbound delivery document.
    - **B**: Relevant for order-related billing. The billing for this item is based on the sales document. The billing status in the sales document item is defined according to the order quantity, for instance, item category REN (returns).
    - **C**: Relevant for order-related billing. The billing for this item is based on the sales document. The billing status in the sales document item is defined according to the target quantity, for instance, item category G2N (credit memo).
    - **D**: Relevant for pro forma. This option allows you to make free-of-charge outbound deliveries.
    - **F**: Relevant for order-related billing. The billing for this item is based on the sales document. The billing status in the sales document item and the quantity in the customer invoice are defined according to the invoice receipt quantity. For example, in a third-party sales order scenario, you can specify whether the system must wait for the vendor invoice to be received before an invoice is issued to the customer, for instance, item category TAS (third party business transaction).
    - **G**: Relevant for order-related billing. The billing for this item is based on the sales document and the quantity that has already been delivered. You can bill for multiple partial deliveries in one transaction and one billing document.
    - **H**: Relevant for delivery-related billing. The billing for this item is based on the outbound delivery document. Items with zero quantity are not included in the billing document.
    - **I**: Relevant for order-relevant billing. The billing for this item is based on the sales document. Billing takes place using the billing plan and its status, for instance, item category WVN (maintenance contract).
  - **Determine Cost**: If selected, the system determines the cost of the sales document item.
  - **Statistical Value**: This field specifies whether the value of the item is used to calculate the total value of the order. The following three options are available:
    - <blank>: The system will pass the item value to header totals.

  - **X**: No cumulation. The item value cannot be used statistically.
  - **Y**: No cumulation. The item value can be used statistically.
- **Billing Block**: If a billing block is selected, the system will automatically set the billing block at the item level, so no billing document can be created until the sales document item is reviewed and the billing block removed.
- **Returns**: This field indicates whether the item is a return item.

The item category is automatically determined by the system based on the sales document type, the item category group of the material, the higher-level item category, and the usage of the item category. Most of the time, a user won't need to worry about the item category. However, in some cases, the item category can be changed manually in the sales document item to deviate from the default business transaction.

You can configure the item category determination procedure by assigning item categories to sales document types and item category groups primarily. However, you can also define the usage and higher-level item category when necessary. SAP S/4HANA comes with a list of predefined item category groups covering the most common business scenarios, but you can always define new ones if needed. As shown in Table 4.7, we've listed some of the main item category groups in the standard system.

Item Category Group	Description
NORM	Standard item
LEIS	Service without delivery
DIEN	Service with delivery
CBOR	Third-party sales order with shipping notification
BANC	Individual PO
ERLA	Structure/material above (sales BOM with pricing at main item level)
LUMF	Structure/material below (sales BOM with pricing at subitem level)
VERP	Packaging

Table 4.7 Sample of Item Category Groups in the Material Master

For example, you would normally assign item category group NORM to materials in stock and item category group DIEN to services and nonstock materials that are relevant for delivery.

Tip

Maintaining the correct item category group in the material master record for the relevant sales organizations is vitally important. For each material type, you can define the default item category group, which will be automatically proposed when a material master record is created. This setting helps ensure the expected item category is determined in the sales order item.

The item category usage also contributes to the item category determination and controls the system's response to special cases, for instance, when the item is the result of a batch split or if the item is a text item. As with the item category group, you can create a new item category usage if the standard offerings don't meet your requirements. An item category usage is only used in certain circumstances. To illustrate few special circumstances where the item category usage plays a key role in item category determination, Table 4.8 shows some common values defined in the standard system.

Item Category Usage	Description
CHSP	Batch Split
CSEL	Cross Selling
FREE	Free Goods
PSEL	Product Selection
TEXT	Text Item

Table 4.8 Sample of Item Category Usage

The next step is to assign the default item category and the item categories that can be entered manually to overwrite the value proposed by the system. These alternative item categories are known as *manual item categories*. Figure 4.4 shows the logic to determine the item category in a simple scenario where the sales item contains a material that is neither a subitem nor a special case that utilizes the item category usage.

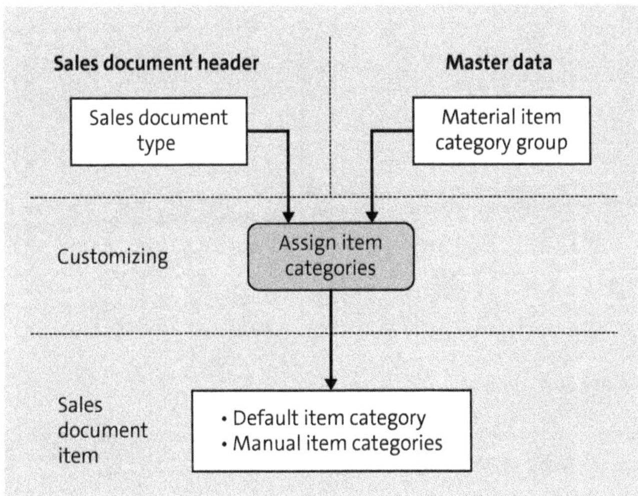

Figure 4.4 Item Category Determination

Now that you understand how the system determines the item category in a sales document, let's take a look at Figure 4.5, which shows a subset of the actual customizing table where default and manual item categories are assigned.

SaTy	ItCGr	Usge	HLevItCa	DfItC	MItCa	MItCa	MItCa
OR	NORM			TAN	TAP	TAQ	TANN
OR	NORM		GTO1	TAN			
OR	NORM		GTO2	TAN			
OR	NORM		GTR2	REN			
OR	NORM		TAC	TAE			
OR	NORM		TAE	TAE			
OR	NORM		TAG	TAN			
OR	NORM		TAM	TAN			
OR	NORM		TAN	TANN			
OR	NORM		TANN	TANN	KBN		
OR	NORM		TAP	TAN			
OR	NORM		TAQ	TAE			

Figure 4.5 Assigning Item Categories

According to the top configuration entry shown in Figure 4.5, if a material with item category group NORM is entered on a sales document type OR and this material is neither a subitem nor a special case, the system will propose item category TAN, which is the item category maintained in the **DfltC** (default item category) column, where both the **Usge** (usage) and **HLevltCa** (higher-level item category) fields are blank. However, business users have the option to manually change the item category to TAP, TAQ, or TANN. Any other item categories are not allowed in this particular case. You should note that all the other configuration entries shown in Figure 4.5 are only relevant for subitems, since the **HLevltCa** column contains the item category of the parent item.

In SAP S/4HANA, you can manually create items in a sales document as subitems or enter a sales BOM, and the system will generate the subitems for you. By definition, a subitem is always allocated to a parent. The item number of the parent can be found in the **HL Itm** (higher-level item) field of the child (subitem). Table 4.9 shows an example of a sales document with three items, where items 20 and 30 are subitems of parent item 10.

Item Number	Material	Quantity	Item Category	Higher-Level Item
10	MAT1	1	TAN	
20	MAT2	1	TANN	10
30	MAT3	1	TANN	10

Table 4.9 Subitems in a Sales Document

Assuming the item category group for materials MAT1, MAT2, and MAT3 is NORM, and the sales document type is OR, the example shown in Table 4.9 is in line with the assignments shown in Figure 4.5. For item 10, the system determines item category TAN (the first entry from the top in the assignment table), whereas item category TANN is proposed for items 20 and 30 (the fourth configuration entry from the bottom).

Another important customizing activity related to item categories allows you to activate enhanced logic to automatically reject sales order items with third-party delivery or individual procurement. If you select the **Enhanced Rejection Logic** checkbox and enable automatic rejection by selecting an option from the **Auto Reject SO Item** dropdown list, as shown in Figure 4.6, you can specify an automatic rejection reason for the sales order item that will be automatically applied when either the **Delivery Complete** or the **Final Invoice** indicator is set in the corresponding PO item.

You can also select the **Update SO Only From Conf. PO** checkbox to update the schedule line data of the sales order item only upon confirmation of the corresponding PO item. In the standard system, the schedule data of a sales order item with individual procurement is updated when the delivery date of the related PO item changes. If selected, this indicator would change the standard behavior, and the system will only update the schedule data of a sales order item when the corresponding PO item has been confirmed.

Additional Settings for Item Categories with Purchasing

Item category	Description	Enhanced Rejection Logic	Auto Reject SO Item	Reason for Rejection	Update SO Only From Conf. PO
CB1	3rd Party with SN	☐	Disable Automati...		☐
CB2	3rd Party w/o SN	☐	Disable Automati...		☐
CB3	3rd Pty with SN free	☐	Disable Automati...		☐
CB4	3rd Pty w/o SN free	☐	Disable Automati...		☐

Figure 4.6 Additional Settings for Item Categories with Purchasing

Example

Let's look at an example of how the automatic rejection functionality works in SAP S/4HANA: You create a sales document using the third-party delivery item category CB1 for 50 units. A PO is generated in the system for 50 units and sent to the vendor.

The vendor has only 45 units available, and there is no plan to manufacture more units as this product has been discontinued. After doing the post goods receipt for 45 units, the buyer sets the **Delivery Complete** indicator in the PO item.

Assuming both the enhanced rejection logic and the automatic sales order update from PO functionalities have been activated in customizing, as shown in Figure 4.6, for item category CB1, the system automatically sets the reason for rejection selected on the sales document item. As a result, even though only 45 out of 50 units have been delivered, the sales order item is marked as closed, and it won't show up in backorder reports.

Note

You can enable automatic rejection and select a reason for rejection only for item categories involving third-party delivery. For example, the automatic rejection works with item category CB1 (third-party delivery), but it cannot be activated for item category TAB (individual procurement).

The functionality to update the sales order item only upon confirmation of the corresponding PO item can be enabled for item categories with third-party delivery or individual procurement.

4.4 Using Bill of Materials in Sales Documents

A BOM is defined as a group of materials that make up a product. In SAP S/4HANA, a sales BOM is formed by the header item and then subitems or components.

Companies often want to sell single products bundled into a sales kit; for example, a company selling consumer electronics can create a sales BOM in SAP S/4HANA, as shown in Table 4.10.

Header Material Number	Component Material Number	Description
KIT11		Kit Smartphone 11 w/ Insurance
	SMPH11	Smartphone Model 11
	CASE11	Model 11 protective case
	400000140	1 Year insurance

Table 4.10 Example of a Sales BOM

Sales kits are commonly used when the products that are part of the sales kits can also be sold independently as single products. However, buying a sales kit is usually more cost effective. If the kit is ordered, when the sales BOM header material KIT11 is entered in the sales document item, the system automatically explodes the BOM adding the three components. Materials SMPH11, CASE11, and 400000140 are added as children of the same parent item. Using SAP terminology, these children are called "subitems of the same higher-level item."

Processing pricing and inventory control at both header and subitem levels wouldn't make sense. Thus, if the components are assembled, then the sales BOM must be defined so only the main item is relevant for pricing and delivery processing. This type of sales BOM is known as a *main item level BOM*. On the other hand, if the components are not assembled, pricing must be maintained at the subitem level, and for those components that are also stock materials, inventory is also controlled at the subitem level. This other type of sales BOM is known as a *subitem level BOM*.

The item category group entered in the **Sales: sales org. 2** view in the material master record defines how the system processes the sales BOM. To create a main item level BOM, you must maintain item category group ERLA for the top-level material, and the system will carry out pricing, delivery control, and delivery processing at the main item level. When the main item is entered in a sales document, the components will be exploded as text items; thus, they won't be relevant for delivery and won't drive any functionality. Table 4.11 shows the item category groups you must assign to each material to create a main item level BOM based on the previous example shown earlier in Table 4.10. Table 4.11 also displays the expected default item category in the sales order.

Material Number	Item Category Group	Item Category Group Description	Item Category in Sales Order	Item Category Description
KIT11	ERLA	Structure/Mat. above	TAQ	Pric.at Header Level
SMPH11	NORM	Standard item	TAE	Explanation
CASE11	NORM	Standard item	TAE	Explanation
400000140	LEIS	Service w/o Delivery	TAE	Explanation

Table 4.11 Main Item Level BOM: Item Category Groups and Item Categories

Note

We're assuming that service item 400000140 should not be delivered, which is why item category group LEIS has been assigned to the insurance item. In the event you want to copy the service item to the outbound delivery, you should maintain the item category group DIEN.

The next step to set up a sales BOM is to create the actual BOM that links the parent material with the components. We'll show you how to create a BOM relevant for sales later in this subsection, but first, let's take a look at the BOM shown in Figure 4.7, which is a main item level BOM in a sales order after entering the main item KIT11. Note that only item 10, the main item, has a price and that all the components have the value 10 in the **HL Itm** (higher-level item) column, which indicates they all belong to item 10.

All Items

Item	Material	Order Quantity	Un	S	Item Description	ItCa	HL Itm	CnTy	Net Price	Crcy
10	KIT11	1	EA		Kit Smartphone 11 w/ Insurance	TAQ	0	PR00	546.000	USD
20	SMPH11	1	EA		Smartphone Model 11	TAE	10	PR00	0.000	USD
30	CASE11	1	EA		Model 11 protective case	TAE	10	PR00	0.000	USD
40	400000140	1	EA		1 Year insurance	TAE	10	PR00	0.000	USD

Figure 4.7 Main Item Level BOM in a Sales Document

Don't forget to check that the item category determination has been set up correctly for the main item and its components. In the example shown in Figure 4.7, we want the system to propose item category TAQ for the main item and item category TAE for all the components (both stock and service items), so you must ensure the entries shown in Figure 4.8 have been maintained in the customizing activity, which can be accessed via **Sales and Distribution • Sales • Sales Documents • Sales Document Item • Assign Item Categories**.

SaTy	ItCGr	Usge	HLevItCa	DfItC	MItCa	MItCa	MItCa
OR	ERLA			TAQ			
OR	LEIS		TAQ	TAE			
OR	NORM		TAQ	TAE			

Figure 4.8 Item Category Assignment Required for Main Item Level BOM

Now, imagine you would like the system to carry out pricing, inventory control, and delivery processing at the component level for the sales BOM in the previous example. For this procedure to happen, you would need to define the BOM as a subitem level BOM by changing the item category group to LUMF for the top-level material KIT11. You don't need to change the item category group for the subitems, however. Table 4.12 shows the item category groups and the expected item categories for all the materials in the example shown earlier in Table 4.10 when the BOM is configured as a subitem level BOM.

Material Number	Item Category Group	Item Category Group Description	Item Category in Sales Order	Item Category Description
KIT11	LUMF	Structure/Mat. below	TAP	Pric.at Item Level
SMPH11	NORM	Standard item	TAN	Standard Item
CASE11	NORM	Standard item	TAN	Standard Item
400000140	LEIS	Service w/o Delivery	TAD	Service

Table 4.12 Subitem Level BOM: Item Category Groups and Item Categories

Figure 4.9 shows a subitem level BOM in a sales order after entering the main item KIT11. This time, item 10 is just a text, and pricing and inventory are managed at the subitem level.

All Items

Item	Material	Order Quantity	Un	S	Item Description	ItCa	HL Itm	CnTy	Net Price	Crcy
10	KIT11	1	EA	☐	Kit Smartphone 11 w/ Insurance	TAP	0	PR00	0.000	USD
20	SMPH11	1	EA	☐	Smartphone Model 11	TAN	10	PR00	525.000	USD
30	CASE11	1	EA	☐	Model 11 protective case	TAN	10	PR00	10.500	USD
40	400000140	1	EA	☐	1 Year insurance	TAD	10	PR00	21.000	USD

Figure 4.9 Subitem Level BOM in a Sales Document

In this case, you should configure the item category assignment by maintaining the entries shown in Figure 4.10, so the system determines item category TAP for the main item, TAN for the components in stock, and TAD for the service item.

SaTy	ItCGr	Usge	HLevItCa	DfltC	MItCa	MItCa	MItCa
OR	LUMF			TAP	TAQ	GT01	GT02
OR	LEIS		TAP	TAD	TAW		
OR	NORM		TAP	TAN			

Figure 4.10 Item Category Assignment Required for the Subitem Level BOM

After maintaining the correct item category assignment in customizing and making sure all materials involved in the BOM have the correct item category group, you still need to create a sales BOM to enable the use of this functionality. So, let's now explore how to create BOMs that are relevant for sales in SAP S/4HANA. To create a BOM, either execute Transaction CS01 or use the SAP Fiori app Create Bill of Material, depending on your user interface (UI). You'll be presented with the screen shown in Figure 4.11, where you must enter the material number of the main item and the BOM usage. On the initial screen, you can restrict the use of the BOM to a particular plant by specifying the plant as well as entering the date from which the BOM is valid.

Create material BOM: Initial Screen

Create variant of...

Material	KIT11
Plant	US20 Long Island City
BOM Usage	5 Sales and Distribution
Alternative BOM	
Validity	
Change Number	
Valid From	02/16/2021

Figure 4.11 Creating a Sales BOM

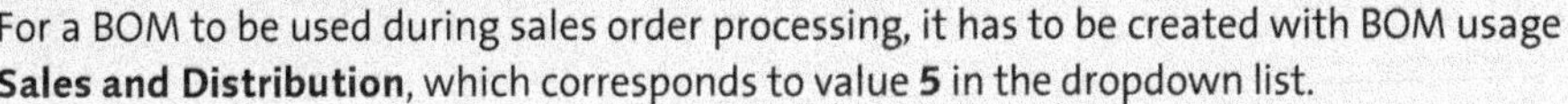

Note

For a BOM to be used during sales order processing, it has to be created with BOM usage **Sales and Distribution**, which corresponds to value **5** in the dropdown list.

On the next screen, you can specify the mandatory information at the component level, such as the material number, the quantity, the unit of measurement (UoM), and the validity period. You also need to pick an item category depending on the nature of the subitem. The following item categories are available:

- **C: Compatible Unit**
- **D: Document item**
- **I: PM structure element**
- **K: Class item**

- **L: Stock item**
- **M: Intra material**
- **N: Non-stock item**
- **R: Variable-size item**
- **T: Text item**

In a sales BOM, you would normally use item category L for goods that are physically in the warehouse and item category N for service items. The rest of the fields are optional when creating a sales BOM, so at this point, the BOM can be saved. Figure 4.12 shows the BOM definition for the example **Kit Smartphone 11 w/ Insurance.**

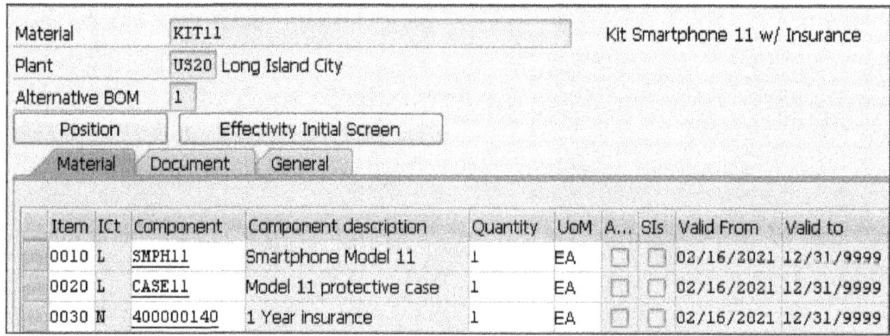

Material KIT11 Kit Smartphone 11 w/ Insurance
Plant US20 Long Island City
Alternative BOM 1

Item	ICt	Component	Component description	Quantity	UoM	A...	SIs	Valid From	Valid to
0010	L	SMPH11	Smartphone Model 11	1	EA			02/16/2021	12/31/9999
0020	L	CASE11	Model 11 protective case	1	EA			02/16/2021	12/31/9999
0030	N	400000140	1 Year insurance	1	EA			02/16/2021	12/31/9999

Figure 4.12 Sales BOM: General Item Overview

When configuring item categories for sales BOMs, a good practice is to think about forming *delivery groups* automatically, which allow you to combine several items in a sales document, so they are delivered together. This functionality is particularly useful for subitem level BOM scenarios, since customers typically want all the components to be delivered at the same time. Going back to the example of the smartphone sales kit, where pricing and inventory are controlled at the subitem level, let's say a customer orders the kit, but the protective case is currently available in stock. However, the smartphone will arrive in the warehouse in 2 days. In this case, you wouldn't want the system to generate a delivery for the case today and another delivery for the smartphone in a few days. To avoid this problem, you can select **A** in the **Create Delivery Group** field for the item category of the sales BOM main item (e.g. TAP), and the system will generate correlated schedule lines provided the **Complete Dlv.** indicator is not checked in the sales document. The system will confirm a quantity on the first day something is available for all the components in the BOM (a quantity greater than zero must be available for all subitems). The same confirmed delivery date will be assigned to all the sales BOM items relevant for delivery, as they are now part of a delivery group. The quantity confirmed depends on the ratio of the BOM. In the example used in this chapter, the BOM ratio is 1:1:1:1, so the system will always confirm the same quantity for all the subitems.

Note

As explained in question two in SAP Note 546719, incomplete delivery groups are not copied to the delivery when using transactions such as VL04, VL10, VL10A, VL10B, or VL10C. In other words, if at least one item of the delivery group is unavailable at the time of the delivery creation via multiple processing, the whole delivery group is rejected. However, the user can use Transaction VL01N to manually copy the available part of the delivery group to the delivery if the customer agrees to receive only the components available. The delivery creation log in Transaction VL01N will inform the user that the delivery group is incomplete.

However, for a delivery group with correlated schedule lines to work, the following prerequisites must be met:

- The delivery group may only contain one BOM.
- It cannot contain any fixed quantities.
- It cannot contain any items that have partial delivery indicator C (one-time delivery).
- It cannot contain any partially delivered items.
- It can only contain items that can be at least partially delivered.
- It can only contain items that have a maximum of one requested schedule line.

4.5 Understanding Schedule Line Categories

The schedule line is the third and last level in a sales document, and how it is processed by the system depends on the schedule line category. Sales document items without schedule lines cannot be delivered, so assigning a schedule line category to the item categories that are relevant for delivery is essential. The system will create deliveries for the dates and confirmed quantities in the schedule lines.

In the standard system, the keys for schedule line categories follow a naming convention, as shown in Table 4.13. The first character of the key represents the type of document, and the second character represents the usage in materials management.

First Character of the Key	Type of Document
A	Inquiry
B	Quotation
C	Sales order
D	Returns

Table 4.13 Schedule Line Category Naming Convention

Second Character of the Key	Usage
T	No inventory management (e.g., services)
X	No inventory management with goods issue
N	No planning
P	Material requirements planning (MRP)
V	Consumption-based planning

Table 4.13 Schedule Line Category Naming Convention (Cont.)

As with the sales document type and item category, if you need to define a new schedule line category according to your requirements, we suggest you copy a similar schedule line category and make the appropriate changes. Also, to avoid confusion, the first character of the new schedule line category should be Z.

To set up sales document scheduling line categories, follow the menu path **SAP Customizing Implementation Guide • Sales and Distribution • Sales • Sales Documents • Schedule Lines**. When creating your own schedule line category, you must define the parameters shown in Figure 4.13. As an example, Figure 4.13 shows the standard schedule line category **CP**, which is used in sales documents and is relevant for MRP.

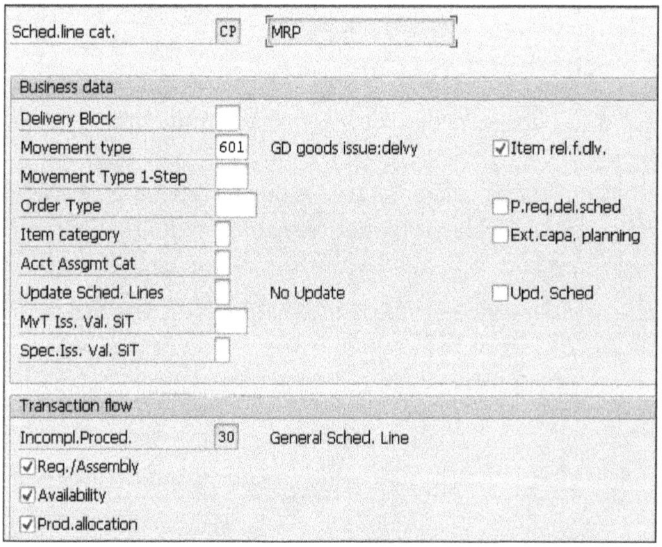

Figure 4.13 Maintaining Schedule Line Categories

The parameters of the schedule line category are divided into two sections, which contain the following fields.

- **Business data**
  - **Delivery Block**: This field allows you to block the schedule line for delivery. This feature could be useful if you want the schedule line to be reviewed before the item is delivered.

- **Movement type**: You can use this parameter to define the movement type to be used when posting a goods issue. No movement type is required for schedule line categories assigned to items with no stock movement (e.g., quotation items).
- **Movement Type 1-Step**: This movement type is used, for example, in stock transfers in two steps.
- **Order Type**: This field is the purchase requisition type to be used by the system to create a purchase requisition when the sales document is saved.
- **Item category**: This field is the item category of the purchase requisition (e.g., service, third-party, standard, stock transfer, subcontracting, etc.).
- **Acct Assgmt Cat** (account assignment category): This field is only relevant for schedule lines that will generate a purchase requisition and defines the account assignment category in the purchase requisition. Some common values include the following:
  - **A: Asset**
  - **K: Cost center**
  - **W: Third-party with shipping notification**
  - **Y: Third-party without shipping notification**
  - **H: Non-stock sales**
- **Update Sched. Lines** (update schedule lines): This field is used in third-party PO scenarios and controls whether the schedule line in the sales document must be automatically updated when a shipping notification is received from the supplier or when a goods receipt is entered for the corresponding individual PO item. The following three values are available:
  - <blank>: No update
  - **1**: Update only for goods receipt
  - **2**: Update for goods receipt and shipping notification
- **MvT Iss. Val. SiT** (movement type for issuing valuated stock in transit): This field specifies the movement type for the follow-on material document for issuing valuated stock in transit. This movement type applies to the following scenarios:
  - Stock transfers with valuated cross-company stock in transit
  - Stock transfers with valuated intracompany stock in transit
  - Delivery to customers with valuated stock in transit
- **Spec.Iss. Val. SiT** (special issuing of valuated stock in transit): This field indicates whether the exact transfer of title occurs during transit or on goods receipt. This field only applies to stock transfers and purchasing scenarios relevant for valuated stock in transit.

- **Item rel.f.dlv.** (item relevant for delivery): Select this checkbox if the schedule line category is relevant for delivery.
- **P.req.del.sched** (purchase requisition delivery schedule): If selected, the system calculates the delivery dates of the schedule line while taking into account the delivery date in the purchase requisition.
- **Ext.capa. planning** (external capacity planning): In a third-party order scenario, if this indicator is selected, the system executes external capacity planning in the SAP Supply Chain Management (SAP SCM) system.
- **Upd. Sched** (update schedule): This checkbox can be used to automatically update the availability date in the schedule line when the delivery date in the corresponding PO item is changed. This functionality only works if the PO is a stock transport order.

- **Transaction flow**
  - **Incompl.Proced.** (incompletion procedure): You can define an incompletion procedure with a set of mandatory fields at the schedule line level. If one or more fields in the incompletion procedure are not entered, the schedule line is considered incomplete. This functionality will be covered in more detail in Chapter 5, Section 5.2.
  - **Req./Assembly** (requirements/assembly): This field specifies whether the transfer of requirements to MRP must be carried out. Note that you can only activate the transfer of requirements at the schedule line level if the transfer of requirements functionality is already activated in the requirements class. We'll cover the transfer of requirements in more depth in Chapter 7, Section 7.3.
  - **Availability**: This checkbox indicates if an availability check must be carried out for the item.
  - **Prod.allocation** (production allocation): This indicator can be used to activate product allocation for the schedule line category.

SAP S/4HANA automatically determines the schedule line category using the item category of the sales document item and the MRP type maintained in the material master record (**MRP 1** view). Therefore, after defining a new schedule line category, you must assign the schedule line category to the relevant item categories and MRP types.

As shown in Figure 4.14, you can specify the proposed schedule line category for each item category and MRP type. If required, you can maintain up to three manual schedule lines, which can be selected by a business user to overwrite the schedule line category determined by the system.

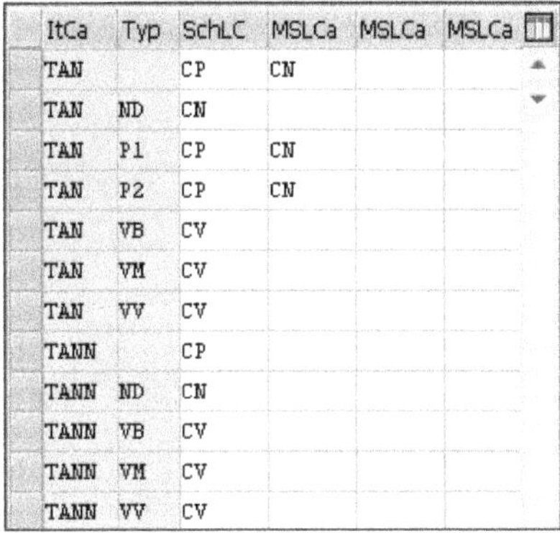

ItCa	Typ	SchLC	MSLCa	MSLCa	MSLCa
TAN		CP	CN		
TAN	ND	CN			
TAN	P1	CP	CN		
TAN	P2	CP	CN		
TAN	VB	CV			
TAN	VM	CV			
TAN	VV	CV			
TANN		CP			
TANN	ND	CN			
TANN	VB	CV			
TANN	VM	CV			
TANN	VV	CV			

Figure 4.14 Assigning Schedule Line Categories

4.6 Understanding Sales Contracts

Many businesses sign long-term agreements for the supply of goods or services at a specific price and quantity over a certain period. These agreements are known in SAP S/4HANA as *sales contracts* or *customer contracts*.

The following four categories of sales contracts have been defined in a standard SAP S/4HANA system:

- Master contract (type GK)
- Quantity contract (type CQ)
- Value contract (types WK1 and WK2)
- Service contract (types WV and MV)

In the following sections, we'll review the key features of these four types of sales contracts.

4.6.1 Master Contract

Master contracts are sales documents used to bundle contracts for the same customer. A master contract only contains general header data that is relevant for all lower-level contracts. You can enter the following data in a master contract:

- Billing plan data
- Partner data
- Business data
- Contract data

The overview screen of a master contract shows the lower-level contracts that refer to it. By using a master contract, you can ensure that the terms in the master contract apply to all lower-level contracts. In SAP S/4HANA, you can group quantity contracts, value contracts, and service contracts under a master contract. This functionality allows you to link sales contracts to a master contract and is known as *contract grouping*. When a lower-level contract is assigned to a master contract, the header information is copied from the master contract into the lower-level contract. However, contract grouping has some limitations since the structures of multiple levels of contracts linked together are not allowed. You can have a master contract with several lower-level contracts, but all lower-level contracts must be the same level. Also, the master contract and all its lower-level contracts must be assigned to the same sales area.

To group sales contracts, you must first define which types of contracts a master contract can refer to, the fields that are relevant for contract grouping, and the rules for copying the data from the master contract into the lower-level contract. You can perform these activities in customizing under **Sales and Distribution • Sales • Sales Documents • Contracts • Master Contract • Define Referencing Requirements**. First, assign the sales contract types that can be linked to each master contract type. This step allows you, for example, to ensure service and maintenance contracts (type WV) are only assigned to master contract type GK, as shown in Figure 4.15.

TarDoc	Description	Source	Description
CQ	Quantity Contract	GK	Master Contract
MV	Rental Contract	GK	Master Contract
WK1	Value Contract- Gen.	GK	Master Contract
WK2	Matl-rel. Value Cont	GK	Master Contract
WV	Service and Maint.	GK	Master Contract

Figure 4.15 Defining Reference Sales Document Types

Then, you can either assign the standard referencing procedure SDGK (SD Sample Reference Procedure) to the sales document type of your master contract, or if the standard procedure does not meet your business requirements, you can define a new referencing procedure by specifying the technical name of the tables and fields to be checked or copied from the master contract. As shown in Figure 4.16, the checkbox in the **Message** column can be used to decide if the system will issue a message in the lower-level contract to inform a business user which changes will be copied from the master contract. The **Copy rule** column is an important parameter that controls how the system treats each field when copying data into the lower-level contract. The following three copy rules are available:

- **A: Identical field**
  These fields must have the same value in the master contract and lower-level contract. If you change a value in this field in the lower-level contract, the system will not allow you to link both contracts. Note that contract data fields (table VEDA) must all have copy rule A to be added to the referencing procedure.
- **B: Copy field**
  The values in these fields are copied from the master contract and cannot be changed in the lower-level contracts.

- **C: Proposal field**
  The system proposes the value from the master contract into the lower-level contract. Changes are allowed in the lower-level contract so business users can overwrite the proposed value. However, the system will only copy changes if both contracts had the same value before the master contract was changed.

Dialog Structure
- Procedures
  - Fields

Group proced. SDGK

Table	Field name	Function	Copy rule	Message
FPLA	AUTTE		B	✓
FPLA	PERIO		B	✓
VBAK	FAKSK		B	✓
VBAK	SPART		A	✓
VBAK	VKORG		A	✓
VBAK	VTWEG		A	✓
VBKD	INCO1		B	✓
VBKD	INCO2		B	✓
VBKD	ZTERM		B	✓
VBPA	KUNNR	SP	A	✓
VBPA	KUNNR	BP	A	✓
VBPA	KUNNR	PY	A	✓
VEDA	VLAUEZ		A	✓
VEDA	VLAUFK		A	✓
VEDA	VLAUFZ		A	✓

Figure 4.16 Defining Referencing Procedures

The referencing procedure is then assigned to the sales document type of the master contract. Also, you can select the **Update low.lev.cont.** indicator if you want the system to update the lower-level contract as soon as the master contract is changed. The update will be carried out via a workflow according to the copy rules specified in the referencing procedure. If the indicator is not activated, the system will only update the lower-level contract when it is called for processing. These two parameters are maintained in the **Contract** section of the definition of the sales document type, as shown in Figure 4.17.

Contract

PricProcCondHeadr		Contract data allwd.	X
PricProcCondItem		FollUpActivityType	
Contract profile	0002 Contract w/ Renewal	Subseq.order type	
Billing request		Check partner auth.	
Group Ref. Procedure	SDGK	✓ Update low.lev.cont.	

Figure 4.17 Contract Data for Sales Document Type GK: Master Contract

Since the master contract is nothing but a document used to group sales contracts and ensure header data is consistent, the master contract is not relevant for delivery or billing. Therefore, you do not need to assign either delivery type or billing type to the sales document type of the master contract.

Note

Master contracts do not have item or schedule lines data to be transferred to a subsequent sales document, so you only need to maintain copy control rules at the header level. We'll review how to set up the copying control in Chapter 6, Section 6.2.

Now, you can create a master contract via Transaction VA41 or the SAP Fiori app Create Contract. Just enter the header information and save the master contract. The next step is to create the lower-level contracts using the same transaction or app, but this time, select the sales document type of the lower-level contract (e.g., WV for a service contract). You'll link the contracts by entering the master contract number in the **Master Contract** field under the **Sales** tab at the header level in the lower-level contract. Figure 4.18 shows the overview screen for a master contract with two service contracts.

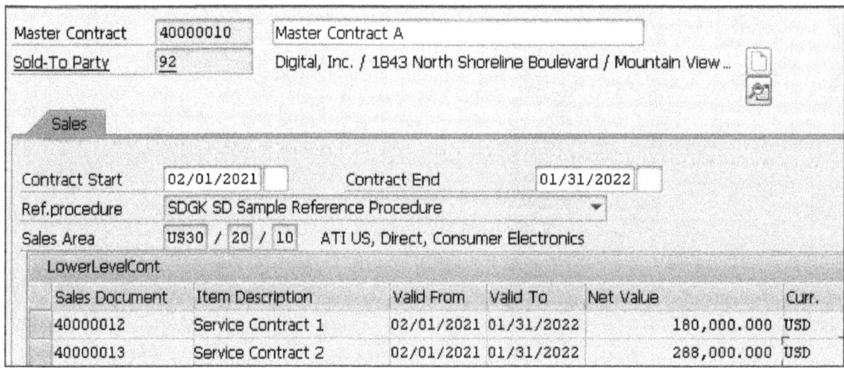

Figure 4.18 Master Contract Overview Screen

4.6.2 Quantity Contract

A quantity contract can be defined as a legally binding agreement where a certain quantity of goods must be supplied to a customer during a specified period. The customer is usually offered a special price in return. The target quantity entered at the item level is not transferred to MRP; thus, it does not create demand to satisfy the contract. The quantity contract has no schedule lines, but it has a start and end date at both header and item level. If required, the validity dates can be different for each item and differ from the start and end dates at header level. The standard sales document type for quantity contracts in SAP S/4HANA is CQ.

Once a sales order is created with reference to a quantity contract, the requirement is placed in MRP. These sales orders are known as *release orders*. Assuming the document type of the subsequent release order is OR, if the stock available on the requested delivery date is less than the quantity in the release order, the system will try to satisfy the release order by either buying more stock or producing more depending on the procurement type of the material.

Each release order automatically decreases the released quantity in the quantity contract. However, this functionality only works if the **Update Document Flow** field is not blank on the contract item category in the copy control rules, as shown in Figure 4.19. Otherwise, the customer can order as much as he would like, and the quantity in the contract won't be a limiting factor.

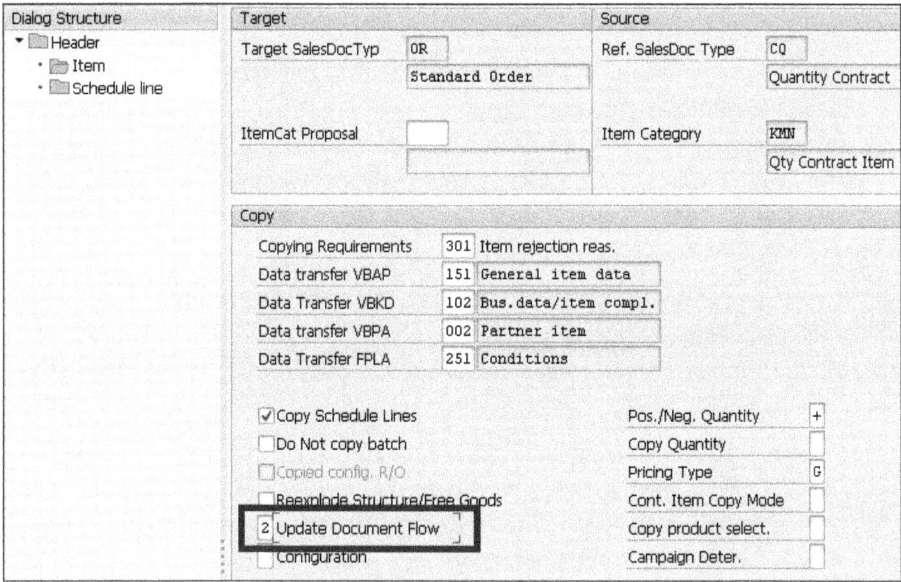

Figure 4.19 Set Copying Control: Quantity Contract to Sales Order

SAP S/4HANA allows you to specify which business partners, besides the sold-to party, can be used to release orders against the quantity contract. When a sales order is created with reference to a contract, the system checks whether the customer number entered is authorized or not by applying one of the following rules specified in the **Check partner auth.** (check partner authorization) field in the sales document type of the contract. This field may have the following values:

- <blank>: No partner check.
- **A**: The customer entered in the sold-to party field in the release order has the partner function sold-to party (AG) or sold-to party for release order (AA) in the contract.
- **B**: The customer entered in the sold-to party field in the release order belongs to the customer hierarchy of the sold-to party (AG) of the contract.

When creating a business partner master record, you can define multiple authorized sold-to and ship-to parties via standard partner functions AA and AW, respectively. These parties will appear as suggestions in the dialog box for creating a contract for the business partner in question.

Also, you can define multiple ship-to parties for a release order in the contract itself with partner function AW. Then, when a release order is created with reference to the contract, the system displays a list of possible ship-to parties from which business users can select. However, you could configure the copying control, so the rest of the partners (other than the sold-to party) are not copied from the contract but instead determined from the customer master record of the sold-to party.

To activate the partner authorization check, you must assign a partner determination procedure containing partner functions AW and AA to the sales document type of the quantity contract. You can use the standard partner determination procedure KAB (Release Order), which comes with the two required partner functions. Also, don't forget to select one of the rules (A or B) in the contract sales document type, in the **Check partner auth.** field.

Quantity contracts can act as lower-level contracts if a master contract number is entered in the master contract field of the quantity contract. The quantity contracts will then appear in the **LowerLevelCont** section of the overview screen of the master contract.

Another feature of sales contracts in SAP S/4HANA is the capability to use cancellation rules to specify the times when a customer can terminate the contract. You can define cancellation rules for a quantity contract in customizing under **Sales and Distribution • Sales • Sales Documents • Contracts • Contract Data • Control Cancellation.**

The first step is to define a cancellation procedure for which you can enter a validity period. You can assign the cancellation procedure to the contract profile, maintained in the **Contract** section of the sales document type. The next step is to define the cancellation rule and assign it to the relevant cancellation procedure. The flowchart shown in Figure 4.20 summarizes how the system uses the allocations made in customizing to determine the cancellation rule in a sales contract.

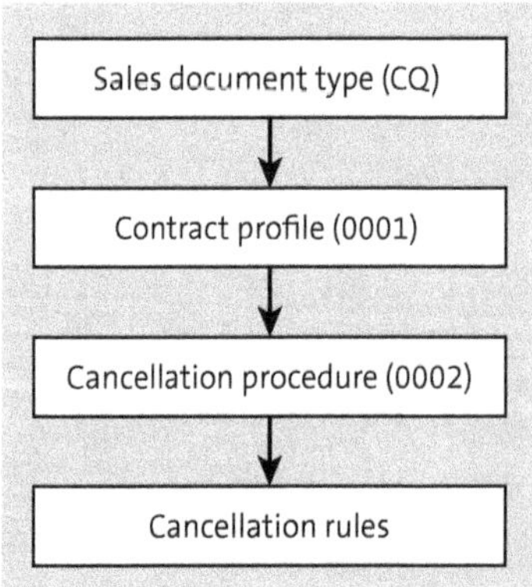

Figure 4.20 Cancellation Rule Determination in a Contract

The cancellation rule is defined by the following four parameters:

- **Canc.date** (cancellation date): This field specifies the rule for calculating the cancellation date, which is the date when the contract will be terminated, provided the cancellation is received in the period of notice.
- **Notice**: This field, alongside the cancellation date, determines the period when the customer must submit a cancellation request. Some examples of periods you can enter in this field include "3 days," "2 weeks," or "1 month."
- **Canc.period** (cancellation period): This field is the period by which the contract is automatically extended if no cancellation is received.
- **Rule valid to**: This field defines the period in which the cancellation rule is valid.

Example

Let's say a sales contract has been created with a start date of January 1st and should adhere to the following cancellation rule:

- Rule for the cancellation date: Start of contract + 1 year
- Period of notice: 2 weeks
- Cancellation period: 1 month

The contract can only be terminated after 1 year, that is, on December 31st, provided the customer submits a cancellation request 2 weeks before the cancellation date. In other words, the cancellation request must be received by December 17th. If the cancellation request is not received by December 17th, the contract is extended by 1 month.

Finally, to ensure that your quantity contracts are fully configured, you should check that a pricing procedure, partners, texts, and item categories have been assigned to the sales contract document type. Also, make sure that copying control rules have been maintained for the target sales document types of the release orders.

4.6.3 Value Contract

Value contracts are similar to quantity contracts in the sense that a customer cannot buy unlimited goods with reference to the contract. In the case of value contracts, the limiting factor is not the target quantity but the total value of the contract (also referred to as the target value).

As with quantity contracts, you can activate the partner authorization check in value contracts to control the additional ship-to and sold-to partners that can release against the contract. Again, partner functions AA (sold-to for contract release order) and AW (ship-to for contract release order) can be used in these scenarios.

In SAP S/4HANA, you can specify a predefined list of materials in the value contract, which could be either an assortment module or a product hierarchy. This type of value contract is often called a *general value contract*. Alternatively, you can create a value contract for only one material, which is typically used for configurable materials. This other type of value contract is known as a *material-related value contract*.

The standard system offers the following two types of value contracts:

- **Document type WK1: General value contract**
  You must select this type of contract if you want to create a value contract for a group of materials belonging to the same product hierarchy or assigned to an assortment module.
- **Document type WK2: Material-related value contract**
  You must select this type of contract if only one material is relevant for release.

To create an assortment module, you can use Transaction WSV2 or the Create Assortment Module – Value Contract app. You can specify an external module number if required; otherwise, the system assigns an internal number. Then, you'll enter the validity period, a short description for the module, and the list of materials. Figure 4.21 shows an example of an assortment module valid for the year 2021.

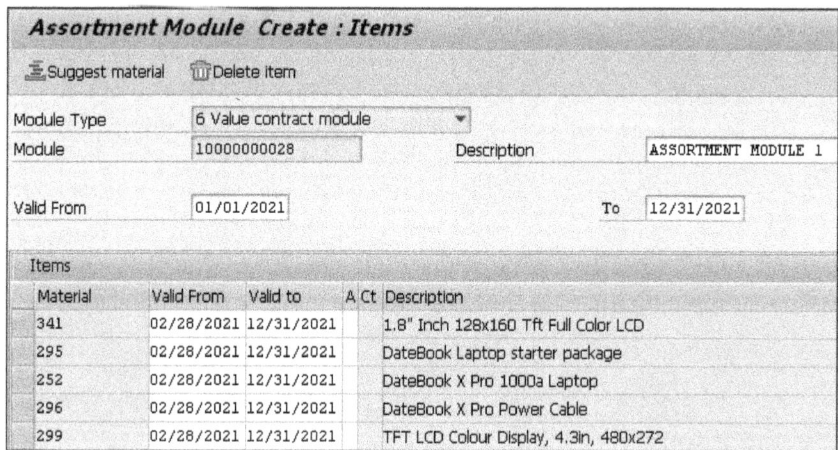

Figure 4.21 Creating an Assortment Module

Note

When creating a release order for a general value contract (WK1), the system will only show those materials in the assortment module that have been maintained in the sales organization and distribution channel of the release order being created.

The released value in the contract is automatically updated every time release orders are created. Once the target value is reached, no more release orders can be created for the contract. To calculate the released value, the system takes into account the amounts in open release orders, deliveries, and billing documents as well as any price adjustments, cancellations, returns, or rejection of order items.

Release orders can be created in any currency. The system will convert the currency in the release order into the currency in the value contract and update the release value accordingly.

In the **Define Item Categories** customizing activity for the item category of the value contract, you can use the **Contract Relase Ctrl** field to control how the system reacts when the value of your release order reaches or exceeds the open target value of the value contract item. The following five options are available:

- <blank>: No response if the target value is exceeded
- **A**: Warning if the target value is exceeded; contract is blocked when a release order exceeds the target value
- **B**: Error message if the target value is exceeded
- **C**: Permitted with a warning only the first time the target value is exceeded
- **D**: As for **A** but the value contract is not blocked

Note
A particular release order can only refer to either a quantity or a value contract. You cannot have an item linked to a quantity contract and another item linked to a value contract in the same release order.

If required, you can assign a cancellation procedure to the contract profile in the sales document type of the value contract, so cancellation rules apply in the event the customer wishes to terminate the contract.

4.6.4 Service Contract

Service contracts are outline agreements between the customer or receiver of the service and the supplier of the service. Release orders are not required to bill the customer since service materials are not relevant for delivery. If the service is billed at regular intervals (e.g., monthly), a billing plan type must be assigned to the sales document type of the contract. We won't go over the details of billing plans in this section since this topic will be reviewed in Chapter 11, Section 11.6. The two types of service contracts in the standard system are rental contracts (type MV) and maintenance contracts (type WV).

As with quantity contracts and value contracts, you can define cancellation rules in service contracts to determine whether a cancellation request from the customer must be processed to terminate the contract.

You can use the service contract to trigger follow-up activities, such as phone calls or quotations, before the contract expires. The type of follow-up activity is defined in the sales document type of the **FollUpActivityType** (follow-up activity type) field of the contract.

Service contracts can also be used to determine which services are covered by the contract and, if the service task is not covered, what price agreement must be applied. For example, a customer can choose between a gold and a silver service contract to cover a machine that is out of warranty. The gold contract is more expensive, but all spare parts are included. In contrast, the silver contract is cheaper but does not include spare parts. Thus, if a machine under a silver contract requires a new spare part, the customer will have to pay for the part. Usually, a special price is defined in the price agreement of the silver contract, so the customer is not charged full price for materials not covered by the monthly service fee. Billing for any additional services or products is initiated by a billing request.

To create a price agreement for a service contract, you must create condition records for the sold-to party in the service contract and the relevant materials or material groups. The standard version of the system provides three condition types for price agreements in service contracts, as shown in Table 4.14.

Condition Type	Description
PKAR	Header price settlement
PPAG	Item price settlement (price for material group)
PPAR	Item price settlement (price for material)

Table 4.14 Pricing Condition Types for Price Agreements in Service Contracts

Pricing condition records for price agreements at both the header and item level can be created directly in the service contract. To create a price agreement at header level, go to the overview screen of the service contract in change mode (via Transaction VA42 or the Manage Sales Contracts app) and select **Extras • Price Agreements • Header**. Then, click on **Other Condition Type**, select the condition type, and maintain the contract-specific pricing condition record, as shown in Figure 4.22.

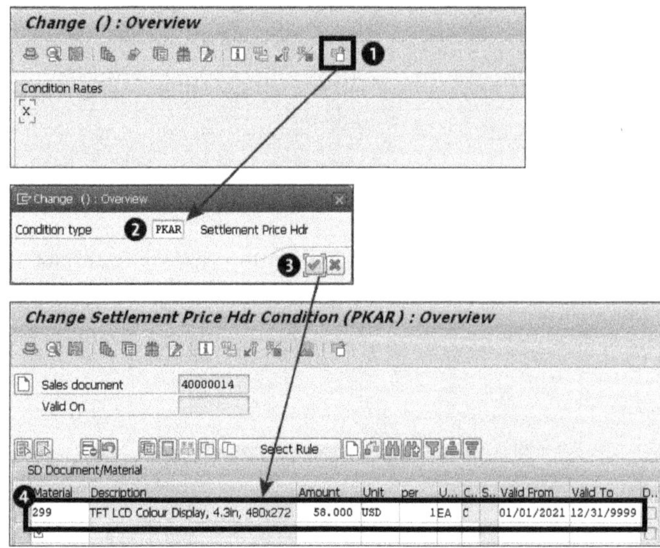

Figure 4.22 Creating a Price Agreement in a Service Contract

Finally, following the example shown in Figure 4.22, let's say the standard price for material 299 is $99.00. However, if a service order is created for service contract 40000014 and material 299 is consumed to repair the device assigned as a technical object in the contract item, the system will determine a price of $58.00 in the debit memo request generated when the service order is processed through resource-related billing as per the price agreement maintained in the service contract.

4.7 Important Terminology

In this chapter, the following terminology was used:

- **Sales document type**
  This key is used to classify sales documents and determines key data in the document such as pricing procedure, partner functions, item categories, incompleteness log, etc.
- **Sales document item category**
  Item categories in conjunction with document types define the business transaction. This key allows for different ways to process items within the same sales document. For instance, item 10 might be sold from local stock, while item 20 must be procured externally from a third-party vendor.
- **Item category group**
  This field in the material master is used to determine the item category in the sales document.
- **Item category usage**
  This field controls the system reaction when processing a sales document in a specific scenario (free goods, text items, batch split, etc.).
- **Third-party delivery item**
  This special category of sales item automatically creates a purchase requisition to buy the goods ordered by the customer from a vendor. The vendor ships directly to the customer. These order items, also known as third-party orders, are needed when the goods ordered by the customer are not in stock and cannot be manufactured in-house.
- **Individual procurement item**
  This special category of sales item automatically creates a purchase requisition to buy the goods ordered by the customer from a vendor. The vendor ships directly to your warehouse. These order items, also known as individual POs, are needed when the goods ordered by the customer are not in stock and cannot be manufactured in-house.
- **Schedule line category**
  This element is used to classify schedule lines and determines how the schedule line is processed by the system. For example, you can decide whether or not a

specific schedule line category passes on requirements to material requirements planning (MRP).

- **Sales bill of materials (sales BOM)**
  A sales BOM defines the different components that together create a product that can be sold to a customer.
- **Sales contract**
  This contractual agreement lays out the terms of a transaction of goods or services within a certain time period. Besides validity dates, a sales contract can contain the target quantity, target amount, cancellation conditions, price agreements, materials, billing plan details, etc.
- **Assortment module**
  This SAP S/4HANA Sales master data object is used to store a list of materials that can be released from a value contract. A validity period can be defined that is checked when a release order is created to ensure the release date lies within the validity period.
- **Master contract**
  Master contracts are sales documents used to bundle contracts for the same customer. The master contract contains only general header data that is relevant for all lower-level contracts.
- **Value contract**
  Value contracts are legally binding agreements specifying that a certain quantity of goods must be supplied to a customer during a specified period. The target value is the limiting factor of the contract.
- **Quantity contract**
  A quantity contract can be defined as a legally binding agreement specifying that a certain quantity of goods must be supplied to a customer during a specified period. The target quantity is the limiting factor of the contract.
- **Service contract**
  Service contracts are outline agreements between the customer or receiver of the service and the supplier of the service.

4.8 Practice Questions

These practice questions will help you evaluate your understanding of the topics covered in this chapter. The questions shown are similar in nature to those found on the certification examination. Although none of these questions will be found on the exam itself, they will allow you to review your knowledge of the subject. Select the correct answers and then check the completeness of your answers in the next section. Remember that, on the exam, you must select all correct answers and only correct answers to receive credit for the question.

1. Which of the following customizing settings are available when defining sales document types in SAP S/4HANA? (There are three correct answers.)

☐ A. Specify whether a credit check must be performed

☐ B. Define whether schedule lines are allowed

☐ C. Specify whether a sales document can only be created with reference to a preceding document

☐ D. Indicate how a BOM is processed in a sales document

☐ E. Decide whether the division entered at the header level takes precedence over the item specifications

2. In the standard system, the first character of the schedule line category refers to the type of sales document where it is used. If "C" is the first character, what document type is it referring to?

☐ A. Quotation

☐ B. Returns

☐ C. Sales order

☐ D. Inquiry

3. Which of the following options are available when configuring the schedule line category to control how the system updates the sales order schedule line in third-party PO scenarios? (There are two correct answers.)

☐ A. Update schedule line when the PO is approved through flexible workflow

☐ B. Update schedule line only for goods receipt

☐ C. Update schedule line for invoice receipt

☐ D. Update schedule line for goods receipt and shipping notification

4. True or False: It is not possible to assign an incompleteness procedure at the schedule line level.

☐ A. True

☐ B. False

5. In the customizing for sales and distribution, where do you define the movement type to be used at post goods issue?

☐ A. In the definition of sales document types

☐ B. In the definition of sales item categories

☐ C. In the definition of schedule line categories

☐ D. In the definition of delivery types

6. When assigning schedule line categories in customizing, how many manual schedule line categories can you maintain for each combination of item category and MRP type?

☐ A. Zero. You can only maintain the proposed schedule line category.

☐ B. One

☐ C. Two

☐ D. Three

7. True or False: You can assign a quantity contract to a master contract by entering the quantity contract number in the **Sales Document** field in the master contract overview screen.

☐ A. True

☐ B. False

8. Which of the following factors can limit the goods or services provided to a customer in a value contract?

☐ A. Both the target quantity and the contract end date

☐ B. The target quantity specified in the contract

☐ C. The target value specified in the contract

☐ D. Both the target value and quantity, whichever occurs first

9. True or False: Quantity and value contract items cannot be processed together in one release order.

☐ A. True

☐ B. False

10. In a standard SAP S/4HANA system, which of the following categories of sales contracts allow the creation of release orders? (There are two correct answers.)

☐ A. Master contract

☐ B. Service contract

☐ C. Value contract

☐ D. Quantity contract

11. When configuring sales documents, which of the following elements can be restricted by the combination of the sales organization and the sales document type? (There are two correct answers.)

☐ A. Order reason

☐ B. Delivery block

☐ C. Billing block

☐ D. Reason for rejection

12. True or False: In SAP S/4HANA, you can specify whether a particular sales document type is only allowed in a specific sales area.

☐ A. True

☐ B. False

13. True or False: All sales documents have data stored at the header, item, and schedule line levels.

☐ A. True

☐ B. False

14. What is the definition of a master contract in SAP S/4HANA?

☐ A. It is an agreement that contains the materials and services that the customer may receive within a time period and up to a target value.

☐ B. It is an agreement that contains the conditions for offering a certain service to the customer.

☐ C. It is an agreement that enables the customer to order a certain quantity of a product during a specified period.

☐ D. It is a document in which you can group contracts together as lower-level contracts.

15. Which of the following statements regarding master contracts are true? (There are two correct answers.)

☐ A. The master contract contains header data only.

☐ B. The master contract and the lower-level contract can be assigned to different sales organizations.

☐ C. Changes in the master contract can be automatically copied into the lower-level contract.

☐ D. Contract grouping supports hierarchical relationships of several lower-level contracts.

16. True or False: In the additional settings for sales order items with individual procurement, item category TAB, you can specify an automatic rejection upon completion of the corresponding PO item.

☐ A. True

☐ B. False

17. Which of the following settings can you configure in the schedule line category? (There are two correct answers.)

☐ A. Credit active

☐ B. Create PO automatically

☐ C. Delivery block

☐ D. Item relevant for delivery

18. Which of the following settings can you configure in the item category? (There are three correct answers.)

☐ A. Shipping condition

☐ B. Pricing relevance

☐ C. Billing relevance

☐ D. Item relevant for delivery

☐ E. Default delivery type

19. Which of the following elements can influence the item category determination for a sales document item? (There are three correct answers.)

☐ A. Usage

☐ B. Sales organization

☐ C. Sales document type

☐ D. Item category group

☐ E. Division

20. What customizing parameter controls whether a sales order item will automatically explode to generate subitems for the BOM components?

☐ A. The business transaction in the item category

☐ B. The item type in the item category

☐ C. The usage in the sales document type

☐ D. The structure scope in the item category

4.9 Practice Question Answers and Explanations

1. Correct answers: **A, C, and E**

   The checkbox to specify whether schedule lines are allowed is a configuration setting for the item category, so answer B is not correct since this setting is not available when defining sales document types. Similarly, the item category controls how a BOM is processed in a sales order, specifically through field structure scope; thus, answer D is also not correct. Thus, only three answers remain that correspond to the three customizing settings that are available when defining a sales document type.

2. Correct answer: **C**

   In the entries shown earlier in Table 4.13, the letter "C" in the first character refers to sales orders, meaning that schedule line categories beginning with C have been defined to be used in sales orders.

3. Correct answers: **B and D**

   The **Update Schedule Lines from External Procurement** field controls how the system updates the sales order schedule line in third-party PO scenarios. If the field is left blank, the system does not perform any update. If the value maintained is 1, then the schedule line is updated only for goods receipts. Finally, if the value is 2, the schedule line is updated for goods receipts and shipping notifications. Therefore, answers B and D are the two correct answers to this question.

4. Correct answer: **B**

   False. As shown earlier in Figure 4.13, you can assign an incompleteness procedure to the schedule line category in the **Incompl.Proced** field in the **Transaction flow** section.

5. Correct answer: **C**

   The movement type to be used at post goods issue is defined in the schedule line category field **Movement Type**, shown earlier in Figure 4.13.

6. Correct answer: **D**

   In the customizing activity **Sales and Distribution • Sales • Sales Documents • Schedule Lines • Assign Schedule Line Categories**, you can maintain the proposed schedule line category and up to three manual schedule lines, as shown earlier in Figure 4.14.

7. Correct answer: **B**

   False. This statement is incorrect. To assign a quantity contract to a master contract, the master contract number must be entered in the **Master Contract** field under the **Sales** tab at the header level in the lower-level contract. This field will create the link, and the quantity contract will automatically appear in the **Sales Document** field in the master contract overview screen, as shown earlier in Figure 4.18.

8. Correct answer: **C**

   For this question, answer C is the obvious choice. In a value contract, the target value is the limiting factor. As release orders are created, the system automatically updates the released value in the contract. Once the target value is reached, no more release orders can be created for the contract. All the other answers mention the target quantity, which is not even available for input in a value contract.

9. Correct answer: **A**

   True. A particular release order can only refer to either a quantity or a value contract. In other words, in the same release order, you cannot have an item created with reference to a value contract and another item created with reference to a quantity contract.

10. Correct answers: **C and D**

    Release orders are created for either value or quantity contracts. Services are invoiced directly from the service contract, usually based on a billing plan, so a release order is not needed since service items are not delivered. Master contracts are used to group the lower-level contracts, so no release orders can be created for the master contract.

11. Correct answers: **A and D**

    SAP S/4HANA introduced the ability to restrict both the order reason and the reason for rejection by sales organization and sales document type, as shown earlier in Figure 4.2. This feature was not available in older versions of SAP.

12. Correct answer: **A**

    True. By default, all sales document types can be used in all sales areas. However, if needed, you can restrict the sales document type per sales area.

13. Correct answer: **B**

    False. Some sales documents, such as contracts, debit memo requests, and credit memo requests do not contain data at the schedule line level. When configuring the item category, you can decide whether schedule lines are allowed.

14. Correct answer: **D**

    Answer A is the definition of a value contract, B is the definition of a service contract, and C is the definition of a quantity contract. Therefore, answer D corresponds to the definition of a master contract, which is a contract that only contains data at the header level and is used to link lower-level contracts.

15. Correct answers: **A and C**

    You can only assign a lower-level contract to a master contract if both contracts have been created in the same sales area. Also, SAP S/4HANA does not support multilevel structures of contracts where a master contract has several lower-level contracts, and each of those has its lower-level contracts, etc. Therefore, answers B and D are incorrect statements. On the other hand, master contracts do contain header data only, and you can configure the system so changes in master contracts are automatically transferred into their corresponding lower-level contracts.

16. Correct answer: **B**

    False. The *automatic rejection upon completion of the PO item* functionality is only available for item categories involving third-party delivery (CB1, CB2, CB3, and CB4), as shown earlier in Figure 4.6. Individual procurement item categories like TAB, where goods are physically received into the warehouse and then shipped to the customer, do not support this feature.

17. Correct answers: **C and D**

    This question concerns settings on the customizing screen for the schedule line category. On that screen, you can specify a delivery block that will be applied automatically as well as indicate whether the item linked to the schedule line is relevant for delivery. Both fields are shown earlier in Figure 4.13. The settings to activate credit management functions and create a PO automatically are not available when defining schedule line categories but instead when defining item categories.

18. Correct answers: **B, C, and D**

    On the customizing screen for the item category, you can specify whether the item is relevant for pricing, billing, and delivery. All these three fields are shown earlier in Figure 4.3. It is not possible to define the shipping condition, nor the default delivery type on the item category. Both the shipping condition and the default delivery type are specified in the definition of the sales document type.

19. Correct answers: **A, C, and D**

    This question tests your understanding of the logic to determine the item category in a sales document. Following the customizing table shown earlier in Figure 4.5, the system uses four fields to determine the item category: sales document type, material item category group, usage, and higher-level item category. Neither the sales organization nor the division are used during item category determination.

20. Correct answer: **D**

    This straightforward question relates to the setting that activates the functionality to automatically explode a sales BOM in a sales document. This feature is controlled by the **Structure Scope** field in the definition of the item category. You can use this field to indicate how a BOM is processed in a sales document.

4.10 Takeaway

In this chapter, you gained a good overview of all basic customizing settings regarding sales document types in SAP S/4HANA. You're also familiar with the important role the item category plays in sales documents, and you learned about its assignment and basic configurations. The same goes for the assignment and configuration options available for the schedule line category.

You've seen how to maintain a BOM that can be used in sales documents and learned about the differences between main item level BOMs and subitem level BOMs when deciding which items are just text items and which are relevant for pricing and delivery. You're also aware of the master data and configuration changes required for each type of sales BOM.

Finally, you got a thorough introduction to the four types of contracts in SAP S/4HANA: master contracts, value contracts, quantity contracts, and service contracts. You should understand the purpose of each type of sales contract, as well as some of the basic settings to enable features such as cancellation rules or rules to copy the data from a master contract into lower-level contracts.

4.11 Summary

The customizing of sales documents is the fundamental building block required for the successful processing of all types of sales documents such as sales orders, quotations, inquiries, contracts, and debit and credit memo requests. By ensuring that document types, item categories, and schedule line categories are configured in full alignment with your customers' requirements, you'll set the course for a successful SAP S/4HANA Sales implementation in your SAP S/4HANA project.

In the next chapter, we'll analyze some of the basic functions within the customizing of sales and distribution, including incompleteness procedures, partner determination, free goods, and material determination.

Chapter 5
Basic Functions Customizing

Techniques You'll Master

- Configure sales incompletion logs
- Set up partner determination in sales documents and customer master records
- Define the rules for material determination in sales documents
- Set up material listing and exclusion to control which materials can be sold to a customer
- Set up automatic free goods determination

In this chapter, we'll explore the most relevant settings for the basic functions in sales and distribution processing. These functions are used across different areas in SAP S/4HANA Sales such as sales, shipping, and billing.

Real-World Scenario

As an SAP S/4HANA Sales consultant, you'll be asked to configure the system in a way that meets specific business requirements, which is particularly important when working with SAP S/4HANA on-premise, which provides the full scope of configuration and functionality. In the current fast-moving and competitive world, companies that process high volumes of data need systems that help minimize errors by validating the information entered by users or by automating certain steps. Therefore, knowing the different options available through standard configuration is critical to building the right system for your customers. A good rule of thumb is to avoid custom developments whenever a requirement can be met via customizing.

You may be required to set up the system so deliveries cannot be created when a required field is not filled in during sales order entry or prevent a list of materials from being sold to a customer. These scenarios are just two examples of basic functions available in SAP S/4HANA, and SAP S/4HANA Sales consultants must know how to configure these functions.

5.1 Objectives of This Portion of the Test

The goal of this portion of the SAP S/4HANA Sales Certification exam is to test your understanding of the following topics:

- Definitions and assignments possible for Incompleteness Procedures
- Configuration of partner determination
- Configuration and maintenance of material determination
- Configuration and maintenance of material listings/exclusion lists
- Configuration and maintenance of free goods determination

Note

The Basic Functions Customizing topic makes up 8%–12% of the total SAP S/4HANA Sales 2020 exam.

The content of this chapter is included in the Sales topic of the SAP S/4HANA Sales 2020 Upskilling exam, which makes up more than 12% of the total certification exam.

5.2 Setting Up Incompleteness Procedures

Sales incompleteness procedures, also known as *incompletion logs*, are one of the most common features in sales document processing. This functionality reminds users when important data is missing, showing the specific fields that must be populated to avoid a block on the document preventing further processing. The sales document is the baseline for any subsequent document, and its data is passed through to the delivery and finally the billing document. Thus, guaranteeing completeness is important before subsequent functions can be carried out. When the document is saved, all the missing data is logged in an incompletion log, and, depending on the configuration settings, the user is prompted to enter the missing information.

A significant amount of information is stored in each sales document at the header, item, and schedule line levels. Thankfully, the system usually proposes the majority of the required data from material and business partner master records. However, users can manually change the proposed data and/or enter additional data manually.

If an incompleteness procedure is set up correctly, the system will inspect specific fields to check if data has been maintained in these fields. In the event data is not maintained, you can tell the system how to respond. For example, when creating a sales order, the purchase order (PO) number must be entered at the header level for the order to be complete. Should the PO number not be maintained, this field would be added to the incompletion log. Via customizing, you can decide to issue a warning message blocking all follow-up activities. Alternatively, you can specify that a warning message should not be issued while allowing all delivery-related activities but blocking billing document creation. Also, you can configure the system so that a specific document type can only be saved if the data is complete. These examples illustrate the flexibility of the incompletion log, a powerful tool that can help your business avoid missing information in customer-facing documents like invoices, avoid processing issues downstream, and ensure data is consistent and reliable.

Figure 5.1 shows how the incompletion log is presented to the user in a sales order. In this example, the PO number (**Cust. Reference**) is missing, and this particular field has been added to the incompleteness procedure.

If the user clicks **Edit**, the system will show all the fields in the incompletion log that must be populated for the document to be considered complete, as shown in Figure 5.2.

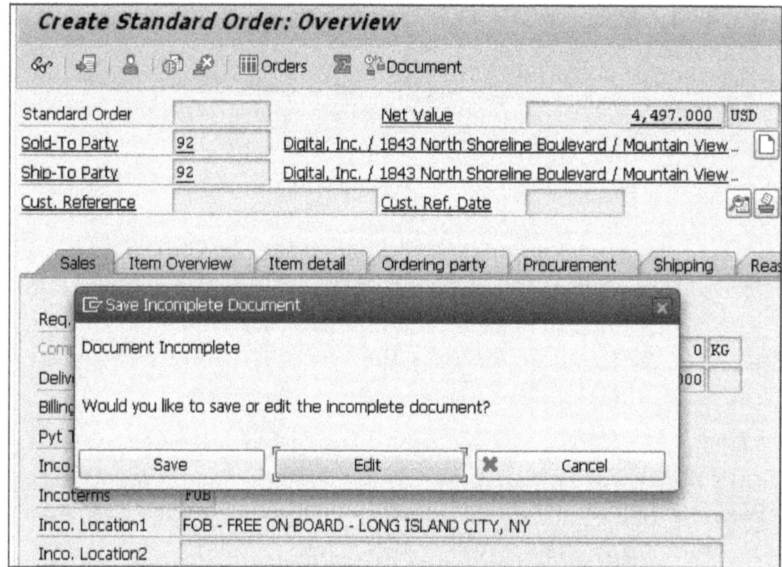

Figure 5.1 Incompletion Log: Popup Message When Saving the Document

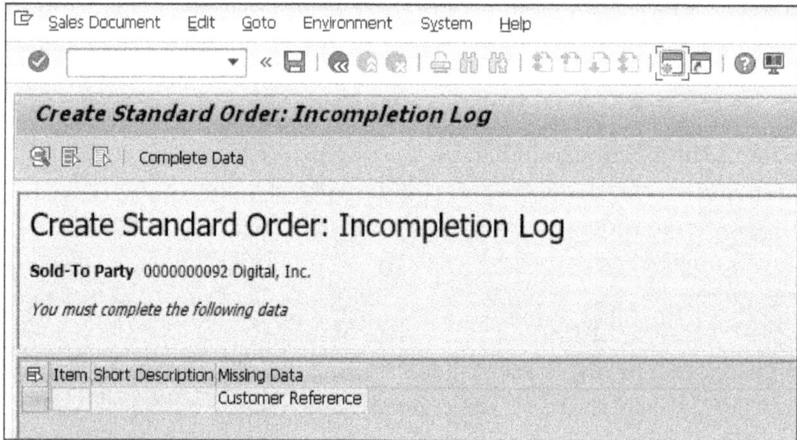

Figure 5.2 Incompletion Log: Missing Data at the Sales Document Header Level

However, if you click **Save**, the sales document will be saved with the completeness status **Incomplete**, as shown in Figure 5.3.

Note that, even if an incomplete document can be saved, the missing data may affect further processing of the sales document. In the example shown in Figure 5.3, an outbound delivery can be created, but the billing function is blocked as indicated in the completeness status for header billing data.

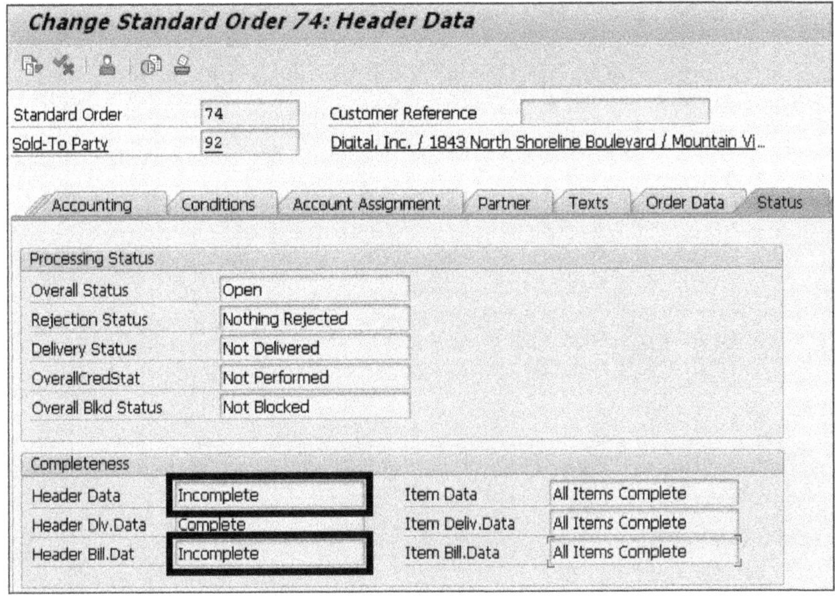

Figure 5.3 Sales Document Completeness Statuses

Let's now take a look at the customizing for incompleteness procedures, which can be found at the menu path **SAP Customizing Implementation Guide • Sales and Distribution • Basic Functions • Log of Incomplete Items.** SAP S/4HANA provides an incompletion log structure for the following data:

- Sales document header data
- Sales document item data
- Sales document schedule line data
- Partner data in sales documents and deliveries
- Delivery header data
- Delivery item data

Note

In SAP S/4HANA, as discussed in Chapter 4, the term *sales document* is used to refer to several types of documents, such as quotations, sales orders, sales activities, and contracts.

First, you need to define the status groups. In a status group, you can define which functions can be performed for incomplete sales and distribution documents or for items where a field entry is missing. You can use this function to block a document for delivery, billing, or pricing. These status groups are eventually assigned to the specific field in the incompletion log.

The table shown in Figure 5.4 shows several status groups where each checkmark ✓ indicates the functions blocked by the status group. If the default status groups defined in the standard system do not meet your requirements, you can amend existing status groups or create new groups.

Tip

We do not recommend changing the standard status groups; instead, you should copy a status group that closely resembles your requirements, change its name, and then select/deselect the relevant statuses to specify which statuses should be blocked.

Change View "Incompletion Control: Status Groups": Overview

New Entries

Incompletion Control: Status Groups

Status grp	General	Delivery	Billing Doc.	Price	Goods movement	Picking/putaway	Pack
00							
01	✓						
02	✓	✓					
03	✓		✓		✓		
04	✓	✓	✓		✓		
05	✓		✓	✓	✓		
06	✓	✓	✓	✓	✓		
16	✓					✓	
30	✓						
32	✓						✓
58	✓		✓		✓	✓	✓

Figure 5.4 Status Groups

Next, we'll explain the effect of selecting each of the statuses shown in Figure 5.4:

- **General**
  With this status, the system will check the fields that have been assigned to the status group in the incompletion procedure. If data is missing in one or more of these fields, the status message will indicate that the general information is incomplete but will allow the document to be processed further.
- **Delivery**
  The system will check the fields that have been assigned to the status group in the incompletion procedure. If data is missing in one or more of these fields, the status message will indicate that the delivery information is incomplete. This status is valid in both sales and shipping documents. For example, the creation of an outbound delivery can be prevented if the shipping point is missing in the sales order or the creation of the billing document can be prevented if the net weight is missing in the delivery.
- **Billing Doc.**
  The system will check the fields that have been assigned to the status group in the incompletion procedure. If data is missing in one or more of these fields, the status message will indicate that the billing information is incomplete. This status is valid in both sales and shipping documents. For example, if the terms of payment are missing in a sales order, invoice creation is not allowed.

- **Price**
  The system will check if pricing has been carried out. If not, the sales document will be flagged as incomplete for further processing.
- **Goods movement**
  The system will check the fields that have been assigned to the status group in the incompletion procedure. If data is missing in one or more of these fields, the status message will indicate that the goods movement data is fully or partially incomplete. This status is only valid for shipping documents.
- **Picking/putaway**
  The system will check the fields that have been assigned to the status group in the incompletion procedure. If data is missing in one or more of these fields, the status message will indicate that the document picking data is fully or partially incomplete. This status is only valid for shipping documents.
- **Pack**
  The system will check the fields that have been assigned to the status group in the incompletion procedure. If data is missing in one or more of these fields, the status message will indicate that the packing data is fully or partially incomplete. This status is only valid for shipping documents.

Now, we can define our incompletion procedures. You'll need to choose the group for which the incompletion procedure will be created. Each incompleteness group represents an object in sales and distribution, as shown in Figure 5.5.

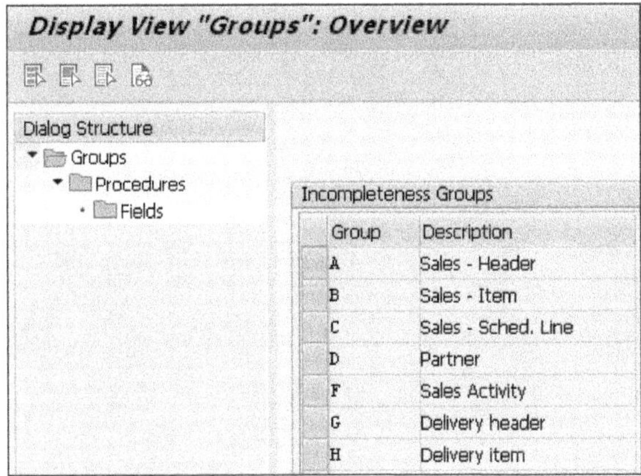

Figure 5.5 Incompleteness Groups

> **Note**
> You cannot create your own incompleteness group, for example, "Billing header."

Once a group has been selected, you can define multiple procedures for each group. Each procedure has a short description that gives you an idea of the purpose of the procedure. Some standard procedures assigned to group **A** (**Sales – Header**) are shown in Figure 5.6.

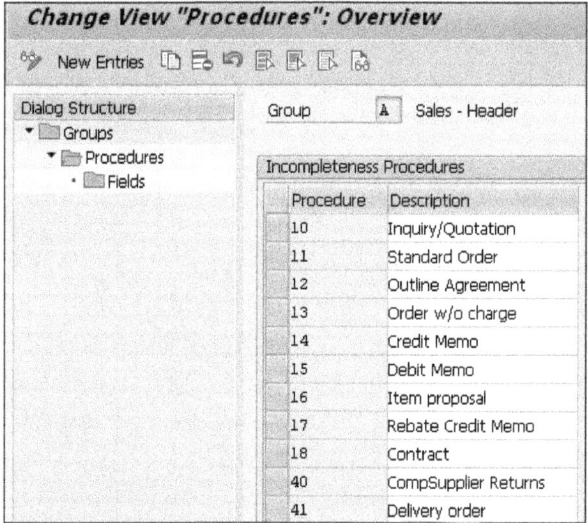

Procedure	Description
10	Inquiry/Quotation
11	Standard Order
12	Outline Agreement
13	Order w/o charge
14	Credit Memo
15	Debit Memo
16	Item proposal
17	Rebate Credit Memo
18	Contract
40	CompSupplier Returns
41	Delivery order

Figure 5.6 Incompletion Procedures

When you create a new incompletion procedure, you should create a copy of a standard procedure, rename it, and then add or remove fields as needed. Only fields that must be checked should be in the procedure. Figure 5.7 shows the fields assigned to procedure **11 (Standard Order)** as well as the screen used to complete the missing data, the status group, the warning indicator, and the processing sequence.

Display View "Fields": Overview

Dialog Structure
Groups
Procedures
Fields

Group A Sales - Header
Procedure 11 Standard Order
Fields of Incompleteness Procedure

Table	Field Name	Description	Screen	Status	Warning	Seq.
VBAK	AUDAT	Document Date	KKAU	01	☐	0
VBAK	WAERK	Document Currency	KKAU	03	☐	0
VBKD	BSTKD	Customer Reference	KBES	01	☑	0
VBKD	INCO1	Incoterms	KDE3	04	☐	0
VBKD	PRSDT	Pricing Date	KKAU	04	☐	0
VBKD	ZTERM	Payment terms	KDE3	03	☑	0

Figure 5.7 Fields of Incompleteness Procedure and Other Options

The first two columns indicate the technical name of the table and field to be checked. These values are available in the ABAP dictionary tool (Transaction SE11). As previously discussed, the **Status** column (status group) is an important setting since it controls whether the system records a status message for this field or whether certain subsequent activities for the document are blocked. By selecting the **Warning** checkbox, you can also specify whether a warning message should be issued during processing if no data has been entered into the relevant field. Finally, the last column refers to the processing sequence, which determines the sequence in which the fields that need to be processed appear (fields with lower values must be maintained first).

Let's now imagine that a sales order is created where the **Customer Reference** field is not filled in at the header level. Based on the incompleteness procedure settings shown in Figure 5.7, status group 01 is assigned to the field in question. According to the status group definition, shown earlier in Figure 5.4, the **General** status is the only checkbox selected for status group 01. Also, the warning indicator is set for this field. As a result, the system will issue a warning message if the **Customer Reference** field is not populated. In this case, the system will indicate the sales document is incomplete and offer an option to fill in the missing data. Should the user decide to save the document without completing the data, the system will save the document as incomplete but still allow any further processing because the **General** status does not block a document from subsequent functions.

The last step to successfully set up an incompletion log is to assign the incompleteness procedure to any of the six objects shown in Figure 5.8.

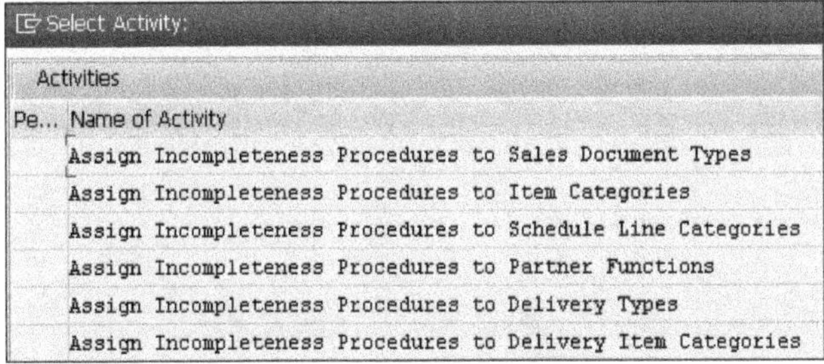

Figure 5.8 Assigning Incompleteness Procedures

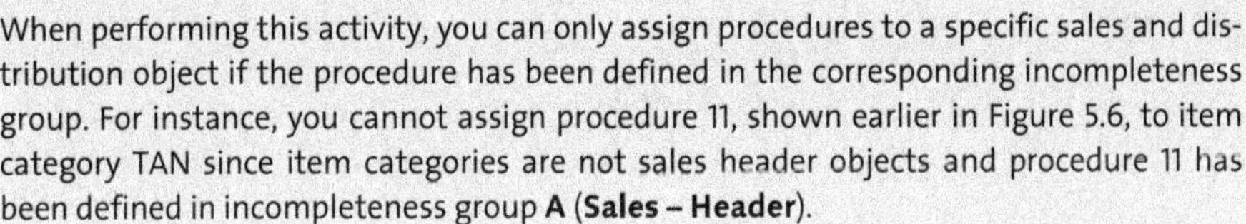

Warning

When performing this activity, you can only assign procedures to a specific sales and distribution object if the procedure has been defined in the corresponding incompleteness group. For instance, you cannot assign procedure 11, shown earlier in Figure 5.6, to item category TAN since item categories are not sales header objects and procedure 11 has been defined in incompleteness group **A** (**Sales – Header**).

Let's look briefly at each of these activities next:

- **Assign Incompleteness Procedures to Sales Document Types**
  You can assign the relevant procedure to one or more sales document types, as shown in Figure 5.9. The incompleteness check at the sales document header level will only be carried out for those sales document types where a procedure has been assigned.

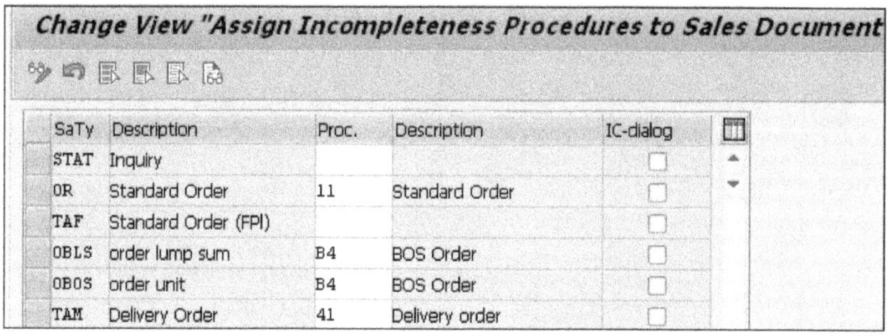

Change View "Assign Incompleteness Procedures to Sales Document

SaTy	Description	Proc.	Description	IC-dialog
STAT	Inquiry			☐
OR	Standard Order	11	Standard Order	☐
TAF	Standard Order (FPI)			☐
OBLS	order lump sum	B4	BOS Order	☐
OBOS	order unit	B4	BOS Order	☐
TAM	Delivery Order	41	Delivery order	☐

Figure 5.9 Assigning Incompleteness Procedures to Sales Document Types

The **IC-dialog** (incompleteness dialog) checkbox has been added to this configuration table to disable the popup window shown earlier in Figure 5.1. If the indicator is selected for the sales document type, when you attempt to save an incomplete document, the dialog box is omitted, and the sales document cannot be saved unless the missing data is entered. In this scenario, the system will take you directly to the incompletion screen without giving you the option to save, which is an easy way to ensure that an incomplete sales document cannot be saved.

- **Assign Incompleteness Procedures to Item Categories**
  You can assign the relevant procedure to one or more sales item categories, as shown in Figure 5.10. The incompleteness check at the sales document item level will only be carried out for those sales item categories where a procedure has been assigned.

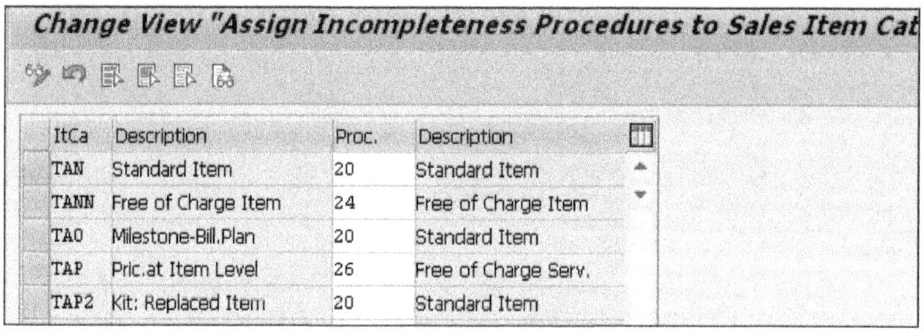

Change View "Assign Incompleteness Procedures to Sales Item Cat

ItCa	Description	Proc.	Description
TAN	Standard Item	20	Standard Item
TANN	Free of Charge Item	24	Free of Charge Item
TAO	Milestone-Bill.Plan	20	Standard Item
TAP	Pric.at Item Level	26	Free of Charge Serv.
TAP2	Kit: Replaced Item	20	Standard Item

Figure 5.10 Assigning Incompleteness Procedures to Item Categories

- **Assign Incompleteness Procedures to Schedule Line Categories**
  You can assign the relevant procedure to one or more sales schedule line categories, as shown in Figure 5.11. The incompleteness check at the sales document schedule line level will only be carried out for those schedule line categories where a procedure has been assigned.

Change View "Assign Incompleteness Procedures to Schedule

SLCa	Description	Proc.	Description
CN	No mat.planning	30	General Sched. Line
CP	MRP	30	General Sched. Line
CS	Leg	31	Sched.Line w/PurReq.
CT	No inv.mgmt/no GdsIs		
CV	Consumption MRP	30	General Sched. Line
CX	No inventory mgmt	30	General Sched. Line
D0	Consignment Returns	30	General Sched. Line

Figure 5.11 Assigning Incompleteness Procedures to Schedule Line Categories

- **Assign Incompleteness Procedures to Partner Functions**
  You can assign the relevant procedure to one or more partner functions, as shown in Figure 5.12. The incompleteness check at the partner function level will only be carried out for those partner functions where a procedure has been assigned.

Change View "Assign Incompleteness Proced

Assign Incompleteness Procedures to Partner Functions

Fu...	Descriptn	Incompl.	Descriptn
SP	Sold-to party	07	Customer
AI	IS-PAM: Cert. owner	07	Customer
A0	Author		
CP	Contact person	06	Contact Person
AR	IS-PAM: Agency	07	Customer
AS	Applicant Warranty		

Figure 5.12 Assigning Incompleteness Procedures to Partner Functions

- **Assign Incompleteness Procedures to Delivery Types**
  You can assign the relevant procedure to one or more delivery types, as shown in Figure 5.13. The incompleteness check at the delivery header level will only be carried out for those delivery types where a procedure has been assigned.

Change View "Assign Incompleteness Procedures to Deliv

Assign Incompleteness Procedures to Delivery Types

DlvTy	Description	InProc	Description
JITD	JIT Outbd. Delivery		
LB	Delivery for Subcon.	K1	Min. Outbound Dely
LB2	DFPS: Reloc. Shpmnts		
LD	Outbound Del.Decentr	K1	Min. Outbound Dely
LF	Outbound Delivery	K1	Min. Outbound Dely

Figure 5.13 Assigning Incompleteness Procedures to Delivery Types

- **Assign Incompleteness Procedures to Delivery Item Categories**
  You can assign the relevant procedure to one or more delivery item categories, as shown in Figure 5.14. The incompleteness check at the delivery item level will only be carried out for those delivery item categories where a procedure has been assigned.

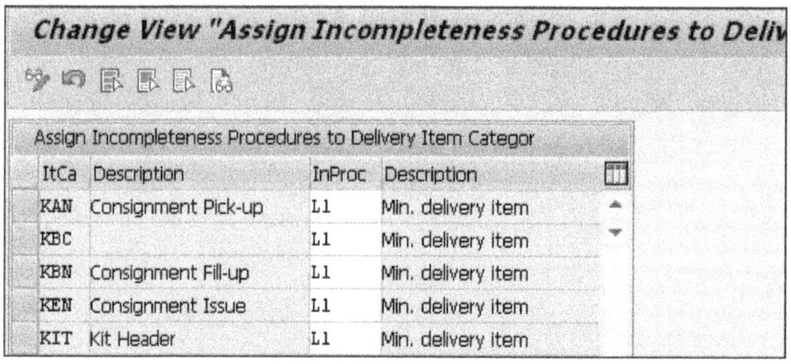

Change View "Assign Incompleteness Procedures to Deliv

Assign Incompleteness Procedures to Delivery Item Categor

ItCa	Description	InProc	Description
KAN	Consignment Pick-up	L1	Min. delivery item
KBC		L1	Min. delivery item
KBN	Consignment Fill-up	L1	Min. delivery item
KEN	Consignment Issue	L1	Min. delivery item
KIT	Kit Header	L1	Min. delivery item

Figure 5.14 Assigning Incompleteness Procedures to Delivery Item Categories

5.3 Setting Up Partner Determination

Partner functions are used in sales documents, deliveries, shipment documents, and billing documents and in the customer master record. The purpose of this object is to define the rights and responsibilities of each business partner in a specific trade transaction. In the following subsections, you'll learn how to set up automatic partner determination in SAP S/4HANA.

Each partner function is identified by an unconverted key in German and an optional language-specific key. Both keys are 2 characters long. Table 5.1 shows some of the most common partner functions defined in the standard system.

Partner Function Key (German)	Name	Converted Key (English)
AG	**Sold-To Party**	SP
WE	**Ship-To Party**	SH
RE	**Bill-To Party**	BP
RG	**Payer**	PY
SP	**Forwarding Agent**	CR
AP	**Contact Person**	CP

Table 5.1 Common Partner Functions in Sales and Distribution

Automatic partner determination is available to assist in the creation of sales and distribution documents as well as customer master records. Once the rules have been defined, the system will propose the allowed partner functions.

To configure automatic partner determination, follow the menu path **SAP Customizing Implementation Guide • Sales and Distribution • Basic Functions • Partner Determination • Set Up Partner Determination.** You can set up the partner determination functionality for the objects shown in Figure 5.15.

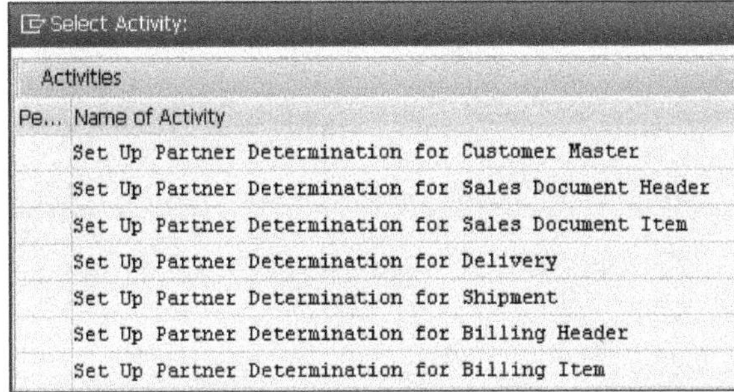

Figure 5.15 Sales Objects That Support Partner Determination

First, you must assign a partner type to each partner function by selecting the **Set Up Partner Determination for Customer Master** option and then selecting **Partner Functions**. You'll be presented with the customizing table shown in Figure 5.16 where partner type KU (customer) has been assigned to partner function **SP** (**Sold-to party**). If required, you can define new partner functions in this table.

Change View "Partner Functions": Overview

Part...	Name	Part...	Error ...	Sup...	Un...	CHT...
SP	Sold-to party	KU	07		☑	
AI	IS-PAM: Cert. owner	KU	07		☑	
AO	Author	US			☐	
CP	Contact person	AP	06		☐	
AR	IS-PAM: Agency	KU	07		☑	
AS	Applicant Warranty	KU			☐	
AU	Author	PE	09		☐	
AW	SH Contract rel. ord	KU	07		☐	

Figure 5.16 Defining Partner Functions

SAP S/4HANA provides 13 partner types to choose from, as shown in Table 5.2. Note that you cannot create new partner types.

Partner Type	Description
A	Work center
AP	Contact Persons
KU	Customer
LI	Vendor

Table 5.2 Partner Types

Partner Type	Description
LS	Logical system
MA	Mail Address
MP	Mail Partner
O	Organizational Unit
PE	Personnel Number
S	Position
US	User
VS	Shipping Point
WK	Plant

Table 5.2 Partner Types (Cont.)

The parameters you can assign to each partner function, as shown in Figure 5.16, from left to right, are as follows:

- Partner type
- Incompleteness procedure
- Higher-level partner function
- Uniqueness in the master data indicator
- Customer hierarchy type

Note

Describing each individual parameter shown in Figure 5.16 is beyond the scope of this book, but you should know that, if the **Unique** indicator is selected for a specific partner function, only one entry for this partner function will be allowed in the customer master data or in the supplier master data.

The **Partner Function Conversion** folder, located at the bottom of the dialog structure shown in Figure 5.16, can be used to maintain language-specific keys for partner functions. As previously mentioned, this setting is optional.

Second, you must assign partner functions to the permitted account groups (second entry from the bottom in the dialog structure shown in Figure 5.16). If you maintain this assignment in one of the activities shown earlier in Figure 5.15, the assignment applies to all sales objects. Figure 5.17 shows some of the account groups that have been assigned to the sold-to party partner function.

Next, you must define a partner determination procedure and assign the relevant partner functions. This activity is slightly different for a customer master record, a sales document header, and the rest of the transactional documents shown earlier in Figure 5.15. You'll need to learn about the configuration parameters available when assigning partner functions to a partner procedure for a customer master and for sales and distribution documents at the header and item levels.

Change View "Account Groups - Function Assignment": Overview

New Entries

Dialog Structure
- Partner Determination P
  - Partner Functions in P
- Partner Determination P
- Partner Functions
- Account Groups - Functi
- Partner Function Conver

Account Groups - Function Assignment

Partn.Fu...	Name	Account Grp	Name
SP	Sold-to party	0001	Sold-to party
SP	Sold-to party	0005	Prospective customer
SP	Sold-to party	0007	Sales partners
SP	Sold-to party	0100	Distribution center
SP	Sold-to party	0110	Branch w/o intercomp.billing
SP	Sold-to party	0120	Branch with intercomp.billing
SP	Sold-to party	0130	Branch with external billing

Figure 5.17 Assigning Partner Functions to Account Groups

Let's review the assignment of partner functions to a partner determination procedure for a customer master record and a transactional document. We've chosen the sales document header as an example of a transactional document, but you can carry out a similar configuration for sales document items, deliveries, shipments, billing document header, and billing document items:

- **Partner procedure for a customer master record**
  To create a new partner procedure key, click on the top folder (**Partner Determination Procedures**) in the dialog structure box shown earlier in Figure 5.16 and maintain the procedure key and description. Then, if you select a specific procedure and click on the **Partner Functions in Procedure** option underneath, you'll be presented with the screen shown in Figure 5.18, where you can assign the required partner functions and decide whether or not the partner is changeable ❶ and/or you want to make the partner mandatory ❷. In this example, we'll show you the assignment table for the standard procedure **AG** (**Sold-To Party**).

Change View "Partner Functions in Procedure": Overview

New Entries

Dialog Structure
- Partner Determination Procedures
  - Partner Functions in Procedure
- Partner Determination Procedure Assig
- Partner Functions
- Account Groups - Function Assignment
- Partner Function Conversion

Partner Functions in Procedure

Part...	Part...	Name	Not Modi... ❶	Mand... ❷
AG	AA	SP Contract rel. ord	☐	☐
AG	SP	Sold-to party	☑	☑
AG	CP	Contact person	☐	☐
AG	AW	SH Contract rel. ord	☐	☐
AG	ED	EDI mail recipient	☐	☐
AG	BU	Buyer	☐	☐
AG	EU	Enduser for F.Trade	☐	☐
AG	ES	External Agent	☐	☐
AG	VN	Vendor	☐	☐

Figure 5.18 Assigning Partner Functions to Procedures: Customer Master Record

- **Partner procedure for a sales document header**
  When assigning partner functions to a procedure for a sales document header, you have five additional configurable options, as shown in Figure 5.19.

Note

As shown in Figure 5.19, parameters ❸, ❹, and ❺ are only available for the sales document header. All the other parameters shown in this screenshot are available for sales document items, deliveries, shipments, billing document headers, and billing document items.

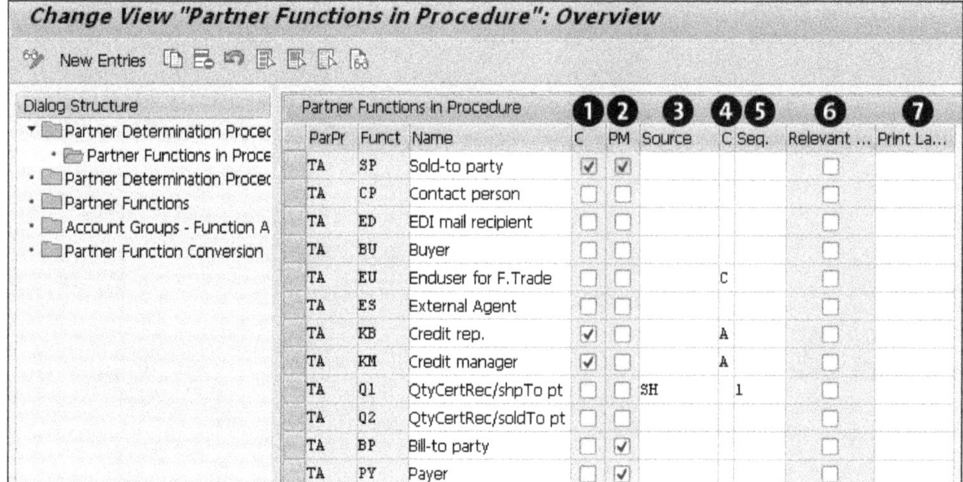

Figure 5.19 Assigning Partner Functions to Procedures: Sales Document Header

Altogether, you have the following seven options:

❶ Partner not changeable: After entry, the partner can no longer be changed in the document.

❷ Partner mandatory: This indicator makes the partner function mandatory.

❸ Source partner function: By default, for each partner function in the procedure, the business partner is automatically copied from the sold-to party master record when creating sales documents. You can use this parameter to ensure the business partner is taken from another partner function (e.g., forwarding agent in the sales document determined from the ship-to party and not the sold-to party).

❹ Origin table: If the field is empty when creating a sales document (default), the partner functions are determined from those partner functions given in the sold-to party customer master. Maintain this field if you want a specific partner to be determined that is not in the partner screen of the sold-to party master.

❺ Sequence: Indicates the priority during partner determination. If the field is blank, the system determines this partner first. If this field contains the value 1, the system determines the partner in the second step. If the field contains the value 2, the system determines the partner in the third step, and so on.

❻ Print-relevant indicator: Indicates that the partner function is relevant to printing the information of the desired business partners.

❼ Print label: Label used to identify partner functions in print forms. This print label field must be unique for each partner determination procedure and must be maintained once the print relevant indicator is enabled.

Once partner functions have been assigned to a partner determination procedure, the final step is to assign the procedure to a partner object key. Depending on the customizing activity you're performing, you may need to assign the procedure to different object keys, as shown in Table 5.3.

Activity: Set Up Partner Determination for...	Assign Partner Determination Procedure to...
Customer master	Account group
Sales document header	Sales document type
Sales document item	Item category
Delivery	Delivery type
Shipment	Shipment type
Billing header	Billing type
Billing item	Billing type

Table 5.3 Partner Determination Procedure Assignment Based on Activity

5.4 Condition Technique Overview

The condition technique is an infrastructure developed to assist with the determination of master data across many areas of SAP S/4HANA. This functionality is a proven and successful mechanism for determining prices (including taxes), table NAST-based outputs, listings/exclusions, texts, batch numbers, product substitutes, and more.

The condition technique is used by the system to make a choice from among a number of alternatives that are stored in condition records. The name *condition technique* comes from the fact that the system makes the choice based on conditions.

The basic elements used in the condition technique include the following:

- **Condition records**
  These records are master records where the values to be determined in the document are stored. The determined values could be a price, a material code, an output record, etc.
- **Condition types**
  The way SAP S/4HANA identifies the various elements in the procedure. For

instance, in a pricing determination procedure, the condition type is used to differentiate between prices of different natures, such as freight charges, discounts, net prices, and so on.

- **Condition table**
  This element determines the combination of the fields required for creating a master record. The field catalog provides you with the fields available for building a condition table.
- **Access sequence**
  Using this search strategy, the system can search for valid condition records for each condition type. Access sequences contain one or more condition tables and the sequence in which those condition tables must be used to find condition records. An access sequence also tells the system when to stop searching.
- **Procedure**
  This element determines the condition types to be used when looking for condition records.

At a high level, the procedure assigned to the sales document dictates which condition types should be considered in the search of condition records. Each condition type is assigned to an access sequence, which has the condition tables that enable the system to find valid condition master records based on the values in the corresponding fields of the sales document.

In the following sections, you'll see how the condition technique is used for material determination, listings/exclusions, and free goods determination. Later on, in Chapter 8, Section 8.2.4, we'll review the condition technique applied to pricing determination in sales documents.

5.5 Setting Up Material Determination

In some scenarios, users may find entering the customer-specific material number or the International Article Number (EAN) easier than entering the material code. Also, when processing electronic data interchange (EDI) orders, the material code available is rarely the same material code that exists in SAP S/4HANA. Material determination, also known as *material substitution*, allows you to set up the system so that, when a specific material key is entered in a sales document, this key is automatically replaced by the material code found based on the master record for material determination. If you use a one-to-one automatic swapping of materials, by using material determination, you don't need to create a material master record for each key entered, thus saving a considerable amount of master data maintenance.

To configure material determination, follow the menu path **SAP Customizing Implementation Guide • Sales and Distribution • Basic Functions • Material Determination**.

Material determination can be particularly useful in the following scenarios:

- You want to present the user with a list of materials to choose from and swap the material entered for the material chosen from the list.
- You want the system to automatically swap a specific material for another (e.g., a customer-specific material is swapped with the actual material code maintained in the system).
- You'd like the system to automatically swap the material entered for another material available, should the availability check for the material entered fail.

Material determination is controlled via the condition technique. The following steps carried out in sequence will enable you to use the condition technique to set up material determination:

1. Maintain field catalog to make sure the fields you need are available.
2. Create condition tables.
3. Maintain access sequence(s) and assign relevant condition table(s).
4. Define condition type(s) and assign an access sequence to each condition type.
5. Maintain procedure and assign the relevant condition type(s).
6. Assign procedures to sales document types.
7. Define substitution reasons.

Note

Steps 1 through 5 are common to other applications that also use the condition technique (e.g., pricing determination). Steps 6 and 7 are specific to material determination.

After maintaining the prerequisites for material determination using the condition technique (steps 1 through 5), you must specify which sales document types will carry out the checks for material substitution by assigning procedures to sales document types, as shown in Figure 5.20, where procedure A00001 is assigned to sales document type OR.

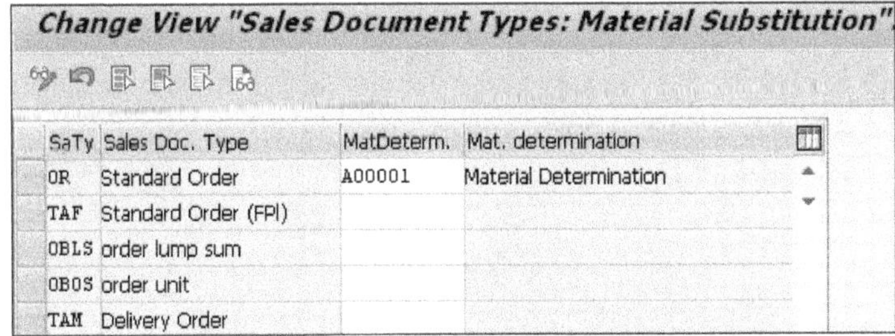
Change View "Sales Document Types: Material Substitution"

SaTy	Sales Doc. Type	MatDeterm.	Mat. determination
OR	Standard Order	A00001	Material Determination
TAF	Standard Order (FPI)		
OBLS	order lump sum		
OBOS	order unit		
TAM	Delivery Order		

Figure 5.20 Assigning Procedures to Sales Document Types

The last configuration step is defining the substitution reasons, which are required in master records for material determination. The substitution reason controls the overall substitution process. As shown in Figure 5.21, five different parameters are available for each substitution reason to define how the substitution should be performed by the system.

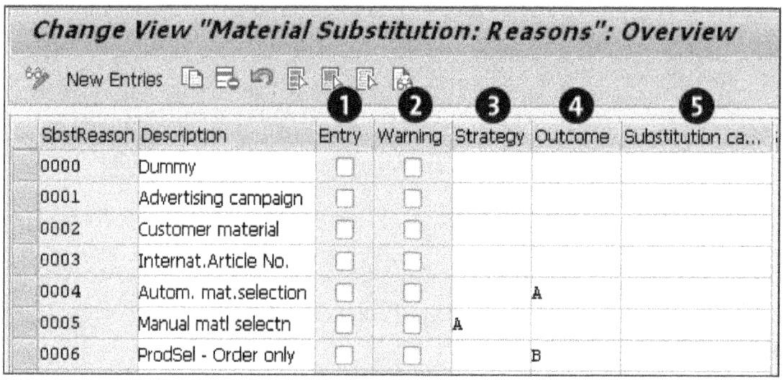

Change View "Material Substitution: Reasons": Overview

New Entries

SbstReason	Description	Entry ❶	Warning ❷	Strategy ❸	Outcome ❹	Substitution ca... ❺
0000	Dummy	☐	☐			
0001	Advertising campaign	☐	☐			
0002	Customer material	☐	☐			
0003	Internat.Article No.	☐	☐			
0004	Autom. mat.selection	☐	☐		A	
0005	Manual matl selectn	☐	☐	A		
0006	ProdSel - Order only	☐	☐		B	

Figure 5.21 Defining Substitution Reasons

As shown in Figure 5.21, you have the following options:

❶ **Entry**: Select this checkbox if you want the original entry to be printed on the corresponding output.

❷ **Warning**: If selected, the system displays a warning message before substituting a material.

❸ **Strategy**: Leave this field blank if you want the system to perform an availability check in the background and select the material that is available based on the quantity entered. If value **A** is selected, substitute products are displayed for selection in a dialog box. The user has to select one of the products offered, and then, an available-to-promise (ATP) check will be performed to inform the user whether the selected material is available. Selecting value **B** will result in a similar outcome as value **A** but with no ATP check.

❹ **Outcome**: This field controls whether or not the proposed item for material substitution should automatically replace the existing material or if the substituted material should be created as a subitem of the original material. Leave this field blank if you want the item to be replaced. Select value **A** if you want the substitution product to be displayed as a subitem in sales orders and deliveries. Selecting value **B** is similar to value **A** but only in sales orders. Note that, if you opt for a one-to-one automatic swap (value blank), the product entered does not need to be maintained in the system. However, when you use subitems, both materials (the main item and the subitem) must be created in the system.

❺ **Substitution category**: This field is used only in repair orders with service products. If you select value **A**, the serviceable equipment in the repair request item is replaced with a suitable service product in the **Material** field. Thus, the service product becomes the main material. This field is useful if you want pricing to be performed for the service product and not for the serviceable equipment.

Once the configuration has been completed, you must maintain the master record for material determination, that is, the condition record required to find the substitute product(s). Condition records have validity dates, which is useful, for example, if you know a product will become obsolete in 3 months and you want to create the condition record now but make sure material determination doesn't happen for 3 months.

Now, let's review the master data setup. Material determination master records are maintained via Transaction VB11. To create a material determination record, first select the condition type. For our example, we'll use standard condition type A001, as shown in Figure 5.22. Then, you must specify the validity dates and the reason for substitution. In the **Material Entered** column, enter the original material (or customer material) and then enter its substitute in the **Material** column. If required, you can maintain more than one substitute in the material determination record. If you maintain multiple substitutes, the checkbox in the **Alternative materials** column will be automatically selected.

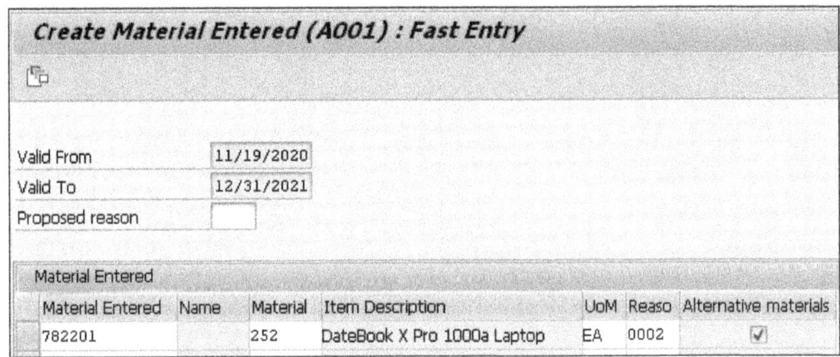

Figure 5.22 Material Determination Condition Record

In the event you maintain multiple alternative materials, the system will determine the substitute product in the sales order based on the sequence in which they have been entered in the material determination record. Figure 5.23 shows two alternative materials for material 782201. In this case, during automatic substitution, the system replaces the entered material 782201 with material 252, if that material is available, or with material 295, if material 252 is not available.

Tip

Enter the original product as the first alternative if you want the system to check this material for availability first, before checking for other alternatives.

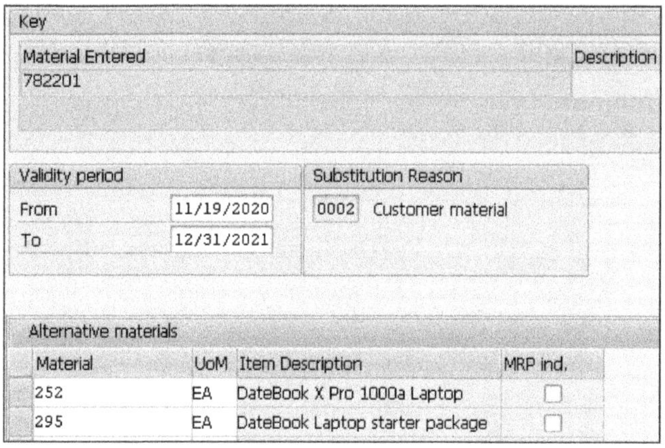

Key

Material Entered: 782201 | Description

Validity period

From: 11/19/2020

To: 12/31/2021

Substitution Reason: 0002 Customer material

Alternative materials

Material	UoM	Item Description	MRP ind.
252	EA	DateBook X Pro 1000a Laptop	☐
295	EA	DateBook Laptop starter package	☐

Figure 5.23 Multiple Alternative Materials

Finally, you must understand the role of the **MRP ind.** checkbox, which activates material requirements planning (MRP) on the alternative materials fast entry screen shown in Figure 5.23. If the **MRP ind.** checkbox is selected for an alternative material, in the event of a shortage for the original material entered in the order, the shortfall will be reported for the alternative material, and the original material will not be replenished. This capability is particularly useful when a material will be discontinued and you want to sell off all your existing stock while ensuring MRP stops creating purchase requisitions or planned orders for the material to be decommissioned. You should also know that only one alternative material can be indicated as relevant for materials planning in the material determination record.

5.6 Setting Up Material Listing and Exclusion

Material listing/exclusion is a functionality that allows you to restrict the materials (products or services) that specific customers can or cannot purchase. For instance, in the consumer electronics industry, a typical legal requirement is to provide a user manual in the local language alongside the product. If same sales organization sells a new product to distributors in several European countries, but the user manual has not been translated into Swedish yet, for example, you may be asked to prevent this product from being sold to distributors in Sweden. This block can be achieved via material exclusion so customers in Sweden cannot buy the new product. Once the user manual is translated into Swedish, the new product can be removed from the exclusion list for your distributors in Sweden.

On the other hand, you can set up a material listing for a customer to ensure that a customer can only buy the materials on the list. If a material listing is created for a customer, the system does not allow any materials to be entered in the sales document if those materials are not contained in the material listing.

To configure listings and exclusions, follow the menu path **SAP Customizing Implementation Guide • Sales and Distribution • Basic Functions • Listing/Exclusion.**

Material listing/exclusion also uses the condition technique described in Section 5.5. Similar to the steps described in material determination, the following customization activities are required to set up material listing and exclusions:

1. Maintain condition tables for listing/exclusion.
2. Maintain access sequences for listing/exclusion and assign condition tables to the access sequence.
3. Maintain listing/exclusion condition types and assign the relevant access sequence.
4. Maintain the determination procedure and assign the relevant condition types to the procedure.

After completing these four condition technique steps, you must activate listing/exclusion by sales document type. Unlike other determination procedures, the access sequences for listing and exclusions don't have an **Exclusion** checkbox, so in this final step, you also must specify the system's response if multiple accesses exist for the same access sequence. Figure 5.24 shows this particular scenario where three accesses have been maintained for access sequence A001.

Dialog Structure
- Access sequences
  - Accesses
    - Fields

Access Sequence A001 Listing

Overview Accesses

No.	Table N...	Description	Requirement
10	001	Customer/Material	
20	001	Customer/Material	1
30	004	Cust./Material w/ Organiz. Units	

Figure 5.24 Maintaining Access Sequences for Listing/Exclusion

Tip

You can assign requirements to specific accesses. Requirements are just conditions written in ABAP code that must be met for the corresponding listing/exclusion to be applied. As shown in Figure 5.24, only access 20 has a value in the **Requirement** column, which in our example screenshot is represented with the value **1**.

To activate listings, you must assign a procedure for material listing to the relevant sales document type. The same requirement is valid for exclusions.

If required, you can assign both listing and exclusion procedures to the same sales document type simultaneously. As shown in Figure 5.25, material listing procedure A00001 and material exclusion procedure B00001 have both been assigned to sales document type OR.

Change View "Sales Document Types: Material Listing and Exclusion":

New Entries

SaTy	Sales Doc. Type	Pro	Listing	Listing	Exclusion	Exclusion
OR	Standard Order		A00001	Listing	B00001	Exclusion
TAF	Standard Order (FPI)					
OBLS	order lump sum					
OBOS	order unit					

Figure 5.25 Activating Listing/Exclusion by Sales Document Type

You must understand the logic behind the **Procedure for material listing** field ❶, which defines how the system should react if multiple accesses are found for the access sequence being processed. You can select one of the following four values:

- **<blank>**
  The first access in the access sequence is used (requirement, if assigned, must be met).
- **A**
  The last access in the access sequence is used (requirement, if assigned, must be met).
- **B**
  At least one access must apply (requirement, if assigned, must be met).
- **C**
  All accesses must apply (requirements, if any, must be met).

Finally, you must create the condition master records for material listing/exclusion. Listing/exclusion master records are maintained in Transaction VB01. First, enter the condition type and select the key combination record if more than one access is maintained for the access sequence assigned to the condition type. Then, you must specify validity dates and maintain the condition record. As an example, Figure 5.26 shows an exclusion condition record where customer 92 is not allowed to buy material 252 during the year 2020.

Create Exclusion (B001) : Fast Entry

Customer	92	Digital, Inc.
Valid From	01/01/2020	
Valid To	12/31/2020	

Customer/Material

Material	Description
252	DateBook X Pro 1000a Laptop

Figure 5.26 Creating Condition Records

Note

Material listing/exclusion is also carried out in delivery documents but only for items added directly into the delivery document and not for items copied from the preceding sales document. However, in deliveries created with reference, material listing/exclusion will be carried out if the functionality is active for the sales document type of the preceding sales document. For standalone delivery documents, this check will be performed if this functionality is active for the document type of the default order type assigned to the delivery type in customizing (e.g., in standard SAP S/4HANA, default order type DL is assigned to delivery type LO).

5.7 Setting Up Free Goods

In some industries, a common practice is to provide goods for free when a certain amount of goods is ordered. SAP S/4HANA supports the following two types of free goods:

- **Inclusive bonus quantity**
  The customer pays for part of the goods ordered, and the rest is offered free of charge. The total quantity supplied matches the quantity purchased, but a portion is not charged to the customer. For example, let's say 1 out of 3 packs of biscuits is free. So, if a customer orders 3 packs, 3 are delivered, but the customer only pays for 2.
- **Exclusive bonus quantity**
  The customer pays for all the goods ordered, but additional goods are offered free of charge. The materials delivered as free goods can be different from the materials ordered. For example, let's say a customer orders one printer, and two ink cartridges are provided as free goods. Therefore, the customer pays for what was ordered but will receive additional goods free of charge.

Typically, free goods are displayed in the sales order as separate subitems of the original item. However, you can configure inclusive free goods without item generation. In this case, the correct discount is determined in pricing for the main item.

Since free goods must be delivered, these items are copied into the delivery document. Via customizing, you can specify whether free goods items must be copied into the billing document, which helps prove that these items are free of charge on the invoice.

Warning

In the following scenarios, free goods determination is not supported:

- Free goods are not supported for deliveries without reference to a sales order.
- Free goods cannot be used in make-to-order production, third-party order processing, or scheduling agreements.

- Free goods are only supported for sales orders with document category C (e.g., not quotations, scheduling agreements, contracts, etc.).
- Free goods are not supported in combinations with sales bills of materials (BOMs).
- Free goods are only supported on a 1:1 ratio. For example, the following agreement is not supported: If material A is purchased, then materials B and C are provided free of charge.

To configure free goods, follow the menu path **SAP Customizing Implementation Guide • Sales and Distribution • Basic Functions • Free Goods.**

First, you'll need to maintain all the relevant objects used by the condition technique, such as condition tables, access sequences, condition types, and procedures. In previous sections of this chapter, we explained how the condition technique works, so we'll jump into configuration steps specific to free goods determination.

You need to activate free goods determination by sales area, document pricing procedure, and customer pricing procedure, as shown in Figure 5.27.

Change View "Determining Free Goods Procedure in Sales":

New Entries

SOrg.	DChl	Dv	DoPr	CuPP	Proc.
US30	20	10	A	1	NA0001
ZRIV	ZR	ZR	A	1	
ZRIV	ZR	ZR	A	2	

Figure 5.27 Activating Free Goods Determination

Next, you'll configure item category determination for free goods items. This setting will allow the system to enter a free of charge item category (e.g., TANN), on the free goods subitem for free goods with item generation. The item category is determined by the following four elements:

- Sales document type of the sales document
- Item category group of the material
- Item usage, which defines the context of the item category determination (For free goods, the item usage is **FREE**.)
- Item category of the higher-level item (original item)

Figure 5.28 shows how free of charge item category **TANN** is determined for free goods in sales documents where the standard document type is **OR**, and the item category group of the material is **NORM**.

Finally, you'll need to create a condition master record according to the business requirement for free goods. Free goods master records are maintained in Transaction VBN1. To illustrate the master data setup, let's go back to the example for inclusive free goods mentioned at the beginning of this section.

Change View "Assign Item Categories": Overview

New Entries

SaTy	ItCGr	Usge	HLevItCa	DfItC	MItCa	MItCa	M
OR	NORM	FREE	TAN	TANN			

Figure 5.28 Determining an Item Category for Free Goods Items

For example, let's say a customer places an order for 6 packs of biscuits. For every 3 packs purchased, 1 pack is delivered for free, so the customer will receive 2 packs free. Figure 5.29 shows the free goods condition master record needed for this example.

Create Free Goods SD (NA00)

Exclusive

Sales Organization	US30	ATI US		
Distribution Channel	20	Direct		
Customer	92	Digital, Inc.		
Valid From	11/28/2020	To	12/31/9999	

Customer / material — Free goods view - INCLUSIVE

Material	Name	Min. Qty	For	Un...	Add. qn...	AddQty...	in %	Calc.Rule	FreeGoods	FGDely...	Add...
299	Pack choc biscuits	1	3	EA	1	EA	33.33	1	1		

Figure 5.29 Free Goods Master Data

The key fields on this screen include the following:

- **Min.Qty (minimum quantity)**
  Indicates the minimum quantity that must be ordered to qualify for free goods. This value must be greater than 0.
- **For**
  This field is the free goods quantity and specifies the amount for which free goods are granted in the order.
- **AddQty (additional quantity)**
  This field specifies the quantity of free goods granted to the customer when the minimum quantity is exceeded. The value is calculated based on the free goods quantity and the calculation rule.
- **Calc.Rule (calculation rule)**
  The calculation rule defines how the amount of free goods is calculated based on the free goods quantity (**For** field), the additional quantity (**AddQty** field), and the document quantity. The following calculation rule options are available:
  - **1**: Proportional (pro-rata)
  - **2**: Related to number of units (unit reference)
  - **3**: Only for whole units

For example, a customer is provided with 10 additional units free of charge for every 100 units purchased. If a customer orders 150 units, based on rule 1, the customer receives 15 units free. Based on rule 2, the customer receives 10 units free, and based on rule 3, the customer receives 0 units free (because 150 is not a multiple of 100).

- **FreeGoods**
  The free goods category categorizes free goods in the following ways:
  - **1**: Inclusive with item generation
  - **2**: Exclusive
  - **3**: Inclusive without item generation
- **FGDely (free goods delivery)**
  The free goods delivery control specifies how free goods are delivered. The following options are available:
  - <blank>: Delivery takes place regardless of the main item
  - **A**: Delivery only if the main item is partially delivered
  - **B**: Delivery only if the main item is fully delivered
  - **C**: Delivery if the main item is part. delivered, only with the main item
  - **E**: Delivery proportional to quantity in the main item

Finally, the screen shown in Figure 5.30 displays the sales order for our inclusive free goods example, where the customer orders 6 packs of biscuits but only 4 packs are charged in the order. The other 2 packs are delivered free of charge.

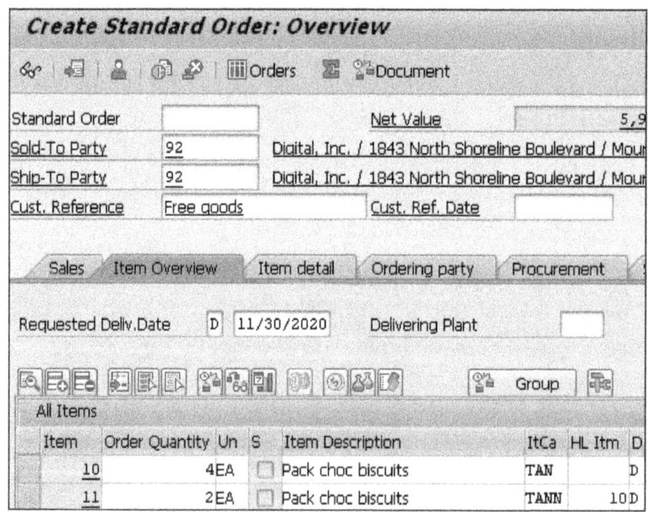

Figure 5.30 Inclusive Free Goods

Note

Item 11 is a subitem of item 10. Item category TANN (free of charge) has been determined on the subitem.

5.8 Important Terminology

In this chapter, the following terminology was used:

- **Partner function**
  This key defines the function that a person or legal entity fulfills in a business process. Sold-to party and ship-to party are two examples of partner functions.
- **Partner type**
  Specifies whether the partner is a customer, supplier, individual person, logical system, etc.
- **Condition technique**
  This infrastructure helps define master data and configure the determination of master data. The master data are called *condition records*, and their structure is defined in condition tables. Four elements are used to configure the determination of condition records: condition types, condition tables, access sequences, and procedures.
- **Status group**
  Defines the completeness status and which functions can be performed on an incomplete sales and distribution document.
- **Sales document type**
  This key represents various business transactions in SAP S/4HANA Sales, such as sales contracts, quotations, sales orders, scheduling agreements, etc. The specifications made for the sales document type control the entire sales order process.
- **Sales document item category**
  In conjunction with document types, item categories define the business transaction. This key allows for different ways to process items within the same sales document. For instance, item 10 might be sold from local stock, while item 20 must be procured externally from a third-party vendor.
- **Schedule line category**
  This key determines how the system processes the schedule line, for example, whether the system passes on requirements to MRP.
- **Item category group**
  This element in the material master can be defined at the sales area level and is used to determine item categories during the processing of sales documents.
- **Account group**
  This key is used to classify customers, determining the number range for the customer and specifying which fields are mandatory or optional in the customer master record.
- **Material requirements planning (MRP)**
  MRP is a component of SAP S/4HANA used to procure or produce the requirement quantities on time both for internal purposes and for sales and distribution. The main goal of MRP is optimizing service levels while minimizing costs and inventory.

- **Available-to-promise (ATP)**
  ATP is a component of SAP S/4HANA used to calculate the quantity that can be allocated to sales orders. Based on the material, plant, quantity, and requested date, the system checks the current stock situation and any future receipts or confirmed issues to generate confirmation proposals.
- **Material listing**
  This function is used to control the materials a customer can buy. In other words, the customer can only buy the materials contained in the list.
- **Material exclusion**
  This function is used to control the materials a customer cannot buy. In other words, the customer is not allowed to buy any of the materials in the list.
- **Incompleteness procedure**
  This feature determines if a sales and distribution document must be marked as incomplete; whether or not an incomplete document can be saved; and, if so, how the system reacts when an incomplete document is processed to create subsequent documents.

5.9 Practice Questions

These practice questions will help you evaluate your understanding of the topics covered in this chapter. The questions shown are similar in nature to those found on the certification examination. Although none of these questions will be found on the exam itself, they will allow you to review your knowledge of the subject. Select the correct answers and then check the completeness of your answers in the next section. Remember that, on the exam, you must select all correct answers and only correct answers to receive credit for the question.

1. Which objects can be assigned to an incompleteness procedure in SAP S/4HANA? (There are two correct answers.)

☐ A. Delivery item categories

☐ B. Business partner account groups

☐ C. Partner functions

☐ D. Billing item categories

2. You've been asked to set up the system to propose a substitute product if the material entered in the sales order is not available. What do you need to do?

☐ A. Set up a material determination record and ensure the MRP indicator is NOT selected

☐ B. Set up a material determination record and ensure the MRP indicator is selected

- ☐ C. Select the **ATP** checkbox in the material determination procedure assigned to the relevant sales document type
- ☐ D. Create a material determination record and assign a substitution rule with the **Strategy** field blank

3. True or False: Free goods determination is supported for quotations.

- ☐ A. True
- ☐ B. False

4. True or False: Free goods determination is not supported in combinations with material structures (for example, product selection, BOMs, variants with BOM explosion).

- ☐ A. True
- ☐ B. False

5. How do you activate material listing and/or exclusion in a sales document?

- ☐ A. By assigning a procedure for material listing/exclusion to the sales document item category
- ☐ B. By assigning a procedure for material listing/exclusion to the sales document type
- ☐ C. By assigning a procedure for material listing/exclusion to the customer account group
- ☐ D. By assigning a procedure for material listing/exclusion to the material item category group

6. In a standard SAP S/4HANA system, which calculation rules can be selected when creating a free goods condition record? (There are three correct answers.)

- ☐ A. Pro-rata
- ☐ B. Unit reference
- ☐ C. Whole units
- ☐ D. Minimum weight
- ☐ E. Factory calendar

7. True or False: An incompleteness procedure at the sales document item level can be used to make specific pricing condition types (like PR00) mandatory.

- ☐ A. True
- ☐ B. False

8. You need to configure partner determination for customer master. Which element can be assigned to the partner determination procedure?

☐ A. Sales document type

☐ B. Delivery type

☐ C. Partner type

☐ D. Account group

9. True or False: The standard system allows you to create incompletion logs to check if the entry in the field assigned to the incompleteness procedure is equal to a specific value.

☐ A. True

☐ B. False

10. A stock material will be discontinued soon, and the equivalent new product is already available in stock. How can you configure the system to sell off all the available inventory of the material to be discontinued while replenishing only the new product?

☐ A. Deactivate the material availability check for the material to be discontinued

☐ B. Set the **End of life** value in the **Distribution-chain-specific material status** field in the material to be discontinued

☐ C. Set up a material determination record and select the MRP indicator for the alternative material

☐ D. Deactivate the material availability check for the new material

11. When setting up incompletion procedures, which statuses can be selected in a status group to block an incomplete sales document from further processing? (There are two correct answers.)

☐ A. Shipment document

☐ B. Billing document

☐ C. Accounting document

☐ D. Price

12. True or False: You cannot assign both listing and exclusion procedures to the same sales document type simultaneously.

☐ A. True

☐ B. False

13. Your project requires that an incomplete sales document cannot be saved in the database. How do you achieve this?

☐ A. Assign a status group with the **General** status to all fields in the incompleteness procedure and assign the procedure to the respective sales document type

☐ B. Set the dialog message in the **Assign Incompleteness Procedures to Partner Functions** customizing activity

☐ C. Create an incompleteness procedure in the **A** (**Sales – Header**) incompleteness group and assign it to the sales document type

☐ D. Set the dialog message in the **Assign Incompleteness Procedures to Sales Document Types** customizing activity

14. Christmas season is approaching, and you've been asked to configure the system so that, for the next 30 days, all customers receive a specially wrapped product when the standard product is entered in the order. What must you do?

☐ A. Assign a calendar to the relevant sales order type and activate "packaging substitution" for the next 30 days

☐ B. Set up material determination and maintain the relevant substitute material and validity dates in the condition record

☐ C. Assign a calendar to the relevant sales item category and activate "packaging substitution" for the next 30 days

☐ D. Ask a developer to find a suitable user exit and implement custom code to swap the material entered if the system date is greater than today's date plus 30 days

15. True or False: One of the benefits of standard material determination is that the material entered in the sales order does not need to be created in the system as it is always replaced by the substitute product.

☐ A. True

☐ B. False

16. Which of the following scenarios do not support free goods determination? (There are two correct answers.)

☐ A. Deliveries without reference to a sales order

☐ B. Exclusive bonus quantity, where additional goods are offered free of charge

☐ C. Quotations

☐ D. Inclusive bonus quantity, where parts of the goods ordered are offered free of charge

17. Which of the following statements is true?

☐ A. The condition technique is an obsolete infrastructure that has been replaced by Business Rule Framework plus (BRFplus).

☐ B. The condition technique cannot be used for pricing determination.

☐ C. The condition technique can only be used for material determination.

☐ D. The condition technique can be used for free goods determination.

5.10 Practice Question Answers and Explanations

1. Correct answers: **A and C**

   As shown earlier in Figure 5.8, you can assign incompleteness procedures to six objects. Delivery item categories and partner functions are two of them. You cannot assign incompleteness procedures to business partner account groups or to billing item categories.

2. Correct answer: **D**

   You must assign a substitution rule where the **Strategy** field is blank to the condition master record if you want the system to perform an availability check in the background and select the substitute product available based on the quantity entered in the sales order.

   The MRP indicator in the material determination condition record is not related to the availability check functionality but instead controls the transfer of material shortage to SAP's planning application.

   No **ATP** checkbox exists in the material determination procedure. Answer C seems to fit but is incorrect since the feature doesn't exist.

3. Correct answer: **B**

   False. Free goods determination is only supported for sales documents with document category C (field VBAK-VBTYP). Quotations have document category B; thus, free goods cannot be determined in quotations.

4. Correct answer: **A**

   True. You cannot use free goods determination with material structures. This limitation is one of the constraints of free goods determination given the number and complexity of sales scenarios concerning BOMs.

5. Correct answer: **B**

   To activate material listing and/or exclusion in a sales document, you must assign a procedure to the relevant sales document type. You cannot assign listing/exclusion procedures to any other elements.

6. Correct answers: **A, B, and C**

   Only three calculation rules are available to choose from when setting up condition master records for free goods:

 - **1**: Pro-rata
 - **2**: Unit reference
 - **3**: Whole units

 The other two answers "minimum weight" and "factory calendar" have nothing to do with free goods determination.

7. Correct answer: **B**

   False. This question is a kind of tricky. You can make net value field (field VBAP-NETWR) mandatory in the incompletion procedure at the item level, which would flag the document as incomplete if no price is maintained. However, if a price has been determined or manually entered, you cannot use incompletion procedures to check the pricing condition used. You can mark a pricing condition as mandatory in the pricing procedure but not in the incompletion procedure.

8. Correct answer: **D**

   When configuring partner determination for a customer master, you must assign the partner procedure to the customer account group, as shown earlier in Table 5.3. The partner type is used in the definition of partner function and not in the assignment of the procedure. The sales document type and delivery type are used in the procedure assignment for the sales document header and delivery, but not in customer master. This question is specific to partner determination for customer masters.

9. Correct answer: **B**

   False. All standard incompleteness procedures do is to check whether certain fields are populated or not. If any of the fields in the incompleteness procedure are blank, the document is marked as incomplete. The system will never perform additional checks if a field in the procedure is populated.

10. Correct answer: **C**

    If you want the system to pass information about the substitute material shortage onto planning, you have to set up a material determination condition record and set the MRP indicator for the alternative material. If the MRP indicator is blank, the original material (material to be discontinued) will be replenished in the event of a shortfall. You can't use the other options to meet the requirement.

11. Correct answers: **B and D**

    If you're familiar with the status groups control table shown earlier in Figure 5.4, you can discard answers A and C since neither "shipment document" nor "accounting document" are available in this table, which leaves you with the two correct answers B and D. The "billing document" status will prevent any invoices from being created if the sales document is flagged as incomplete. Also, if the "price" status is selected, the system will block the document for further processing since pricing hasn't been carried out.

12. Correct answer: **B**
    False. You can assign both listing and exclusion procedures to the same sales document type simultaneously, as shown earlier in Figure 5.25. This approach is needed when the same document type (e.g., OR) is used to sell to customers that can only buy specific products (listing) and customers that are not allowed to buy certain products (exclusion).
13. Correct answer: **D**
    By selecting the **IC-dialog** checkbox in the **Assign Incompleteness Procedure to Sales Document Type** customizing activity, you'll omit the popup window that gives the option to save the document as incomplete, thus forcing the user to complete the missing data to save the document. Both answers A and C describe steps for setting up incompleteness procedures, but they are not enough to prevent an incomplete document from being saved. The dialog checkbox is the only standard way to achieve this functionality, which is only available when assigning incompleteness procedures to sales document types, which invalidates answer B.
14. Correct answer: **B**
    You can set up automatic product substitution via material determination, and the validity dates in the condition master record allow you to activate this functionality during the desired period. Answers A and C describe options that may sound right but are not even available in SAP S/4HANA. Answer D requires implementing custom code and thus cannot be the correct answer. Remember that the objective of the certification exam is to test your knowledge of the standard configuration; thus, any answers that mention custom code must be discarded.
15. Correct answer: **B**
    False. While this statement might be true in certain scenarios (e.g., when the material entered in the order is the customer's material number), in some cases the material entered in the sales order must be maintained in the system. For example, you can use material determination to carry out an ATP check, so if the material entered in the order is not available, it is automatically swapped for another available material. In this scenario, both materials (original and substitute) must be created in the system.
16. Correct answers: **A and C**
    Some constraints exist on free goods determination. The standard system only supports free goods for sales documents with document category equal to C (order); thus, quotations are not supported. Also, free goods determination is not supported for deliveries without reference to a sales order. The other two scenarios mentioned support free goods determination.

17. Correct answer: **D**
    We've described several ways to use the condition technique in this chapter, and free goods determination is one of them. The condition technique is NOT an obsolete infrastructure since this capability is widely used in SAP S/4HANA to determine batches, prices, texts, material listings and exclusions, substitute products, etc. Therefore, answers A and C are incorrect. The condition technique is used for pricing determination, which invalidates answer B. In Chapter 8, Section 8.2.4, we'll review this topic extensively.

5.11 Takeaway

You now understand how to create and assign incompleteness procedures to ensure mandatory fields are populated in sales and distribution documents as well as how to use partner functions. You learned the steps to configure partner determination in sales and distribution to enable the system to determine the correct partner automatically when creating or processing sales documents.

We also introduced you to the condition technique, a special mechanism used in pricing, material determination, NAST-based output control, free goods, and more—basically anywhere you have a condition record.

You also understand the basic settings to enable material determination and how this functionality can be used in real-life scenarios to propose alternative products during a limited period or to help your business sell off the remaining stock of a product soon to be discontinued.

We then discussed the deployment options for material listings and exclusions, which lets you control which products specific customers may or may not buy.

Finally, we covered how free goods determination is configured to supply the customer with a portion of goods free of charge or other goods to be offered for free in addition to the quantity ordered.

5.12 Summary

In this chapter, we provided a brief overview of the most frequently used basic functions within sales and distribution. SAP S/4HANA allows you to configure incompleteness logs and the automatic determination of several objects in different ways. Knowing the range of options provided as standard is important to zero in on the right approach for your specific business scenario. You also can set up master records for the central functions that take advantage of the condition technique.

In the next chapter, we'll cover the customizing setting relevant for the whole process chain such as copying control, text control, and NAST-based output determination.

Chapter 6
Cross-Functional Customizing

Techniques You'll Master

- Set up copy control
- Identify text sources
- Configure text control
- Adjust output determination based on table NAST
- Use enhancement technologies
- Add new fields to communication structures

In this chapter, we'll explore the customization settings required to perform cross-functional business processes. We'll go over the most important settings to map requirements in the area of sales and distribution that require some level of understanding of cross-application customizing functions, such as the configuration of copy control, text control, and output control based on table NAST (Message Status) as well as system modification options and personalization techniques.

Real-World Scenario

Working as an SAP S/4HANA sales consultant, you must have a good knowledge of cross-application customizing options that help you to fulfill specific business requirements, such as the management of texts in sales documents and specifying what data is copied from source documents to subsequent documents.

For instance, you might be asked to ensure that changes to a specific sales order text are also visible in the delivery and billing document texts. Terms of payment are a classic example of a text that should be consistent in the document header from the quotation up until billing.

You also should have an understanding of how to perform system modifications using the enhancement technologies provided by SAP S/4HANA. Some companies, for example, have a designated storage area for samples that can be offered to customers free of charge. Thus, you may be required to determine a default storage location to serve as a specific location defined for free goods in sales document items for all sales orders where the sales document type is **Deliv.Free of Charge** and make this storage location dependent on the delivering plant. Storage location determination in sales documents is not supported by the customizing (IMG), so you'll require an enhancement to the standard sales order processing logic.

The ability to implement enhancements and make modifications to the standard software is also an extremely useful skill when the options for customizing are insufficient and using enhancement technologies is the only way to streamline business processes.

6.1 Objectives of This Portion of the Test

The goal of this portion of the SAP S/4HANA Sales Certification exam is to test your understanding of the following topics:

- Configuring the copy control
- Defining and assigning text determination procedures

- Referencing and duplicating texts
- Setting up table NAST-based output determination
- Configuring output types
- Adding new fields
- Performing system modifications

Note

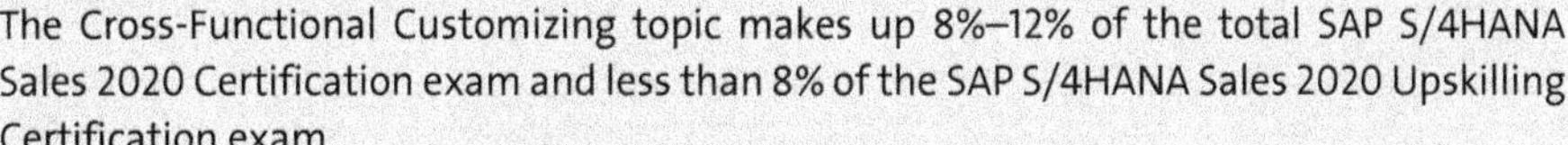

The Cross-Functional Customizing topic makes up 8%–12% of the total SAP S/4HANA Sales 2020 Certification exam and less than 8% of the SAP S/4HANA Sales 2020 Upskilling Certification exam.

6.2 Setting Up Copy Control

Arguably one of the most important customizing settings in sales and distribution, copy control is used by the system to create subsequent documents from a source document and to update the document flow. A document will only be created if copy control is maintained for the source and target document types. Any item categories to be copied must also be specified in the copy control; otherwise, the item won't be transferred. Therefore, copy control plays a key role in controlling the types of documents that can be used and the information passed on to the next document when processing sales transactions in the system, from quotation to billing.

Example

Let's say your project requires a new type of quantity contract ZCQ, which has already been defined in the system. When testing the new contract type in the development client, you can successfully create contracts, and everything within the new quantity contract seems to work as expected. However, the system returns the error message **Copying from sales document type ZCQ to OR is not supported** when you try to create a release order type OR for the contract. This step would work fine if the release order (OR) had been created with reference to a standard quantity contract type CQ.

This scenario is a classic example of missing copy control for a newly defined sales document type. Since the release order inherits information from the quantity contract, the system checks whether it can use sales document type ZCQ as the source document and OR as the target document. Since the combination ZCQ → OR has not been maintained yet, the system is unable to proceed.

Copy control entries for standard document types are already preconfigured, but you must remember to check if additional copy control settings are required whenever new sales document types, item categories, or schedule lines are defined. However, when you create new document types or item categories by copying existing ones, the system will ask if the new entry is relevant for copy control. If you click on **Yes**, the system will automatically generate all the copying rules based on the rules of the object used as a reference.

The customizing activities to maintain the copy control for sales documents can be accessed in the IMG menu under **Sales and Distribution • Sales • Copying Control for Sales Documents**. For deliveries, the menu path is **Logistics Execution • Shipping • Copying Control**, and for billing documents, **Sales and Distribution • Billing • Billing Documents • Maintain Copying Control for Billing Documents**.

To illustrate how often the system makes use of copy control rules, Figure 6.1 shows an example of an end-to-end sales process flow where the product sold is invoiced and then returned by the customer. Each box represents a new document in the document flow, and each arrow symbolizes a copy control check carried out by the system.

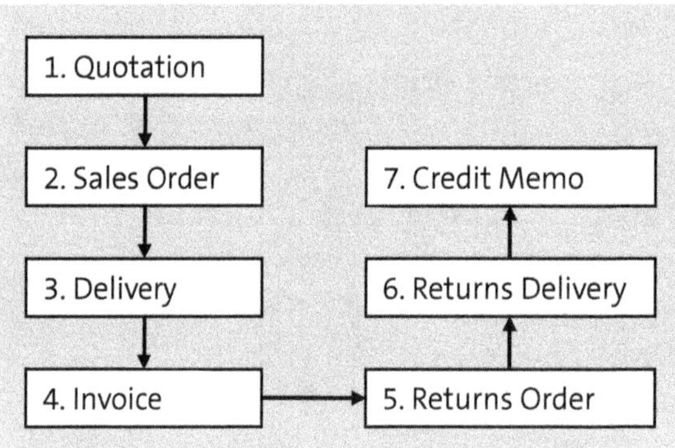

Figure 6.1 Sales and Distribution Documents in a Typical Sales Flow with Returns

You can also use copy control to create a sales document with reference to another document of the same type (e.g., OR into OR). This capability could be useful if a business user needs to create a sales order similar to an existing one. By using the **Create with Reference** option, the user won't have to re-create all the items—all the information will be copied automatically.

You can specify copying rules for sales documents, deliveries, and billing documents at both the header level and at the item level. In addition, when configuring copy control for sales documents, you can set up copying rules at the schedule line level. Next, we'll explain the copy control settings available at the header level for sales documents. Figure 6.2 shows the standard copying rules from quantity contract (type CQ) to standard sales order (type OR).

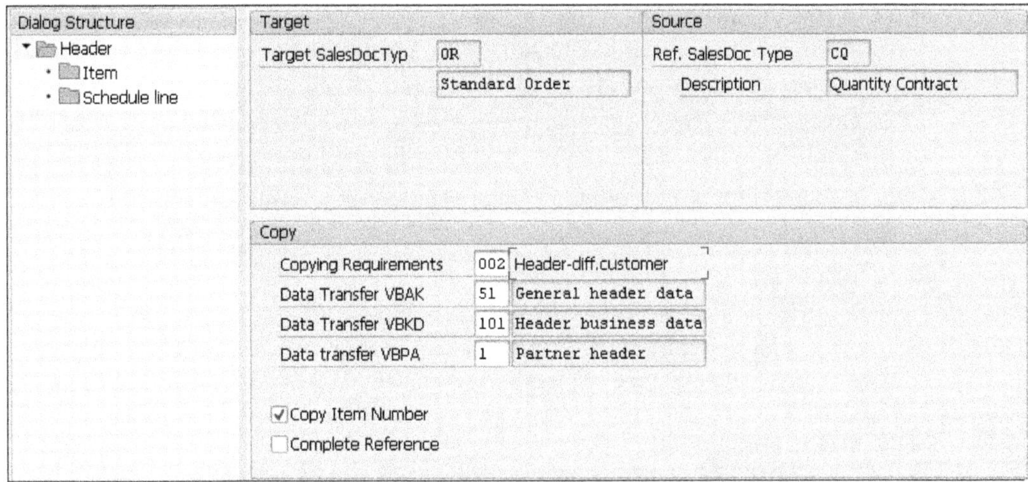

Figure 6.2 Sales Document Copy Control Settings at the Header Level (CQ to OR)

The important fields on this screen include the following:

- **Copying Requirements**
  This field contains the identification number of a routine where the copying requirement is coded in ABAP. The copy procedure will only take place if the condition is satisfied. SAP S/4HANA provides many standard routines that should cover most business requirements, but you can develop a customized routine with the help of an ABAP programmer if needed. Some standard copying routines include the following:
  - **001** (Header same customer): This routine checks whether the currency, customer (sold-to), and sales area are the same in both the reference document and the target document at the header level as well as whether the reference document is currently valid and complete. The standard include is LV45C001.
  - **002** (Header different customer): This routine checks whether the currency and sales area are the same in both the reference document and target document at the header level as well as whether the reference document is currently valid and complete. The standard include is LV45C002.
- **Data Transfer VBAK**
  This ABAP routine specifies the header data from the source document that is copied into the target document. Custom header data routines can also be developed to accommodate specific business requirements.
- **Data Transfer VBKD**
  This ABAP routine specifies the business data from the document header of the source document that is copied into the target document. Business data custom routines can also be developed to accommodate specific business requirements.

- **Data Transfer VBPA**
  This ABAP routine specifies the partner data from the header of the source document that is copied into the target document. Item custom routines can also be developed to accommodate specific business requirements.
- **Copy Item Number**
  If this checkbox is selected, the system copies item numbers from the source document into the target document.
- **Complete Reference**
  If selected, the system copies the whole source document into the target document and returns a message if the quantity in the target document is changed or an item is deleted later.

Now, let's look at the copy control for sales documents at the item level. As shown in Figure 6.3, you can make a few more settings at the item level compared to the header level.

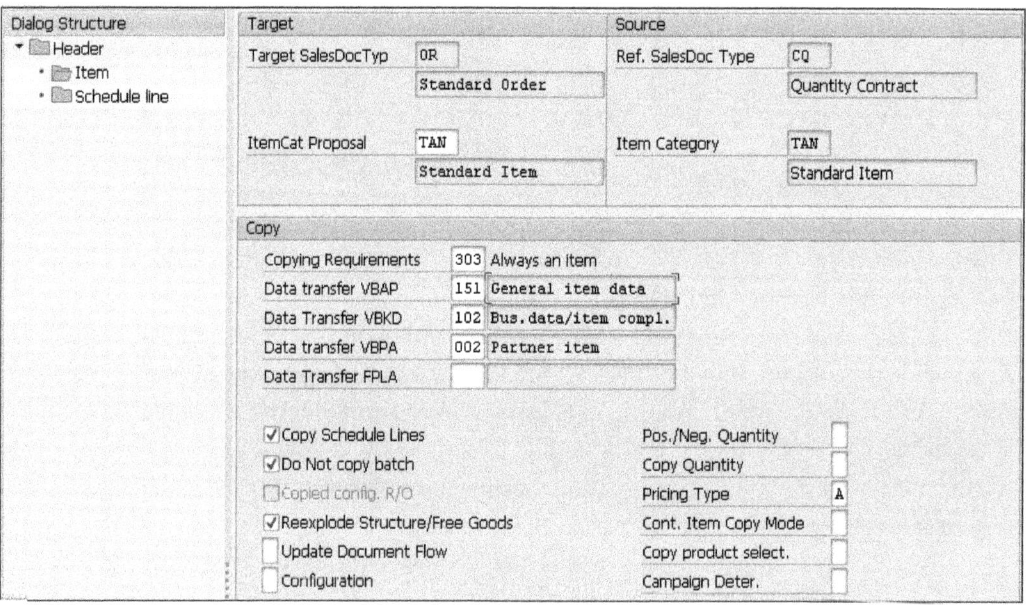

Figure 6.3 Sales Document Copy Control Settings at the Item Level (CQ to OR and TAN)

As in the header copy control settings, in this case, we also have copying requirements and data transfer routines that apply at the item level. The following tables are used for item data transfer:

- Table VBAP: Sales Document: Item Data
- Table VBKD: Sales Document: Business Data (item)
- Table VBPA: Sales Document: Partner (item)
- Table FPLA: Billing Plan (This routine is used to check if the billing plan data must be copied.)

Other important copying rules at the item level include the following:

- **Copy Schedule Lines**
  This indicator must be selected if you need schedule lines to be copied to the target document. If the indicator is set, you must maintain the copy control at the schedule line level.
- **Do Not copy batch**
  This checkbox dictates whether the batch number of the material must be copied into the subsequent document.
- **Reexplode Structure/Free Goods**
  This indicator controls how sales bill of materials (BOM) items are processed when a sales document is created with reference. New orders might be created a long time after the source order was created; thus, the BOM may have changed. If this checkbox is selected for the BOM main item, the system does not copy the subitems but instead reexplodes the BOM using the new date. If you don't select this indicator, the subitems are copied into the following document.
- **Update Document Flow**
  This field determines whether entries are automatically generated in the document flow table when a new sales document is created with reference. The information in the document flow might be needed since this information is used by the system to determine the status of the source document. You can select one of the following three values:
  - <blank>: The system does not create document flow records. You may want to select this option when the source and the target documents share the same document type, that is, if you decide to create a document with reference to a similar one, you can avoid entering in all the lines in the target document and thus save time.
  - **X**: The system creates document flow records. Let's say the originating document is a quotation. When a sales order is created with reference to a quotation, the sales order creates a document flow record for the quotation. Similarly, deliveries and invoices create document flow records for both the sales order and the quotation.
  - **2**: The system creates document flow records except for deliveries, goods issues, and invoices. In the event the source document is a quantity contract, the release sales order creates a document flow record for the contract, but the delivery, goods issue, and billing documents only create a document flow entry for the sales order. As a result, for example, you won't be able to see the contract in the document flow of the delivery, and you won't see the delivery, goods issue, or invoice in the document flow of the contract.
- **Pos./Neg. Quantity**: This field defines the effect of the quantity in the target document on the open quantity in the source document. The following three options are available:

  - <blank>: The quantity on the following document does not affect the quantity to be completed on the source document.
  - **+**: The quantity on the following document has a positive effect on the quantity to be completed on the source document. For example, let's say a contract item has been created with a quantity of 100 units. A release sales order is created with reference to the contract for 40 units. The open quantity in the contract is updated to 60 units.
  - **-**: The quantity on the following document has a negative effect on the quantity to be completed on the source document. For example, let's say a contract item has an open quantity of 20 units. A returns order is created with reference to the contract for 10 units. The open quantity in the contract is updated to 30 units.
- **Copy Quantity**: This field is used to specify the quantity to be copied to the target document. The following three options are available:
  - <blank>: Automatic quantity determination. For instance, if you were creating a sales order from a quotation, the system would copy the open quotation quantity. However, when copying from a billing document to a credit memo, the system will copy the billed quantity.
  - **A**: The system copies the order quantity in the source sales document.
  - **B**: The system copies the target quantity in the source sales document.
- **Pricing Type**: This important field controls which condition types are redetermined and which are copied unchanged when creating sales documents with reference. SAP S/4HANA offers over 20 different pricing types to choose from. Details about each pricing type are beyond the scope of the certification exam. In this bullet point, we'll review some of the most common pricing types. To learn more about other pricing types, refer to the SAP Note 24832.
  - **A**: The system copies pricing elements unchanged, but scales are redetermined.
  - **B**: The system ignores any manual pricing conditions in the source document and carries out new pricing. The result is equivalent to create a new item.
  - **C**: This option copies manual pricing elements and redetermines the others.
  - **D**: The system copies pricing elements unchanged without redetermining scales.
  - **G**: The system copies pricing elements unchanged and redetermines taxes, rebates, cost conditions, and intercompany billing conditions.
  - **H**: This option redetermines freight conditions and delivery costs.
- **Cont. Item Copy Mode** (contract item copy mode): This field only applies to value contract items. The following three options are available:
  - <blank>: No restrictions regarding the material code.

 - **A**: The material code in the value contract item cannot be copied into the release order. A different material must be selected in the release order.
 - **B**: Only the material code in the value contract can be used in the release order.
- **Copy product select.** (copy product selection): This field defines whether the results of product selection are automatically copied from a source document to a target document. The following options are available:
 - <blank>: The main item is copied, and a new product selection is carried out for the subitems.
 - **A**: The main item and its subitems are copied into the target document. If you make changes to the order that triggers the available-to-promise (ATP) check, the system redetermines product selection.
 - **B**: The main item and its subitems are copied into the target document. If you make changes to the order that triggers the ATP check, the system does not redetermine product selection.

Next, you must maintain copy control settings at the schedule line level for each combination of target document type and source document type where information is stored in the schedule lines of the reference document. Figure 6.4 shows the copying rules when copying from a quotation (type QT) into a standard order (type OR) and the schedule line category in the quotation is category BP (MRP).

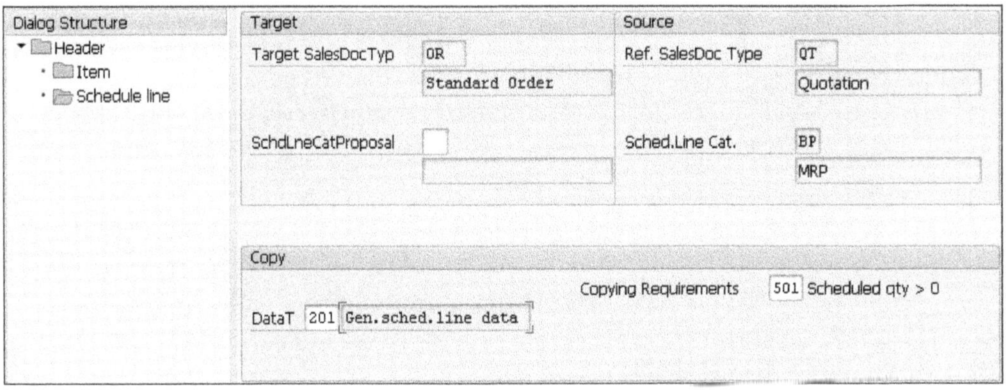

Figure 6.4 Sales Document Copy Control Settings at the Schedule Line Level (QT to OR and BP)

The copy control for the schedule line presents fewer options than the item category, but you can still make use of data transfer and copying requirement routines.

So far, we've shown you the screens for setting up the copy control when copying sales documents to sales documents. However, you should know that the layout of these copy control screens, and the fields available, vary slightly depending on the nature of the source and target documents. For instance, when copying from a billing document into a sales document, the copy control settings at the header level present exactly the same options as those between sales documents, as shown earlier in Figure 6.2. Differences arise at the item level when copying a billing item into a sales document item in which case the following four settings are not available:

- Contract item copy mode
- Reexplode structure/free goods
- Copy product selection
- Copy quantity

Also, note that deliveries do not have schedule line information; thus, you cannot configure copying rules for schedule line data when copying from a sales document into a delivery document. In this scenario, only header and item data are transferred to the delivery. The same principle applies when copying data from a sales document into a billing document since billing documents don't have schedule lines.

When configuring the header copy control from a sales document or a delivery into a billing document, you'll be presented with a different screen where no data transfer routines are available. However, SAP S/4HANA provides a copying requirements routine that can be used, for example, to check whether the delivery or sales document is blocked for billing. Besides the copying requirements routine, the following three parameters are only available on the header copy control for billing documents:

- **Determ.of Export Data** (determination of export data): You can use this indicator to control how the export data must be obtained for the billing document. You can decide whether to redetermine the export data in the billing document, copy the export data from the delivery into the billing document, or refer to the export data in the delivery document.
- **Assignment Number**: This parameter plays an important role in billing split criteria and controls how the assignment number field (field VBRK-ZUONR) is populated in the billing document. The assignment number is set in the header of the billing document and transferred to the financial journal entry. The following options are available:
  - <blank>: The field remains empty
  - **A**: Customer purchase order number
  - **B**: Sales order number
  - **C**: Delivery number
  - **D**: External delivery number
  - **E**: Actual billing document number
  - **F**: External delivery number if available, otherwise use the delivery number

- **Reference Number**: Like the assignment number, this parameter also affects the split criteria during billing and controls how the reference document number field (field VBRK-XBLNR) is populated, which is stored at the billing document header level. The reference document number field contains the document number with the business partner and is also transferred to the financial journal entry. The following options are available:
  - <blank>: Current billing document number
  - **A**: Customer purchase order number
  - **B**: Sales order number
  - **C**: Delivery number
  - **D**: External delivery number
  - **E**: Current billing document number
  - **F**: External delivery number if available, otherwise use the delivery number

Figure 6.5 shows the header copy control settings when copying from a delivery document (type LF) into a billing document (type F2).

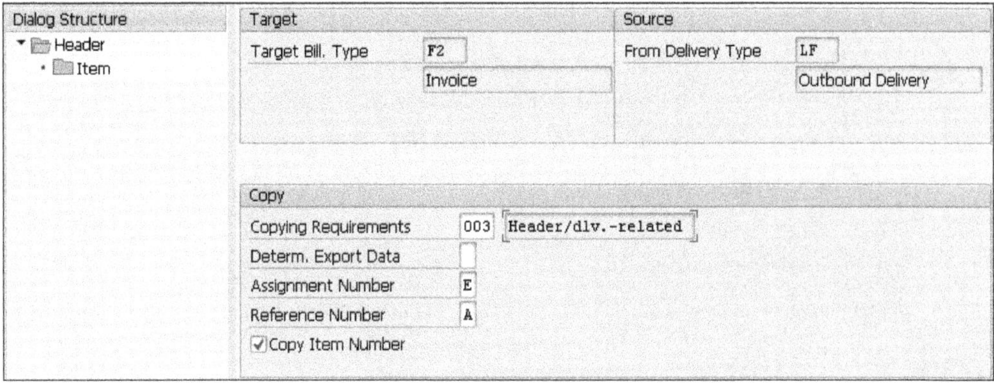

Figure 6.5 Billing Document Copy Control Settings at the Header Level (LF to F2)

Finally, let's take a look at the copy control settings at the item level when copying deliveries or sales documents into a billing document. As in other copy control screens, SAP S/4HANA provides routines for copying requirements and data transfer. In this case, the same data transfer routine is used for both header data (table VBRK) and item data (table VBRP), and this routine can contain the logic for the split criteria when creating billing documents. For example, when combining multiple sales documents into one invoice, the split criteria could be used to create separate billing documents if either the distribution channel or the division in the source sales document is different.

Some of the parameters shown in Figure 6.6 have already been covered in the item copy control for sales documents.

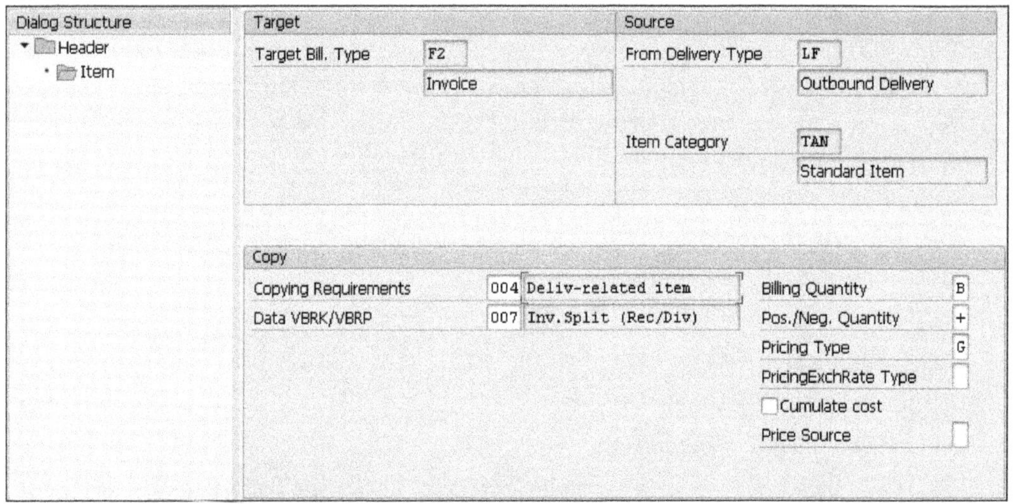

Figure 6.6 Billing Document Copy Control Settings at the Item Level (LF to F2 and TAN)

However, the following parameters are specific to the billing document copy control:

- **Billing Quantity**: This field specifies the quantity to be copied into the billing document item. The following values are available:
  - **A**: Order quantity minus invoiced quantity
  - **B**: Delivery quantity minus invoiced quantity
  - **C**: Order quantity
  - **D**: Delivery quantity
  - **E**: Goods receipt quantity minus invoiced quantity
  - **F**: Invoice receipt quantity minus invoiced quantity
  - **G**: Cumulative batch quantity minus invoiced quantity
  - **H**: Cumulative batch quantity
  - **I**: Purchase quantity minus quantity already invoiced
- **PricingExchRate Type** (pricing exchange rate type): This field determines the date used as the basis to calculate the exchange rate.
- **Cumulate cost**: If selected, the cost of the subitems is added up and passed to the main item. This function works with sales BOMs, free goods, and batch split items. Of particular importance is selecting this indicator for those item categories where the components are not relevant for billing.
- **Price Source**: This field dictates the document and sequence used to get the pricing information that will be copied into the invoice.

6.3 Setting Up the Text Control

Playing an important role in sales and distribution documents, texts are often an efficient communication channel with customers, suppliers, and business users. For example, a text explaining that free goods have been added to the sales order item can be printed on an order confirmation. Also, when terminating a service contract, the user can enter an internal text at the header level and provide details about contract cancellation for future reference. These examples are just two basic applications of texts in sales documents, but texts and text determination are flexible functions that can be tailored to meet all kinds of business requirements.

You can set up texts in customizing under **Sales and Distribution • Basic Functions • Text Control • Define and Assign Text Determination Procedures.** First, select one of the following text objects available in SAP S/4HANA:

- Customer
  - Central texts
  - Contact person
  - Sales and distribution
- Info record
  - Customer-material info record
- Pricing conditions
  - Agreements
  - Conditions
- Sales document
  - Header
  - Item
- Delivery
  - Header
  - Item
- Billing document
  - Header
  - Item
- Shipment
  - Header
- Settlement management
  - Header
  - Item
- Trading contract
  - Header
  - Item

Note

Each text object is referenced by a 4-character key. To make the relevant settings for text determination in SAP S/4HANA Sales, you must be familiar with the various text objects:

- **KNA1**: Customer master central texts in the **General Data** screen.
- **KNVK**: Contact person texts in the **General Data** screen.
- **KNVV**: Customer master sales texts in the **Sales and Distribution** screen.
- **KNMT**: Customer-material information record texts.
- **KONA**: Texts defined for agreements.
- **KONP**: Texts defined for pricing condition records.
- **VBBK**: Text defined in the preceding document at the header level. The preceding document could be a sales document, a delivery, or a billing document. For example, if a contract is copied to a sales order, then this text object refers to the header text in the contract.
- **VBBP**: Text defined in the preceding document at the item level. The preceding document could be a sales document, a delivery, or a billing document.
- **VTTK**: Text defined in the shipment document at the header level.
- **WBRK**: Text defined in the settlement management document at the header level.
- **WBRP**: Text defined in the settlement management document at the item level.
- **WBHK**: Text defined in the trading contract at the header level.
- **WBHI**: Text defined in the trading contract at the item level.
- **TEXT**: This standard standalone text is created via Transaction SO10 or an SAP Fiori app (the Administration of Form Texts app). For this text object, the text ID is ST, and you must specify the name of the standard text in the access sequence.

The next step is defining the permitted text IDs, also known as *text types*, for the chosen text object by clicking the **Text types** button. Figure 6.7 shows some standard text types for text object KNA1 (Customer Central Texts). If needed, you can create your own text types.

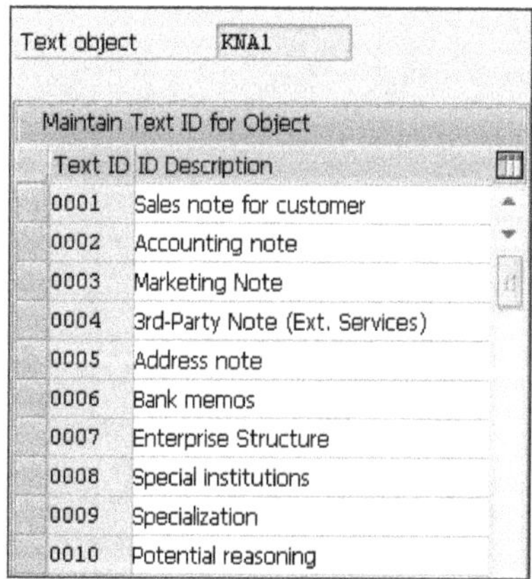

Text object KNA1

Maintain Text ID for Object

Text ID	ID Description
0001	Sales note for customer
0002	Accounting note
0003	Marketing Note
0004	3rd-Party Note (Ext. Services)
0005	Address note
0006	Bank memos
0007	Enterprise Structure
0008	Special institutions
0009	Specialization
0010	Potential reasoning

Figure 6.7 Text Types for Text Object Customer Central Texts

To configure text determination for the customer master record, first you must define a text determination procedure. Then, you'll need to specify which text IDs will be part of the text procedure. Finally, you can assign the text procedure to the customer account group.

Note

You cannot assign access sequences to text IDs in the text procedure for the customer master record—neither for customer central texts nor for customer sales and distribution texts (text objects KNA1 and KNVV, respectively). You cannot define access sequences for pricing condition texts and shipment texts.

Once text types have been defined and text procedures are assigned to customer account groups, you can maintain the texts in the customer master record. For example, the customer central texts shown in Figure 6.7 can be maintained in the business partner master record, for BP role FLCU01 (Customer), under the **Customer: Texts** tab in the **General Data** view, as shown in Figure 6.8.

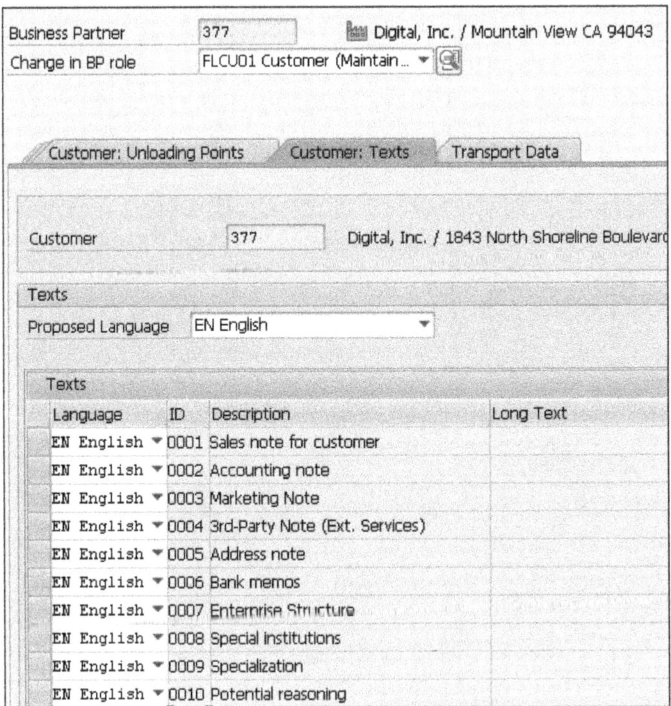

Figure 6.8 Customer Central Texts in the Business Partner Master Record

Via text determination, the system can propose text IDs from customer master records for inclusion in sales documents, deliveries, and billing documents. You can also copy one or more text IDs automatically from the preceding sales and distribution document or determine a standard text created through Transaction SO10.

To configure text determination for the sales document header, the first step is to define a text determination procedure. Then, for each of the text IDs in the procedure, you must specify the parameters shown in Figure 6.9.

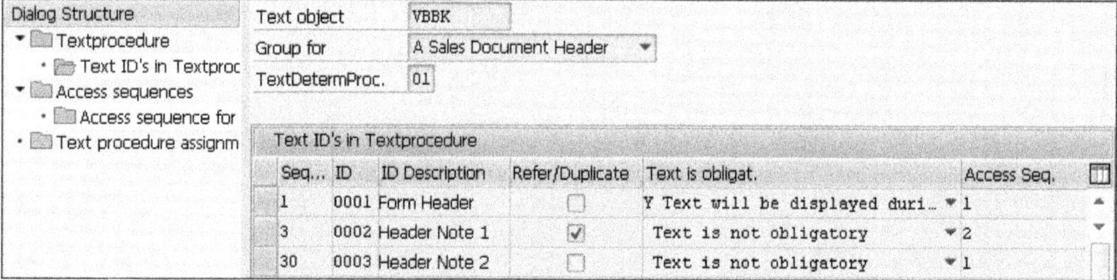

Figure 6.9 Text IDs in Text Procedures for Sales Documents (Header)

For each text ID, you can configure the following parameters in the text determination procedure for sales documents:

- **Refer/Duplicate**
  This checkbox determines how a text ID is transferred from a master record or from a sales and distribution document to a subsequent sales and distribution document. If this indicator is selected, the text is duplicated from the original. In other words, the system creates a copy of the text, and any changes on the original text do not affect the copied text. Likewise, a change to the copied text has no effect on the original text. On the other hand, if the indicator is blank, the text is referenced, meaning that only one version of it exists on the database. In this case, if you change the original text, the change appears in all sales and distribution documents that reference it. But if you change a text that refers to an original text, the text is then duplicated and becomes an independent text. For more information on this indicator, including some examples, refer to SAP Note 817101.
- **Text is obligat. (text is obligatory)**
  This parameter allows you to select one of the following four values:
  - <blank>: The text is not obligatory.
  - **X**: The text is obligatory. If the text is missing, the document will be marked as incomplete.
  - **Y**: The text is displayed in a popup window during copying.
  - **Z**: The text is obligatory and appears in a popup window during copying. If the text is missing, the document will be marked as incomplete.
- **Access Seq. (access sequence)**
  This field specifies the number of the access sequence used by the system to copy a text from a preceding document or a master record. You can define access sequences for sales documents. Basically, this field determines the text object, the text IDs for each text object, and the sequence in which the texts are checked. When defining the access sequence, we recommend using the same

text ID in both the source and destination text objects. For example, text ID 0001 in the customer master should be copied into the text ID 0001 of the sales document header. If the texts are copied from a customer master record, you can specify the partner function of the customer to be used to find the texts. You can also specify the language (or languages) of the text to be copied. With regard to the language, the following options are available:

- All texts that meet the criteria are copied regardless of the language.
- Only the texts in a specific language are copied.
- Only the texts in the language of the partner function specified are copied.
- Only the texts in the language of the sales organization are copied.

Finally, you need to assign the text determination procedure to the relevant sales document types or item categories. Table 6.1 shows the type of sales and distribution element to which a text procedure must be assigned, depending on the text object being configured.

Text Object	Text Procedure Assigned to...
Sales document header	Sales document type
Sales document item	Item category
Delivery header	Delivery type
Delivery item	Item category
Billing document header	Billing document type
Billing document item	Billing document type
Pricing agreement	Agreement type
Pricing condition	Pricing condition type
Shipment header	Shipment type
Settlement management header	Settlement document type
Settlement management item	Settlement document type
Trading contract header	Contract type
Trading contract item	Item category

Table 6.1 Text Determination Procedure Assignment

6.4 Output Control Based on Table NAST

In this section, we'll review the configuration of the output control based on table NAST using the condition technique. Traditionally, output management based on NAST was the only standard option to determine outputs in SAP ERP. However, with SAP S/4HANA, a new output management approach is also in place that utilizes a technology known as Business Rule Framework plus (BRFplus). Therefore, SAP S/4HANA offers two output management frameworks in sales and distribution:

- New output management for SAP S/4HANA
- Output determination based on table NAST

Note

Each output management framework belongs to a specific SAP component. The details of the two output-related components can be summarized in the following ways:

- New output management for SAP S/4HANA (CA-GTF-OC)
  - CA: Cross-Application Components
  - CA-GTF: General Application Functions
  - CA-GTF-OC: SAP S/4HANA Output Management
- Output determination based on table NAST (SD-BF-OC)
  - SD: Sales and Distribution
  - SD-BF: Basic Functions
  - SD-BF-OC: Output Determination

You should be aware that, in SAP S/4HANA, output management based on table NAST is no longer available in some billing apps, like the Manage Billing Documents app, so if you choose NAST-based output determination for billing, you won't be able to issue outputs from apps that don't support this framework. Also, you cannot combine output management frameworks in billing. Therefore, you must choose which output management framework will be used by all apps in billing. Table 6.2 shows which output management framework each billing app can use.

SAP Fiori App	Output Management Framework
Create Billing Documents	New output management for SAP S/4HANA (CA-GTF-OC)
Manage Billing Documents	
Schedule Billing Output	
Create Billing Document – VF01	Either new output management for SAP S/4HANA (CA-GTF-OC) or NAST-based output determination (SD-BF-OC)
Change Billing Document – VF02	
Display Billing Document – VF03	
Create Billing Documents – VF04	
Create Invoice Lists – VF21	
Change Invoice List – VF22	
Display Invoice List – VF23	
Create Invoice Lists – VF24	
List Invoice Lists – VF25	
Cancel Invoice Lists – VF26	

Table 6.2 Output Management Framework versus Billing App

Note

In this chapter, we won't discuss the new output management for SAP S/4HANA (CA-GTF-OC), which will be covered, however, in Chapter 12, Section 12.5, when we review the main changes introduced by SAP S/4HANA 2020 in the sales and distribution area.

For more information about choosing your output management framework, refer to SAP Note 2267376 for SAP S/4HANA up to release 1709 or SAP Note 2790427 if you're implementing SAP S/4HANA 1809, 1909, or 2020.

In NAST-based output determination, output control is used to send information to internal and external partners. Output control can be divided into two main functions: output determination and output processing. In sales and distribution documents, output control is carried out separately in the header and item levels.

The output determination component allows you to propose output records for sales, shipping transportation, and billing. The system uses the condition technique described in Chapter 5, Section 5.4, to determine the output that is directly linked to the corresponding sales transaction. If needed, you can change an output record in the sales and distribution document.

You can maintain the settings for output determination based on NAST in customizing under **Sales and Distribution • Basic Functions • Output Determination**. The following customizing activities must be performed in sequence:

1. Create condition table
2. Define access sequences
3. Define output types
4. Define the output determination procedure

We already went through a similar process when covering the configuration for other functions that also use the condition technique. First, a condition table must be defined with the selected fields that form the criteria to determine the output. Figure 6.10 shows standard condition table 001, which contains two fields in the condition key: the sales organization and the customer number.

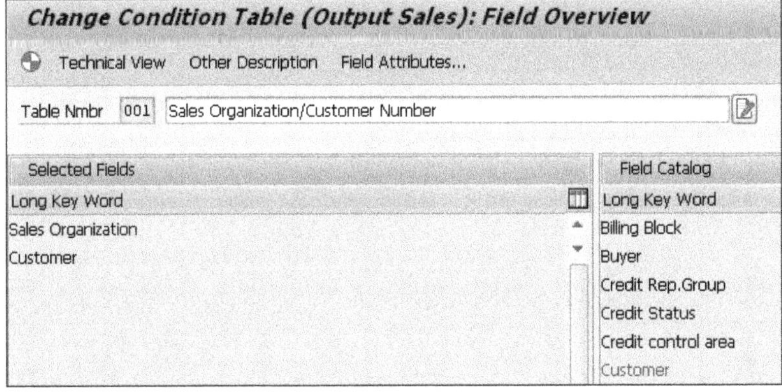

Figure 6.10 Output Condition Table 001

The second step is to define an access sequence by allocating one or more condition tables to each access, or entry, in the access sequence. The system reads the selected fields of the condition table, gets the values from the sales and distribution document, and tries to find a valid output condition record. Then, it moves to the next access in the access sequence and repeats the process. You can assign ABAP requirements to each condition table in the access sequence. If the requirement is fulfilled (SY-SUBRC = 0), the relevant access is executed. You can also set the **Exclusive** indicator for each condition table, so the system stops searching for messages in the access sequence if a valid output condition record was found for a condition table marked as exclusive.

Next, you'll need to define the output type that is used to represent a specific form of output in SAP S/4HANA. The following are some examples of output types in SAP S/4HANA Sales:

- Quotation
- Order confirmation
- Packing list
- Advance shipping notification
- Invoice

Each output type is assigned to a specific application depending on the area of the IMG menu where you're setting up the output control. Table 6.3 shows the main applications for message conditions in sales and distribution.

Application	Description	Examples of Sales and Distribution Documents
V1	Sales	Quotations, sales contracts, and sales orders
V2	Shipping	Deliveries
V3	Billing	Invoices
V6	Handling Units	Handling units
V7	Transport	Shipments

Table 6.3 Applications for Message Conditions in SAP S/4HANA Sales

Before we discuss the configuration options within the output type definition, you should be familiar with some key concepts, such as dispatch time, transmission medium, and communication strategy.

The *dispatch time* refers to the time when the message will be processed. Four options are available:

- Time 1: The output message is processed through program RSNAST00, which can be scheduled periodically or started manually.
- Time 2: Similar to time 1, but in this case, program RSNAST00 will ignore the message before the requested date and time specified in the output message.

- Time 3: The output message is processed with the application's specific transactions, for example, Transaction VF02 if you want to print out one invoice or Transaction VF31 for a set of invoices.
- Time 4: The message is processed immediately when the application document is saved.

The *transmission medium* specifies the message is produced. For instance, you can specify if the output is printed or generated internally and sent by email. In sales and distribution, the following ten options are available for transmitting outputs:

- **1: Print output**
  This option outputs to the spool, and depending on the **Print immediately** indicator in the output condition record, printing can occur immediately or the message can stay in the spool. Print parameters, such as the output device or the **Print immediately** indicator, can be set in the condition record. However, if no output device is specified in the condition record, the print parameters can also be determined based on the output type and the current application.
- **2: Fax**
  This option sends the message as a fax to the fax number (communication method FAX) maintained in the partner data.
- **4: Telex**
  This option sends the message as a telex to the telex number (communication method TLX) maintained in the partner data.
- **5: External send**
  This option sends the message using the preferred communication method for contacting the business partner. This method is determined by the communication strategy maintained in the output condition record. If a communication strategy is defined in the output type in customizing, the system will propose it when creating a condition record.
- **6: EDI**
  This option is used for electronic data interchange (EDI). The system generates the relevant Intermediate Document (IDoc) and sends it to the middleware application, for example, SAP Integration Suite. The IDoc is then converted and forwarded to the recipient.
- **7: Simple mail**
  This output is controlled by the Business Communication Interface. You can define a standard text and a title for the message both in condition records and in the output type customizing. The message is sent to an internal distribution list or to external internet addresses during output processing.
- **8: Special function**
  This option can be used to call your own custom programs.

- **9: Events**
  This transmission medium can be used to trigger an event to start a workflow.
- **A: Distribution**
  This method can be used to exchange data between two SAP systems through Application Link Enabling (ALE). The partner is derived from the distribution model.
- **T: Tasks**
  A task is initiated during output processing, which generates a work item.

The *communication strategy* indicates the sequence in which the system searches for communication methods such as fax, email, etc. in the case of transmission medium 5 (external send) only. This function is not available with transmission mediums other than 5. The first communication method found for the specified address is selected. This field allows you to ensure the preferred communication method is used to contact the business partner. You can define a communication strategy in customizing under **Sales and Distribution • Basic Functions • Output Determination • Determine Communication Strategy**. The following communication methods are available:

- FAX: Fax
- X40: X.400
- LET: Post (letter)
- INT: Email
- RML: Remote mail

Now that you know more about the basic elements in output determination, let's take a look at some of the main settings you can maintain when defining output types in SAP S/4HANA:

- You can assign an access sequence. This step is not mandatory unless you plan to use output condition records.
- If you select the **Access to conditions** indicator, the system determines the output by searching for valid output condition records.
- You can specify that the output cannot be changed manually during processing by selecting the **CannotBeChanged** indicator.
- You can decide whether to allow multiple outputs for the same partner. For example, if new items are added to an order after the order confirmation has been sent to the sold-to party, you may want to send the same output (including the new items) to the same partner function.
- You can indicate whether the output type is partner independent or whether the output can only be created for the partner functions assigned to the output type.

- You can set default values for the following parameters:
  - Dispatch time
  - Transmission medium
  - Partner function
  - Communication strategy
- You can also determine whether any of the four dispatch times should be excluded.
- You can select the indicator that contains the criteria used to determine the print parameters. Print parameters contain information about the output device, the **Print immediately** indicator, the release after output indicator, etc. The possible values for the print parameter indicator depend on the application. For example, for application V1, the following two indicators are allowed:
  - Sales organization
  - Sales organization/distribution channel/division
- You can define the email subject line and text for the body of the email to be used with transmission mediums **5: External send** and **7: Simple mail.**
- You can define the ABAP program and which form routine within the program is called depending on the transmission medium. This information is stored in the configuration table TNAPR.
- You can specify the partner functions allowed for each transmission medium.

Once the output type has been defined, the next logical step is to maintain the output determination procedure and assign all the relevant output types. Only the output types in the procedure will be considered when output determination is carried out for the sales and distribution document. In the procedure, you can use requirements for each output type to restrict access to the output type, unless the criteria in the requirement are met. As shown in Figure 6.11, requirement **2** (order confirmation) has been assigned to output **BA00** in output procedure **V10000**. This particular requirement checks the sales document for completeness, and output BA00 is only generated if the document is complete. On this screen, you can also indicate if an output type is to be entered manually in the sales and distribution document.

Dialog Structure
- Procedures
  - Control Data

Procedure V10000 Order Output

Reference Step Overview

Step	Counter	CTyp	Description	Requiremnt	Manual only
1	0	ESYM	Internal Output	22	☐
10	1	BA00	Order Confirmation	2	☐
20	1	MAIL	Internal Output	0	☑
30	1	KRML	Credit Processing	9	☐
40	1	BAV0	Order confirm. VMI	0	☐
50	0	BSOP	Order Confirmation	0	☐
50	1	MAND	SEPA Mandate	460	☐

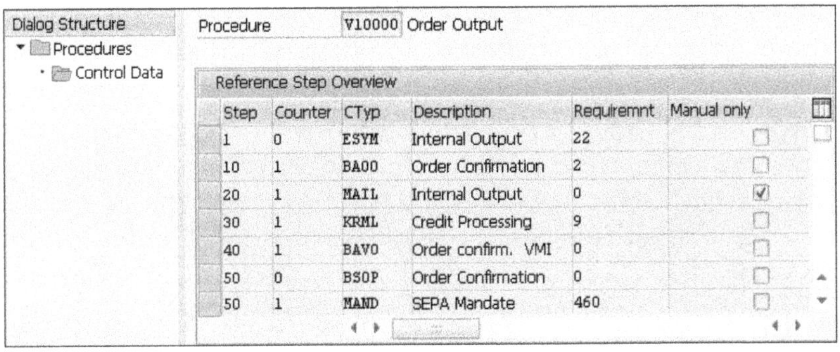

Figure 6.11 Maintaining Output Determination Procedure

An output determination procedure must be allocated to sales and distribution documents, which can be done both at the header level and at the item level. To determine the output at the header level, you must assign the output procedure to the sales and distribution document type. If you want the output to be determined at the item level, then you can assign the output procedure to the relevant item categories.

Finally, you can proceed and create the output conditions records required for output determination, either via Transaction NACE (Conditions for Output Control) and selecting the application concerned or by calling the specific transactions to create or change output condition records for each application. When creating output condition records, you specify the output type and the values for the fields defined in the condition table assigned to the access sequence. Then, you'll enter the partner function, the transmission medium, the dispatch time, and the language. If the transmission medium is 1 (print output), you can click on **Communication** and assign the output device, specify whether the message is to be issued immediately, and specify the number of messages.

6.5 Enhancements and Modifications

When working on a project, you're may be presented with specific customer requirements that are not supported by the standard system. To enable consultants to meet these requirements, the standard software allows for the development of system adaptations using logic specific to the customer without changing the original repository of objects, thus ensuring that changes will not be lost in subsequent system upgrades. The following enhancement techniques are commonly used in the area of sales and distribution:

- **New data fields**
  - **Using condition technique**
    In sales and distribution, the condition technique is used in pricing, output determination, account determination, material listings/exclusions, and material determination. The fields allowed for defining condition tables and accessing condition records are contained in communication structures. Each function has its own communication structures. For example, structures KOMK and KOMP are used in pricing, KOMKBV1 and KOMKBV2 are used in output control, etc. In the event you need to use a particular field in your condition table that is not available in the field catalog for the application concerned, first you must add the required field to the field catalog. However, this step is not enough since the value in the source field will not be transferred to the relevant communication structure, and the system will never find the corresponding condition record. Instead, you'll need to

append the new field to the communication structure and implement the logic to populate the new field in the include used by the system to fill the structure.

Example

Let's say your project requires an output for outbound deliveries that must be determined depending on the document type of the preceding document. You must add the field AUART (sales document type) to the field catalog for outputs for deliveries, then you can create a new condition table for output determination in outbound deliveries containing the sales document type.

Second, you must append communication structure KOMKBV2 (output determination communication area header application V2) and add AUART to the standard structure.

Finally, you must find the standard include reserved for user modifications where additional fields can be moved into the communication structure. When working with structure KOMKBV2, you can implement your ABAP code in include RVCOMFZZ, form USEREXIT_KOMKBV2_FILL, to populate field AUART with the sales document type of the reference sales order.

- **Without condition technique**

  Adding new fields for printing documents, copying customer master fields into sales documents, adding new fields for the billing interface, and adding new fields for the transfer to accounting are some system adaptations when it comes to creating or copying new fields. Covering all these system adaptions is beyond the scope of this book, but we'll explain the basics of system modification in the area of document printing to provide an example of these types of enhancements.

  Depending on the scope of the modification, we can distinguish between two types of modifications when printing documents: first-level modifications and second-level modifications.

  First-level modifications are simple enhancements that do not require appending new fields to the communication structure since the field to be added to the form is already included in the communication structure. In this case, the new field is assigned to an existing field in the form.

  Second-level modifications are more complex modifications where the new field is neither in the standard communication structure nor in the form. First, you must enter the new field in the relevant communication structure for additional fields. Then, if the new field doesn't match the name of the SAP field, you must add an ABAP statement like MOVE X TO Y in the user exit designed for this purpose, where X is the technical name of the SAP field in the sales and distribution document and Y is your new field. Table 6.4 summarizes the communication structures and user exits available for printing documents.

Type of Field	Communication Structure	Include Communication Structure (New Fields)	User Exit (Include)
Sales document header	VBDKA	VBDKAZ	V05DZZAK
Sales document item	VBDPA	VBDPAZ	V05DZZAP
Delivery header	VBDKL	VBDKLZ	V050ZZLK
Delivery item	VBDPL	VBDPLZ	V050ZZLP
Billing document header	VBDKR	VBDKRZ	V05NZZRK
Billing document item	VBDPR	VBDPRZ	V05NZZRP

Table 6.4 Communication Structures and User Exits for Printing Documents

- **Routines**
  Routines are usually short programs used to perform checks during document processing. SAP S/4HANA provides several standard routines, but you can create your own routines via Transaction VOFM to fulfill your customers specific requirements. Standard routines are delivered within the number range 001–599, whereas custom routines are created within the number range 600–999. This naming convention ensures that future upgrades won't overwrite your own routines. Four types of routines are available in sales and distribution:
  - **Copying requirements**
    These routines determine whether the whole source document or only specific items can be copied, based on the result of a check carried out during the copying of documents.
  - **Data transfer**
    Routines for data transfer are useful to dictate which fields are copied by default and which are only copied if certain conditions are met.
  - **Requirements**
    Requirements can be assigned in the access sequences for both pricing and output control. You can also assign these routines to pricing conditions in the pricing procedure. For example, you can control the determination of a particular pricing condition type, based on the value of the item category or any other field in the sales document.
  - **Formulas**
    Formulas are used in pricing to influence the scale base value, condition base value, condition amount, rounding rules, and group key. For example, you can use a formula to distribute a header discount among items in an order according to weight.

- **Customer exits**
  Customer exits allow you to add your own functionality to SAP's standard business applications without having to modify the original source code. SAP delivers customer exits for specific programs and screens, which just act as placeholders where you can implement bespoke logic. By using customer exists, you won't change the source code of standard programs since your enhancements are encapsulated in objects that exist separately from SAP's standard software. Also, when you use customer exits, your customer objects follow strict naming conventions, which allows the system to keep them safe and unchanged during an upgrade to a new system release. SAP organizes its exits in packages called SAP enhancements.

  To locate the exits available in the SAP system, execute Transaction CMOD (Project Management of SAP Enhancements) and choose **Utilities • SAP Enhancements**. On this screen, you can search enhancements by development class or by enhancement package. The report returns the list of enhancement packages that meet the selection criteria. Select one package and click on [icon] to display the function module exits and other enhancement objects. On this screen, for some enhancements, you can also access useful documentation by clicking on the [icon] icon.
- **Business Add-Ins (BAdIs)**
  BAdIs are mechanisms for modifying the functionality of a business function without making changes to the standard source code. BAdIs are object-oriented enhancements; thus, they are a most sophisticated option than customer exits for implementing custom code. SAP guarantees that future upgrades of the original business function won't affect these customer-specific enhancements. In SAP S/4HANA, two types of BAdIs are available: classic and new BAdIs. New BAdIs are significantly faster, are integrated into the kernel, and are switchable. Therefore, we recommend that you define only new BAdIs and migrate all existing classic BAdIs to new BAdIs using the semiautomated procedure delivered by SAP. Enhancing the automatic search for a reference document at the item level when you create a sales document is an example of a BAdI available in the area of sales and distribution.

6.6 Important Terminology

In this chapter, the following terminology was used:

- **Text ID or text type**
  This key defines different types of texts allowed for a particular text object. For example, ST is used for individual standard texts, and 0001 refers to form headers in sales documents.

- **Text determination procedure**
  This value identifies a group of text IDs that can be used, for example, in a customer master record or in a sales document header. This procedure also defines the sequence in which texts are shown in a sales and distribution object.
- **Electronic data interchange (EDI)**
  EDI is a technology for delivering data messages to a receiving system. In SAP terms, we refer to EDI when data is exchanged between an SAP and a non-SAP system. Several EDI standards are available to address the needs of specific industries.
- **Output type**
  This value specifies the kind of output to be produced. Outputs can be sent by email, by mail, by fax, electronically via EDI, or even used to call an ABAP program. In the context of sales and distribution, examples of common output types include order confirmations, shipping notifications, or invoices.
- **Application for output conditions**
  This value identifies the applications from which an output can be sent, for example, sales order processing or invoice processing.
- **Output message**
  When output determination is carried out successfully, an output message is generated for the sales and distribution object and stored in table `NAST`. An output message record can also be created manually and contains information regarding the language of the output, processing status, processing time, transmission medium, partner information, etc.
- **Output management framework**
  This framework controls the components to exchange information with internal and external partners via outputs. This output could be an email, an EDI message, or a hard copy of an invoice sent by mail. Two output management frameworks are available in sales and distribution, and you can choose between the new SAP S/4HANA output control (CA-GTF-OC) or the classic output determination procedure (SD-BF-OC). Some billing apps, however, can only use the new SAP S/4HANA output control.
- **User exits**
  User exits are an obsolete enhancement method that allows you to add your own functionalities. User exits are implemented as subroutines within standard includes. These subroutines usually start with the statement `FORM` or `PERFORM`. Changes to user exits in SAP S/4HANA Sales are modifications. Before you can add your custom logic, first, you need access to change the standard include that calls the user exit via SAP Software Change Registration (SSCR). User exits have no defined interface; that is, a lot of data is available, and a lot of program-internal data can be changed. Therefore, you must be careful when using user exits for enhancements. All SAP sources are replaced during an upgrade. As a result, Transaction SPAU displays all user exits changed by the customer, and you can adjust these user exits if you want to retain your custom enhancements.

- **Customer exits**
  Customer exits are improved versions of user exits and consist of placeholders in the standard code that do not contain any logic but instead allow you to add your own functionalities. Unlike the user exits described earlier, customer exits have a defined interface; thus, you can only transfer certain fields to the interface, and you can also only change certain fields in a customer exit. Customer exits can be switched to be made active or inactive and are implemented as function modules that call an include in the customer space (Z include). Thus, no SSCR access key is required to change them. Customer exits neither affect standard SAP source code nor software updates.
- **Copy control**
  The copy control functionality can be defined as all the configuration parameters that control which fields are copied to a newly created sales and distribution document when the new document is to be created with reference to a preceding document. With copy control, you can also specify the requirements that must be met for a specific field to be copied.

6.7 Practice Questions

These practice questions will help you evaluate your understanding of the topics covered in this chapter. The questions shown are similar in nature to those found on the certification examination. Although none of these questions will be found on the exam itself, they will allow you to review your knowledge of the subject. Select the correct answers and then check the completeness of your answers in the next section. Remember that, on the exam, you must select all correct answers and only correct answers to receive credit for the question.

1. True or False: In SAP S/4HANA, within billing, you cannot combine the output determination framework based on NAST with the new output management framework for SAP S/4HANA.

☐ A. True

☐ B. False

2. Which of the following settings are available in the copy control at the item level when copying from a billing document into a sales document? (There are two correct answers.)

☐ A. Reexplode structure/free goods

☐ B. Update document flow

☐ C. Copy quantity

☐ D. Do not copy batch

3. True or False: When configuring the copy control from a sales document to a delivery document, you can maintain copying rules at the header, item, and schedule line levels.

- ☐ A. True
- ☐ B. False

4. When creating a billing document for a debit memo request, what configuration setting will allow you to copy pricing elements unchanged and redetermines taxes?

- ☐ A. Maintain the relevant price source in the copy control at the header level
- ☐ B. Maintain the relevant pricing type in the copy control at the item level
- ☐ C. Assign a custom calculation routine to the tax condition type in the pricing procedure
- ☐ D. Maintain the relevant price source in the copy control at the item level

5. Your project requires setting up copying rules for billing documents, specifically when copying delivery documents to billing documents. Which of the following settings are available in customizing?

- ☐ A. Maintain copying rules at the header, item, and schedule line levels
- ☐ B. Maintain copying rules at the header and item levels
- ☐ C. Maintain copying rules at the header level only
- ☐ D. Maintain copying rules at the item level only

6. In a standard SAP S/4HANA system, which of the following text objects allow you to define access sequences for text determination? (There are three correct answers.)

- ☐ A. Customer central texts
- ☐ B. Pricing condition texts
- ☐ C. Sales document header texts
- ☐ D. Delivery header texts
- ☐ E. Billing document header texts

7. Let's say you've defined a specific header text type in a quotation, and you've configured the text control so the text is transferred into a follow-up sales order. Where do you maintain the settings to ensure that changes in the quotation text are automatically updated in the sales order text?

- ☐ A. Text IDs in the text determination procedure
- ☐ B. Access sequence

☐ C. Definition of text types

☐ D. Text procedure assignment

8. You've been asked to set up the system so a text ID in the customer master is automatically copied to a header text in quantity contracts for that customer. Which settings enable this? (There are two correct answers.)

☐ A. Define a text determination procedure for the customer master record

☐ B. Define a text determination procedure at the sales document header level

☐ C. Assign a text determination procedure to the relevant sales document type

☐ D. Assign a text determination procedure to the account group of the customer

9. When creating a sales order with reference to a quotation, you want the system to copy the quotation into the sales order completely and send a message if a sales order item is deleted. What must you do?

☐ A. Select the **Complete Reference** indicator in the copy control at the item level

☐ B. Select the **Complete Reference** indicator in the copy control at the header level

☐ C. Select the **Complete Reference** indicator in the copy control at the schedule line level

☐ D. Select the **Copy in full** checkbox in the document type of the quotation

10. True or False: A text ID in the text determination procedure can be set to be displayed in a popup window during copying.

☐ A. True

☐ B. False

11. What are characteristics of NAST-based output determination? (There are two correct answers.)

☐ A. You can assign ABAP requirements to each condition table in the access sequence.

☐ B. You can specify which transmission mediums are not allowed for an output type.

☐ C. You can set the **Access to conditions** indicator for each condition table in the access sequence.

☐ D. You can set the default dispatch time for an output type.

12. True or False: In output management based on NAST, a communication strategy to determine the preferred communication method can only be used with transmission medium 5 (external send).

☐ A. True

☐ B. False

13. You're working on a project that requires adapting the standard SAP system to suit your company's specific requirements. Which of the following enhancement techniques ensure upward compatibility? (There are three correct answers.)

☐ A. Customer exits

☐ B. Modification to SAP dictionary object

☐ C. Routines

☐ D. Modification to SAP source code

☐ E. BAdIs

6.8 Practice Question Answers and Explanations

1. Correct answer: **A**
   True. In SAP S/4HANA, the following three apps can only use the SAP S/4HANA output control (CA-GTF-OC) framework to send outputs:
   - Create Billing Documents
   - Manage Billing Documents
   - Schedule Billing Output
2. Correct answers: **B and D**
   This question is a kind of tricky. Answers A and C are incorrect since **Reexplode Structure/Free Goods** and **Copy Quantity** are two of the four settings that are available in the copy control between sales document items, as shown earlier in Figure 6.3. However, these settings are not available when copying billing document items into sales document items, which is the scenario described in the question. This question highlights the fact that copy control settings may vary depending on the nature of the source and target documents.
3. Correct answer: **B**
   False. This question is a straightforward knowledge question. In this case, you cannot maintain copying rules at the schedule line level since deliveries don't have schedule line information.
4. Correct answer: **B**
   This specific question concerns settings on the customizing screen for the copy control from sales documents to billing documents at the item level. The **Pricing Type** parameter, shown earlier in Figure 6.6, controls how pricing data is

treated when copying documents. If value **G** is selected, the system copies price elements unchanged and redetermines taxes.

5. Correct answer: **B**

   When copying delivery documents to billing documents, you cannot make settings at the schedule line level because neither deliveries nor billing documents have schedule line information. Therefore, answer A is incorrect. You can maintain copying rules both at the header and item levels, as shown in the dialog structure shown earlier in Figure 6.5 and Figure 6.6.

6. Correct answers: **C, D, and E**

   Answer A is incorrect since you cannot assign access sequences to text IDs in the text procedure for the customer master record. Similarly, answer B is also incorrect because, in SAP S/4HANA, you cannot assign access sequences for pricing condition texts. Therefore, the remaining answers are the three correct answers to the question.

7. Correct answer: **A**

   If the **Refer/Duplicate** field in the text determination procedure, shown earlier in Figure 6.9, is blank for a particular text ID, the text is referenced. As a result, only one version of the text exists on the database. Thus, if you change the original text, the change appears in all sales and distribution documents that reference it. Therefore, answer A is the only correct answer.

8. Correct answers: **B and C**

   To set up text determination for sales documents at the header level, you'll need to define a text determination procedure for text object VBBK (sales header texts) then define an access sequence for the concerned text IDs. Finally, you'll assign the text determination procedure to the corresponding sales document type. A text determination procedure for the customer master record and assigning the procedure to the customer account group do not affect text determination in sales documents, such as quantity contracts.

9. Correct answer: **B**

   The **Complete Reference** indicator in the copy control at the header level allows you to copy the source document (quotation) into the target document (sales order) completely. If an item is deleted after the target document has been created, the system sends a message. This indicator is neither available in the copy control at the item level nor at the schedule line level; thus, answers A and C are incorrect. The setting described in answer D does not even exist in SAP S/4HANA.

10. Correct answer: **A**

    True. This scenario can be achieved by selecting the values **Y** or **Z** in the **Text is obligat.** field, which is available in the text determination procedure for each text ID, as shown earlier in Figure 6.9.

11. Correct answers: **A and D**

    This specific question concerns settings in output determination using the condition technique. When defining an output type, you can specify which dispatch times are not allowed. You cannot specify which transmission mediums are not allowed; thus, answer B is incorrect. Answer C is also incorrect since the **Access to conditions** indicator can be set for the output type and not for the condition table in the access sequence. Thus, only A and D remain as the two correct answers to the question.

12. Correct answer: **A**

    True. The function communication strategy is used to determine the preferred communication method to contact the customer and is only available with transmission medium 5 (external send).

13. Correct answers: **A, C, and E**

    Any kind of modification to SAP standard objects, whether source code or dictionary objects, should be avoided because, during updates, older SAP objects are overwritten by the delivered objects. In summary, the update process overwrites all modifications of SAP objects. Therefore, answers B and D are incorrect, leaving you with the only three correct answers.

6.9 Takeaway

In this chapter, we discussed how to set up copy control within sales and distribution. You learned how to control which fields should be copied from the source document to the target document, at the different levels where the information is stored. We then explored the rules used to perform text determination to make a text flow from the customer master record to the sales and distribution document or from a reference document to another sales and distribution document.

Next, we covered the output determination component used for output control. We reviewed the customizing steps to adjust the output determination and output types, which allows the system to automatically propose outputs for sales and distribution documents. This fundamental function helps your company exchange information with internal and external partners.

Finally, we discussed several common enhancement techniques available in the area of sales and distribution.

6.10 Summary

Knowing the settings available with regard to cross-application customizing functions is critical for the success of any SAP S/4HANA implementation project. This knowledge enables you to map requirements in the area of sales and distributions and configure special functions like text control, copy control, and output control.

You now also understand the differences between the various system modification options used in sales.

In the next chapter, you'll learn about the material availability check and the transfer of requirements in sales documents.

Chapter 7
Availability Check and Advanced ATP

Techniques You'll Master

- Set up the transfer of requirements in sales and distribution
- Maintain settings for availability checks through checking groups
- Configure the scope of the availability check
- Configure default settings by sales area and configure the system's response if the availability check shows a shortage
- Execute an available-to-promise (ATP) check and explain the calculation of available quantities
- Describe backorder processing in SAP S/4HANA for advanced ATP
- Understand plant substitution with alternative-based confirmation in advanced ATP

The focus of this chapter is on explaining the availability check in sales and distribution processing, also known as the *product availability check*. We'll discuss the main configuration steps and functionality of the ATP check, the transfer of requirements to materials planning, and some of the key features of SAP S/4HANA for advanced ATP, which is a new system functionality available since SAP S/4HANA 1809.

Real-World Scenario

When working on a project for a customer, let's say you're asked to explain, to managers and key users in the sales department, how SAP S/4HANA calculates the quantity available and confirms the delivery of goods against required delivery dates and quantities. For this task, you must understand the difference between concepts like on-hand inventory and ATP stock, as well as understand all the factors that can influence the confirmed date proposed by the system during the creation of a sales order.

A common business requirement is to set up the system so that the availability check considers different system components depending on the material. For example, for material A, perhaps only on-hand inventory must be used whereas for material B other incoming stock, like purchase orders (POs), must be taken into account. The scope of the availability check allows you to define which parameters must be considered or ignored by the system during an ATP check, based on the availability checking group assigned to the material. Manufacturing companies may require special rules for make-to-order materials; for example, you may be asked to ensure that only released production orders are included in the availability check. To provide good service to your customers, you should be able to maintain the relevant settings so that the ATP check behaves as expected in all cases and the correct quantities and dates are confirmed by the system and communicated to customers. This confirmation capability is of great importance to allow your business to plan for demand accordingly, with reliable information.

7.1 Objectives of This Portion of the Test

The goal of this portion of the SAP S/4HANA Sales Certification exam is to test your understanding of the following topics:

- Backward and forward scheduling
- Availability overview screen and calculation of available quantities
- Configuration of transfer of requirements
- Configuration of the scope of availability check
- Configuration of default settings by sales area

- Delivery agreements in customer master records
- Using fixed quantities and dates in sales documents
- Advanced ATP features in SAP S/4HANA

Note

The Availability Check and Advanced ATP topic makes up less than 8% of the total SAP S/4HANA Sales 2020 exam.

The content of this chapter is included in the Sales topic of the SAP S/4HANA Sales 2020 Upskilling exam, which makes up more than 12% of the total certification exam.

7.2 Product Availability Check

The product availability check refers to what was previously known as the ATP check in legacy SAP ERP systems and is now part of SAP S/4HANA for advanced ATP (sometimes referred to as aATP). SAP S/4HANA for advanced ATP contains six major processes:

- Product availability checks
- Backorder processing
- Controller and central functions
- Product allocations
- Supply protection
- Alternative-based confirmations

In this section, we'll review the basic configuration and functionality of the product availability check process.

First, you must understand that an ATP check is not just a check of your current on-hand inventory, but a complex check that can be performed for any future date. ATP checks involve many other variables, including the current warehouse stock and any planned movements into and out of the warehouse. In addition to unrestricted stock, and depending on how the capability is configured, ATP can also take into account safety stock, stock in transfer between plants, stock in quality inspection, restricted-use stock, and blocked stock. Table 7.1 shows the different requirements and supply elements that can be considered by the availability check.

Future Supply (+)	Requirements (–)
POs	Dependent requirements
Purchase requisitions	Reservations
Shipping notifications	Sales requirements

Table 7.1 Planned Elements in ATP Checks

Future Supply (+)	Requirements (–)
Planned orders	Deliveries
Production orders	Dependent reservations
	Stock transport orders

Table 7.1 Planned Elements in ATP Checks (Cont.)

An availability check can be carried out at the plant level or, if material requirements planning (MRP) areas have been defined, at the MRP area level. In any case, a product availability check is always executed for a single material.

If your business produces make-to-order goods or uses consignment stock, the ATP check must be performed against the special stock E for make-to-order or special stock W for customer consignment stock. This function is controlled by the checking rule, which we'll discuss later in Section 7.2.

You can also set up the ATP check so that replenishment lead times are factored in. The replenishment lead time is the time needed to either produce or procure the requested stock. For a given material, the replenishment lead time is specified in the material master record. By definition, any (reasonable) quantity requested can be confirmed beyond the end of the lead time. Figure 7.1 shows how the available quantity can change over time in a simplified scenario where packing, picking, and transportation times are not taken into account. In this example, the current day is marked as day 1, and the customer has requested 170 units of a particular material to be delivered on day 5. The replenishment lead time for this hypothetical material is 10 days.

In this case, the system cannot confirm the quantity requested on day 5 because the quantity available on day 5 is below 170 units. However, 170 units are confirmed on day 6 because a planned receipt of 260 units will take place on day 6, which increases the quantity available to 290 units on day 6. From day 10 onwards, any quantity can be confirmed since that is the end of the replenishment lead time.

An important task is to understand the role of scheduling in product availability checks because, to confirm the requirement, a material may need to be available days before the requested delivery date to allow time for transportation planning, picking/packing, loading, and transporting to the destination location. The following business functions must be carried out before a delivery date can be met:

- Transportation planning: The time required to arrange for transportation (e.g., obtain space on an aircraft or truck).
- Picking and packing the materials: The time that takes to find the items in the warehouse, pack them, and move them to the loading bay.
- Procuring or producing materials: If materials are not available, this time is the time needed to manufacture or obtain the required materials from the vendor.

- Loading the materials: The time required to place the goods into a container, truck, etc.
- Transporting the materials to the customer: The time required to transport the materials to the destination location.

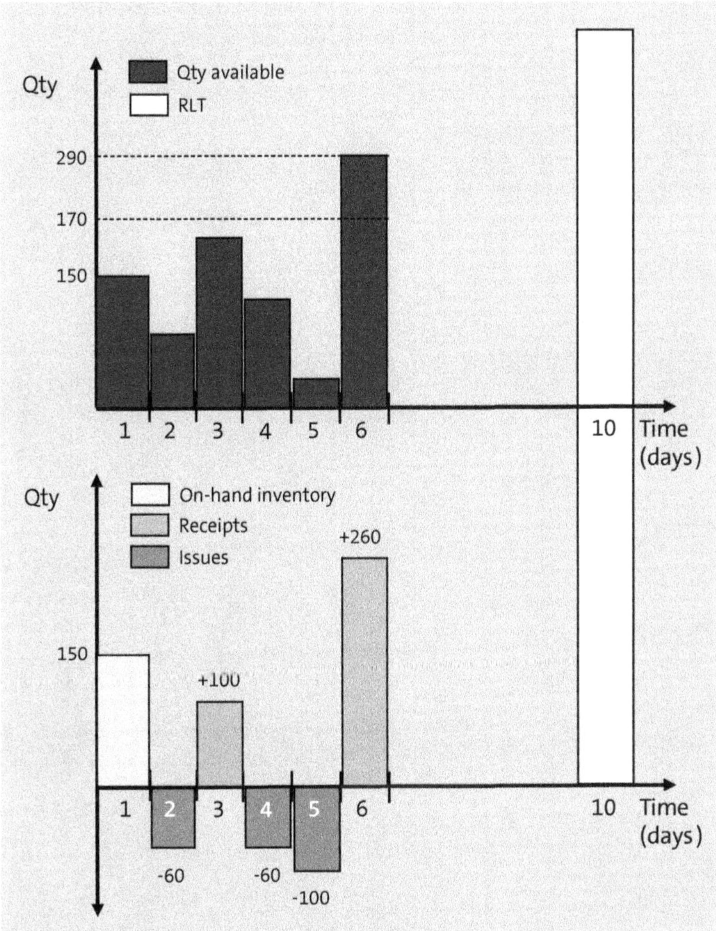

Figure 7.1 Available Quantity versus Time

The following two scheduling scenarios are supported by the ATP check functionality:

- **Backward scheduling**
  The system first carries out backward scheduling to determine if the requested delivery date is achievable. If the date can be met, then the system confirms the requirement for the requested delivery date, and no further analysis is needed.
- **Forward scheduling**
  If backward scheduling returns a transportation planning start date, procurement start date, or material availability date in the past, the customer requested

delivery date is not achievable. Therefore, the system must perform forward scheduling to calculate the date when the quantity requested can be delivered.

Note

Forward scheduling is only carried out if the delivery date requested by the customer is *not* achievable.

Figure 7.2 shows the backward scheduling process in a scenario where the requested delivery date cannot be met since the start of the procurement date is in the past.

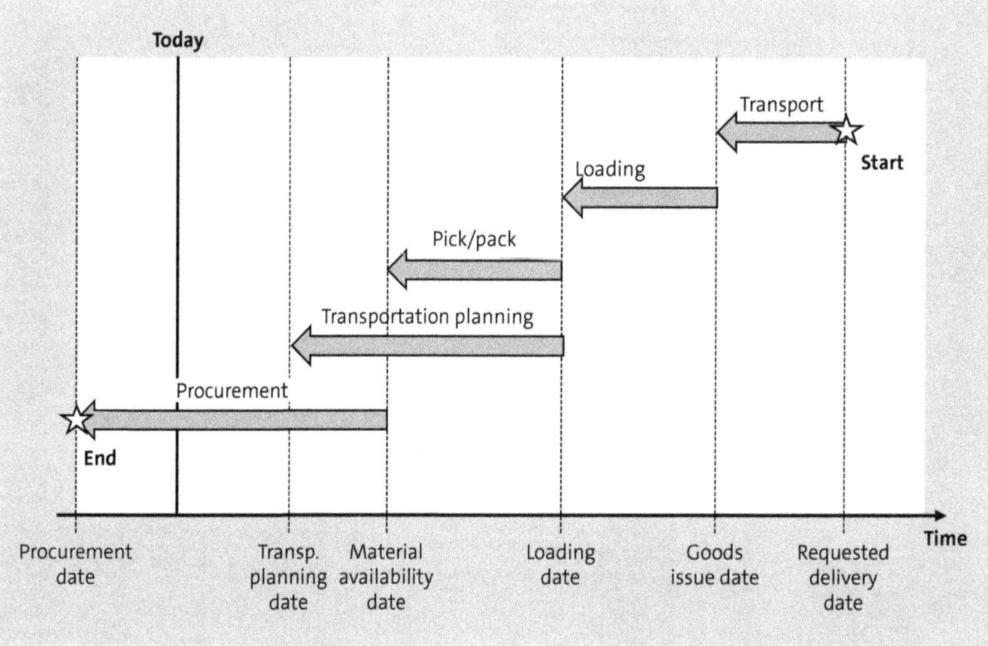

Figure 7.2 Backward Scheduling

When this event happens, the system performs forward scheduling to determine the confirmed delivery date, as shown in Figure 7.3.

The **Schedule lines** tab of the sales document provides the result of the ATP check. If backward scheduling determines that the requested delivery date cannot be met, the confirmed quantity is set to zero on the original schedule line. Then, forward scheduling creates a new line for the confirmed date where the quantity requested is zero, but the confirmed quantity matches the quantity ordered. This scenario is shown in Figure 7.4. In this example, the customer wants 2 units on 11/10/2020. However, this date is not achievable, and the system can only confirm 2 units on 11/12/2020.

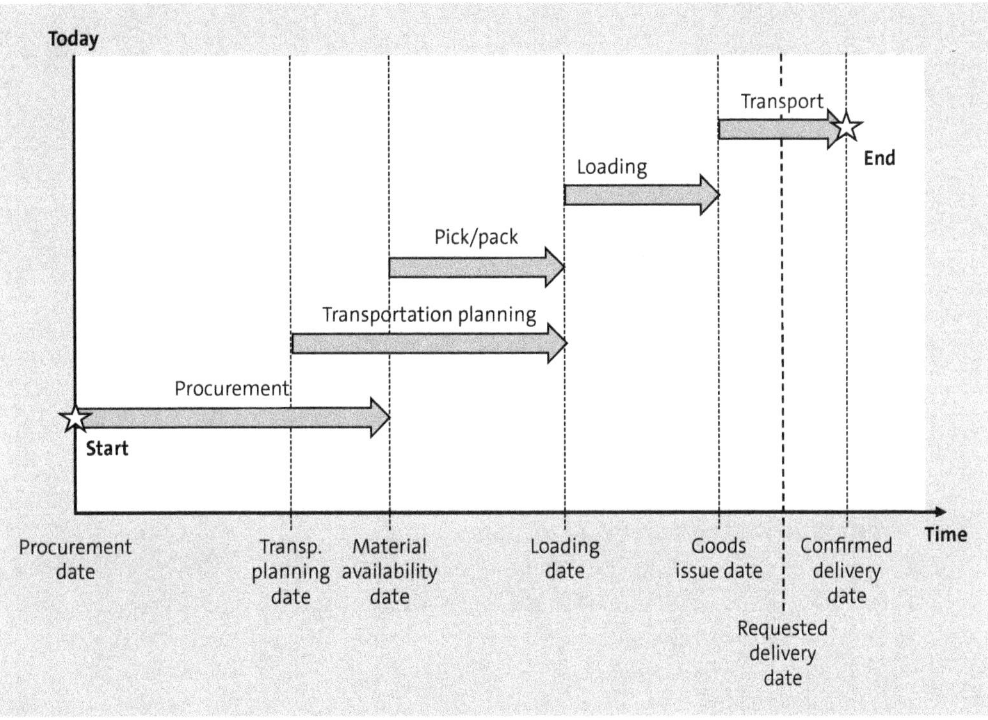

Figure 7.3 Forward Scheduling

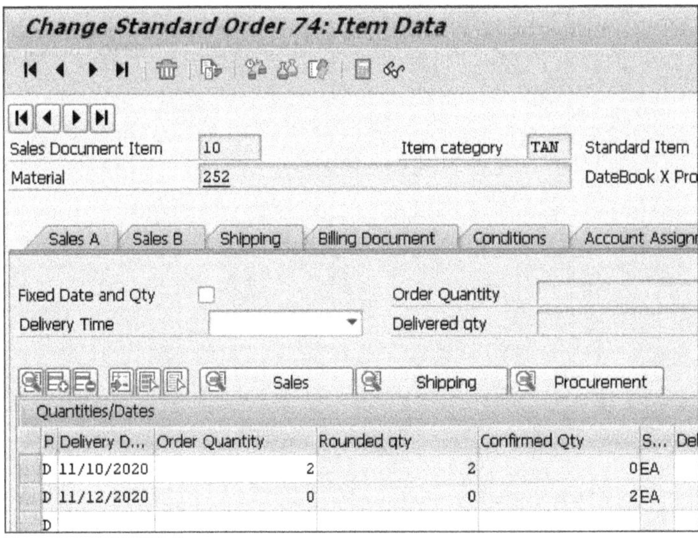

Figure 7.4 Sales Order Item: Schedule Lines Tab

You can perform an availability check manually in a sales document at any point in time. This feature is particularly useful when stock levels or planned requirements have changed since the last check. However, the system will automatically perform a new availability check in the following situations:

- Change of plant
- Change of quantity
- Change to the requested delivery date or material availability date
- Change of storage location or batch
- Change of route or shipping point
- Change to the fixed data
- Removal of reason of rejection
- Removal of blocks that affect requirements

In some cases, a customer may order a quantity that cannot be fully delivered on the requested date. Customer preferences vary from one customer to another. Some customers would like all the items to be delivered at a later date, while others might ask you to allow a certain number of partial deliveries for each item. You can use the customer record to define the way in which goods should be delivered when the requested date for all the items cannot be met. As shown in Figure 7.5, the **Complete Delivery** indicator ❶ is copied to the sales document header from the customer master record and would ensure the sales order is delivered completely in a single delivery. Note also the **Max.Part.Deliveries** (maximum partial deliveries) field ❷ and the **Part.dlv./item** (partial delivery) indicator ❸. These two fields dictate the number of partial deliveries for each item and are copied into the **Shipping** view of the sales document item.

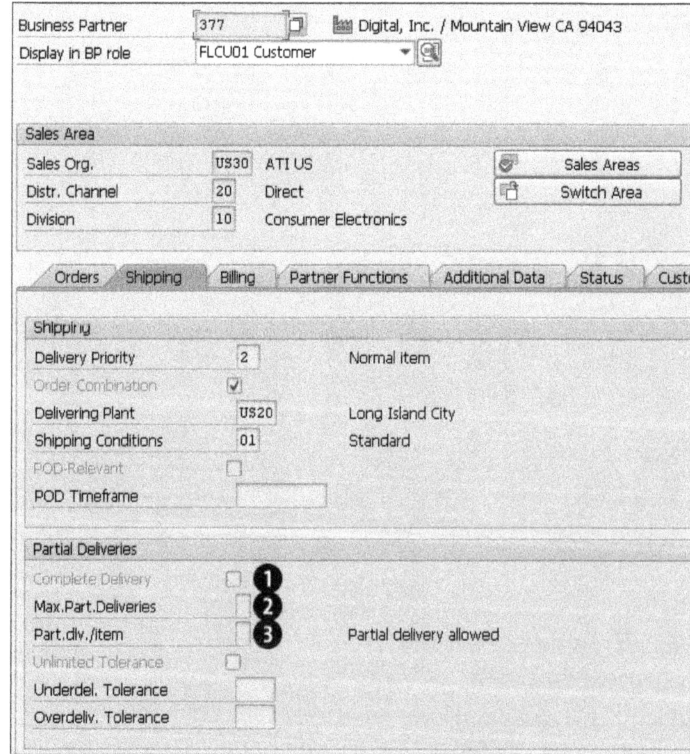

Figure 7.5 Delivery Agreements in the Customer Master Record

You can select any of the following five values for the **Part.dlv./item** indicator:

- <blank>: Partial delivery allowed
- **A**: Create a delivery with a quantity greater than zero
- **B**: Create only one delivery (also with quantity zero)
- **C**: Only complete delivery allowed
- **D**: No limit to subsequent deliveries

Tip

If the **Complete Delivery** indicator is not selected in a sales document, partial deliveries may be allowed depending on the number of items, dates, material availability, etc. However, this scenario could potentially lead to a high number of deliveries, thus incurring unwanted transportation costs. However, you can ensure as few deliveries as possible by selecting value C (only complete delivery allowed) in the **Part.dlv./item** indicator for all items while leaving the **Complete Delivery** indicator unselected at the header level. As a result, the system won't create more than one delivery per item.

Another interesting feature of the availability check is the ability to fix quantities and dates in a sales document item. This feature is useful when an item cannot be delivered fully on the requested date, but the customer accepts the schedule line proposal. As shown in Figure 7.6, the **Fix Qty/Date** (fixed quantity and date) indicator is displayed both on the **Schedule lines** tab and on the availability control screen. Once this indicator is selected, the system transfers the requirements to MRP, and even if later the requested quantity becomes available before or on the original requested date, the rescheduling job won't change the confirmed schedule lines on the item. The **Fix Qty/Date** indicator must only be selected if your customer will accept the delivery dates and quantities proposed by the system. SAP S/4HANA will consider a fixed item completed if the confirmed quantity has been fully delivered. The confirmed quantity could be smaller than the quantity originally ordered by the customer, so you must be sure the customer understands and agrees to this scenario before fixing the date and quantity of an item.

Note

In third-party items (item categories TAS or CB1, for example), the meaning of the **Fix Date and Qty** indicator changes slightly. In this case, the indicator determines which material availability date is transferred to the purchase requisition and subsequently to the purchase order item. If the indicator is set, the material availability date from the confirmed schedule line is transferred to the purchase requisition. However, if the indicator is not set, the material availability date from the requested schedule line is transferred to the purchase requisition. For further information on the role of the **Fix Date and Qty** indicator in connection with third-party or individual purchase order items, you can refer to SAP Note 210455.

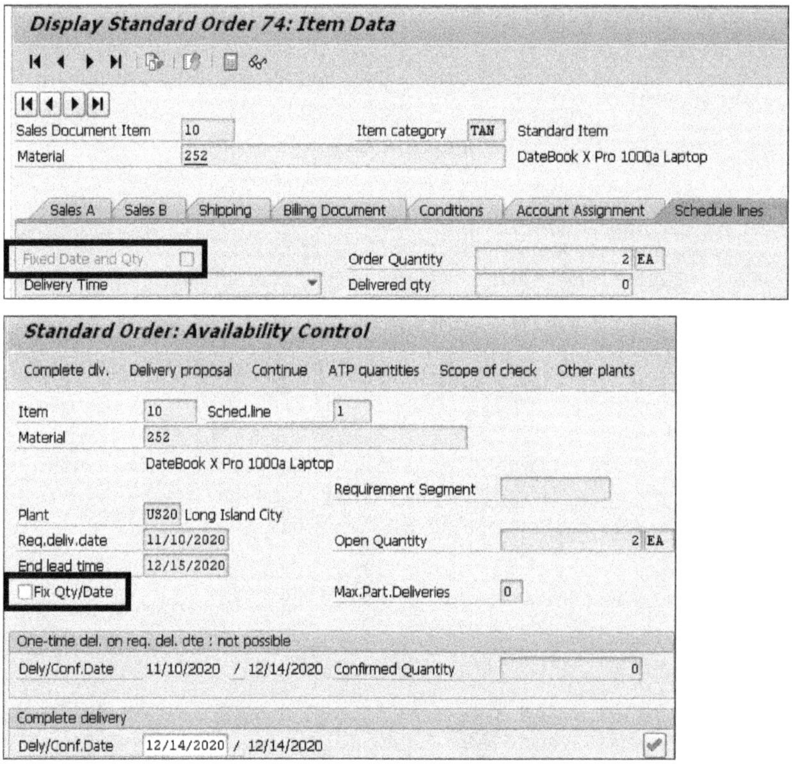

Figure 7.6 Fix Date and Quantity Indicator

For your system to confirm requirements on time, the quantity requested must be available on the material availability date. The method to calculate available quantities is quite simple: The system starts with the current stock, adds any future receipts, and subtracts any future requirements on their corresponding dates. This calculation is performed for every day up to the material availability date. If the cumulated available quantity is greater than or equal to the quantity requested on the material availability date, the system confirms full delivery on the date requested by the customer. Otherwise, the system will make the following three proposals to deliver the goods:

- **One-time del. on req. del. dte (one-time delivery on the requested delivery date)**
  Only the quantity available on the material availability date is delivered. No further deliveries will take place if the quantity available is less than the quantity requested.
- **Complete delivery (complete delivery in one shipment)**
  The system uses forward scheduling to calculate the material availability date when the full quantity requested will be available, which becomes the baseline for suggesting a complete delivery date.
- **Dely proposal (delivery proposal)**
  Multiple dates and quantities may be proposed so that goods are shipped as

soon as they become available. Regardless of partial deliveries allowed, the system will propose multiple dates on the availability control screen. However, if the **Complete delivery** indicator is selected on the order, the **Schedule lines** tab will show one schedule confirmed for complete delivery even if the multiple delivery proposal option is chosen.

Let's now analyze the availability control, schedule lines, and availability overview screens with the following example:

- Imagine today's date is 12/15/2020, and a customer requests 50 units of material *TFT LCD Colour Display, 4.3in, 480x272* to be delivered on 12/17/2020. To not complicate things, let's assume the warehouse team is extremely quick processing deliveries, and the customer is located "down the road." Thus, the transportation planning time, pick/pack/loading time, and transportation time are each zero days.
- The current inventory on hand is 26 units. A confirmed requirement exists for 10 units for another order on 12/16/2020, and a PO for 20 units has a planned goods receipt date of 12/18/2020.
- The end of the lead time for this material is 12/21/2020.

Based on this scenario, the system can only confirm 16 units on the requested delivery date (12/17/2020). Alternatively, the system can confirm the 50 units on 12/21/2020, which is the end of the lead time, that is, the date when any quantity can (in theory) be confirmed. A third and final option is to make three partial deliveries and start shipments as soon as the cumulated available quantity is greater than zero. The availability control screen, shown in Figure 7.7, displays these three delivery proposals.

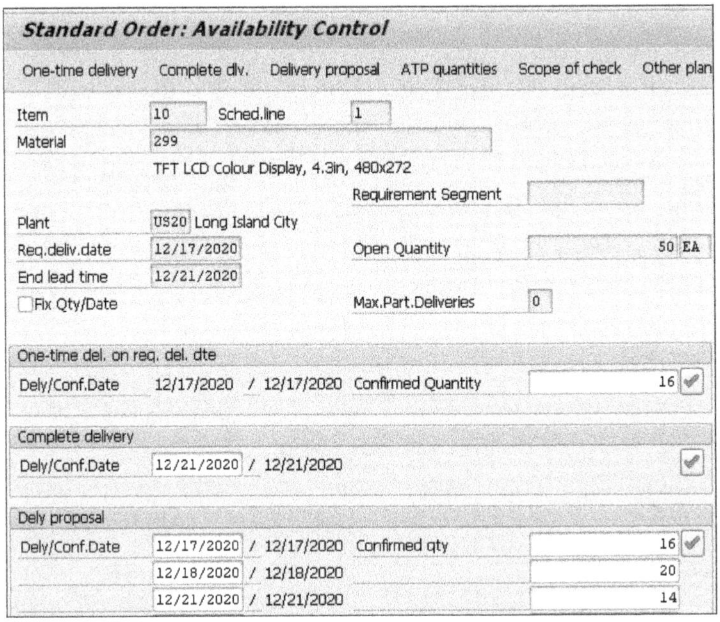

Standard Order: Availability Control

One-time delivery Complete dlv. Delivery proposal ATP quantities Scope of check Other plan

Item 10 Sched.line 1
Material 299
TFT LCD Colour Display, 4.3in, 480x272
Requirement Segment
Plant US20 Long Island City
Req.deliv.date 12/17/2020 Open Quantity 50 EA
End lead time 12/21/2020
Fix Qty/Date Max.Part.Deliveries 0

One-time del. on req. del. dte
Dely/Conf.Date 12/17/2020 / 12/17/2020 Confirmed Quantity 16

Complete delivery
Dely/Conf.Date 12/21/2020 / 12/21/2020

Dely proposal
Dely/Conf.Date 12/17/2020 / 12/17/2020 Confirmed qty 16
12/18/2020 / 12/18/2020 20
12/21/2020 / 12/21/2020 14

Figure 7.7 Availability Control Screen

If your customer wants the goods to be delivered as soon as possible in multiple shipments, you must enter multiple delivery proposals in the **Dely proposal** section, shown at the bottom of Figure 7.7. In this case, the **Schedule lines** tab for the item would have three schedule lines for three partial deliveries, as shown in Figure 7.8.

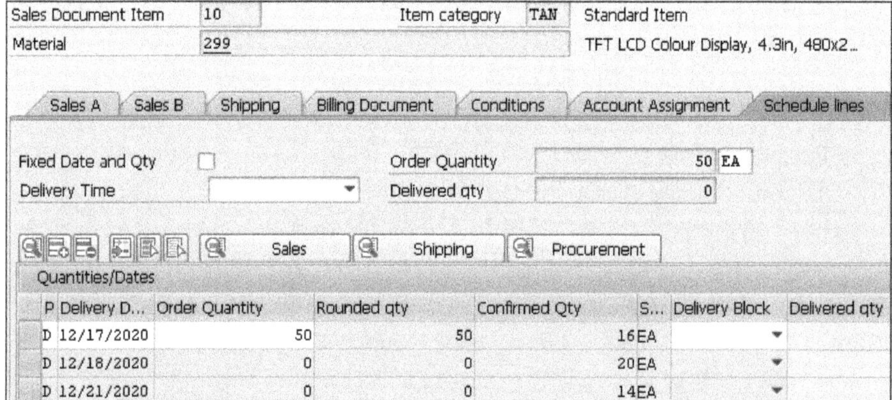

Figure 7.8 Schedule Lines

The goods would be delivered in the following way:

1. 16 units would be delivered on 12/17/2020 (original requested date). A confirmed order of 10 units to be delivered on 12/16/2020 is subtracted from the current on-hand inventory (26 units), resulting in 16 units available on 12/17/2020.
2. 20 units would be delivered on 12/18/2020. These 20 units are expected to be received through the PO on 12/18/2020.
3. 14 units (the remaining quantity) would be delivered on 12/21/2020, when the replenishment lead time ends. The system assumes that MRP will generate a purchase requisition to ensure the material in question is available by that date.

Finally, to understand this example even better, Figure 7.9 shows the availability overview screen where you can see how the cumulated available quantity or the ATP quantity evolves over time and the different MRP elements that impact the ATP quantity over time. You can also see the end of the lead time date, as well as the total receipts (46 units including 26 units of current stock on hand) and the total confirmed goods issues (10 units).

The different MRP elements shown in Figure 7.9 include the following:

- **Stock**: Plant stock
- **CusOrd**: Customer order
- **SimReq**: Simulated requirement from availability check (in this case, the current order being created)

- **POitem**: PO item
- **SLocSt**: Storage location stock

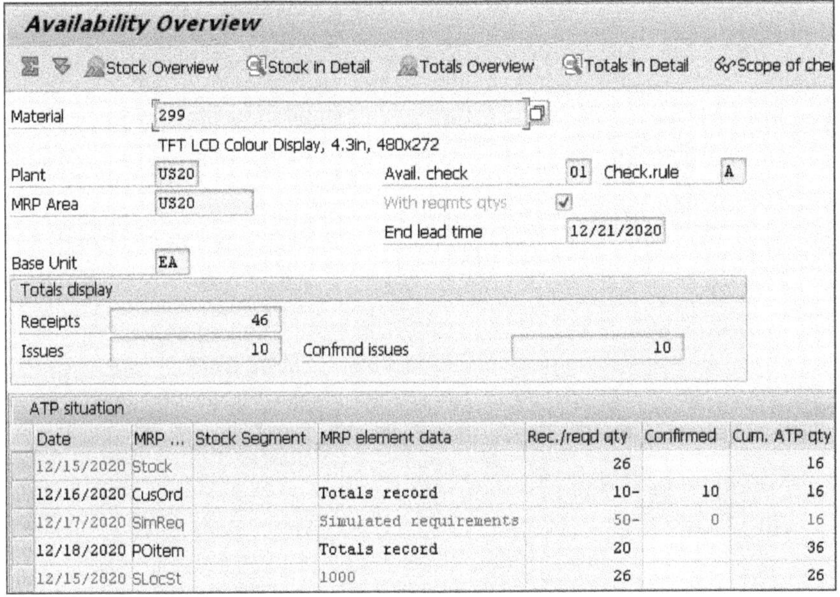

Figure 7.9 Availability Overview Screen

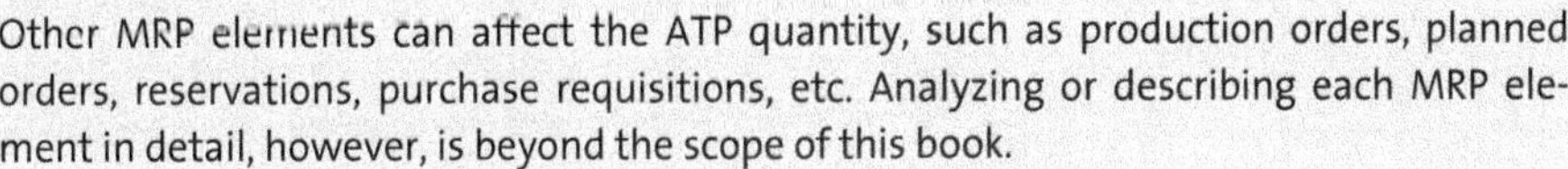

Note

Other MRP elements can affect the ATP quantity, such as production orders, planned orders, reservations, purchase requisitions, etc. Analyzing or describing each MRP element in detail, however, is beyond the scope of this book.

7.3 Transfer of Requirements

The main objective of the transfer of requirements is to ensure that goods are available in time for delivery. Information from the schedule line is transferred to MRP, which determines if enough quantity is available for the requested date. If necessary, MRP will create either production orders or purchase requisitions to fulfill the expected demand.

You can set up the transfer of requirements for both individual and collective requirements. Individual requirements transfer demand at the schedule line level, whereas with collective requirements the system consolidates requirements into groups and transfers these groups to MRP either daily or weekly. This functionality is controlled by **Checking Group for Availability Check** field in the material master record, in the **Sales: General/Plant** view, as shown in Figure 7.10.

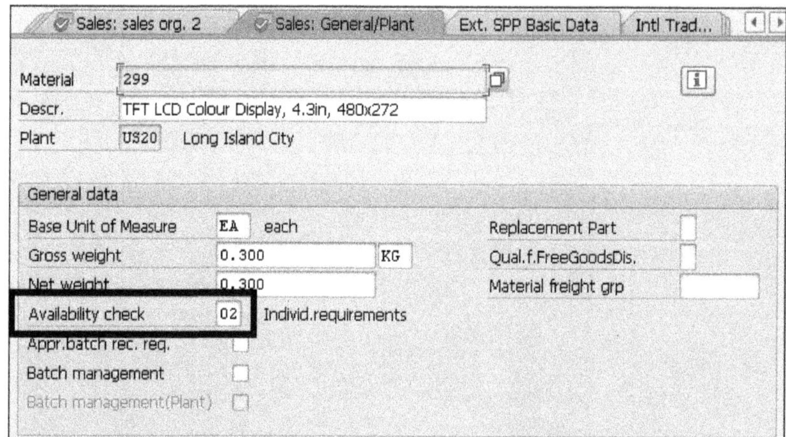

Figure 7.10 Checking Group for Availability Check in the Material Master Record

The transfer of requirements depends on the following customizing elements:

- Requirements class
- Requirements type
- Checking group
- Schedule line category

A *requirements class* controls the transfer of requirements for all sales document types, but if needed, fine-tuning is possible at the schedule line level. This element determines how the stock available is consumed as well as the replenishment strategy. A requirements class is based on another element called a *requirements type*. We'll explain how to configure both requirements classes and requirements types in this section. To access the customizing activities related to the transfer of requirements, follow the menu path **SAP Customizing Implementation Guide • Sales and Distribution • Basic Functions • Availability Check and Transfer of Requirements • Transfer of Requirements.**

Before we dive into the configuration steps, note that a transfer of requirements cannot be carried out unless the following criteria are met:

- The transfer of requirements must be switched on at the requirements class level.
- A plant must be entered in the sales document item.
- The transfer of requirements must be switched on at the schedule line category level.
- You must define a checking group and assign it in the material master record.

Tip

Prior to creating a new requirements class, check if any of the requirements classes available in the standard version meet your customer's demands.

As an example, the standard requirements class 030 is shown in Figure 7.11.

Reqmts class 030 Sale from stock

Requirements	
Avail. Check	✓ ❶
Req. transfer	✓ ❷
Allocation ind.	❸
Prod.allocation	
Ind.req.reductn	✓ ❹
No MRP	1 ❺

Configuration	
Configuration	
Cons.of config.	

Costing	
Costing	
Costing ID	
Costing Method	
Costing Variant	
Costing Sheet	
Copy cstg sheet	
CndTypLineItems	
CondTypLinItFix	

Assembly	
Assembly type	
Order costing	
Automatic plnng	
Special Stock	
Order Type	
Avail.components	
Type comp.check	
Online assembly	
Capacity check	
No update	
OCM	

Account assignment	
Acct Assgmt Cat	
Valuation	
W/o Val. Strat.	
Settlmt Profile	
Strategy Seq.	
Changeable	0
RA Key	
Consumption	
Functional Area	

Figure 7.11 Requirements Class 030

We won't go into the detail about each indicator, but let's briefly explain some main parameters you must understand at the requirements class level:

❶ **Avail. Check**: If selected, the order requires an availability check.

❷ **Req. transfer**: If this indicator is selected, the transfer of requirements is carried out for the order. You can set the indicator for the requirements class and decide whether or not the transfer of requirements is required at the schedule line level. However, if the indicator is not set for the requirements class, the transfer of requirements to MRP is not carried out regardless of the settings made at the schedule line level.

❸ **Allocation ind.**: This indicator defines the consumption of customer requirements with planned independent requirements. The following values are available:

- <blank>: No consumption with customer requirements
- **1**: Consume planning with assembly
- **2**: Consume planning without assembly
- **3**: Consume planning material (without assembly)

❹ **Ind.req.reductn**: The requirements reduction indicator controls if the order is considered part of independent requirements planning in a make-to-stock scenario.

❺ **No MRP**: This indicator specifies whether customer requirements are included in the planning run. Three options are available:

- <blank>: Requirements planned and displayed
- **1**: Requirements not planned but displayed
- **2**: Requirements neither planned nor displayed

The next step is to define the requirements types. A requirements type can only be assigned to one requirements class, but a requirements class can be assigned to several requirements types. Figure 7.12 displays requirements types in the first column from the left, and the corresponding requirements class.

RqTy	Requirements type	ReqCl	Description
011	Delivery requirement	011	Delivery requirement
021	Unchecked order/delivery	021	Unchecked order/dlv
031	Order requirement	031	Order requirements
041	Order/delivery requirement	041	Order/delivery reqmt
110	Dely of mat. to subcon/reserv.	110	Delivery w/o MRP
BSF	Gross planned indep. reqmts	102	Gross reqmts plnning
ELVV	Make-to-ord.variant + plngMat.	061	Mk->O. MatVar.PlgMat

Figure 7.12 Configuring Requirements Types

Once your requirements types have been defined, you must configure the assignment of requirements types to the respective sales transaction. In this customizing activity, you'll define the search strategy for determining the requirements type based on the value set in the **Q** field, which contains the origin of the requirements type, as shown in Figure 7.13. In this screenshot, the first column starting from the left is the item category, the second column is the MRP type, and the third column is the requirements type. Then, you have the origin of requirements type (**Q**).

You must select one of the following three values for the **Q** column:

- **0** or <blank>: The system first tries to get the requirements type from the material master. For this process, the strategy group is read for the relevant material and plant, and then the system determines the requirements types from the strategy assigned to the strategy group. If no strategy group is maintained for the material, the system uses the MRP group. If this step also fails, the system tries using the material type. Finally, if none of the previous attempts were successful, the system uses the sales document item category and the material MRP type, according to the configuration table shown in Figure 7.13. If no specific entry exists for the MRP type in the material, the system uses the entry where the MRP type is <blank>.

- **1**: This search strategy uses the item category and the material MRP type.
- **2**: Same as search strategy 1 with an additional check of the allowed requirements type.

Assignment of Requirement Types to Transaction

ItCa	Typ	RqTy	Q	Requirements type description
TAN		011		Delivery requirement
TAN	ND	011		Delivery requirement
TAN	P1	041		Order/delivery requirement
TAN	P2	041		Order/delivery requirement
TAN	VB	011		Delivery requirement
TAN	VM	011		Delivery requirement
TAN	VV	011		Delivery requirement

Figure 7.13 Determining the Requirements Type by Item Category and MRP Type

Note

Typically, field **Q** has value 0 (or blank), but in some scenarios, such as a goods issue from customer consignment stock, you may want the system to ignore the requirements type derived from the strategy group in the material master since doing so may trigger an availability check against unrestricted stock, which won't work for consignment processing. However, by selecting an alternative search strategy in the **Q** field (like 1 or 2), and based on the consignment issue item category, you can determine a requirements type that triggers an availability check against customer consignment stock (special stock W).

The last key step in setting up the transfer of requirements is to define whether or not an availability check and/or transfer of requirements must be carried based on the schedule line of the sales document. As mentioned earlier in this section, you can only activate an option at the schedule line level if this option is already activated at the requirements class level. Figure 7.14 shows the fine-tuning of the availability check and transfer of requirements at the schedule line level, where the **AvC** and **Rq** fields correspond to the fields indicated by ❶ and ❷ shown earlier in Figure 7.11, respectively. The **SLCa** column contains the schedule line category code.

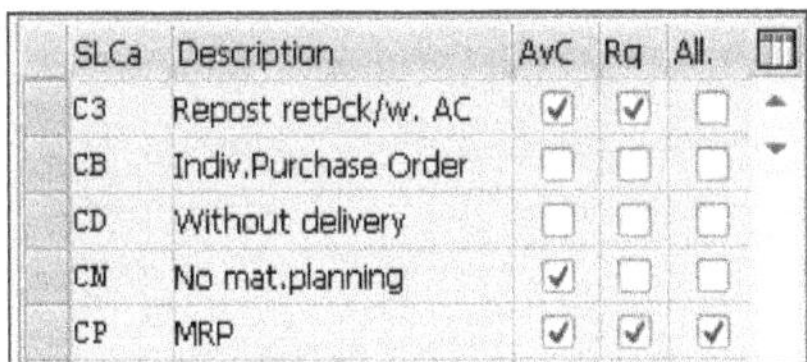

SLCa	Description	AvC	Rq	All.
C3	Repost retPck/w. AC	☑	☑	☐
CB	Indiv.Purchase Order	☐	☐	☐
CD	Without delivery	☐	☐	☐
CN	No mat.planning	☑	☐	☐
CP	MRP	☑	☑	☑

Figure 7.14 Defining Procedures for Each Schedule Line Category

7.4 Availability Check with Available-to-Promise Logic

The customizing activities related to availability checks with ATP logic (or to product availability checks) can be found at the menu path **SAP Customizing Implementation Guide • Sales and Distribution • Basic Functions • Availability Check and**

Transfer of Requirements • Availability Check • Availability Check with ATP Logic or Against Planning.

The first step is to define the checking group for availability check, which, alongside the checking rule, controls the scope of the availability check.

In previous versions of SAP, the checking group could be used to define whether the system should record individual or collective requirements in sales and shipping documents. However, since SAP S/4HANA was released, the use of collective requirements is no longer supported, and only individual requirements are used. This change makes sense because, with individual requirements, each demand is transferred immediately to MRP, which provides visibility into the order and delivery items in both the availability check and stock requirements list screens. Individual requirements also allow for backorder processing. Previously, the only disadvantage of using individual requirements is its impact on system performance, but with the introduction of the SAP HANA database, this impact is no longer a concern—hence the decision to only support individual requirements going forward. As a consequence, when you define checking groups in SAP S/4HANA, the three fields in the box shown in Figure 7.15 are displayed for reference and cannot be edited.

Av	Description	TotalSales	TotDlvReqs	Block QtRq	No PAC	Accumul.	RelChkPlan	Advanced ATP
01	Daily requirements	A	A	☑	☐			A Active
02	Individ.requirements	A	A	☑	☐			A Active
CH	Batches	A	A	☐	☐			Inactive
DR				☐	☐			Inactive
KP	No check			☐	☑			Inactive
Z1	Daily-Advcd ATP Accu	A	A	☑	☐	1		A Active
Z2	Daily-No Advancd ATP	A	A	☑	☐	1		Inactive

Figure 7.15 Defining the Availability Checking Group

The **TotalSales** and **TotDlvReqs** fields define whether the system must create individual or collective requirements for sales documents and deliveries, respectively. Value **A** means individual requirements (single records), and this value cannot be changed. The **Block QtRq** (material block in availability check with transfer of quantities) field is also not editable and, if selected, the material is blocked only during the availability check. The quantity reserved is recorded in a blocking table, and if another user carries out an availability check for the same material in the same plant, the system takes the blocked quantity into account, thus providing more accurate information about the quantity available. If you create a new checking group, the **Block QtRq** indicator is selected by default, and you cannot change it.

Four other editable fields are shown in Figure 7.15:

- The **No PAC** indicator is used to disable the standard product availability check when executing advanced ATP check.
- The **Accumul.** field controls whether the system must:

- Take the sum of all previously confirmed quantities into account when calculating the cumulated ATP quantity when creating/changing sales orders.
- Take the sum of all open requirements quantities into account when calculating the cumulated ATP quantity when creating sales orders and no cumulation when changing sales orders.
- Take the sum of all open requirements quantities into account when creating sales orders and the previously confirmed quantities when changing sales orders.
- Use the old checking logic without cumulation.

- The **RelChkPlan** field indicates whether or not the component of a product is checked against planning.
- The **Advanced ATP** field controls whether the system uses the advanced product availability check or the standard product availability check for the checking group concerned.

The checking group for the availability check is specified in the material master record at the plant level in the **Sales: General/Plant** and **MRP 3** views. Based on the material type and the plant, you can define the default checking group proposed by the system when new material master records are created, as shown in Figure 7.16.

Scope of Availability Check (Default Values)

Matl Type	Plant	Descriptn	Name 1	AvailCheck	Descriptn
FERT	US20	Finished Product ...	Long Island City	02	Individ.requirements
HAWA	US20	Trading Goods	Long Island City	02	Individ.requirements

Figure 7.16 Configuring the Default Values for Availability Check Group

The next step is to tell the system which stocks, issue elements, and supply elements must be considered by the availability check by configuring the scope of the availability check. As shown in Figure 7.17, these settings are based on two parameters: the checking group (**Avail** field) determined by the material master record and the checking rule (**CRl** field) assigned to the sales transaction.

Avail	Description	CRl	Checking Rule
02	Individ.requirements	A	SD order
02	Individ.requirements	AE	SD order; Make-to-Order stock
02	Individ.requirements	AQ	SD order; Project stock
02	Individ.requirements	AR	Supply Assignment (ARUN)
02	Individ.requirements	AV	SD order; Returnable Packaging
02	Individ.requirements	AW	SD order; Consignment
02	Individ.requirements	B	SD Delivery
02	Individ.requirements	BB	
02	Individ.requirements	BE	SD Delivery; Make-to-Order stock
02	Individ.requirements	BQ	SD delivery; Project stock
02	Individ.requirements	BV	SD delivery; Returnable packaginç
02	Individ.requirements	BW	SD delivery; Consignment

Figure 7.17 Configuring the Scope of the Availability Check: Overview

For each combination of checking group and checking rule, you must define the scope, as shown in Figure 7.18.

Availability Check 02 Individ.requirements
Checking Rule A SD order

Stocks
With Safety Stock
With Stock in Transfer
With Quality Inspection Stock
With Blocked Stock
With Restricted-Use Stock

Requirements
With Sales Requirements
With Delivery Note
With Stock Transport Reqts Exclude
With Reservations
With Dependent Requirements
With Dependent Reservations Exclude

Future Supply
With Purchase Requisitions
With Purchase Orders X Include (for STO, use order qua...
With Shipping Notifications
With Planned Orders Exclude
With Production Orders Exclude

Replenishment Lead Time
Without Replenishment Lead Time

Special Scenarios
Without Storage Location Check
Without Subcontracting

Delayed Supply
Without Receipts in Past
Show Message for Delayed Supply

Missing Parts Processing
Checking Period: Goods Receipt

Figure 7.18 Configuring the Scope of the Availability Check: Details

The system always considers unrestricted-use stock when performing an availability check. In addition, you can indicate to the system whether any of the following five types of stocks must also be taken into account:

- Safety stock
- Stock in transfer between plants
- Quality inspection stock
- Blocked stock
- Restricted-use stock

For example, if you select the **With Safety Stock** checkbox, both unrestricted-use and safety stock are included in the product availability check.

You can also configure the scope of the check to include or exclude the requirements and future supply elements shown earlier in Table 7.1.

Warning

The scope of the availability check must be defined by your business, but you should explain the consequences of certain configurations that may result in unwanted outcomes of the ATP check. For example, if both sales and delivery requirements are taken into account for checking rule A (sales documents) but only delivery requirements are considered for checking rule B (deliveries), a potential danger exists that quantities already confirmed in the sales documents are considered available stock by the availability check in the delivery. As a result, the system would allow for deliveries to be created, thus pushing out the dates in the sales orders since those quantities are now assigned to the newly created deliveries.

The following settings provide even more flexibility when configuring the scope of the availability check:

- **Special scenarios**
  You can decide to select the **Without Storage Location Check** checkbox to perform the ATP check at the plant level only. Also, if you select the **Without Subcontracting** checkbox, then subcontractor stock is not considered available stock.
- **Replenishment lead time**
  If the **Without Replenishment Lead Time** flag is set, the replenishment lead time is ignored, and the ATP check will only confirm schedule lines based on planned future receipts. In other words, if no future receipts are planned or are not considered in the scope of the check, the ATP check will not confirm schedule lines where the delivery date falls after the end of the replenishment lead time.
- **Delayed supply**
  You can select the **Without Receipts in Past** indicator if you want the ATP check to ignore supply elements with a delivery date in the past assuming that they won't be delivered. The **Show Message for Delayed Supply** flag allows you to display a message to inform the user whether receipts in the past have been included or ignored.
- **Missing parts processing**
  The **Checking Period: Goods Receipt** field determines how many days in the future the system checks for the missing components of a production order. This functionality is not used in sales and distribution.

We've already covered most of the main configuration steps of the product availability check, but you should know about one more useful setting that allows you to specify certain defaults for the availability check by sales area. In SAP S/4HANA, you can specify whether the **Fix Qty/Date** flag is set by default for a specific sales area. Also, you can specify, for a particular sales area, the system behavior when the availability check cannot confirm the ordered quantity on the requested date. This behavior is controlled by the value in the **Avail.check rule** field, shown in Figure 7.19.

Sales Org.	Distr. Chl	Division	Fixed Date and Qty	Avail.check rule
US30	20	10	☐	
Z001	00	00	☐	
Z001	00	Z1	☐	

Figure 7.19 Configuring Default Settings by Sales Area

Setting the **Fix Qty/Date** flag will result in this checkbox being selected by default in all sales documents created in the sales area concerned.

The **Avail.check rule** field is a bit more complex, as you're presented with several options to choose from, as shown in Table 7.2.

Availability Check Rule	Short Description	System Reaction When the Order Quantity Cannot Be Satisfied
A	**One-Time Delivery**	The system automatically opts for one-time delivery on the requested date. The availability control screen is not shown.
B	**Full Delivery**	The system automatically opts for a complete delivery. The availability control screen is not shown.
C	**Delivery Proposal**	The system automatically opts for a delivery proposal. The availability control screen is not shown.
<blank>	**Dialog Box for Shortages (Batch/BAPI Full Delivery)**	The availability control screen is displayed in foreground processing. In backorder processing, system behavior matches availability check rule B.
D	**Dialog Box for Shortages (Batch/BAPI One-Time Delivery)**	The availability control screen is displayed in foreground processing. In backorder processing, system behavior matches availability check rule A.
E	**Dialog Box (Batch/BAPI Delivery Proposal)**	The availability control screen is displayed in foreground processing. In backorder processing, system behavior matches availability check rule C.

Table 7.2 Rules for Transferring the Results of the Availability Check

7.5 Advanced Available-to-Promise

SAP S/4HANA gives you the option to activate the advanced product availability check functionality, also known as SAP S/4HANA for advanced ATP, for the checking group. In this way, you can specify for which materials the system will use SAP S/4HANA for advanced ATP and for which materials the system will use the classic ATP check. You can activate SAP S/4HANA for advanced ATP by choosing **A: Active** from the **Advanced ATP** dropdown list shown earlier in Figure 7.15.

Note

The features activated by SAP S/4HANA for advanced ATP are not part of the standard SAP S/4HANA enterprise license. Your company will require a dedicated license for SAP S/4HANA for advanced ATP if this switch is set to **Active**.

However, the features activated by SAP S/4HANA for advanced ATP are part of the standard SAP S/4HANA Cloud license.

SAP S/4HANA for advanced ATP is particularly useful for analyzing sales orders with many items or analyzing production orders comprising multiple components. This functionality has been designed to process large volumes of data. SAP S/4HANA for advanced ATP also offers innovative features, such as product allocations, backorder processing, supply assignment, alternative-based confirmations, supply protection, and more. Covering all these features is beyond the scope of this book, but in this section, we'll provide an overview of two main features: backorder processing and alternative-based confirmations.

7.5.1 Backorder Processing

Imagine a scenario where a sales order item was confirmed to be delivered 4 days after the customer requested delivery date due to an unexpected high demand, which has led to shortages. A PO was placed a few days ago to replenish stock with the delivery date next week. However, the buyer has asked the vendor to expedite the shipment, and the goods have been received in the warehouse today. Now, enough stock is available to deliver the sales order on time, but to update the confirmed date and quantity on the sales order item, a new availability check must be carried out. This ATP check after the initial confirmation is known as *backorder processing*. This example illustrates a change in supply, but the availability of a certain product can also be affected by changes in demand, for instance, if a customer cancels an order that was scheduled to be delivered before the order in question.

Backorder processing can be scheduled to run periodically, and its main objective is to reconfirm sales orders and stock transport orders by checking material availability when the supply or demand situation has changed.

SAP S/4HANA for advanced ATP has increased flexibility in backorder processing to allow your company to control which orders get the limited available stock and in which priority. This functionality is achieved by using the six confirmation strategies shown in Table 7.3, which can result in requirements being confirmed earlier, later, partially, or not confirmed at all.

Confirmation Strategy	Check Sequence	Description
Win	1	Requirements are fully confirmed on the requested date.
Gain	2	Requirements will keep their original confirmations or improve if stock becomes available.

Table 7.3 Confirmation Strategies in Backorder Processing

Confirmation Strategy	Check Sequence	Description
Redistribute	3	Requirements can be updated with a better, equal, or worse confirmed date. For example, this group may lose some original commitments to the previous two groups, but if stock is available after the Win and Gain groups have been processed, then this group will gain additional quantity or get an earlier confirmation date.
Fill	4	Requirements get an equal or worse confirmation.
Lose	5	Requirements lose their confirmations to higher-ranked groups.
Skip	Not applicable	Requirements are excluded from the ATP check, thus keeping current confirmation, which would be the case if the **Fix Qty/Date** flag is set in the sales order item.

Table 7.3 Confirmation Strategies in Backorder Processing (Cont.)

Now, you must define your backorder processing segments for each of the confirmation strategies. When defining backorder processing segments, you'll specify the selection criteria and how to prioritize requirements. You can only create backorder processing segments in the SAP Fiori launchpad via the Configure BOP Segment app. Figure 7.20 shows an example of a segment created for requirements with delivery priority 01 (high) and item category TAN in sales organization US30.

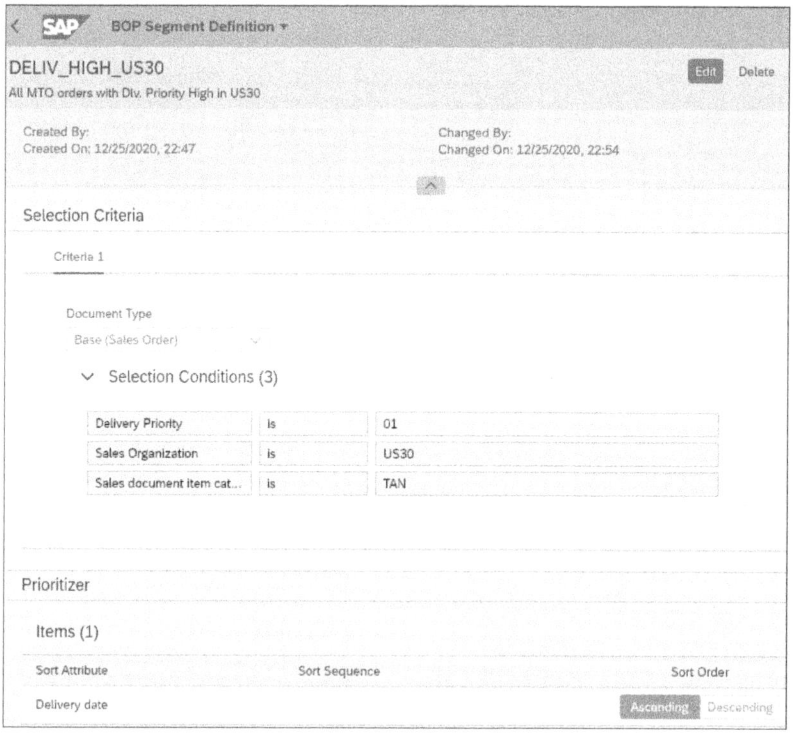

Figure 7.20 Backorder Processing Segment Definition

The entries selected will be sorted by the delivery date in ascending order as defined in the **Sort Attribute** field at the bottom of the screen.

Once your backorder processing segments have been defined, you must assign them to the relevant confirmation strategies via the Configure BOP Variant app. In this example, backorder processing segment DELIV_HIGH_US30 has been added to strategy **WIN**, as shown in Figure 7.21. As a result, orders that meet the selection criteria shown in Figure 7.20 will always get better confirmations should the materials become available.

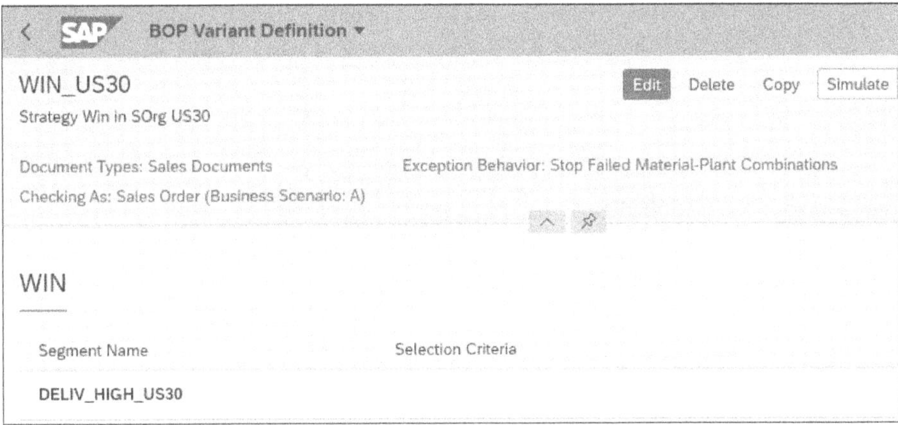

Figure 7.21 Backorder Processing Variant Definition

The next step is to schedule the backorder processing run for the variant created by using the Schedule BOP Run app. SAP S/4HANA provides the Monitor BOP Run app to analyze the result of the backorder processing run.

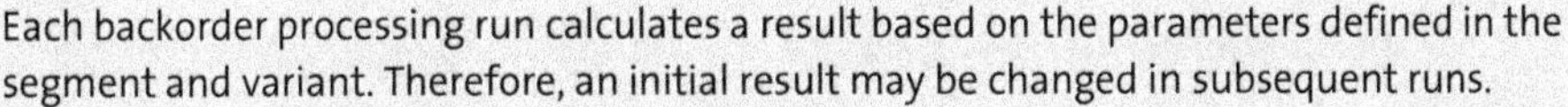

Note

Each backorder processing run calculates a result based on the parameters defined in the segment and variant. Therefore, an initial result may be changed in subsequent runs.

Another useful feature of backorder processing in SAP S/4HANA for advanced ATP is the ability to define responsibility groups for materials with shortages. You can then assign these responsibility groups to users who can then use the SAP Fiori app Release for Delivery to manually determine which requirements must be unconfirmed to free up stock for other orders, which will take the available stock, etc.

In summary, backorder processing through SAP S/4HANA for advanced ATP gives you full flexibility to define your own selection and sorting criteria by creating backorder processing segments, which you can then assign to different confirmation strategies. In this way, you can ensure the system prioritizes orders or customers that are crucial to the success of the business.

7.5.2 Alternative-Based Confirmation

As mentioned earlier, a sales document item that cannot be confirmed on the requested date may be due to limited availability in the delivering plant. However, in some cases, you can still fulfill the order on time if the goods are delivered from an alternative plant. This scenario is precisely when alternative-based confirmations through SAP S/4HANA for advanced ATP comes in handy. With alternative-based confirmations, SAP S/4HANA evaluates all possible options and automatically swaps the original delivering plant with the alternative plant that provides a sufficient quantity or confirms an earlier delivery date.

Note

You can only use the alternative-based confirmation feature to process requirements from sales documents of the sales order document category.

For a plant to be considered an alternative plant, this plant must be assigned to the same sales area as the default plant.

To enable plant substitutions for sales orders in alternative-based confirmation, you must create a substitution strategy and define an alternative control via the following two SAP Fiori apps for alternative-based confirmation in SAP S/4HANA for advanced ATP:

- Configure Substitution Strategy
- Configure Alternative Control

When creating a substitution strategy, you can choose between one of the following three substitution item types:

- **Inline substitution**
  Substitutions are made at the item level. For example, a sales order for 50 units might be delivered from plant 0001, but only 20 units are available there. The system checks the availability in all the alternative plants and finds that plant 0002 has the full quantity available. Plant 0001 is changed to plant 0002 in the sales order item.
- **Subitem generation**
  Subitems are generated for each of the alternative plants that can fulfill the requested quantity. For example, a sales order created for 50 units might be delivered from plant 0001, but only 20 units are available there. However, currently, 10 units are available in plant 0002, and 20 more units are available in plant 0003. The system will generate two subitems, as shown in Table 7.4, to fulfill the quantity of 50 units on the requested date.

Item Number	Plant	Order Quantity	Higher-Level Item
10	0001	20	
20	0002	10	10
30	0003	20	10

Table 7.4 Subitem Generation in Alternative-Based Confirmation

- **Combined strategy**
  If the full quantity cannot be fulfilled by the default plant, the system will try to find an alternative plant that can fulfill the full item quantity. If found, the system performs an inline substitution; otherwise, subitems are generated. This option is the default substitution item type.

In addition, you must select one of the following three alternative determination rules in the **Building Rule** section:

- **Full Confirmation**: This option selects the plant that can confirm the full quantity requested. If fulfillment not possible, no confirmation is generated. If more than one alternative plant can confirm the full quantity, the system chooses the plant that minimizes the delay for the final confirmation.
- **Maximum Confirmation Earlier**: This option selects the plant that can confirm the maximum quantity. If more than one alternative plant can confirm the same maximum quantity, the system chooses the plant that minimizes the delay for the final confirmation.
- **On-Time Confirmation**: This option selects the plant that can confirm the maximum quantity on the requested delivery date only.

In the event that multiple plants can confirm equally, for instance, two alternative plants can fulfill the full quantity on time, the system will apply the following tie-breaker logic to determine the best option:

1. Choose the alternative plant that is in the same country as the ship-to party
2. Choose the alternative plant that is in the same region as the ship-to party
3. Choose the alternative plant with the shortest transit time
4. Choose the alternative plant with the shortest overall delivery time (including picking and packing)
5. If steps 1 through 4 don't provide a unique result, then:
   - Choose the originally requested plant
   - Choose the plant according to alphabetical sorting (numbers are sorted before characters)

To illustrate how substitution strategies appear once saved, Figure 7.22 shows strategy **INLINESUB_MAXCONF**, which has been created with the substitution item type **Force Inline Substitution (I)** and the alternative determination **Maximum Confirmation Earlier**.

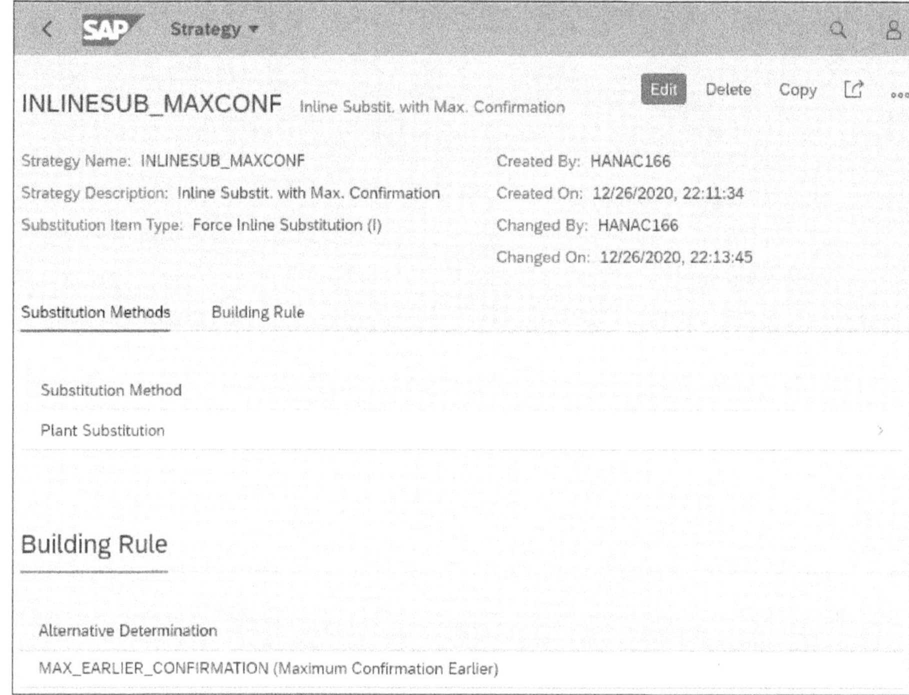

Figure 7.22 Defining a Substitution Strategy

Next, you'll need to configure an alternative control and define the criteria that will determine your substitution strategy. Figure 7.23 shows an example of an alternative control where the previously created substitution strategy **INLINESUB_MAXCONF** is determined based on the sales document type **OR**, sales organization **ZRIV**, and sold-to party **42**.

Configure Alternative Control

Sales Order

Alternative Controls (1)

Search

Edit

Sales Document Type	Created By
OR	HANAC166

Sales Organization: ZRIV

Sold-to party - Customer Number: 0000000042

Substitution Strategy: INLINESUB_MAXCONF

Execute ABC: ☑

Figure 7.23 Configuring Alternative Control

If plant substitution for sales orders in alternative-based confirmation has been activated successfully, an availability check is executed with alternative-based confirmation when creating or changing a sales document. The delivery process, however, does not currently support alternative-based confirmation.

7.6 Important Terminology

In this chapter, the following terminology was used:

- **Available-to-promise (ATP)**
  This business function provides the quantity available of a material in a specific plant on the required date so goods can be delivered to the customer on time. If the quantity ordered is not available on the customer's requested date, this function generates confirmation proposals for the corresponding material and plant.
- **Availability check**
  This function, also known as the product availability check, calculates the confirmed quantity based on the customer requirement. Not a simple stock check, this functionality considers the scheduling for picking or packing times, the time taken to procure or manufacture the goods, and any relevant planned receipts/planned issues as defined in the scope of the availability check.
- **Requirement**
  The quantity requested of a certain material to be delivered to the customer on a specific date.
- **Confirmation**
  A combination of the committed delivery date and quantity for a customer requirement. A confirmation is the outcome of the availability check.
- **Confirmation strategy**
  Defines how a requirement must be handled by the system. In SAP S/4HANA for advanced ATP, with backorder processing, a requirement can be assigned to one of five possible confirmation strategies (win, gain, redistribute, fill, and lose) that will eventually determine the priority of the requirement and help the system to decide the sequence in which the requirements receive the available quantity.
- **Alternative-based confirmation**
  This function within advanced ATP in SAP S/4HANA uses a heuristic approach to calculate all possible options to deliver goods from an alternative plant if the material is not available in the originally requested delivery plant. The aim is to meet the customer's requested date and quantity by exchanging the plant in the sales document for a plant with better material availability.
- **Replenishment lead time**
  This value is the duration, in working days, to procure or manufacture any quantity of a product. Any requirement with a material availability date after

the replenishment lead time can be confirmed since (in theory) enough time is available to produce or procure the material, even if the material is not currently available. The replenishment lead time is taken into account in the product availability check and can be disabled in the **Scope of check** (scope of availability check).

- **Backorder**
  This term refers to sales documents that cannot be delivered on time based on the calculations carried out by the availability check. This situation occurs when backward scheduling returns a material availability date in the past.
- **Backorder processing**
  This feature is available in SAP S/4HANA for advanced ATP and allows you to react to changes in the demand or supply of materials in sales order or stock transport order items already confirmed. By assigning one of the five confirmation strategies (win, gain, redistribute, fill, or lose) to each backorder processing segment, you can manage business priorities and protect customer interest. Periodic jobs can be scheduled to execute backorder processing runs in the background based on the variants defined.
- **Backorder processing segment**
  This term is used in backorder processing in advanced ATP to refer to a saved selection of parameters that determines which requirements are selected and the sequence in which the requirements are prioritized.
- **MRP group**
  This field, specified in the **MRP 1** view in the material master record, is used to group materials that can be planned in the same way from the MRP perspective. Materials with the same MRP group share common control parameters, such as the strategy group, consumption mode, and planning period. A strategy group can be determined from the MRP group if not maintained in the **MRP 3** view in the material master.
- **Strategy group**
  This field, specified in the **MRP 3** view in the material master record, defines the main strategy, as well as up to seven alternative strategies that can be used for the particular material. If the system cannot find a strategy group in the material master for a given plant, it reads the MRP group for the material (if maintained) and uses the strategy group assigned to the MRP group.
- **Checking rule**
  The checking rule of the availability check indicates the final procedure for checking material availability in different business transactions, such as sales orders, deliveries, make-to-order stock in sales orders, make-to-order stock in deliveries, consignment in sales orders, consignment in deliveries, and more. In SAP S/4HANA Sales, the checking rule is specified within the system and cannot be changed.
- **Sales area**
  The combination of sales organization, distribution channel, and division.

- **MRP area**
  The organizational unit where MRP is carried out. Three types of MRP areas exist: at the plant level, at the storage location level, or in MRP areas for subcontractors.

7.7 Practice Questions

These practice questions will help you evaluate your understanding of the topics covered in this chapter. The questions shown are similar in nature to those found on the certification examination. Although none of these questions will be found on the exam itself, they will allow you to review your knowledge of the subject. Select the correct answers and then check the completeness of your answers in the next section. Remember that, on the exam, you must select all correct answers and only correct answers to receive credit for the question.

1. In which of the following situations is a new availability check executed automatically? (There are two correct answers.)

☐ A. Change of shipping type

☐ B. Change of plant

☐ C. Change of shipping point

☐ D. Change of Incoterms

2. True or False: A customer requests 50 units to be delivered in 6 days. The material replenishment lead time is 3 days. Picking/packing, loading, and transportation times all add up to 1 day. The 50 units are not available today, and no future receipts are currently planned for this material. However, the ATP check will confirm the full quantity on the required date if the **Without Replenishment Lead Time** flag is NOT set in the configuration of the scope of availability check.

☐ A. True

☐ B. False

3. In a sales order with three items, each item has a different material availability date, so the customer has accepted partial deliveries on this order. Which of the following options will minimize the number of partial deliveries and help reduce transportation costs?

☐ A. Carry out an availability check twice a day through backorder processing to confirm each item as soon as the material is available

☐ B. Leave the **Part.dlv./item** field blank for each item

☐ C. Select value C (only complete delivery allowed) in the **Part.dlv./item** field for each item

☐ D. Set the **Complete delivery** indicator in the customer master record

4. A customer typically prefers partial deliveries to get the products as soon as they become available. However, in a particular order, all the materials have to be delivered together to meet country-specific regulations in the destination country. What do you need to do?

☐ A. Select the **Complete delivery** indicator in the sales order

☐ B. Leave the **Part.dlv./item** field blank for each item

☐ C. Select the **Complete delivery** indicator in the customer master record

☐ D. Set a delivery block in the sales order at the header level and remove the block once all items are available

5. Which of the following statements regarding the **Fix Date/Qty** indicator in a sales document are true? (There are two correct answers.)

☐ A. A fixed item is considered complete once the entire confirmed quantity has been delivered.

☐ B. If the **Fix Date/Qty** indicator is selected, the system then transfers the requirements for the schedule lines to MRP.

☐ C. If the **Fix Date/Qty** indicator is selected, the requested date and quantity can no longer be changed for the item.

☐ D. The **Fix Date/Qty** indicator is only displayed on the schedule line overview screen.

6. Select the answers that represents a true statement regarding the availability overview screen in SAP S/4HANA. (There are two correct answers.)

☐ A. The availability overview screen shows the available quantity over a period of time for multiple materials in the selected plant

☐ B. The availability overview screen shows the available quantity over a period of time for a single material in the selected plant

☐ C. The availability overview screen shows the available quantity over a period of time for a single material in the selected company code

☐ D. The availability overview screen shows the available quantity over a period of time for a single material in the selected MRP area

7. During an ATP check, which of the following objects can be considered as future receipts? (There are three correct answers.)

☐ A. Deliveries

☐ B. Planned orders

☐ C. Reservations

☐ D. POs

☐ E. Purchase requisitions

8. True or False: SAP S/4HANA uses backward scheduling first to find out if the requested delivery date is achievable. Only if the requested date cannot be met will the system then perform forward scheduling to propose a viable delivery date.

☐ A. True

☐ B. False

9. If forward scheduling is carried out, the availability control screen may offer several options for delivery. Select the valid options. (There are three correct answers.)

☐ A. Delivery from an alternative plant

☐ B. Complete delivery

☐ C. One-time delivery on requested delivery date

☐ D. Delivery proposal

☐ E. Partial deliveries of the same quantity

10. True or False: When configuring the transfer of requirements, if the **Req. transfer** indicator is not set at the requirements class level, you can still decide whether or not to activate the transfer of requirements to MRP by setting the indicator at the schedule line level.

☐ A. True

☐ B. False

11. You need to configure the availability check with ATP logic in an SAP S/4HANA system. Which of the following settings is NOT possible when defining a new checking group?

☐ A. Select the **Material block in availability check with transfer of quantities** indicator

☐ B. Set value A (Single records) in the **Total sales order requirements** field for individual requirements

☐ C. Set value B (Totals records per day) in the **Total sales order requirements** field for collective requirements

☐ D. Set the **Advanced ATP** field to **Active**

12. Where do you configure whether the stock in transfer between plants is added to the available inventory in the receiving plant?

☐ A. **Define Availability Check Group**

☐ B. **Activate Stock Transfer Between Storage Locations**

☐ C. **Configure Requirement Classes**

☐ D. **Configure Scope of Availability Check**

13. In a standard SAP S/4HANA system, which element influences the rules for transferring the results of the availability check and the default value of the **Fix Qty/Date** field?

☐ A. The plant in the sales order item

☐ B. The item category in the sales order item

☐ C. The sales area

☐ D. The sales document type

14. True or False: When performing an availability check, in the event of shortages, the system always presents the user with the availability control screen to choose the preferred delivery option.

☐ A. True

☐ B. False

15. Which of the following statements regarding SAP S/4HANA for advanced ATP are true? (There are two correct answers.)

☐ A. SAP S/4HANA for advanced ATP is activated or deactivated at the plant level.

☐ B. SAP S/4HANA for advanced ATP can be activated for a specific group of materials only.

☐ C. SAP S/4HANA for advanced ATP features are part of the standard SAP S/4HANA Cloud license.

☐ D. SAP S/4HANA for advanced ATP features are part of the standard SAP S/4HANA enterprise license.

16. When carrying out an ATP check for a delivery item, which of the following elements may be taken into account? (There are three correct answers.)

☐ A. Purchase requisitions

☐ B. Inbound Intermediate Documents (IDocs)

☐ C. Credit memos

- ☐ D. Dependent requirements
- ☐ E. Production orders

17. When configuring backorder processing through SAP S/4HANA for advanced ATP, to which of the following confirmation strategies can you assign a backorder processing segment? (There are three correct answers.)

- ☐ A. Secure
- ☐ B. Lose
- ☐ C. Expedite
- ☐ D. Fill
- ☐ E. Redistribute

18. You're creating a new substitution strategy for alternative-based confirmation in SAP S/4HANA for advanced ATP. Which of the following substitution types can be selected? (There are two correct answers.)

- ☐ A. Subitem generation
- ☐ B. Alternative confirmation
- ☐ C. Inline substitution
- ☐ D. Partial substitution

7.8 Practice Question Answers and Explanations

1. Correct answers: **B and C**

   The two fields from the four options given that would trigger a new availability check when changed are the shipping point and the plant. Plant is an obvious answer since stock checks and planning are carried out at the plant level. Thus, a change in the delivering plant requires a new ATP check. The shipping point is used to determine both the loading time and the pick/pack time. Different shipping points may result in different times, which could affect the confirmed delivery date; thus, a new ATP check is also required if the shipping point is changed. Incoterms and shipping types don't trigger automatic ATP checks since those two fields are not used in scheduling or to calculate the inventory available.

2. Correct answer: **A**

   True. If the material is currently out of stock, backward scheduling will fail, and the system will perform forward scheduling to calculate the confirmed delivery date. The system takes the replenishment lead time into account since the **Without Replenishment Lead Time** checkbox is NOT checked in the scope of the availability check. Assuming no planned future receipts are expected, the material availability date will be the end of the replenishment lead time (any quantity is, in theory, available on this date). As a result, the material availability date is today plus 3 days. Then, the system will add 1 day for picking/packing, loading, and transportation, so the calculated delivery date will be today plus 4 days (material availability date plus 1 day). Since the calculated delivery date through forward scheduling (today plus 4 days) is before the customer requested date (today plus 6 days), the ATP check confirms the full quantity.

3. Correct answer: **C**
   The customer has accepted partial deliveries, so answer D can be discarded. Answer B may result in multiple deliveries per item, so answer B isn't the correct answer either. The only answer that will guarantee the minimum number of partial deliveries is to select value C in the **Partial Delivery** field for each item. This value guarantees one delivery per sales document item, so if three items exist in the order, a maximum of three deliveries will be created.

4. Correct answer: **A**
   In this case, the customer has specific delivery requirements on a particular order, so the change must be made in the order and not in the customer master record. By selecting the **Complete delivery** checkbox in the sales order at the header level, you can ensure the order in question is delivered completely in a single delivery.

5. Correct answers: **A and B**
   As shown earlier in Figure 7.6, the **Fix Date/Qty** indicator is displayed on both the schedule line overview and the availability control screens; thus, answer D is not correct. Answer C is also false since you can still make changes to the quantity ordered and to the requested delivery date regardless of whether the **Fix Date/Qty** checkbox has been selected. If selected, the rescheduling job will not update the confirmations on the relevant item, but you still can change the requested quantity and/or date, thus leaving you with only two correct answers.

6. Correct answers: **B and D**
   As shown earlier in Figure 7.9, the availability overview screen shows the quantity available for a single material in the selected plant and MRP area only. You cannot display the availability at the company code level (which might include several plants) or for multiple materials at the same time.

7. Correct answers: **B, D, and E**
   In SAP S/4HANA, outbound delivery notes are typically referred to as deliveries. Both deliveries and reservations of stock are requirements from the materials planning perspective; in other words, they are both outward movements of stock. Only inward movements of stock can be considered as future receipts. Therefore, both answers A and C are false, which leaves you with the three correct answers.

8. Correct answer: **A**

   True. The statement describes exactly how backward and forward scheduling works in SAP S/4HANA. The system only performs forward scheduling if backward scheduling returns a start of transportation planning or procurement date in the past, as shown earlier in Figure 7.2. Then, forward scheduling is carried out to calculate an achievable delivery date.

9. Correct answers: **B, C, and D**

   As shown earlier in Figure 7.7, the three options offered by the availability control screen are:

   - Complete delivery
   - One-time delivery on requested delivery date
   - Delivery proposal

   The other two answers are made-up options.

10. Correct answer: **B**

    False. The fine-tuning of the availability check for sales documents that you carry out at the schedule line level depends on their global configuration at the requirements class level. You can deactivate the transfer of requirements at the schedule line level if it's active at the requirements class level. You cannot activate the transfer of requirements at the schedule line level if it's NOT active at the requirements class level.

11. Correct answer: **C**

    This rather specific question concerns settings on a customizing screen. When defining a new checking group, you cannot set value B (Totals records per day) in the **Total sales order requirements** field since the use of collective requirements is no longer supported. In SAP S/4HANA, only individual requirements are used.

12. Correct answer: **D**

    In the **Configure Scope of Availability Check** customizing activity, you can select the **With Stock in Transfer** checkbox to take into account any stock that is being transferred into the plant where the ATP check is carried out.

13. Correct answer: **C**

    This question tests your knowledge of the **Configure Default Settings by Sales Area** customizing activity. As shown earlier in Figure 7.19, you can set up rules for transferring the results of the availability check and the default value of the **Fix Qty/Date** field based on the sales area.

14. Correct answer: **B**

    False. You can decide whether to hide or show the availability control screen based on the sales area via customizing. As shown earlier Table 7.2, when configuring default setting by sales area for the availability check, if you select values A, B, or C in the **Avail.check Rule** field, the availability control screen is not shown.

15. Correct answers: **B and C**

    As an SAP S/4HANA Sales consultant, you must know if certain features require an additional license or are included in the standard package. SAP S/4HANA for advanced ATP is part of the standard SAP S/4HANA Cloud license. However, this functionality is NOT included in the standard SAP S/4HANA enterprise license. Thus, answer C is a true statement. As shown earlier in Figure 7.15, you can activate SAP S/4HANA for advanced ATP for a specific checking group, which you can then maintain in the material master record for a group of materials. Therefore, answer B is also a true statement.

16. Correct answers: **A, D, and E**

    This question tests your knowledge of the scope of the availability check. The only two elements from the five answers provided that do not exist within the scope of the availability check, as shown earlier in Figure 7.18, are credit memos and inbound IDocs. On the other hand, production orders, purchase requisitions, and dependent requirements are all elements within the scope of the availability check, which can be enabled or disabled through configuration.

17. Correct answers: **B, D, and E**

    All the confirmation strategies are listed in Table 7.3. Lose, fill, and redistribute are valid confirmation strategies for backorder processing. The other two answers (secure and expedite) are made-up strategies that don't exist in SAP S/4HANA.

18. Correct answers: **A and C**

    When creating a substitution strategy for alternative-based confirmation, you can choose between one of the following three substitution types:

    - Inline substitution
    - Subitem generation
    - Combined strategy

    Answers B and D are made-up substitution types that don't exist in SAP S/4HANA.

7.9 Takeaway

In this chapter, you learned how SAP S/4HANA uses backward scheduling to determine if the requested delivery date is achievable. If not achievable, forward scheduling calculates a realistic delivery date. You also understand the options offered by the system on the availability control screen and how schedule lines are created in the event of shortages. You now know how the available quantity of a particular material changes over time and how this information is presented in the availability overview screen.

You're familiar with the delivery agreement indicators in both the customer master record and the sales document and how these fields control the number of deliveries allowed for a sales document, which can save on shipping costs and ensure customer requirements are met. You were also introduced to the effect of the **Fix Qty/Date** indicator, especially the important role this checkbox plays when the rescheduling job is run periodically.

You can configure the transfer of requirements to MRP and the availability check with ATP logic. You now know the features introduced by SAP S/4HANA for advanced ATP and how to activate SAP S/4HANA for advanced ATP for the selected checking groups. Finally, you've gained a good overview of the benefits of backorder processing and alternative-based confirmation—two of the main functionalities within SAP S/4HANA for advanced ATP.

7.10 Summary

Availability checks and the transfer of requirements are two important integrated components that need to be understood together. If set up coherently, the system will generate accurate confirmation proposals and ensure materials are procured on time. You've learned that the availability check is configurable to a highly detailed level, thus allowing you to decide, for example, which elements must be taken into account from a supply and requirements perspective, what stock types must be considered, and even what default values certain parameters should take by sales area.

SAP S/4HANA for advanced ATP is a new tool that includes the standard ATP functionality and, if activated, provides additional features, such as prioritizing overdue or unconfirmed requirements via backorder processing or checking availability in alternative plants in the event of shortages in the requested delivery plant.

In the next chapter, we'll analyze the pricing determination functionality in sales documents, pricing condition records, and condition contracts.

Chapter 8
Pricing

8

Techniques You'll Master

- Understand the condition technique as used in pricing and learn ways to ensure optimal performance
- Configure pricing in sales and distribution
- Apply special pricing functions
- Conduct pricing analysis
- Use special condition types
- Manage condition contracts in SAP S/4HANA Sales

The focus of this chapter is on explaining the functions and customizing settings for pricing within the area of sales in SAP S/4HANA. We'll discuss the condition technique in sales and distribution documents; the design and configuration of the various elements used in pricing determination; special pricing functions; special condition types, such as discounts and statistical condition types; and condition contract management. Master pricing condition records play a critical role in pricing determination; however, we won't discuss this topic again, which has already been covered in Chapter 3, Section 3.4.

Real-World Scenario

In SAP S/4HANA, *pricing* refers to the calculation of the amount a customer must pay in exchange for goods or services. Pricing also helps a company meet legal requirements according to the tax law of the relevant countries where it operates; thus, pricing can be considered one of the most important components in sales. In this often complex process, the system must consider many factors to arrive at a specific price for a customer. As an SAP S/4HANA Sales consultant, your responsibility is to make price determination as automated as possible, while ensuring the correct price is determined in all sales documents.

For example, your project may require a pricing procedure that has more than one discount condition type. One discount condition type could be a customer discount where VIP customers get a 5% discount when ordering consumables, and the other could be a 10% discount that only applies to certain materials on sale during a limited time. The sales manager asks you to set up the system so only the most favorable discount is determined in the sales order. You should be equipped with the technical skills to configure condition exclusion groups, so if a VIP customer orders a consumable that is currently on sale, he doesn't get a 15% discount but only the most favorable of the two discounts is applied, which in this example is the temporary 10% discount on consumables that are on sale.

Finally, you must be able to use the condition technique with confidence, including describing its components and explaining how pricing condition records are created. You must know how to set up different pricing conditions in the pricing procedure to meet your company's pricing requirements. You should also have knowledge of special pricing functions and special condition types.

8.1 Objectives of This Portion of the Test

The goal of this portion of the SAP S/4HANA Sales Certification exam is to test your understanding of the following topics:

- Configuration of condition tables and field catalogs
- Define pricing condition types
- Set up access sequences for pricing in sales
- Define and assign pricing procedures
- Set up exclusion procedures for groups of conditions
- Understand special pricing condition types
- Initiate a pricing analysis on a sales document
- Understand condition contract settlement management in SAP S/4HANA Sales

> **Note**
> 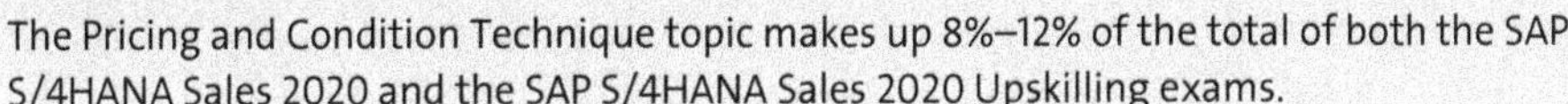
> The Pricing and Condition Technique topic makes up 8%–12% of the total of both the SAP S/4HANA Sales 2020 and the SAP S/4HANA Sales 2020 Upskilling exams.

8.2 Configuring Pricing

Pricing in SAP S/4HANA can be described as the methodology used by the system to determine the prices of any products or services sold to customers. Within sales and distribution, pricing is carried out in quotations, inquiries, sales orders, debit memos, credit memos, contracts, and schedule agreements. The actual mechanism that enables the system to find the correct price given a specific set of variables is known as the *condition technique*. In Chapter 5, Section 5.4, you learned about the basic elements used by the condition technique. In this section, we'll expand on its different uses and settings in pricing, as well as what can you do to optimize system performance during price determination.

> **Note**
> In Chapter 3, Section 3.4, you learned how to maintain pricing condition records in SAP S/4HANA, so even though this topic is part of this portion of the exam, we won't review this topic again in this chapter. We encourage you, however, to revisit Chapter 3, Section 3.2, regarding pricing condition records before diving into the following key concepts.

Five elements are used to determine prices via the condition technique, and we suggest setting them up in the following sequence:

1. Create the condition tables.
2. Set up access sequence and assign condition tables.
3. Create the condition type and assign the access sequence.

4. Define and assign the pricing procedure.
5. Create master pricing condition records.

Example

Let's say a customer wants to buy material SMPH11 "Smartphone Model 11." However, two pricing condition records exist in the system for this material, both under condition type PR00:

- $650 is the list price valid for all customers.
- $580 is a special price that only applies to a specific group of retailers that usually purchase this material in bulk.

The condition record with the list price of $650 is stored in the standard condition table A304 (Material with release status), whereas condition records with the special price of $580 are stored in the standard condition table A305 (Customer/material with release status) where the customer number is one of the selected fields.

The access sequence has two steps, and the most specific condition table (the one including the customer number) is the first step:

1. Table A305: Customer/material with release status
2. Table A304: Material with release status

The access sequence is assigned to condition type PR00, and this condition type is contained in the standard pricing procedure RVAA01.

A sales order is created for material SMPH11, and the customer entered is one of the retailers with a special price. Procedure RVAA01 is the pricing procedure on the order, so the condition technique tries to find a price for condition PR00. First, the system sees the access sequence for condition type PR00 and then searches for a condition record using the first step in the access sequence, which corresponds to condition table A305. Since a special price of $580 has been maintained for this customer and material, the system stops searching and assigns $580 to condition type PR00 in the **Conditions** tab of the sales document item.

If the system fails to find a valid condition record in condition table A305, it will go to the next condition table A304, which is not customer dependent, and the price of $650 will be applied to the item. The **Exclusive** indicator on the access sequence step controls whether the system must stop searching once a valid condition record has been found. This indicator will be discussed later in Section 8.2.2.

Now that you understand the elements and configuration steps involved in pricing determination, let's discuss each element in detail.

8.2.1 Condition Tables

Pricing condition tables define the fields used to find a master pricing condition record. All the pricing condition records for a given set of attributes or fields are stored in the corresponding condition table. SAP S/4HANA offers several pricing condition tables, but if these tables are not sufficient for your requirements, you can create custom condition tables. Figure 8.1 shows the technical view of the standard pricing condition table A305 mentioned in the previous example.

Change Condition Table (Pricing Sales/Distribution): Technical View

Dictionary Elements Other Description Field Attributes...

Table 305 Customer/material with release status

With validity period ☑

with release status ☑

Selected Fields

Short Description	Key	Item Fld	Text Field	Field Name	Data element	Domain name	No...	Da...
Sales Organization	☑	☐	○	VKORG	VKORG	VKORG	4	C
Distribution Channel	☑	☐	○	VTWEG	VTWEG	VTWEG	2	C
Customer	☑	☐	○	KUNNR	KUNNR_V	KUNNR	10	C
Material	☑	☑	◉	MATNR	MATNR	MATNR	40	C

Figure 8.1 Pricing Condition Table A305

You can maintain condition tables and change the field catalog in the IMG menu under **Sales and Distribution • Basic Functions • Pricing • Pricing Control • Condition Tables and Field Catalog.**

When creating a new condition table, you can choose the key fields from a list known as the *field catalog*. In some cases, the provided list may not be enough, and you may need to add fields to the standard field catalog for pricing. Adding fields is possible via the **Change Field Catalog** customizing activity. On the main screen, if you click on the **New Entries** button, you'll be presented with an empty column where you can use the matchcode tool to select, from a list, the field you want to add to the field catalog. Then, click on **Save**, and the field catalog will be updated with the new field. You can now create a condition table with the new field.

However, if the field you need in your condition table is not in the list of allowed fields, you can still use the field, but the field must be added to the list of allowed fields first. This process is regarded as an enhancement and is complex since you'll need the help of an ABAP developer to make changes to communication structure KOMPAZ (Pricing communication item: Customer modifications) if it's an item field or to structure KOMKAZ (Pricing communication header: Customer modifications) if it's a header field. Structures KOMPAZ and KOMKAZ are subsets of the pricing communication structures KOMP and KOMK, respectively. Changes to the user exit designed to fill these structures are also required. Since we provided an overview on adding new data fields to communication structures, including an example, in Chapter 6, Section 6.5, we won't discuss this topic again in this chapter.

Tip

Avoid creating condition tables with fields that don't add value in the search of condition records. For example, a condition table to search by product hierarchy, material group, and material number is not an optimal choice. If you really need to go down to the level of material number, then the material group and the product hierarchy are redundant fields in the condition table. The system will find the same price with a condition table with just one field—the material number.

Now that we've covered condition tables, let's discuss access sequences.

8.2.2 Access Sequences

To define access sequences, follow the menu path **Sales and Distribution • Basic Functions • Pricing • Pricing Control • Access Sequences.**

An *access sequence* refers to the search strategy used by the system to fetch a valid pricing condition record. Only condition types for which the condition amount is determined via condition records require an access sequence. For example, you should not assign an access sequence to manual condition types, such as discounts where the amount or percentage is entered manually in the sales document.

In the access sequence, you'll define the order of condition tables that must be searched to find a condition record that meets the criteria—in other words, to find a valid price for a particular condition type. You must configure the steps in the access sequence from the most specific condition table (usually the condition table with more key fields) to the most generic condition table. Figure 8.2 shows the standard access sequence PR02 (Price with Release Status), which has been defined with four steps and uses three condition tables. The three condition tables are tables A305, A306, and A304.

Access Sequence PR02 Price with Release Status

Overview Accesses

No.	Table Number	Description	Requirement	Exclusive
10	305	Customer/material with release status		☑
20	306	Price list category/currency/material with release status		☑
30	306	Price list category/currency/material with release status	3	☑
40	304	Material with release status		☑

Figure 8.2 Access Sequence PR02 (Price with Release Status)

The following elements are maintained to set up an access sequence:

- **No.**: This column contains a sequential number that indicates the sequence in which the condition tables must be accessed. In the access sequence shown in Figure 8.2, the system starts with table A305 assigned to the first step and ends with table A304 at the bottom. First, the system goes to access "10" and reads condition table A305, which is **Customer/material with release status**; if a valid condition record is not found, the system goes to access "20" and reads table

A306, which has the description **Price list category/currency/material with release status**. This process continues until the system finds a condition record or until it reaches the last step in the sequence.

- **Table Number**: This value is the number of the condition table that contains the key fields required to create a pricing condition record. The first letter of database condition tables is A, which is omitted in this column.
- **Requirement**: This value represents a condition written in ABAP code that must be met to access the condition table in the step. In access sequence PR02, shown in Figure 8.2, requirement "003" is assigned to access "30." If the requirement is fulfilled, the relevant step is executed.
- **Exclusive**: If selected, the system stops if a condition record is found. In our example, if a condition record is found in step "10," then it doesn't go to step "20" to continue the search.

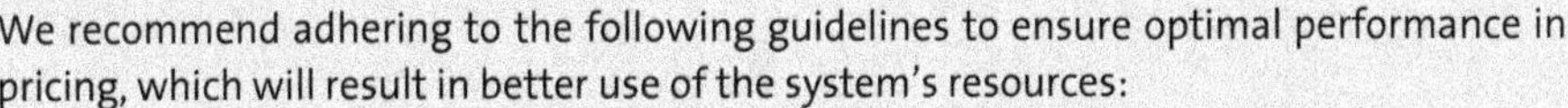

Tip

We recommend adhering to the following guidelines to ensure optimal performance in pricing, which will result in better use of the system's resources:

- Access sequences should not have redundant or unnecessary access steps. When copying a standard access sequence, you must remove any steps that are not needed.
- Remember to use the **Exclusive** indicator in the access sequence, so the system stops searching for further condition records after finding a valid one.

Next, you'll need to set up a pricing condition type and assign it to the relevant access sequence.

8.2.3 Pricing Condition Type

Pricing elements such as discounts, gross prices, surcharges, and taxes are represented by pricing condition types. The condition type controls the properties of pricing conditions. Figure 8.3 shows some of the key control elements for the standard pricing condition type PR00.

You can define condition types by following the menu path **Sales and Distribution • Basic Functions • Pricing • Pricing Control • Define Condition Types**.

At the top of the screen, maintain the condition type description and the access sequence that will be used to find condition records for the concerned condition type.

Note

You can assign the same access sequence to more than one condition type.

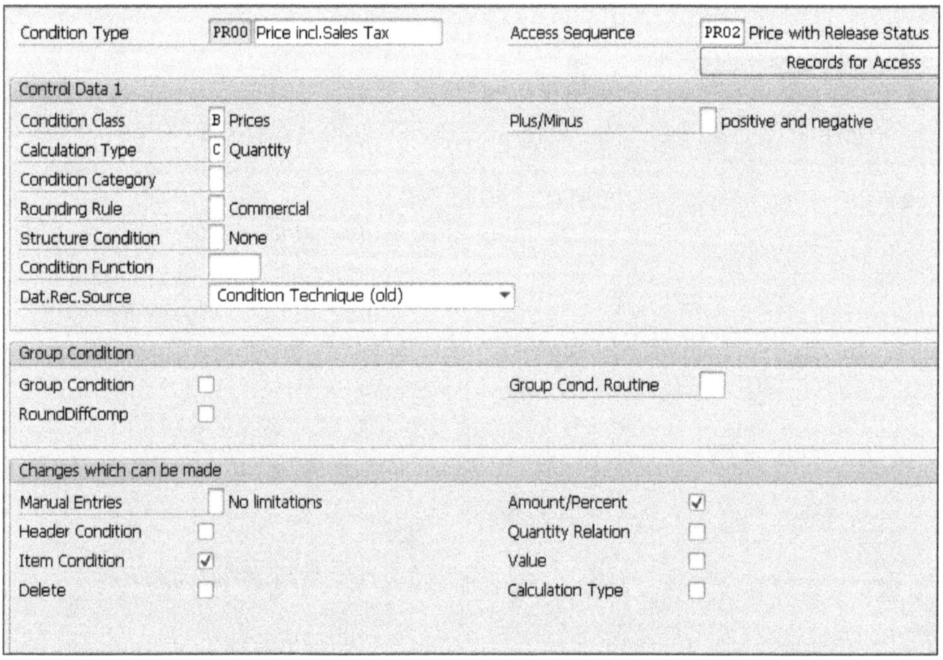

Condition Type PR00 Price incl.Sales Tax
Access Sequence PR02 Price with Release Status
Records for Access

Control Data 1
Condition Class B Prices
Plus/Minus positive and negative
Calculation Type C Quantity
Condition Category
Rounding Rule Commercial
Structure Condition None
Condition Function
Dat.Rec.Source Condition Technique (old)

Group Condition
Group Condition
Group Cond. Routine
RoundDiffComp

Changes which can be made
Manual Entries No limitations
Amount/Percent
Header Condition
Quantity Relation
Item Condition
Value
Delete
Calculation Type

Figure 8.3 Pricing Condition Type PR00

For the purposes of this book, we'll only explain the most relevant attributes of a condition type. First, let's look at the fields in the **Control Data 1** section, as follows:

- **Condition Class**: This field specifies how the pricing condition type is categorized. The following values are the most common:
  - **A: Discount or surcharge**
  - **B: Prices**
  - **C: Expense reimbursement**
  - **D: Taxes**
- **Calculation Type**: This field determines how the system calculates the condition type. You can set the calculation type in the pricing condition record too. If the calculation type is left blank in the condition record, the value maintained in the condition type is valid for the condition record. For instance, if a condition type for a freight surcharge is marked as a header condition and the calculation type "D" (gross weight) is selected, the system distributes the amount entered at the header level among the items based on the gross weight. The most commonly used calculation types are the following:
  - **A: Percentage**
  - **B: Fixed amount**
  - **C: Quantity**
  - **D: Gross weight**
  - **F: Volume**

- G: Formula
- M: Quantity – Monthly Price
- **Condition Category**: This field is used to group pricing conditions according to predefined categories, such as freight costs, transfer prices, base amounts, taxes, etc.
- **Rounding Rule**: This field determines the rounding rule used by the system to round up or down the condition type value, so the amount is converted into the smallest significant subdivision of the currency. You have the following three options are available:
  - <blank>: This option uses commercial rounding, which is the rule often used for currency conversions and for rounding prices (e.g., $43.675 is rounded up to $43.68, whereas $43.674 is rounded down to $43.67).
  - **A**: This rule always rounds values up (e.g., $43.671 is rounded up to $43.68).
  - **B**: This rule always rounds values down (e.g., $43.679 is rounded down to $43.67).
- **Structure Condition**: This control is often used for bills of materials (BOMs) or configurable materials. If you leave this field blank, the parameter is deactivated. The other two values have the following effects:
  - **A**: The condition is duplicated into all the subitems (or components).
  - **B**: The price of the subitems is added to determine the value of the condition.
- **Plus/Minus**: This field determines the sign of the condition amount. If you leave the field blank, the condition amount can be either positive or negative. If **A** is selected, the condition value is treated as a positive amount, which is always added to the gross price, for example, a freight charge. If **X** is selected, the amount is always subtracted from the gross price, for example, a discount.

In the **Group Condition** section, you'll maintain the following fields:

- **Group Condition**: This checkbox controls how the system calculates the scale base value. If selected, the system takes into account all related items in the sales order to calculate the scale base value. This feature is reviewed in detail in Section 8.3.
- **RoundDiffComp** (rounding difference comparison): If this indicator is selected, the system compares the condition value at the header level with the total of the condition values at the item level. If the values don't match, the difference is then added to the largest item.
- **Group Cond. Routine**: This field can be used to specify a routine that calculates the basis for the scale value when a group condition occurs in pricing.

In the **Changes which can be made** section, you'll maintain the following fields:

- **Manual Entries**: This setting controls the priority in the event a pricing condition can be determined automatically but is entered manually. If value **B** is selected, the automatically derived entry has priority over the manual entry. If value **C** is selected, the manual entry has priority, and the system won't search for condition records. If value **D** is selected, the condition cannot be processed manually.

- **Header Condition**: If this indicator is selected, the condition type can be entered in the header condition screen. A header pricing condition can be distributed to all of the individual items. The basis of the distribution to the items is controlled by the **Calculation Type** field mentioned earlier.
- **Item Condition**: If this indicator is selected, the condition type can be entered in the item condition screen.
- **Delete**: This checkbox controls whether the condition type can be deleted from the document.
- **Value**: If this indicator is selected, the value of the condition type can be changed on the document.
- **Calculation Type**: This checkbox controls whether you can change the calculation type during document processing.

In the **Master Data** section, you'll maintain the following fields:

- **Pricing procedure**: The pricing procedure entered in this field controls the use of condition supplements in pricing condition records for the condition type concerned. For example, let's say you want to automatically determine two discounts (RA00 and RB00) every time a specific pricing condition record is found for condition PR00 (gross price). These two discounts have been assigned to pricing procedure PR0000. By entering pricing procedure PR0000 in this field for condition type PR00, you can then define a value for discounts RA00 and RB00 when maintaining a pricing condition record for PR00. During pricing, if the PR00 condition record is found, the system will also apply discounts RA00 and RB00. When creating or changing a condition record, you can maintain condition supplements from the top menu **Goto • Condition Supplement**, provided the pricing procedure has been entered in the condition type in customizing. Figure 8.4 shows discounts RA00 and RB00 have been assigned to a PR00 condition record as condition supplements.
- **Proposed valid-from**: This field is used to propose a valid-from date when a rebate agreement is created. The default value can be changed in the rebate agreement if needed. The following default values are available:
  - <blank>: Today's date
  - **1**: First day of the week
  - **2**: First day of the month
  - **3**: First day of the year
  - **4**: No proposal

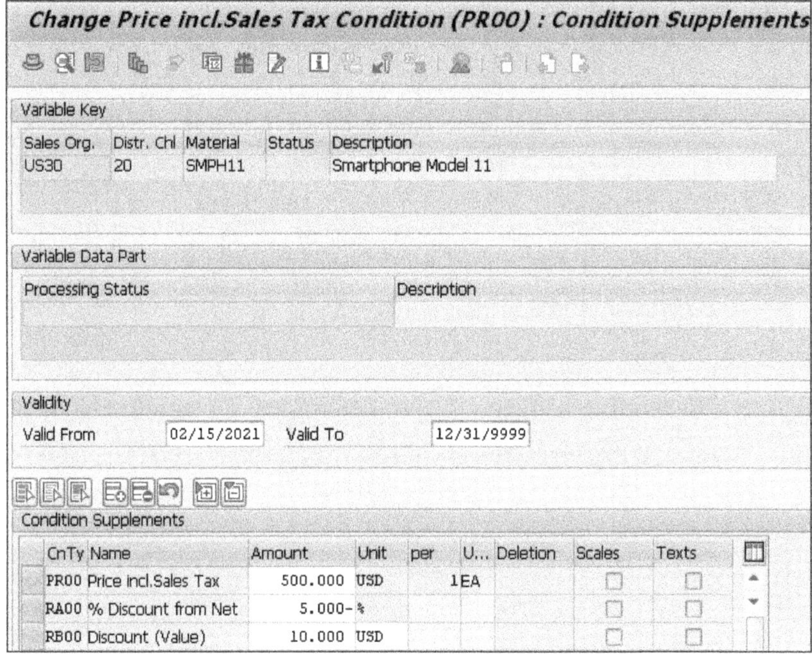

Figure 8.4 Maintaining Condition Supplements

- **Proposed valid-to**: This field is used to propose a valid-to date when a rebate agreement is created. The default value can be changed in the rebate agreement if needed. The following default values are available:
  - <blank>: December 31st, 9999
  - **0**: Today's date
  - **1**: End of the current month
  - **2**: End of the current year
  - **5**: No proposal
- **Ref. condition type**: In some cases, you may want to use two condition types for the same condition record value, where the only difference between the two condition types is the description, access sequence, or calculation type. By defining a reference condition type in one of the condition types, you only need to maintain condition records for the reference condition type, thus creating half of the condition records than would have been needed otherwise.
- **Ref. application**: If you want to reference condition records from other applications, you can maintain the reference application in this field.
- **Delete from DB**: The value maintained in this parameter controls how the system processes the action to delete pricing condition records. You can select one of the following values:
  - <blank>: The deletion flag is set for the pricing condition record so it can no longer be used during pricing.

- **A**: The user receives a popup message asking whether the condition record should be deleted from the database or whether the deletion indicator should be set.
- **B**: The condition record is deleted from the database. The user only receives a popup if there are condition supplements for the condition record.

- **Condition update**: This indicator specifies if the condition record must be determined depending on the maximum value defined. For example, you can create a condition record for a 10% discount that only applies to a particular product but then is deactivated once the cumulative discount value reaches \$1,000. If this indicator is selected, the system keeps track of the cumulative discount value, so if the condition record for the 10% discount is used in two orders of \$2,000 each, the system knows that the cumulative value is *2 × 0.1 × \$2,000 = \$400*, which means that the value left to be discounted via this condition record is \$600. In addition to the maximum value, you can use this function to track when the maximum quantity or the maximum number of orders is reached. This functionality is described in more detail in Chapter 3, Section 3.4.

In the **Scales** section, you'll maintain the following fields:

- **Scale basis**: This field specifies the basis for scales. For instance, the scale can be based on weight, volume, or quantity.
- **Check scale**: This parameter indicates whether the scale for this condition type must be entered in ascending order or in descending order.
- **Scale Type**: This indicator controls the validity of the scale based on the scale value. You can select from the following values:
  - **A (Base-scale)**: You can define a *base scale*, as shown in Figure 8.5. In this example, if an order is received for 55 units, the entire 55 units are valued at \$400; thus, the total order value is \$22,000.

Scales

Scale Type	Scale quantity	UoM	Amount	Unit	per	UoM
From	1	EA	500.000	USD	1	EA
	25		450.000			
	50		400.000			

Figure 8.5 Example of a Scale Type: Base-Scale

  - **B (To-scale)**: You can define a *to-scale*, as shown in Figure 8.6. In this example, if an order is received for 55 units, the entire 55 units are valued at \$400, and the total order value is \$22,000. We get the same result as the base scale shown in Figure 8.5; however, since the scale type is different, the values in the **Scale Type** and **Scale quantity** columns have changed.

Scales						
Scale Type	Scale quantity	UoM	Amount	Unit	per	UoM
to	25	EA	500.000	USD	1	EA
	50		450.000			
	75		400.000			

Figure 8.6 Example of a Scale Types To-Scale or Graduated

- D (Graduated-to interval scale): You can define a graduated scale, as shown in Figure 8.6. In this example, if an order is received for 55 units, the first 25 units are valued at $500 per unit, the next 25 units are at $450, and finally, the last 5 units are at $400. The total gross value, in this case, is $25,750 as displayed in the pricing elements for the item shown in Figure 8.7.

Pricing Elements									
Inactive	CnTy	Name	Amount	Crcy	per	UoM	Condition Value	Curr.
			0.000	USD	1	EA	0.000	USD
▢	PR00	Price incl.Sales Tax	500.000	USD	1	EA	12,500.000	USD
▢	PR00	Price incl.Sales Tax	450.000	USD	1	EA	11,250.000	USD
▢	PR00	Price incl.Sales Tax	400.000	USD	1	EA	2,000.000	USD
		Gross Value	468.182	USD	1	EA	25,750.000	USD

Figure 8.7 Item Pricing Elements in Sales Order Pricing: Graduated Scales Example

- <blank>: The default scale type is **A**, but this default can be changed in the **Details** screen of the pricing condition record. However, you cannot change this field to **D** since interval scales can only be chosen if the condition type is provided with scale type **D**.

In the **Control Data 2** section, you'll maintain the following fields:

- **Exclusion**: This parameter, also known as the *condition exclusion indicator*, controls whether discounts must be excluded during pricing. For example, let's say a large-scale customer who receives a particularly good price should not also get additional discounts that have been assigned to the same pricing procedure. If you select the **Exclusion** indicator in customizing when defining the pricing condition type, the exclusion applies to all condition records for the condition type. However, as mentioned in Chapter 3, Section 3.4, you can maintain this indicator in an individual condition record so that it only applies if the concerned condition record is determined.

Note

The **Exclusion** indicator does not exclude condition supplements that are part of the condition record for the price. Therefore, if condition supplements are found, they will always be taken into account despite the **Exclusion** indicator.

- **Pricing date**: This field dictates the date used by the system to find the available price. Depending on the value selected, the price is determined based on the following dates:

- <blank>: Pricing date on the sales document, but taxes and rebates are determined based on the date of services rendered
- **A**: Date of services rendered
- **B**: Pricing date on the sales document
- **C**: Billing date on the sales document
- **D**: Sales document creation date
- **E**: Date when the sales order was received

Now, let's move on to discuss the definition and assignment of pricing procedures.

8.2.4 Pricing Procedures

Pricing procedures are required so the system knows which condition types are to be taken into account and in which sequence when pricing is carried out in a sales document. You should not change the pricing procedures contained in the standard system; instead, we recommend you create your own pricing procedures that contain only the condition types needed in a specific business context. A common practice is to define separate pricing procedures if your company sells in several countries, as taxes are calculated differently in each country. To define a pricing procedure, follow the IMG menu path **Sales and Distribution • Basic Functions • Pricing • Pricing Control • Define and Assign Pricing Procedures.**

For each pricing procedure, you can maintain two general settings: the **Specific Procedure** field and the **Pricing Type** field, as shown in Figure 8.8.

Dialog Structure
Procedures
Procedures - Control

Usage: A
Application: V

Procedures

Procedure	Description	Specific Procedure	Pricing Type
RVAA01	Standard	☐	
RVAA02	Standard with Price Book	☐	
RVAA03	Standard w. Fixed Down Paym...	☐	
RVAAAR	Standard - Argentina	☐	
RVAACA	Standard - Canada	☐	
RVAACN	Standard	☐	
RVAAED	Standard	☐	
RVAAFR	Standard - France	☐	

Figure 8.8 Pricing Procedures: General Settings

Let's briefly review each setting:

- **Specific Procedure**: If this checkbox is selected, the procedure is considered a special pricing procedure. Special pricing procedures are used for intercompany billing and rebate credit memos. Also, this indicator can be used to redetermine the condition class and the statistical condition indicator when copying from a reference document.

- **Pricing Type**: The value selected in this field determines how pricing conditions are treated when copying sales documents. Some of the values possible for pricing types include the following options:
  - **A**: Copy price elements and redetermine scales
  - **B**: Carry out new pricing
  - **C**: Copy manual price elements and redetermine the others
  - **D**: Copy price elements unchanged
  - **E**: Copy price elements and fix values
  - **F**: Same as pricing type **E**, with the only difference being that the condition value (field KWERT) of the individual condition types is inverted during the transfer
  - **G**: Copy price elements unchanged and redetermine taxes
  - **H**: Redetermine freight conditions

All the condition types of a pricing procedure, such as gross prices, discounts, surcharges, taxes, etc. are contained in the **Procedures - Control Data** screen. Figure 8.9 shows this screen for standard pricing procedure RVAA01.

Procedure RVAA01 Standard

Procedures - Control Data

Reference Step Overview

Step	Co...	Co...	Description	Fro...	To ...	Ma...	R...	St...	Rel...	Print T...	Subtotal	Requir...	Alt. Ca...	Alt. Cn...	Accou...	Accruals
8	0	EK01	Actual Costs			✓					B	2			ERL	
11	0	PR00	Price incl.Sales Tax				✓					2			ERL	
13	0	PB00	Price (Gross)			✓						2			ERL	
20	0	VA00	Variant Price							X		2			ERL	
100	0		Gross Value							X	1		2			
101	0	KA00	Sales deal							X		2			ERS	
102	0	K032	Price Group/Material							X		2			ERS	
103	0	K005	Customer/Material							X		2			ERS	
104	0	K007	Customer Discount							X		2			ERS	
105	0	K004	Material							X		2			ERS	
106	0	K020	Price Group							X		2			ERS	
107	0	K029	Material pricing grp							X		2			ERS	
108	0	K030	Customer/Mat.Pr.Grp							X		2			ERS	

Figure 8.9 Control Data for Standard Pricing Procedure RVAA01

Two types of price elements exist in a pricing procedure: price condition types and subtotals. During pricing, the pricing procedure is read from top to bottom, and all the automatic condition types are determined by the condition technique. When a subtotal pricing element is reached, the system reads the **From** and **To** columns to identify the step numbers that define the range for the subtotal.

Note

The **From** field in the pricing procedure is also used to define the basis for a calculation. For example, if a discount condition is defined as a percentage, the **From** field tells the system which step in the pricing procedure must be used as the basis to calculate the discount amount.

Typically, all gross prices are followed by the **Gross Value** subtotal, followed by all possible conditions for discounts and surcharges. Then, the **Sum Surcharges/Discounts** subtotal is calculated, which is used to calculate the **Net Amount** subtotal. Finally, the price condition for taxes is determined and then added to the **Net Amount** subtotal to calculate the **Total Amount** subtotal, which corresponds to the final price including taxes. Table 8.1 shows, for example, how the different subtotals are calculated in a simple pricing procedure.

Step	Condition Type	Description	Amount	From	To
10	PR00	List Price	$50.00		
20	PMP0	Manual Price			
100		Gross Amount	$50.00		
110	DCM1	Customer Discount	$ -10.00		
120	HAN1	Handling Charges			
130	PAL1	Pallet Charges			
200		Sum Surcharges/Discounts	$ -10.00	101	199
210	PNP0	Net Price			
300		Net Amount	$40.00		
310	TTX1	Output Tax	21%		
400		Total Amount	$48.40		

Table 8.1 Example Calculation of Subtotal Amounts in a Pricing Procedure

On the screen shown in Figure 8.9, you can maintain settings for each price element. For instance, you can decide whether the condition type is determined automatically or added manually, whether the condition is mandatory, and whether the condition type is a statistical condition. If needed, you can also define the requirements that must be fulfilled for a condition type to be considered during pricing. In addition, you can specify whether the price element should be printed at the item level, in the total block, or not at all when printing sales documents.

Tip

For pricing conditions defined as manual in the pricing procedure, we recommend leaving the access sequence unassigned. Otherwise, the system will waste resources trying to find a valid pricing condition records that does not exist.

Once the pricing procedure has been defined, you can now configure the pricing procedure determination. First, you must define customer pricing procedures and maintain these procedures in the customer master record at the sales area level. Second, you'll specify the document pricing procedure for each sales document type and billing document type. Figure 8.10 shows various sales document types and their corresponding document pricing procedures.

SaTy	Sales Document Type	DoPr	Doc. Pricing Proc.
SO	Rush Order	A	Standard
SOR		A	Standard
SRVO	Sales Order (Srvce)	S1	Service Order
SRVP	Order in Soln Qtan	S2	Solution Quotation
STAT	Inquiry	A	Standard
OR	Standard Order	A	Standard
TAF	Standard Order (FPI)	F	DP90 W. Billing Plan
OBLS	order lump sum	S	Bill of Services
OBOS	order unit	S	Bill of Services
TAM	Delivery Order	A	Standard
TAV	Standard Order (VMI)	A	Standard
TBSS	Entry of Srvc Prfrmd	S	Bill of Services
TSA	Telephone Sales	A	Standard

Figure 8.10 Assigning Document Pricing Procedures to Sales Document Types

To complete the setup for pricing procedure determination, you must allocate the customer pricing procedure and the document pricing procedure to a pricing procedure within a specific sales area, as shown in Figure 8.11.

Sales Org...	Distri...	Division	Doc. Pricing Pr...	Cust.Pric.Pr...	Pricing Proced...	Pricing Procedure
US30	20	10	A	1	RVAA01	Standard
US30	20	10	V	1	PSER01	Service

Figure 8.11 Pricing Procedure Determination

Another important customizing activity within pricing control consists of defining, for each item category, whether or not the system should carry out pricing. This step involves maintaining one of the following values in the **Pricing** field, as shown in Figure 8.12:

- <blank>: No pricing is carried out.
- **X**: Standard pricing is carried out.
- **A**: Pricing for empties.
- **B**: Pricing for free goods (100% discount).

On the other hand, the **Statistical Value** field specifies whether the item value is used for statistical purposes or whether it should be included in determining the total value. You can select one of the following three options:

- <blank>: The system will copy the item value to the header totals.
- **X**: No cumulation. Values cannot be used statistically.
- **Y**: No cumulation. Values can be used statistically.

To activate pricing for item categories, follow the IMG menu path **Sales and Distribution • Basic Functions • Pricing • Pricing Control • Define Pricing by Item Category • Activate Pricing for Item Categories**. As shown in Figure 8.12, standard pricing has been activated for item categories TAN and TAO, whereas, for item category TANN, pricing for free goods is carried out.

ItemCat	Description	Pricing	Statistical Value
TAN	Standard Item	X	
TANN	Free of Charge Item	B	
TAO	Milestone-Bill.Plan	X	
TAP	Pric.at Item Level		X
TAP2	Kit: Replaced Item		
TAPA	Standard Sourcing		
TAPB	Pric.at Item Level		X

Figure 8.12 Activating Pricing for Item Categories

Note

Finally, master pricing condition records must be created for the condition technique to find the relevant prices, discounts, surcharges, taxes, etc. To refresh your memory on how to work with pricing condition records, refer to Chapter 3, Section 3.4.

Tip

We recommend adhering to the following guidelines to ensure optimal performance in pricing, which will result in better use of the system's resources:

- Standard pricing procedures may have pricing condition types that are not required by your project. If so, make a copy of the standard pricing procedure and remove any unnecessary pricing conditions.
- Use pricing requirements in the pricing procedure to make sure a condition type is only accessed if the requirement is met. Thus, no reading of the underlying access sequence and condition tables is carried out unless needed.

8.3 Applying Special Pricing Functions

You now have a good understanding of how to design and configure sales pricing conditions, access sequences, condition tables, and pricing procedures. In this section, we'll cover a few common special pricing functions and how they are used in SAP S/4HANA.

8.3.1 Group Conditions

Group conditions enable the system to read the quantity scale cumulatively, with the cumulated quantity of all materials in the sales order that belong to the same material pricing group. For this feature, you must maintain the condition type as a group condition in customizing.

Note

The material pricing group field (field MVKE-KONDM) is available in the material master record, in the **Grouping terms** section of the **Sales: sales org. 2** view. This field is maintained at the sales organization and distribution channel levels.

For example, imagine your company sells consumer electronic devices such as smartphones, tablets, and laptops. A new discount (discount condition type ZDIS) has been introduced, and depending on the number of laptops ordered, a customer might get a 0%, 15%, or 25% discount. Let's say all laptops are in material pricing group 01. In this example, the pricing condition record for condition type ZDIS is maintained for material pricing group 01 only with the scale shown in Table 8.2.

Number of Laptops Ordered	Discount (ZDIS)
1–5	0%
6–10	15%
11+	25%

Table 8.2 Quantities Eligible for Various Discounts

Now, let's say a customer places an order with three lines (two items for laptops and the third item for a tablet). If the **Group Condition** indicator is selected for the discount condition type, the system calculates the discount for each line, as shown in Table 8.3.

Item	Material	Quantity	Unit of Measurement (UoM)	Material Pricing Group	Discount (ZDIS)
10	Laptop model A	4	PC	01	15%
20	Laptop model B	3	PC	01	15%
30	Tablet model X	4	PC	02	0%

Table 8.3 Example of Sales Order Using Group Conditions

Note that neither item 10 nor 20 qualify for a discount individually, since neither item exceeds 5 PCs. However, since both items are of the same material pricing group, they are considered together by the group condition type and the quantity used to calculate the scale base value is 7 PCs. As a result, a 15% discount is applied to both items. For item 30, no condition record is found for condition type ZDIS since the material pricing group is not equal to "01."

You can also use the group conditions functionality with different condition keys. In other words, rather than maintaining one condition record with scales at the material pricing group level, you can maintain several condition records at the material level, and the system reads each condition record with the cumulated quantity for all the items with the same material pricing group. For group conditions with different condition keys to work, you must maintain the following settings for the condition type:

- Select the **Group Condition** indicator.
- Assign standard routine 3 (Mat.Pricing Group) in **Group Cond. Routine** field.

Now, let's illustrate this alternative use of group conditions with another example. Let's say your company wants to promote the sale of two materials by offering a special discount to one of your biggest customers:

- Material 299: TFT LCD Colour Display, 4.3 in, 480x272
- Material 341: 1.8" Inch 128x160 TFT Full Color LCD

Both materials 299 and 341 have material pricing group 01 in sales organization US30. According to the discount policy communicated by your company, the quantities ordered for both materials must be added up to calculate each of the material discounts.

The condition type used for the customer/material discount is K005. As shown in Figure 8.13, standard condition type **K005** has been configured for group condition with routine **3**.

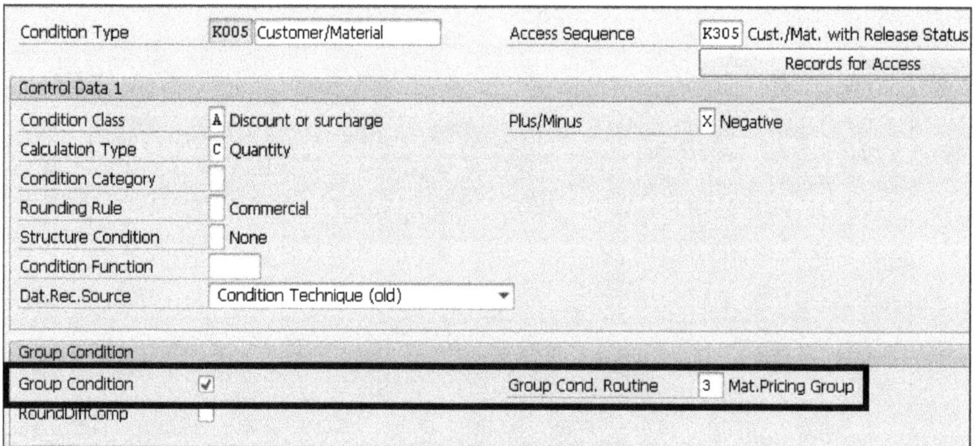

Figure 8.13 Settings for a Group Condition with a Different Condition Key

As mentioned earlier, the discount given to the customer depends on the quantity purchased of both materials, but the amount is different for each material. Figure 8.14 shows the scales in the condition records created for condition type K005 for both materials.

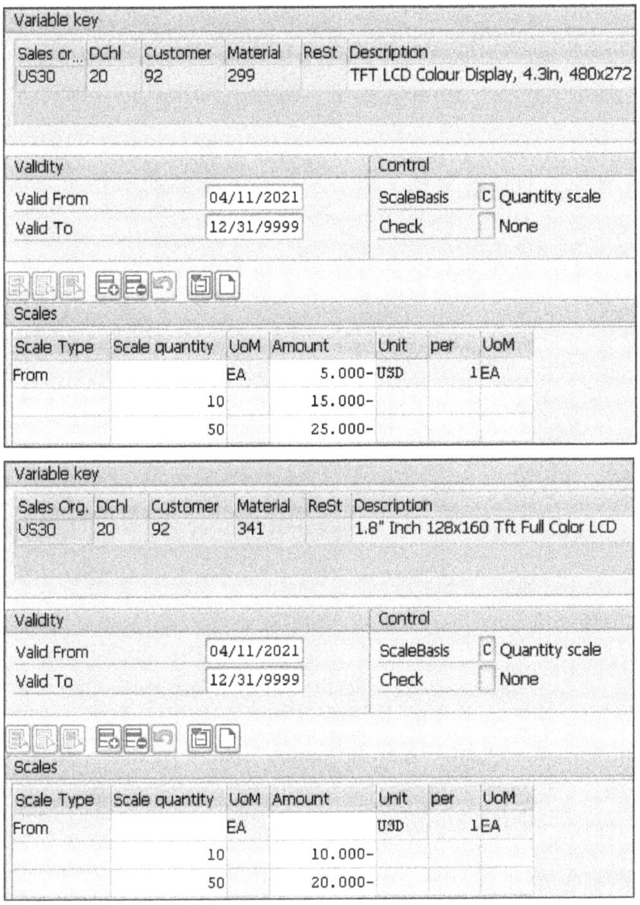

Variable key

Sales or...	DChl	Customer	Material	ReSt	Description
US30	20	92	299		TFT LCD Colour Display, 4.3in, 480x272

Validity		Control		
Valid From	04/11/2021	ScaleBasis	C	Quantity scale
Valid To	12/31/9999	Check		None

Scales

Scale Type	Scale quantity	UoM	Amount	Unit	per	UoM
From		EA	5.000-	USD	1	EA
	10		15.000-			
	50		25.000-			

Variable key

Sales Org.	DChl	Customer	Material	ReSt	Description
US30	20	92	341		1.8" Inch 128x160 Tft Full Color LCD

Validity		Control		
Valid From	04/11/2021	ScaleBasis	C	Quantity scale
Valid To	12/31/9999	Check		None

Scales

Scale Type	Scale quantity	UoM	Amount	Unit	per	UoM
From		EA		USD	1	EA
	10		10.000-			
	50		20.000-			

Figure 8.14 Scales for Discount K005: Materials 299 and 341

Based on the scales shown in Figure 8.14, if an order is created for 6 units of material 299, a discount of $5 is applied to the item. Similarly, if an order is placed for 5 units of material 341, no discount is determined for the item. However, if the customer orders 6 units of material 299 and 5 units of material 341 in the same order, based on the group condition settings for condition type K005, the system uses the quantity of 6 + 5 = 11 to read the scales in both condition records. Thus, the discount amounts of $15 and $10 are applied to materials 299 and 341, respectively. The values determined for condition type K005 for both materials are shown in Figure 8.15.

Note

Group conditions are not limited to discounts; they can also be useful when applied to surcharges such as freight charges. For example, you can define a condition type for a freight charge that reads the scale of the condition record based on the gross weight of the order items where the materials belong to the same material pricing group, and a valid condition record has been found for the freight charge.

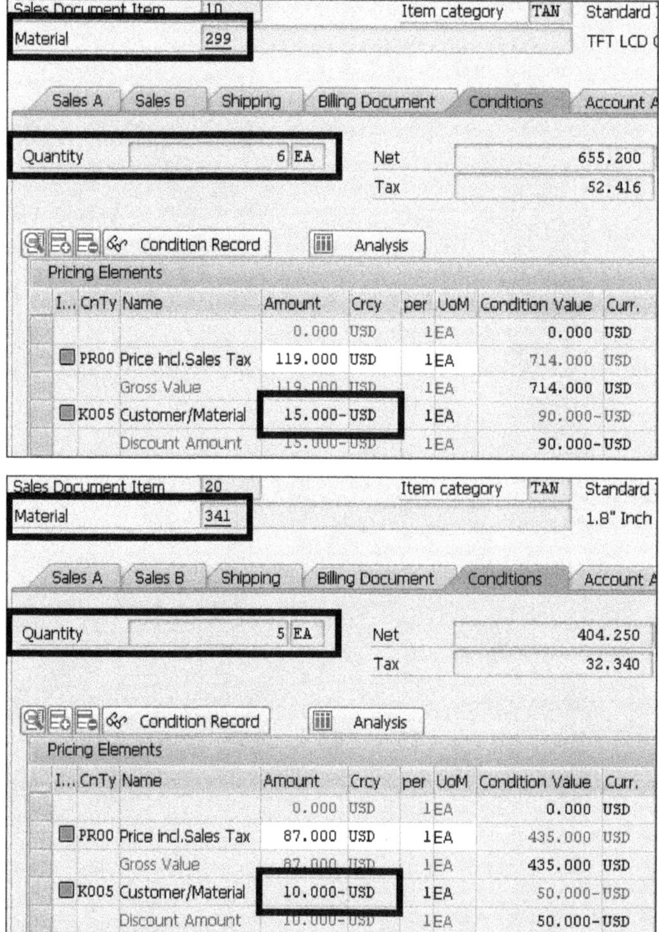

Figure 8.15 Condition Type K005 Discount Determination in a Sales Order Using Group Conditions

8.3.2 Condition Exclusion

During pricing, the system can find more than one valid condition record for a given sales document item, which may result in customers receiving two discounts when, in fact, they should only receive the most favorable price from the two discounts. By using the condition exclusion process, you can compare condition types with each other and determine the best price for the customer. All other condition types within the exclusion group are deactivated.

To customize exclusion groups, follow the IMG menu path **Sales and Distribution • Basic Functions • Pricing • Condition Exclusion • Condition Exclusion for Groups of Conditions.**

The following steps must be executed to enable condition exclusion for groups of conditions:

1. Define condition exclusion groups.
2. Allocate condition types to the condition exclusion groups.
3. Enter condition exclusion groups in the pricing procedure and specify a procedure (or rule) for determining the condition types to be excluded.

When comparing condition types during pricing, depending on how you configure the condition exclusion procedure, the system may exclude an entire group or just individual condition types within a group. SAP S/4HANA provides the following six predefined exclusion rules:

- **A**: The system selects the most favorable condition type within a condition exclusion group.
- **B**: The system selects the most favorable condition record of a condition type if several condition records exist. For example, let's say the access sequence assigned to the condition type has two or more accesses (condition tables), and the **Exclusive** indicator is not selected. In this case, the system may find two valid condition records for the same condition type, and this exclusion rule will choose the best price.
- **C**: The system selects the most favorable of the two condition exclusion groups. First, the system sums the amounts found for each group, the totals are compared, and only the conditions in the group with the overall best price remain active.
- **D**: This rule is known as an *exclusion procedure*. If any of the condition types of the first group exists for the item, all of the condition types of the second exclusion group are excluded.
- **E**: Similar to **B**, but with the most unfavorable condition record.
- **F**: Similar to **C**, but with the more unfavorable condition exclusion group.

You should be aware that condition records excluded by the system are not deleted from the sales order but are simply deactivated. Excluded records are indicated on the pricing screen with a yellow triangular icon. Within the pricing screen, if you double-click on an inactive condition record, you'll navigate to the **Conditions Details** screen where you'll see the **Inactive** indicator with the value "A" (Condition exclusion item).

Tip

In the standard system, conditions with the condition value zero do not participate in the exclusion via exclusion groups. However, you can change the system behavior so that conditions with the condition value zero also participate in the exclusion. In this case, enter the standard calculation formula 038 in the pricing procedure for one of the conditions to be compared.

Let's illustrate how you can use condition exclusion groups to ensure that only the best discount and the lowest freight charge are applied to the sales order item. You'll start by defining the following two exclusion groups:

- G001: Discount Group
- G002: Freight Group

For the purpose of this example, imagine you have two active discounts for the material ordered by the customer. One discount is a promotional 5% discount based on the material group (condition type ZD01), and the second is a 10% discount (condition type ZD02) that is exclusive for the customer and material entered in the sales order. Two freight charges (condition types ZF01 and ZF02) are determined in the same order.

Next, you'll assign the required condition types to the exclusion groups, as shown in Table 8.4.

Condition Exclusion Group	Description	Condition Type	Description
G001	Discount Group	ZD01	Discount 1
G001	Discount Group	ZD02	Discount 2
G002	Freight Group	ZF01	Freight 1
G002	Freight Group	ZF02	Freight 2

Table 8.4 Assigning Condition Types to Exclusion Groups

The next step is to maintain the condition exclusion for pricing procedures. Since we want the system to select the most favorable condition type within the exclusion group for both discount and freight, you must specify exclusion rule A in the **ConditExclusionProc** (condition exclusion procedure) column, as shown in Figure 8.16.

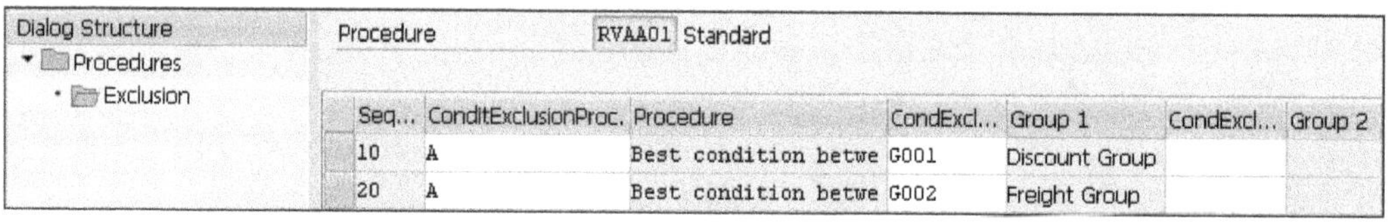

Dialog Structure
- Procedures
  - Exclusion

Procedure RVAA01 Standard

Seq...	ConditExclusionProc.	Procedure	CondExcl...	Group 1	CondExcl...	Group 2
10	A	Best condition betwe	G001	Discount Group		
20	A	Best condition betwe	G002	Freight Group		

Figure 8.16 Maintaining Condition Exclusion for Pricing Procedures

Four valid pricing condition records are found for the condition types in this example: ZD01, ZD02, ZF01, and ZF02. Thus, the preliminary pricing result shown in Table 8.5 is generated internally before the execution of exclusion. This pricing is not visible to the user but helps to illustrate this example.

Condition Type	Description	Condition Amount	Condition Value	Inactive
PR00	List Price	$500 per 1 PC	$500	
ZD01	Discount 1	5%	$25	
ZD02	Discount 2	10%	$50	

Table 8.5 Preliminary Pricing Result before the Exclusion Has Been Executed

Condition Type	Description	Condition Amount	Condition Value	Inactive
	Subtotal		$425	
ZF01	Freight 1	$6	$6	
ZF02	Freight 2	$12	$12	
	Total		$443	

Table 8.5 Preliminary Pricing Result before the Exclusion Has Been Executed (Cont.)

In this example, exclusion group G001 deactivates condition type ZD01 since the ZD02 condition record is the best discount. With regard to freight charges, exclusion group G002 deactivates condition type ZF02 since the condition record found for ZF01 has the lowest freight cost. Table 8.6 shows the final price calculation after the exclusion has been executed. This result is the final pricing result shown to the user on the screen. Note that the excluded condition types now have the value "A" in the **Inactive** column.

Condition Type	Description	Condition Amount	Condition Value	Inactive
PR00	List Price	$500 per 1 PC	$500	
ZD01	Discount 1	5%	$25	A
ZD02	Discount 2	10%	$50	
	Subtotal		$450	
ZF01	Freight 1	$6	$6	
ZF02	Freight 2	$12	$12	A
	Total		$456	

Table 8.6 Final Pricing Result after the Exclusion Has Been Executed

The correct, final price where the best discount and the lowest freight is applied to the sales order item is calculated in the following way:

$500 – $50 + $6 = $456

Now, let's look at the condition types for a minimum order value, which enables you automatically adjust the net value of the order to the minimum value allowed.

8.3.3 Minimum Order Value

Your company might need to set a minimum order amount to ensure that the cost of shipping an order is worthwhile for the business or if you have a set cost for each business transaction. In essence, you're ensuring a minimum profit amount per order. In SAP S/4HANA, you can specify a minimum order value using the following two standard condition types:

- AMIW: Minimum order value
- AMIZ: Minimum value surcharge

Let's consider an example: For material 296 (DateBook X Pro Power Cable), your company has established a minimum order value of $30 for all customers assigned to customer price group 01 (bulk buyer).

First, you must create a condition record for condition type AMIW for the minimum order value. In this example, a pricing condition record is created for the corresponding customer price group, the material, and the amount of $30, as shown in Figure 8.17.

Create Minimum SalesOrdrVal Condition (AMIW) : Fast Entry

Sales Organization	US30	ATI US
Distribution Channel	20	Direct
Price Group	01	Bulk buyer

Select Rule

Price group/Material

Material	Description	Amount	Unit	per	UoM	C..	S..	Valid From	Valid To
296	DateBook X Pro Power Cable	30.000	USD			B		04/14/2021	12/31/9999

Figure 8.17 Condition Record for Minimum Order Value

The list price for each unit is $10, so when a customer of price group 01 orders 2 units of material 296, the value of the sales order ($20) falls below the minimum order value ($30). At this point, the system automatically calculates the required surcharge to meet the minimum order value. As shown in Figure 8.18, the surcharge ($10) is applied in the pricing elements of the order item for condition type AMIZ in the **Condition Value** column. After applying the surcharge, the net amount for the item is $30, which matches the minimum order value.

Sales Document Item	10	Item category	TAN	Standard Item
Material	296			DateBook X Pro Pow

Sales A | Sales B | Shipping | Billing Document | Conditions | Account Assignment

Quantity	2 EA	Net	30.000 USD
		Tax	0.000

Condition Record | Analysis

Pricing Elements

I...	CnTy	Name	Amount	Crcy	per	UoM	Condition Value	Curr.
	PR00	Price incl.Sales Tax	10.000	USD	1	EA	20.000	USD
		Gross	10.000	USD	1	EA	20.000	USD
		Discount Amount	0.000	USD	1	EA	0.000	USD
		Rebate Basis	10.000	USD	1	EA	20.000	USD
		Net Value for Item	10.000	USD	1	EA	20.000	USD
	AMIW	Minimum SalesOrdrVal	30.000	USD			0.000	USD
	AMIZ	Minimum ValueSurchrg	30.000	USD			10.000	USD
			15.000	USD	1	EA	30.000	USD
		Net Value 2	15.000	USD	1	EA	30.000	USD
		Total	15.000	USD	1	EA	30.000	USD

Figure 8.18 Surcharge Added to the Order to Meet the Minimum Order Value

Condition types AMIW and AMIZ have been defined in the standard system for processing minimum value requirements. In this respect, the following settings have already been made for you:

- Condition type AMIZ references condition type AMIW with reference application "V." Therefore, you don't need to create a condition record for condition AMIZ.
- Neither AMIW nor AMIZ can be processed manually.
- Condition type AMIW is defined as a group condition.
- Standard pricing procedure RVCEU1 includes condition type AMIW with the following parameters:
  - **Statistical** = "X"
  - **Subtotal** = "D"
  - **Requirement** = "2"
  - **Account Key** = <blank>
- Standard pricing procedure RVCEU1 includes condition type AMIZ with the following parameters:
  - **Requirement** = "2"
  - **Alt. Calc. Cndn Amnt** (commonly called the *calculation formula*) = "13"
  - **Account Key** = "ERS"

8.3.4 Pricing Analysis

At this point, you've seen how pricing determination is a complex function within SAP S/4HANA. In some cases, you may not be able to tell why a sales document item shows a particular value for a condition type in the pricing procedure.

Pricing analysis is a standard function that can be called from any sales document to help the user understand how the system determined the different pricing elements in the document. You can execute a pricing analysis on a sales document item by clicking the **Analysis** button in the item's **Conditions** tab. You'll see the screen shown in Figure 8.19.

This screen is divided into three areas. On the left, the screen shows an overview of the pricing procedure with all its pricing elements. For each condition type, you can expand the three areas to see the various accesses as well as any condition records found during pricing. In the example shown in Figure 8.19, condition type PR00 has four accesses, and a condition record was found for access 040 (Material with release status) with a value of $119.

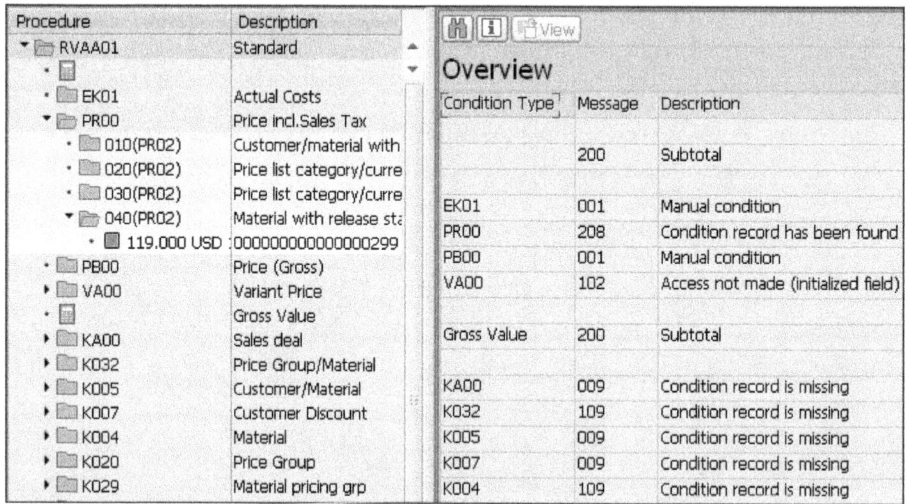

Figure 8.19 Analysis Pricing

In the top-right corner, the screen displays different information depending on which level you select in the overview tree area, as follows:

- If you select the top-level pricing procedure, the screen shows all the pricing elements and a short message with information about whether or not a condition record has been found, whether the condition type is manual, or whether the access could not be made due to a missing value in the document, to mention a few examples.
- If you select a condition type, the screen will display information regarding each of the relevant accesses. Based on this information, you can tell whether the access was not executed because a requirement was not fulfilled, a condition record was found, a condition record is missing, etc.
- If you select one condition access, the screen shows the fields in the condition table for that particular access and the values in the document for each field. By clicking on the **View** button, you can switch to the technical field names. In addition, you can see the pricing date used to fetch the condition record.
- If the system finds a condition record, if you click on the condition record level, the system will take you to a screen displaying the pricing condition record, where you can access all the information for that particular condition record, such as scales, validity periods, condition supplements, etc.

Finally, the lower-right corner of the pricing analysis screen is reserved for additional information on accesses and condition types. This area is particularly useful when a pricing requirement is assigned to either the access or the condition in the pricing procedure.

8.4 Special Condition Types

In this section, we'll cover some of the most relevant special condition types delivered by a standard system. You should be familiar with each of these condition types, which are all preconfigured to serve specific purposes in sales pricing:

- **Statistical condition type VPRS**
  Condition type VPRS (Internal price) is used to capture the standard cost of the material. This condition type is defined in the standard pricing procedure as a statistical condition, meaning this value does not alter the total value of the item. Condition type VPRS is defined with condition category G (internal price); thus, the system receives the cost from the valuation segment of the material master record. Standard pricing procedures containing condition VPRS usually have a subtotal called "Profit Margin," which uses formula 011 to subtract the cost in VPRS from the net value of the item. You cannot manually process condition type VPRS.
- **Manual conditions HM00 and PN00**
  You can use the manual condition HM00 (Order value) to specify the total value of the order at the header level. The system will distribute the HM00 condition value proportionally between the items based on the previous net value of each item. If you wish to enter the net price for an item manually, you can use manual condition PN00 (Price (net)).
- **Minimum price (condition type PMIN)**
  Condition type PMIN (Minimum Price) is a surcharge determined by the system when the minimum price is not met during pricing. First, you'll create a condition record for condition type PMIN and the corresponding material. Then, you'll enter the amount for the minimum price, for example, $150. When a sales order is created, the system checks whether the net price of the item falls below the minimum price and, if so, calculates the difference and assigns the value to condition PMIN. Going back to our earlier example, let's assume the list price of the material in question is $200 and a discount of $60 has been applied. The net value is $140, which is less than the $150 set in the condition record for the minimum price. In this scenario, the system adds the condition PMIN with the value of $10 as a surcharge to ensure the minimum price ($150) is met. Condition type PMIN has been defined so the manual entry has priority. Therefore, you can manually adjust this condition type in a sales order.
- **Cash discount (condition type SKTO)**
  Condition type SKTO (Cash discount) is a statistical condition used to display the cash discount after taxes as defined in the payment terms of the item. For example, let's say a sales order item uses payment term Y059 (10 Days 2% Disc, 20 Days Net). If payment is made within 10 days, the customer is eligible for a 2% cash discount. In this case, and assuming the **Cash Discount** indicator has been selected in the material master record (in the **Sales: sales org. 1** view) for the relevant sales organization, the system reads the 2% discount defined in payment terms Y059 and passes the discount to price element SKTO.

- **Electronic data interchange customer expected (price conditions EDI1 and EDI2)**
  If your company receives orders via electronic data interchange (EDI), you can pass the customer expected unit price and the customer expected item value from the inbound Intermediate Document (IDoc) to condition types EDI1 and EDI2, respectively. You can then use conditions types EDI1 and EDI2 to manage small price differences on the order line item level. The system compares the value in condition EDI1 to the unit price determined by the system. Usually, calculation formula 9 is assigned to condition type EDI1 in the pricing procedure. This formula checks whether the variance is greater than 0.05 currency units, and if so, the order is considered incomplete and must be reviewed and corrected. Similarly, calculation formula 8 is usually assigned to condition type EDI2 and tests for a maximum variance of 1.0 currency units between the calculated item value and the value stored in price element EDI2.

8.5 Condition Contract Management

The traditional rebate solution available in the SAP ERP 6.0 system was not flexible and suffered from major blockages every time the table containing rebate entries (table VBOX) had to be rebuilt due to changes in a customer's scenario. For this reason, with the release of SAP S/4HANA, a new central and standardized solution, called *condition contract management*, was introduced to create and settle all types of conditions for both suppliers and customers. The new condition contract management functionality in SAP S/4HANA has numerous advantages over traditional rebate agreements in SAP ERP 6.0, such as a single transaction code (or SAP Fiori app) to create and maintain condition contracts for customers and suppliers and real-time access to actual business volume data directly from transactional documents (e.g., billing document header and item tables VBRK and VBRP). Therefore, running a program periodically to update the sales volume to be settled in the rebate agreement is no longer required.

For the certification exams, you must have a high-level understanding of condition contracts with a customer as the contract partner. Condition contracts for suppliers are beyond the scope of this book.

Condition contracts allow you to work with the following two types of conditions:

- **Conditions accompanying an invoice**
  These conditions are all the condition types that are valid with immediate effect, which can alter the value of a sales order item. Prices, discounts, and surcharges are examples of conditions that accompany an invoice. You can store and edit these conditions in the condition contract, which acts as the central object where you can define validity periods, mark specific conditions as active, and add notes in the available text elements.

- **Conditions requiring subsequent settlement**
These conditions are settled at a later date. When creating a condition contract with condition types that require subsequent settlement, you must provide settlement dates, the information needed to calculate the business volume, which determines the settlement amount and validity dates. Accruals can be posted when invoices are created. Alternatively, if you don't want to impact the billing process, you can configure the creation of accruals for customer rebates to be carried out separately. For example, accruals can be calculated on a daily or monthly basis. You can settle condition contracts manually or automatically, and you can create condition contracts with a start date in the past. This feature is a nice addition in the SAP S/4HANA solution since the condition contract can be created after the transactional documents used to calculate business volume are posted.

In terms of configuration, the customizing settings related to condition contract management can be found in the IMG menu path **Logistics - General • Settlement Management • Condition Contract Management.**

The main three configuration areas are as follows:

1. **Configure pricing including condition contract conditions**
This configuration step concerns two areas: general settings for sales-related pricing and specific settings for contract conditions related to condition contracts with a customer as contract partner.
First, as in pricing for sales orders, you must define condition tables, access sequences, condition types, and pricing procedures.
The next step is to define sales condition types and activate them for condition contract management. In the standard system, the following condition types are available:
   - RES1: Rebate
   - REA1: Rebate Accruals
Then, you'll need to define sales condition type groups and assign sales condition types to a sales condition type group.
2. **Configure condition contract maintenance**
In this second configuration step, you must define and configure parameters for condition contract types, such as number ranges, condition contract categories, field status groups, status profiles, action profiles, and text determination procedures.
Then, you'll define condition contract types with the configured parameters to determine how a condition contract will be processed. When defining condition contract types, the **Type of Contract Partner** field is mandatory since this field controls whether the contract partner is a customer or a supplier. The **Type of Eligible Partners** field is also mandatory and specifies whether or not eligible partners play a role for the condition contract type.

3. **Configure condition contract settlement**
   Finally, you must maintain the following settings to allow the system to process the condition contract settlement and calculate the settlement amount for a given period:
   - Configure business volume determination: This value is usually based on the total amount of customer invoices.
   - Specify settlement calendar: You must define the factory calendars to be used as settlement calendars and assign them to the relevant condition contract types. The calendar allocated to the condition contract type will be the default calendar, but you can overwrite this value in the actual condition contract if needed.
   - Configure the settlement document creation and settlement amount calculation: In this step, you'll maintain settings for the settlement document creation, such as specifying the condition contract settlement type, the settlement type customer, and the settlement process type. You'll also define the pricing procedure used for the settlement amount calculation and maintain the pricing procedure determination, which depends on the settlement document type of the settlement document.
   - Specify posting and reversal of accruals: This configuration enables the system to reverse accrual postings and replace them with the actual amounts determined during the condition contract settlement.

Let's now review the processing of condition contracts and the settlement process. For condition contracts requiring subsequent settlement, you can perform the following business activities in SAP S/4HANA:

1. **Create or change a condition contract**
   The condition contract management solution revolves around the condition contract document. In this document, you can manage conditions, specify administrative data, define the criteria for business volume selection, specify eligible partners, and maintain settlement calendar information. To create or change a condition contract, use Transaction WCOCO or, from the SAP Fiori launchpad, use the Maintain Contract – Condition Contracts app. Figure 8.20 shows a condition contract as displayed in Transaction WCOCO.

   Once all the required fields have been maintained, you can save a condition contract.

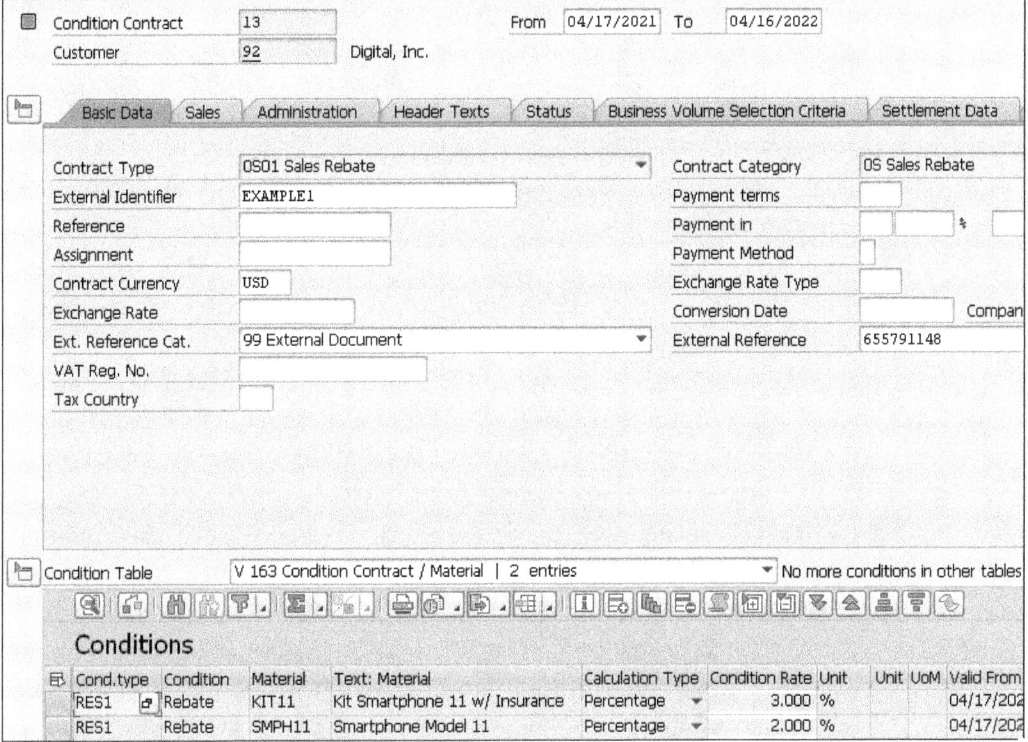

Figure 8.20 Condition Contract: Transaction Code WCOCO

2. **Release a condition contract**
   When a condition contract is created, you'll see the status icon in the header line, which means the contract is inactive for operative use. To release a condition contract, switch to *change mode* by selecting the menu option **Condition Contract • Change** in Transaction WCOCO and click on the function from the toolbar. The status icon in the header changes to . You can also release (or lock) condition contracts in mass via the transaction codes and SAP Fiori apps shown in Table 8.7.

Transaction Code (SAP GUI)	Transaction Code Description	SAP Fiori App
WCOCO_ACTION	Perform Business Transaction for Condition Contracts	Set Status – Condition Contracts
WCOCOALL	Condition Contract List	Display Contracts – Condition Contracts

Table 8.7 Ways to Release/Lock Condition Contracts in Mass

3. **Perform settlement of a condition contract**
   As mentioned earlier, by executing the settlement process, you create the corresponding amounts receivable and payable in financial accounting for a specific period and the relevant business partners.

You can carry out the settlement of a condition contract in SAP S/4HANA with a customer as the condition contract partner via Transaction WB2R_SC or the Settle Condition Contracts app. The key figure used for the settlement, also known as *business volume base*, is determined from the transactional document (e.g., billing document) and constitutes the basis for the settlement. For instance, you can use weight, quantity, volume, or monetary amount as the business volume base.

In the condition contract header, under the **Settlement Calendar** tab, you can also specify the settlement dates and which of the following four settlement types must be used when settlement is performed:

- **<blank>: Final Settlement**: With this settlement type, the conditions of a condition contract are permanently settled for a specific time period. You can have several final settlements in a single condition contract. If necessary, final settlements can be corrected via a delta settlement.
- **1: Partial Settlement**: With this settlement type, the conditions of a condition contract are provisionally settled for a specific time period. Typically, the final settlement is used to settle the partial settlements for a final time.
- **2: Delta Settlement**: Delta settlements can be used to settle subsequent changes in the business volume base, for example, invoice cancellations or returns posted after a final settlement has been processed for a specific time period.
- **3: Delta Accruals**: Delta accruals can be used to make corrections to accruals based on changes in the condition data.

4. **Control the condition contract management process**
   In addition to Transaction WCOCOALL (Condition Contract List), the standard reports shown in Table 8.8 are also available to review condition contract conditions and settlement-related data for condition contracts.

Transaction Code (SAP GUI)	Transaction Code Description	SAP Fiori App
WCOLI	Condition Overview – Condition Contract	Display Conditions – Condition Contracts
WB2R_SETTL_CAL	Settlement Calendar Data for Condition Contract	Display Settlement Calendars – Condition Contracts
WB2R_BUSVOL	Business Volume for Condition Contracts	Display Business Volume – Condition Contracts
WB2R_AB_DOCS	Settlement Management Documents for Condition Contract	Display Settlement Documents – Condition Contracts
WB2R_BVDETAIL	Detailed Statement for Condition Contract Settlement	Display Detailed Statement – Condition Contracts

Table 8.8 Standard Reports for Condition Contract Management

8.6 Important Terminology

In this chapter, the following terminology was used:

- **Field catalog**
  In the context of pricing, the field catalog refers to the list of pricing fields used in condition tables. If the list of allowed fields is not sufficient, you can create a new Z-field and include it in the pricing structure KOMKAZ for header fields and KOMPAZ for line-item fields.
- **Condition table**
  This table defines the combination of fields that can be used for each step of an access sequence. By reading a condition table, the system can identify an individual condition record.
- **Access sequence**
  Stores the condition tables and the sequence in which they are searched to find a valid condition record. Each condition type that uses condition records must have an access sequence, and you can assign the same access sequence to multiple condition types.
- **Condition type**
  Each pricing condition type represents a different kind of price in a sales document. Discounts, surcharges, automatic list prices, manual prices, and taxes are some examples of condition types. Pricing condition types are applicable to either the document header or to line items.
- **Pricing condition record**
  The master data record that contains the pricing value that you want to store and retrieve during pricing.
- **Condition supplement**
  A supplement included every time you apply a material price. For example, let's say you want to apply both a special discount and a customer discount every time the system accesses the pricing condition record for condition PR00 where the condition supplement has been included. When a sales order is created, if the condition record is found, the system automatically calculates the PR00 gross price, the special discount, and the customer discount.
- **Price element**
  An item in the pricing procedure. Each step in the pricing procedure is a price element, which can be either a subtotal or a pricing condition type.
- **Document pricing procedure**
  The key that determines the pricing procedure for each type of sales document together with the sales area and the customer classification for pricing procedure determination in the customer master record.
- **Customer pricing procedure**
  This field, also known as the *customer classification for pricing procedure determination*, is a field in the customer master record in the **Sales and Distribution**

view. This field determines which pricing procedure the system should apply when you create a sales document for the customer.

- **Condition exclusion group**
  A list of condition types that are compared with each other during pricing and result in the exclusion of an entire group or of individual condition types within a group.
- **Condition exclusion procedure**
  A field in the customizing for condition exclusions that controls the exclusion rule within an exclusion group or between two exclusion groups for a pricing procedure. The standard system supports six options.
- **Rebate**
  A rebate is a special discount that is paid retroactively to a customer. This discount is based on the customer's sales volume over a specified time period.
- **Condition contract management**
  Part of settlement management in SAP S/4HANA that allows you to handle the settlement of customer and supplier rebates through a standardized solution. Condition contract management uses conditions that require subsequent settlement as well as conditions that accompany an invoice. The latter are valid with immediate effect and include prices, value conditions, and quantity conditions. These conditions can be maintained in a sales document called the *condition contract* and form the basis for settling claims or rebates on suppliers or customers.

8.7 Practice Questions

These practice questions will help you evaluate your understanding of the topics covered in this chapter. The questions shown are similar in nature to those found on the certification examination. Although none of these questions will be found on the exam itself, they will allow you to review your knowledge of the subject. Select the correct answers and then check the completeness of your answers in the next section. Remember that, on the exam, you must select all correct answers and only correct answers to receive credit for the question.

1. True or False: When creating a new custom condition table in pricing, if the fields in the field catalog are not sufficient, you can add new fields as long as they are in the list of allowed fields.

☐ A. True

☐ B. False

2. You've set up a condition type as a group condition discount. You've created a condition record for this condition type and material pricing group 01 with a scale, so a 10% discount is applied if the quantity ordered is greater than

5 PCs. In which of the following scenarios will the 10% discount be determined for all the items in the order?

☐ A. A sales order with two items: Each item has a quantity of 3 PCs, the materials are different, but both belong to material group 01.

☐ B. A sales order with two items: Each item has a quantity of 3 PCs, the materials are different, and they belong to different material groups.

☐ C. A sales order with one item where the quantity ordered is 5 PCs and the material group is 01.

☐ D. A sales order with two items: Each item has a quantity of 6 PCs, the materials are different, and they belong to different material groups.

3. In pricing for sales, where do you define a reference condition type?

☐ A. In the pricing condition record

☐ B. In the pricing procedure

☐ C. In the pricing condition type

☐ D. In the access sequence

4. What scale types can you select when defining pricing condition types? (There are three correct answers.)

☐ A. Base scale

☐ B. From-scale

☐ C. To-scale

☐ D. Graduated-to interval scale

☐ E. Linear scale

5. Where can you set the condition exclusion indicator? (There are two correct answers.)

☐ A. In the pricing procedure

☐ B. In the pricing condition type

☐ C. In the access sequence

☐ D. In the pricing condition record

6. True or False: To be able to use group conditions in SAP S/4HANA, the condition type must have the **Group Condition** indicator selected, and the materials must belong to the same material pricing group.

☐ A. True

☐ B. False

7. Which of the following customizing steps must be executed to enable condition exclusion for groups of conditions in pricing? (There are three correct answers.)

- ☐ A. Define condition exclusion groups
- ☐ B. Define exclusion rules
- ☐ C. Assign condition types to exclusion groups
- ☐ D. Assign pricing condition records to exclusion groups
- ☐ E. Maintain condition exclusion for pricing procedures

8. Your customer wants to specify a minimum order value for sales order processing. To meet this requirement, you're planning to use standard condition types AMIW and AMIZ. Where do you specify the minimum order value?

- ☐ A. In the customer master record
- ☐ B. In the pricing condition record for condition type AMIZ
- ☐ C. In the pricing condition record for condition type AMIW
- ☐ D. In the pricing condition records for condition types AMIW and AMIZ

9. How do you maintain the settings for pricing procedure determination? (There are two correct answers.)

- ☐ A. You allocate the sales document type to the customer pricing procedure.
- ☐ B. You assign the sales document type to the document pricing procedure.
- ☐ C. You allocate the pricing procedure to the sales document type.
- ☐ D. You assign the pricing procedure to the document pricing procedure and customer pricing procedure within a sales area.

10. When creating a condition contract in SAP S/4HANA with conditions requiring subsequent settlement, which of the following settlement types can be used? (There are three correct answers.)

- ☐ A. Complete settlement
- ☐ B. Final settlement
- ☐ C. Delta settlement
- ☐ D. Future settlement
- ☐ E. Delta accruals

11. Which of the following are the main customizing steps involved in the configuration of condition contract management in SAP S/4HANA? (There are three correct answers.)

☐ A. **Activate Rebate Processing**

☐ B. **Account Determination For Rebates**

☐ C. **Configure Pricing including Condition Contract Conditions**

☐ D. **Configure Condition Contract Maintenance**

☐ E. **Configure Condition Contract Settlement**

12. True or False: In condition contract management, standard condition type RES1 is used to calculate the rebate accrual value.

☐ A. True

☐ B. False

13. Which of the following standard pricing condition types cannot be processed manually? (There are two correct answers.)

☐ A. PMIN – Minimum Price

☐ B. AMIW – Minimum SalesOrdrVal

☐ C. VPRS – Internal price

☐ D. PN00 – Price (net)

14. In a sales document, let's say you want header condition ZDOO (Freight) to be distributed among the items based on the volume. Which of the following fields do you need to maintain for the condition type concerned?

☐ A. Condition category

☐ B. Calculation type

☐ C. Condition class

☐ D. Structure condition

8.8 Practice Question Answers and Explanations

1. Correct answer: **B**
   False. If the required field is not in the list of allowed fields, you can still add it to the field catalog by appending the new field to the communication structures KOMPAZ (Pricing Communication Item: Customer Modifications) or KOMKAZ (Pricing Communication Header: Customer Modifications), depending on whether the new field is an item field or a header field. This change is regarded as an enhancement and requires ABAP code in the corresponding user exit and thus must be implemented with the help of an ABAP developer.

2. Correct answer: **A**

   This question tests your knowledge of the **Group Condition** indicator. The condition type in question has this indicator selected because this condition type has been set up as a group condition; thus, the system will consider all the related items in the order to calculate the scale base value. The pricing condition record for the discount has been created for material pricing group 01 and has a scale. In scenario A, the system considers both items since both materials have the material group 01 despite being different materials, so the quantity used to calculate the condition value is 3 + 3 = 6 PCs. Since 6 is greater than 5, the 10% discount is determined for both items. Answer A is, therefore, the correct answer.

   Answer B is incorrect because only one of the items can have the material pricing group 01. Thus, the system will only consider the quantity for one item, which is 3 PCs, and thus does not qualify for a 10% discount. Answer C is also incorrect because only one item exists in the order and the quantity ordered is not greater than 5 PCs. Finally, answer D is incorrect too because only one of the two items can have the material pricing group 01, meaning the other item does not qualify for the 10% discount despite both items having a quantity greater than 5 PCs. The condition in the statement is that all items in the order must qualify for the discount.

3. Correct answer: **C**

   A reference condition type is defined in the **Master Data** section of the pricing condition type. If the condition type to be used as a reference is entered into the **Ref. Condition Type** field, you only need to maintain condition records for the reference condition type.

4. Correct answers: **A, C, and D**

   The scale types available in SAP S/4HANA are base scale, to-scale, and graduated-to interval scale. These three scale types and the price elements for a graduated-to interval scale example are shown in Figure 8.5, Figure 8.6, and Figure 8.7.

5. Correct answers: **B and D**

   The condition exclusion indicator is available in the definition of the condition type in the **Control Data 2** section as well as in the details view of the pricing condition record in the **Control** section. If maintained for the condition type, the indicator applies to all condition records. However, you can maintain this indicator in an individual condition record so that it only applies if the concerned condition record is determined.

6. Correct answer: **A**

   True. As discussed at the beginning of Section 8.3, group conditions enable the system to read the quantity scale cumulatively with the cumulated quantity of all materials in the sales order that belong to the same material pricing group, provided the condition type has been marked as a group condition in customizing.

7. Correct answers: **A, C and E**

   To enable condition exclusion for groups of conditions, you must perform the following three customizing steps, which can be carried out in the IMG under the path **Sales and Distribution • Basic Functions • Pricing • Condition Exclusion • Condition Exclusion for Groups of Conditions**:

   - Define condition exclusion groups
   - Assign condition types to exclusion groups
   - Maintain condition exclusion for pricing procedures

8. Correct answer: **C**

   When using standard condition types AMIW (Minimum SalesOrdrVal) and AMIZ (Minimum ValueSurchrg) to set up a minimum order value, you must specify the minimum order value in the pricing condition record for condition type AMIW. Condition type AMIZ has already been defined with AMIW as the reference condition type and calculation formula 13 in the pricing procedure. This setup allows the system to determine a surcharge for condition type AMIZ if the order net value falls below the minimum order value specified in the AMIW condition record.

9. Correct answers: **B and D**

   To configure the pricing procedure determination, first, you must assign the sales document type to the document pricing procedure, as shown earlier in Figure 8.10. Then, you'll assign the pricing procedure to the document pricing procedure and customer pricing procedure within a sales area, as shown earlier in Figure 8.11. You cannot assign the sales document type to the customer pricing procedure, which invalidates answer A. Answer C is also incorrect as you cannot directly assign the pricing procedure to the sales document type.

10. Correct answers: **B, C, and E**

    In the condition contract, you can specify which settlement type is to be used. Four settlement types are available in SAP S/4HANA:

    - Partial settlement
    - Final settlement
    - Delta settlement
    - Delta accruals

    Complete settlement and future settlement are not valid settlement types in SAP S/4HANA. Therefore, answers A and D are incorrect.

11. Correct answers: **C, D, and E**

    Answers A and B correspond to settings for rebate processing, which is the legacy SAP ERP 6.0 solution for rebates. With the launch of SAP S/4HANA, rebate processing was replaced by condition contract management. Therefore, answers A and B are both incorrect answers. The main customizing steps to set up condition contract management are:

- **Configure Pricing including Condition Contract Conditions**
- **Configure Condition Contract Maintenance**
- **Configure Condition Contract Settlement**

12. Correct answer: **B**

    False. In condition contract management, the standard condition type used to calculate the rebate accrual value is REA1.

13. Correct answers: **B and C**

    Neither standard condition type AMIW nor VPRS can be processed manually. On the other hand, PN00 is a condition type used to enter the net price for an item manually, and condition PMIN has been defined so the manual entry has priority.

14. Correct answer: **B**

    The **Calculation Type** field determines whether the system calculates a price as a fixed amount or as a percentage based on quantity, volume, or weight. If calculation type "F" (volume) is selected for a freight condition type entered at the header level, the surcharge amount will be distributed among the items based on the volume of each item.

8.9 Takeaway

In this chapter, you learned about the condition technique as used in pricing and how to maintain these settings to ensure optimal performance. You're familiar with the configuration items to determine prices via the condition technique, like creating condition tables, setting up access sequences, defining pricing condition types, and defining and assigning pricing procedures. You've seen how to apply special pricing functions, such as group conditions, exclusion procedures for groups of conditions, and minimum order values and how to conduct pricing analysis.

You can explain the purpose and how to use some of the special pricing condition types delivered by SAP S/4HANA.

Finally, you gained insight into the sales side of condition contract management, the innovative SAP S/4HANA solution to create and settle all types of conditions for both suppliers and customers that replaces the old rebate mechanism.

8.10 Summary

Pricing is a complicated subject, so one challenge was to explain, in a single chapter, the most relevant concepts that will ensure you're prepared to answer a good number of pricing-related questions correctly when taking the certification exams. We're confident the goal has been achieved.

You should now have a good understanding of how to design and configure pricing condition types, access sequences, condition tables, and pricing procedures. You should also understand condition exclusions and how you can use them to determine the best prices for a customer. We discussed group conditions, special pricing condition types, and condition contracts.

In the next chapter, we'll go over the sequence of typical sales processes, and you'll refresh your knowledge on the document flow of sales processes, pre-sales documents, make-to-order production, and complaints and third-party order processing.

Chapter 9
Sales Process

Techniques You'll Master

- Explain the order-to-cash process in SAP S/4HANA
- Use sales inquiries and quotations
- Understand the origin of data in sales documents
- Create sales orders with reference to a preceding document
- Explain rush orders and cash sales
- Understand special business processes in sales
- Handle customer complaints

The focus of this chapter is on explaining basic sales concepts, like the origin of data in sales documents, copying rules, and special sales order types. In addition, we'll discuss sales processes in SAP S/4HANA, such as pre-sales business processes, the make-to-order production process, and third-party order processing. Other topics reviewed include customer claims, returns, and refund management.

Real-World Scenario

SAP S/4HANA delivers, out of the box, the majority of the standard business functions required in sales and other areas. From recording pre-sale activities in the form of an inquiry or quotation to creating the sales order, picking, packing, shipping, and billing, all are steps in the order-to-cash process that can be performed and controlled in the system.

As an SAP S/4HANA Sales consultant, you'll need a good understanding of the more specialized business processes in sales supported in SAP S/4HANA. For example, your company may require you to ship certain products to your customer directly from a supplier rather than from your own company. Once the goods are delivered to the customer, the customer invoice is triggered based on the actual quantity delivered. You'll need to use third-party order processing to manage this scenario efficiently, thus saving on shipping costs and transactions.

Also, to respond to customer complaints (for example, receiving an invoice with an incorrect amount), you must be familiar with the claims and refund management tools that will enable you to implement the corresponding processes to resolve these situations promptly, thus helping to maintain (or even improve) customer satisfaction.

9.1 Objectives of This Portion of the Test

The goal of this portion of the SAP S/4HANA Sales Certification exam is to test your understanding of the following topics:

- Logical sequences for standard order-to-cash process
- Sales inquiries and quotations
- Origin of data in sales documents
- Reference status and copying rules
- Rush orders and cash sales
- Make-to-order production
- Third-party order processing
- Complaint and returns processing

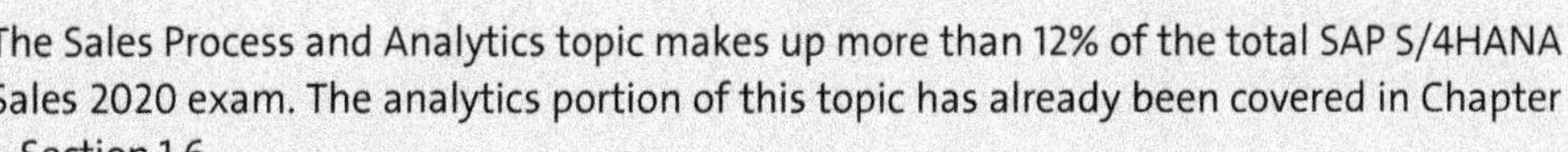

Note

The Sales Process and Analytics topic makes up more than 12% of the total SAP S/4HANA Sales 2020 exam. The analytics portion of this topic has already been covered in Chapter 1, Section 1.6.

The content of this chapter is included in the Sales topic of the SAP S/4HANA Sales 2020 Upskilling exam, which makes up more than 12% of the total exam.

9.2 Order-to-Cash Process

The order-to-cash process, representing the most generic business cycle in sales, is used by every company that delivers goods or provides services to customers. A basic order-to-cash process consists of four stages, usually performed in the following sequence:

1. Pre-sales activities
2. Sales order processing
3. Shipping
4. Billing

Not all four stages are mandatory to complete a sales cycle. For example, when selling services, you can use an item category that is relevant for order-related billing like standard item category TAD. In this scenario, you can skip the shipping stage and go from order processing to billing directly. Pre-sales activities involve the creation of inquiries, quotations, customer contracts, and scheduling agreements. Pre-sales documents are all optional documents. However, from a business perspective, in some cases, pre-sales documents may be required (for instance, contracts and scheduling agreements are key documents in the healthcare and automotive industries). Figure 9.1 shows the two basic order-to-cash processes for goods and services. In these diagrams, each box represents a particular document in SAP S/4HANA.

To deliver goods, additional steps are carried out in the areas of warehouse and transportation management, but these steps are not the focus of this chapter. Let's look at an example to illustrate the general four stages. Imagine a customer calls your company asking for the characteristics and availability of a particular product. At this point, an inquiry is created in the system by the sales representative. Later on, the same customer calls again and requests a quotation, which can be created with reference to the inquiry. The customer agrees with the price and other terms and conditions in the quotation and places an order. The sales representative can now create a sales order with reference to the quotation. The sales order is the contractual agreement between your company and the customer for the supply of the product requested. We're now entering the shipping stage of the order-to-cash process where the outbound delivery must be created and transportation arranged with the carrier.

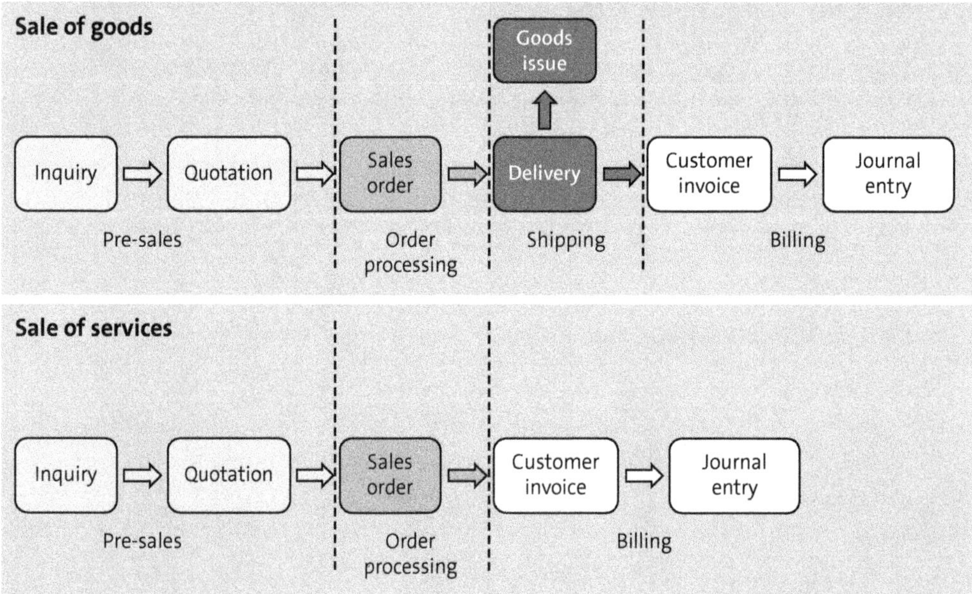

Figure 9.1 Basic Order-to-Cash Processes in SAP S/4HANA

The next step is to pick, pack, and ship the goods from the warehouse. Once the goods leave the docks, the post goods issue is performed from the outbound delivery, and stock levels are adjusted accordingly. Usually, a billing batch job bills the outbound delivery overnight and creates the corresponding billing document, which is sent to the customer via email. Finally, the billing document is released to accounting, which creates a journal entry in SAP S/4HANA Finance. From this point onwards, the accounts receivable team must make sure the payment is collected on time. The payment transaction itself, however, is not considered part of the sales and distribution process.

As we progress through the steps shown in Figure 9.1, several interrelated documents are generated. The sales document flow is a nice feature that displays the entire chain of documents, thus allowing you to visualize all relevant preceding and subsequent documents. You can access the sales document flow at any time by selecting the top menu option **Environment • Display Document Flow** from the relevant transaction code or SAP Fiori app used to display the sales document. The sales document flow can be seen at the header and item levels depending on whether the cursor is positioned in a header field or an item field in the sales document overview screen. Figure 9.2 shows the sales document flow from the sales order, at the header level, for the order-to-cash scenario we've just described.

In fact, the order-to-cash process does not always end with the customer invoice. Customer claims may result in subsequent documents being added to the sales document flow. Let's assume the customer requested all products to be part of the same batch. However, upon inspection of the goods received, the customer realizes that some products belong to a different batch and calls customer service to initiate a returns process. A returns order is created in the system with reference to the customer invoice, followed by the returns delivery. The post goods receipt is performed via the returns delivery once the goods returned by the customer arrive in the warehouse. Then, a credit memo is issued for the quantity received.

Document Flow

Status Overview | Display Document | Service Documents | Additional Links

Business Partner 0000000092 Digital, Inc.

Document	On	Time	Status
Inquiry 0007000005	03/02/2021	16:44:01	Completed
Quotation 0020000012	03/02/2021	16:45:45	Completed
Standard Order 0000000254	03/02/2021	16:46:42	In Process
Outbound Delivery 0080000196	03/02/2021	16:47:14	Completed
Picking Request 20210302	03/02/2021	16:47:23	Completed
GD goods issue:delvy 4900001074	03/02/2021	16:47:33	Complete
Invoice 0090000146	03/02/2021	16:48:06	Completed
Journal Entry 0090000013	03/02/2021	16:48:09	Not Cleared

Figure 9.2 Sales Document Flow at the Header Level

Regarding services, a customer may ask for a refund if the net amount on the invoice does not match the net amount on the quotation. In this case, a different type of sales document, called a *credit memo request*, can be created with reference to the customer invoice. No delivery is required for services, so the credit memo request is billed to generate a credit memo and compensate for the price discrepancy. These new stages of the order-to-cash process are shown in Figure 9.3.

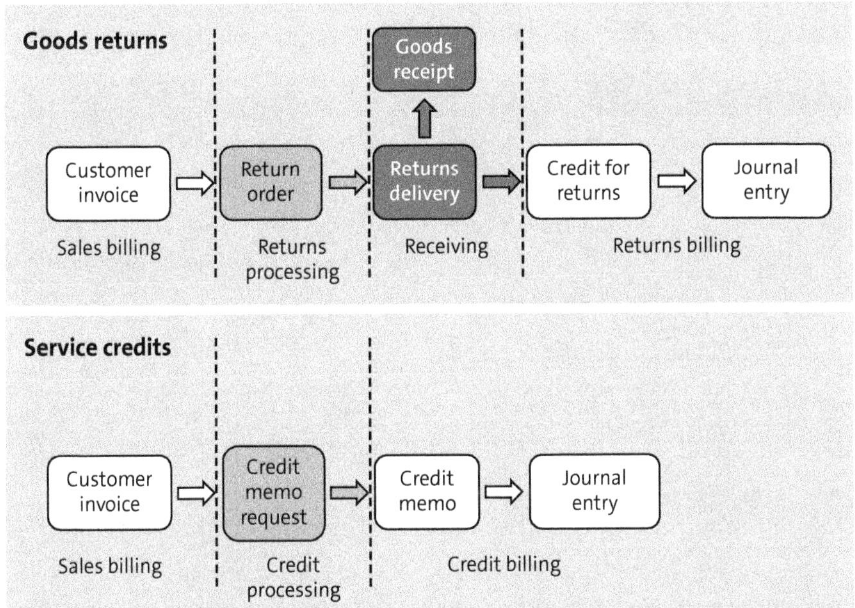

Figure 9.3 Return and Refund Processes in SAP S/4HANA

As mentioned earlier, if subsequent documents are created with reference to existing documents, they will be connected to the preceding document and will be added to the sales document flow, giving business users a good end-to-end overview of the process.

9.3 Using Sales Inquiries and Quotations

Pre-sales documents, such as inquiries and quotations, are not mandatory steps from the order-to-cash process perspective, but these documents can be used to represent the business process that takes place before the actual sale. These documents also help your company record information that can be monitored and used to analyze the products or services customers are interested in as well as the success rate of the formal offers made to customers. Let's briefly define both sales inquiries and sales quotations:

- **Sales inquiry**
  A formal, non-binding inquiry from a customer regarding products or services offered by the company.
- **Sales quotation**
  A legally binding offer valid within a certain timeframe that contains details about conditions of the sale, including price, before goods are delivered or services provided to a customer. In SAP S/4HANA, a quotation can be created with reference to an inquiry, and sales orders can be created with reference to a quotation.

In SAP S/4HANA, you can use the transaction codes or SAP Fiori apps shown in Table 9.1 to manage inquiries and quotations.

Transaction Code	SAP Fiori App	Alternative SAP Fiori App
VA11 – Create Inquiry	Create Inquiry	Manage Sales Inquiries
VA21 – Create Quotations	Create Quotation	Manage Sales Quotations
VA12 – Change Inquiry	Change Inquiry	Manage Sales Inquiries
VA22 – Change Quotations	Change Quotation	Manage Sales Quotations
VA13 – Display Inquiry	Display Inquiry	Manage Sales Inquiries
VA23 – Display Quotations	Display Quotation	Manage Sales Quotations
VA15 – List of Inquiries	List Inquiries	Manage Sales Inquiries
VA25 – List of Quotations	List Quotations	Manage Sales Quotations

Table 9.1 List of Inquiry and Quotation Transaction Codes and SAP Fiori Apps

As mentioned earlier, provided document types and copying requirements have been maintained in customizing, inquiries and quotations can be used to form the basis of subsequent sales documents.

You can define a validity date for both inquiries and quotations. This feature is useful from a reporting point of view since you can monitor both types of documents according to the validity date. A validity date, however, is particularly important for quotations since they are legally binding offers; thus, you should always define an expiry date while taking into consideration price fluctuations. This approach will prevent unwanted situations where the cost of the goods sold (COGS) has gone up substantially, but your company still must honor the price on the offer, which is still as valid today as when sent to the customer a long time ago.

Let's now explore two helpful SAP Fiori apps that you can use to display all the details relevant for a sales inquiry or quotation in one place. These apps are the Sales Inquiry and the Sales Quotations factsheet apps, which can be accessed from the transactional Manage Sales Inquiries app and Manage Sales Quotations app, respectively. Both apps present you with several tabs of data regarding the relevant business partners, related sales documents, a list of the items to be sold, and general information about the document.

Some key features of these two SAP Fiori apps include the following:

- As shown in Figure 9.4 for the Sales Inquiry app, both apps display relevant key facts such as the order reason, the sold-to party, the overall status and the rejection status of the document, the net value, and the order probability (likelihood that the deal will close).
- You can navigate to the Customer – 360° View app for an overview of one specific customer and, for example, to display a list of sales orders or contracts for this customer.
- You can see a list of all the items in a sales document and navigate to the material master page. By selecting an item, you can display relevant key facts about the item, navigate to the Customer – 360° View app, display and navigate to related sales orders, display item-specific business partners, and display product information.
- You can display details about business partners at the header level.
- In the **Process Flow** area, you'll see a graphical representation of the process steps and related documents. You can see the status and valid-to date of your quotation and any preceding or subsequent documents, such as deliveries or billing documents. As shown in Figure 9.5, if the actual delivery and billing documents have not been created yet, the app uses the confirmed delivery date and the billing date in the sales document to display planned documents.
- If the approval status of the quotation is **In Approval**, you can display the workflow steps and find the user responsible for approving the quotation.

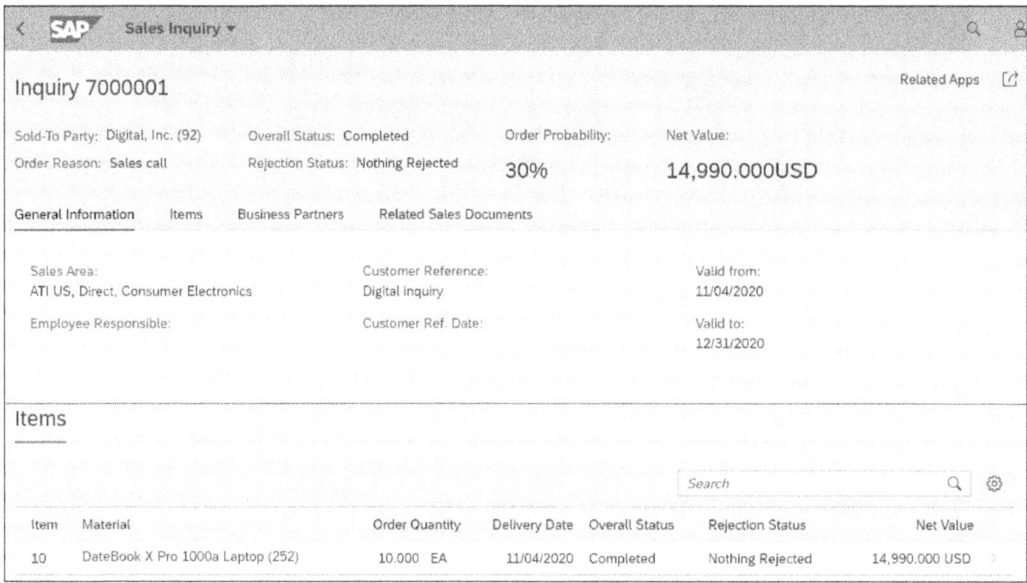

Figure 9.4 Sales Inquiry Factsheet App: General Information

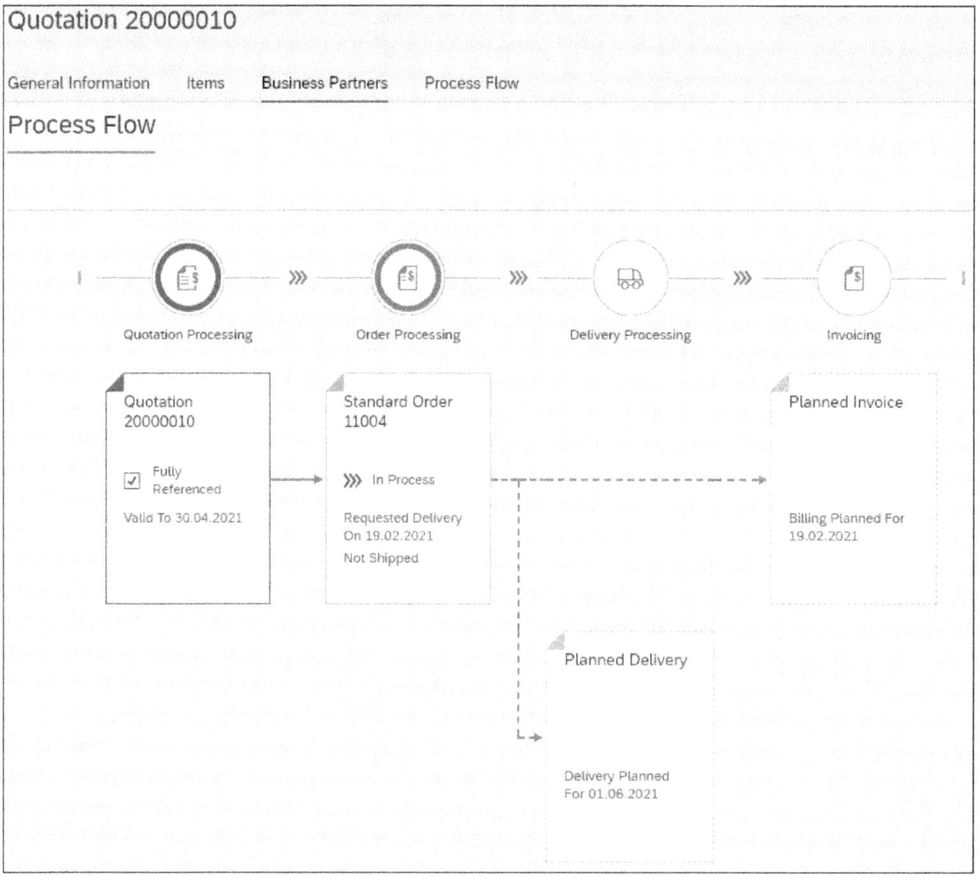

Quotation 20000010
General Information
Items
Business Partners
Process Flow
Process Flow
Quotation Processing
Order Processing
Delivery Processing
Invoicing
Quotation 20000010
Fully Referenced
Valid To 30.04.2021
Standard Order 11004
In Process
Requested Delivery On 19.02.2021
Not Shipped
Planned Invoice
Billing Planned For 19.02.2021
Planned Delivery
Delivery Planned For 01.06.2021

Figure 9.5 Sales Quotations Factsheet App: Process Flow

Tip

The process flow in SAP S/4HANA uses color-coding to indicate the status of a document, and by clicking on the documents, you can navigate to the relevant applications to resolve problems.

When creating an inquiry or a quotation, you can offer your customer optional alternative materials by entering them directly below the item requested. If accepted by the customer, the alternative material can be copied from the inquiry to the quotation or from the quotation to the sales order. To use alternative items, follow these steps:

1. Enter the material requested by the customer in your inquiry or quotation. This material is the main item.
2. Enter the alternative item directly under the main item to be replaced and populate the following fields:
   - **Material** number of the alternative item
   - In the **AltItm** (alternative item) field, enter the item number of the main item you entered in step 1.
   - If you can offer multiple alternative products for the same main item, add them to the document one after the other and enter the same item number (item number of the main item) in the **AltItm** field.

Note

By default, the system copies only the main item when copying a sales document with alternative items into a subsequent document. To select the alternative item, click the **Selection list** button in the dialog box that appears when creating a sales document with reference. Then, select the alternative items you want in your target document.

9.4 Origin of Data in Sales Documents

Every time a sales document is created, most fields are automatically populated by the system. This capability saves business users a lot of time and helps prevent human errors in data entry. For documents created without reference, the two main sources of data are the following:

- Master records
- Customizing tables

Let's discuss first how the system copies data from the master records, more specifically from the business partner, material master, and customer-material information record, as follows:

- **Data from the business partner record**
  Some of the key fields copied from the customer master record into the sales document include the following:

- Complete delivery indicator
- Shipping conditions
- Delivery priority
- Order combination indicator
- Maximum number of partial deliveries allowed per item
- Partial deliveries at the item level
- Delivery plant
- Incoterms
- Payment terms

If a field copied from the business partner should have a different value on a specific sales document, you can always override the default value manually.

Also, note that some fields are copied into the document at the header level that automatically apply to the items too. These fields are referred to as *general business data*, and changing the value in the header field will update the items accordingly. Payment terms and Incoterms are two examples of general business data fields. If you change the value of one of these fields in an item, the item is shown in the *own data* display option when you navigate to the menu path **Edit • Display Range • Own Business Data** from the overview screen. For example, in a sales order with four items, the header and all items have payment terms AFPT. You can change the payment terms for item 30 to CP00. This step creates a deviation from general business data in the header, and item 30 is shown on the own data item list, as shown in Figure 9.6. As a consequence, if you now change the payment terms at the header level from AFPT to TCPT, item 30 won't be updated and will remain with payment terms CP00. At this point, if you change the Incoterms at the header level, all the items will be updated with the new Incoterm entered in the header, except for item 30.

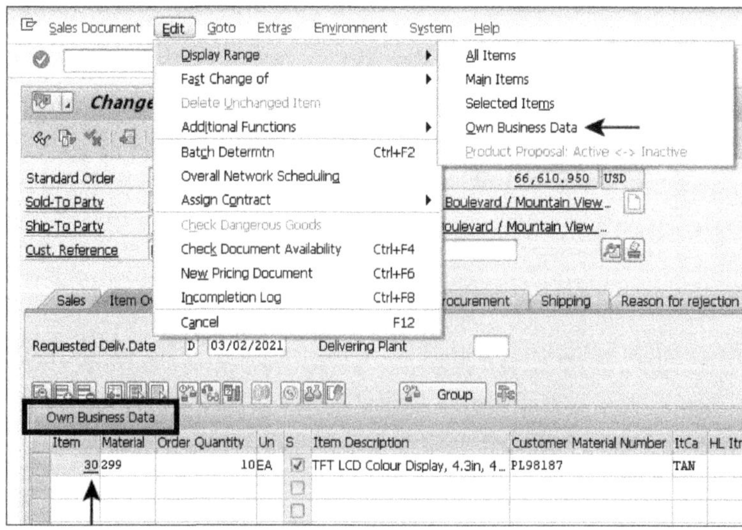

Figure 9.6 Sales Order Item: Own Business Data

As described in previous chapters, a sales document can have a different business partner for each partner function. For instance, the sold-to, ship-to, and payer can all be different, which influences the source of the data being copied into the document since each partner function is used by the system to determine specific information in the following way:

  - Sold-to → Incoterms, pricing conditions, and shipping conditions
  - Payer → Credit limits and terms of payment
  - Ship-to → Taxes, ship-to addresses, and delivering plants

- **Data from the material master record**

  The data from the material master record is copied into the sales document items, but as with the business partner, you can manually change the data in the document if required. Some sales-relevant fields copied from the material master include the following:

  - Gross weight
  - Net weight
  - Volume
  - Delivering plant

  The delivering plant is a special case, since this information can be copied from three master data objects: the material, the customer (ship-to), and the customer-material information record. If the delivering plant has been maintained in the customer-material information record, this value has priority and is copied into the sales document item. In this case, the delivering plants specified in the material and customer master are ignored by the system. But if no customer-material information record exists or it exists but the delivering plant is not maintained, the system follows the logic shown in Table 9.2.

Where Plant Is Defined	Material Maintained for the Plant in the Customer Master?	Delivering Plant Copied From
The material master only	Not applicable	Material master
The customer master only	Yes	Customer master
Both material master and customer master	Yes	Customer master
Both material master and customer master	No	Material master

Table 9.2 Delivering Plant Determination in the Sales Document Item

- **Data from the customer-material information record**

  As described earlier in Chapter 3, Section 3.3, in the customer-material information record, you can maintain the customer material code and description, as well as some shipping-related fields such as delivery priority, delivering plant, underdelivery and overdelivery tolerances, partial delivery at the item level, and the maximum number of partial deliveries per item. The data defined in the customer-material information record has priority over data in the material and/or the business partner master record. Therefore, if a different delivering plant is maintained in the material, in the ship-to party and in the customer-material information record, the delivering plant in the customer-material information record will be copied into the sales document item.

You now understand how the system copies data from several master records into the sales document, However, some fields are determined via specific configuration entries defined in the IMG in combination with the values of certain fields in the sales document, the customer master, and the material master. Two examples of these fields are the shipping point and the route, which can be defined in the following ways:

- **Shipping point**
  A *shipping point* is a required field to create outbound deliveries. This value is determined for all items relevant for delivery in the sales document based on the combination of the **Shipping Condition** from the header of the document, the **Loading group** in the material master, and the **Plant** on the sales document item. Shipping point determination rules are maintained in the configuration table TVSTZ (Organizational Unit: Shipping Points per Plant), which we'll cover in the next chapter.
- **Route**
  A *route* is typically used for the collective processing of deliveries. Routes are determined by the system depending on the following criteria:
  - Country of departure
  - Departure zone of the shipping point
  - Country of destination (as defined in the ship-to party address)
  - Receiving zone of the ship-to party
  - Shipping conditions in the sales order (header data)
  - Transportation group of the material
  - Weight group (optional)

  Figure 9.7 shows an example of route determination without weight group for Germany, which is both the country of departure and the destination.

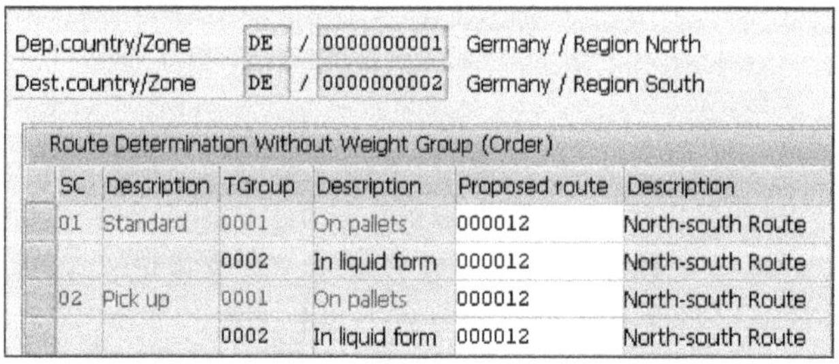

Dep.country/Zone DE / 0000000001 Germany / Region North
Dest.country/Zone DE / 0000000002 Germany / Region South

Route Determination Without Weight Group (Order)

SC	Description	TGroup	Description	Proposed route	Description
01	Standard	0001	On pallets	000012	North-south Route
		0002	In liquid form	000012	North-south Route
02	Pick up	0001	On pallets	000012	North-south Route
		0002	In liquid form	000012	North-south Route

Figure 9.7 Route Determination Without Weight Group: Germany

Note

In SAP S/4HANA, you can also specify whether the system must carry out another route determination when the delivery is created. This capability is configured by delivery type in the customizing activity **Logistics Execution • Shipping • Basic Shipping Functions • Routes • Route Determination • Define New Route Determination by Delivery Type**. In addition, you can define whether an extra check must be performed to confirm that the route found for the delivery is one of the actual routes allowed for the original proposed route, as specified in **Define Allowed Actual Route by Proposed Route**. This topic will be covered in more detail in Chapter 10.

9.5 Reference Status and Copying Rules

The reference status of a sales document tells you whether or not subsequent documents have been created and is updated when a source document is copied. The reference status field is available both at the header level and at the item level. A reference status at the item level is found under the item's **Status** tab and can contain one of the following three values:

- **Not Referenced**: No subsequent documents exist for the item.
- **Partially Referenced**: At least one subsequent document exists for the item but there's still quantity left to be referenced.
- **Fully Referenced**: At least one subsequent document exists for the item, and the item can no longer be copied again. This status is automatically set when the entire quantity has been copied or based on the completion rule defined for the item category.

The reference status at the header level is displayed in the header's **Status** tab and is determined by the reference status of all the items in the sales document in the following way:

- If all the items in the document have the reference status **Not Referenced**, then the reference status at the header level also shows the **Not Referenced** status.
- If all the items in the document have the reference status **Fully Referenced**, then the reference status at the header level also shows **Fully Referenced** status.
- In all other scenarios, the reference status at the header level shows the **Partially Referenced** status.

Note

Note that the rejection reason field is also taken into account when a document is copied. Items that have been rejected can no longer be copied since they are treated as completed items regardless of the value in the reference status field.

From the sales document overview screen, you can click on the icon to access the status overview screen. On this screen, you'll see the reference status of the document and all items at a glance, as shown in Figure 9.8.

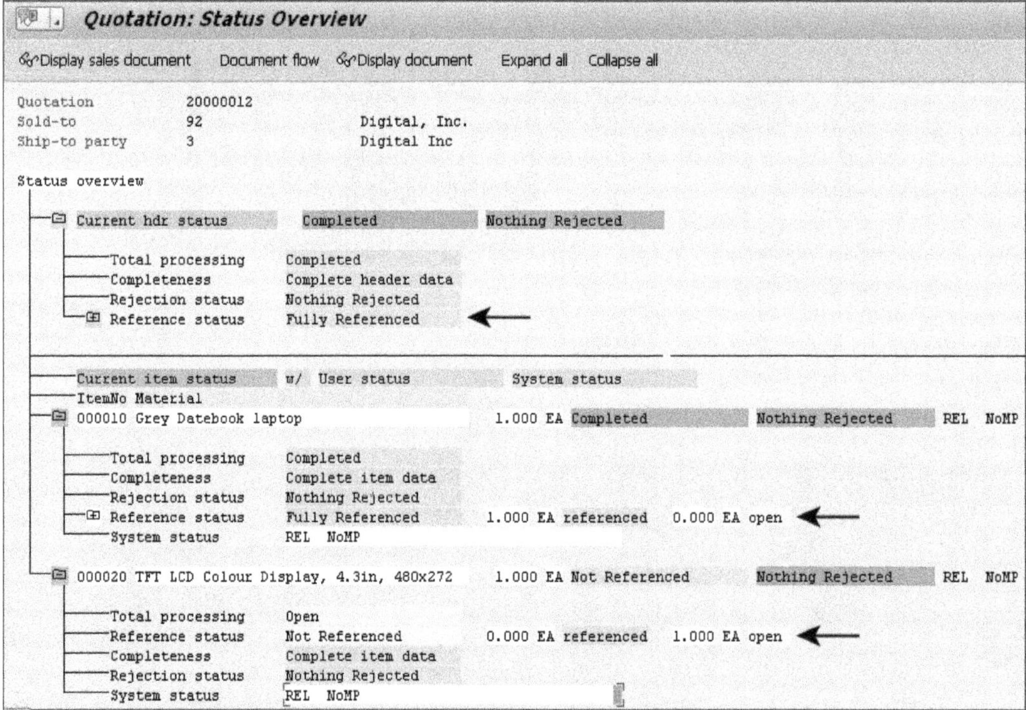

Figure 9.8 Reference Status Fields in the Status Overview Screen

As shown earlier in Figure 9.1, inquiries and quotations are some examples of sales documents that require a subsequent document to move through the order-to-cash process. You can specify whether the source document must be copied completely by selecting the **Complete Reference** checkbox in the header parameters of the copy control. If you select this indicator, you won't be able to change or delete any items when copying from the selection list. The full source document is copied into the target document; thus, the source document is classed as **Fully Referenced** at both the header and item level. This topic was described in more detail when we discussed copy control in Chapter 6, Section 6.2.

Another customizing parameter that influences the reference status of an item is the **Completion Rule** field, which can be maintained when defining the item category. We covered this field in Chapter 4, Section 4.3, but to summarize, the following options are available:

- As soon as an item is referenced once, the item is considered **Fully Referenced** even if the quantity copied is less than the item quantity.
- An item is not classed as **Fully Referenced** until the full quantity has been copied into one or more subsequent documents.
- An item is not classed as **Fully Referenced** until the full target quantity has been copied into one or more subsequent documents. This setting can be used in quantity contracts, for example.

- An item is not classed as **Fully Referenced** until the full target value has been referenced. This setting can be used in value contracts.

Note

In the standard system, item categories for inquiries have been set up so the item reference status is set to **Fully Referenced** after the first reference, regardless of the quantity in the target sales document item. However, if you copy a quotation into a sales order, the quotation is not complete until the full quotation quantity has been referenced by one or more subsequent documents.

Like quotations, contracts are considered **Fully Referenced** only if the full target quantity (or value) of the contract has been referenced.

9.6 Using Special Sales Orders

In some industries, a customer can place an order, pick up the goods, and pay in cash all at the same time—a common scenario in the retail industry, for example. To support this scenario, SAP S/4HANA provides a special sales order type called *cash sales* via sales document type BV. With this sales document type, the system uses the current date as both the delivery date and the billing date. Therefore, as soon as the sales order is saved, an outbound delivery (type BV) is automatically created in the background, and a cash sale invoice can be printed out and given to the customer. The cash sale invoice is an output generated from the sales order and not from the billing document. Since the customer pays in cash, he or she is entitled to receive an invoice right after the payment is made. However, at that time, the billing document may not exist in the system yet (since billing documents are usually created by a periodic batch job). Thus, the cash invoice is always printed out directly from the sales order.

In a cash sales scenario, a document of billing document type BV is created by standard processing from the sales order (order-related billing), and the invoice amount is posted directly to a cash account.

The sales transaction is considered complete once the customer has paid and received the goods. If the customer has the goods at the time of payment, we recommend scheduling a batch job to perform the post goods issue in the background, which will help improve inventory management and keep stock levels up to date.

Warning

When using the cash sale process, a billing document can only be created if the quantity in the sales order (type BV) matches the goods issue quantity. The delivered quantity and billed quantity must be the same.

A standard cash sale process is shown in Figure 9.9.

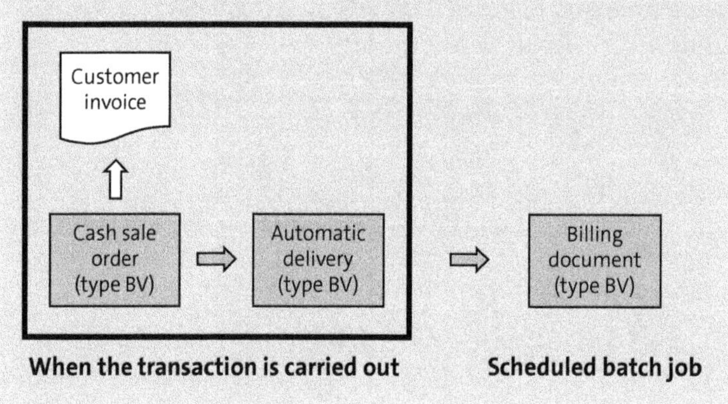

Figure 9.9 Standard Cash Sales Process

The *rush order* process is similar to the cash sale in the sense that the customer either picks up the goods or asks for the goods to be delivered the same day the order is placed. In a rush order scenario, the outbound delivery (type LF) is automatically created when the sales order (type SO) is saved, but both the invoice and payment take place later. Unlike the cash sale, in a rush order transaction, the billing document is generated with reference to the delivery since the billing document uses a delivery-related billing item category (e.g., TAN).

The goods issue is posted as soon as the goods are physically picked and shipped to the customer. The diagram shown in Figure 9.10 represents the rush order process in SAP S/4HANA.

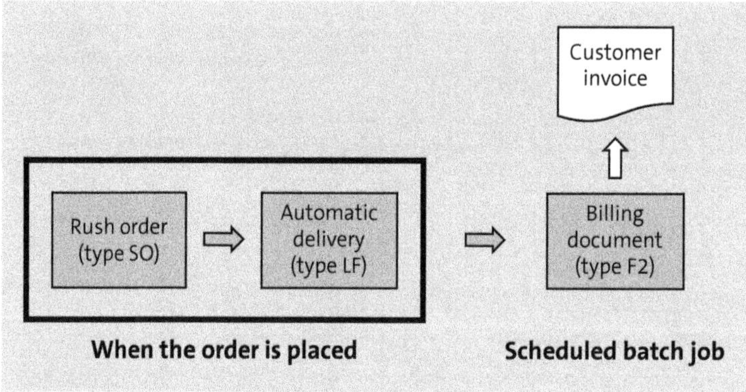

Figure 9.10 Standard Rush Order Process

9.7 Executing Special Business Processes in Sales

Some companies may require you to go beyond the basic sales process and implement special processes to save costs, generate the right level of information, and operate efficiently. In other words, often you'll find yourself activating specific sales business processes, such as processes related to consignment stock, make-to-order items, configurable products, third-party orders, or individual purchase orders (POs) to mention just a few cases. All these special processes require only a little configuration to make them work since these processes are delivered by SAP S/4HANA as part of the standard package.

A detailed discussion of each individual special process is beyond the scope of this book, but in this subsection, we'll provide an overview of a make-to-order production process and a third-party sales order, which are two special processes used extensively across many industries.

9.7.1 Make-To-Order Production

Let's first discuss the make-to-order production process. Unlike mass production where the same finished goods are repeatedly made to stock, in a make-to-order production process, a limited quantity is produced for a specific sales order. This process is typically used in industries where a relatively low volume of configurable goods is manufactured. The fact that the material is configurable may result in a large number of possible permutations of the same base product; thus, you must ensure each variant is allocated to the correct sales order. For example, make-to-order production is common in the automotive industry (especially in luxury car brands) to allow customers to select the characteristics of the vehicle and build the vehicle against a single sales order according to the customer's specifications.

SAP S/4HANA offers you two ways to track the costs associated with make-to-order production:

- Cost management using the sales order
- Cost management using Project System

For the purpose of this certification, we'll only cover cost management using the sales order, where all costs and revenues are allocated to the sales item.

In Chapter 7, Section 7.3, we discussed how the transfer of requirements to production planning is controlled by the requirements type determined by the search strategy used by the system in a sales order item. If the material is defined as a make-to-order product, the sales order item is planned as a customer-individual requirement, and the produced materials are received into sales order-specific stock.

To define a make-to-order material, first, you must identify a requirements class that has settlement profile SD1 (sales order make-to-order prod), special stock indicator E (orders on hand), and accounting via sales order. In addition, both the availability check and transfer of requirements indicators must be selected. As an example, standard requirements class 040 meets the criteria described and is assigned to standard requirements type KE (Indiv.cust.ord. w/o consumpt.), as shown in Figure 9.11.

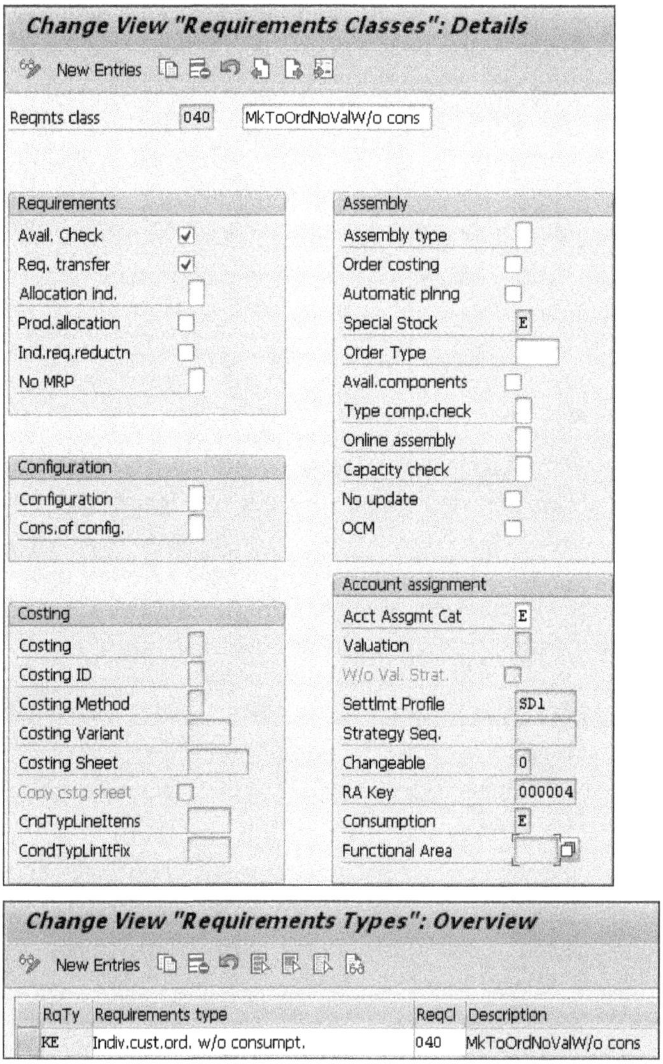

Figure 9.11 Requirement Class 040: Make-to-Order Production without Consumption

You must now find a planning strategy that is assigned to the make-to-order requirements type KE, like for instance, standard planning strategy 20, which is also the only planning strategy in strategy group 20 (Make-to-order production).

Next, you must maintain strategy group 20 in the material master record, in the **MRP 3** view. Alternatively, you can assign strategy group 20 to a material requirements planning (MRP) group in customizing activity **Production • Material Requirements Planning • Master Data • Independent Requirements Parameters • Planning Strategy • Assign MRP Group to Strategy Group**. Then, you can allocate the relevant MRP group to the material (**MRP 1** view).

To finish defining a make-to-order material from a sales perspective, you should specify the in-house production time in the **MRP 2** view and choose an availability check group in the **Sales: General/Plant** view that creates individual requirements, for instance, checking group 02 (Individ.requirements).

When creating make-to-order items in a sales order, you must ensure the plant is populated for the item. Then, if the material has been defined correctly, the system automatically determines the requirements type for make-to-order production.

Note

In early versions of SAP R/3, requirements types were determined using the document item category. However, in SAP S/4HANA, the item category does not take priority to control the requirements type. The system first tries to find a requirements type using the strategy group in the material master, and if a strategy group has not been maintained, the system will use the MRP group. Therefore, we don't recommend using special item categories for controlling make-to-order production. The item category can be, for example, TAN, and still, the item can transfer make-to-order requirements to production planning based on the strategy group or the MRP group in the material master record.

You can confirm that the sales item will be planned as a customer-individual requirement by navigating to the availability overview screen while creating the order and doing the following checks:

- Checking rule **AE** (sales order; make-to-order stock) is used in the availability check.
- The simulated requirement is shown under the MRP element **CustSt** (individual customer stock) where the MRP element data is the sales order number and item number.

The example availability overview screen shown in Figure 9.12 meets these two conditions.

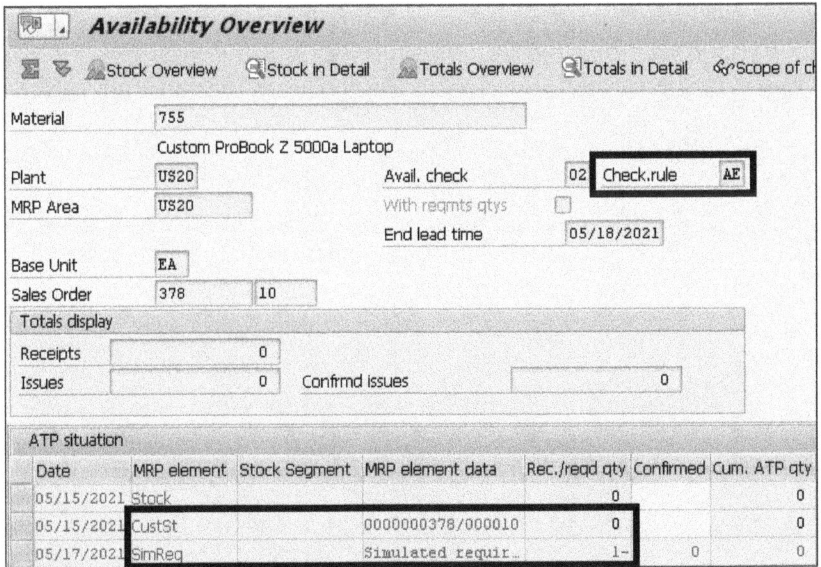

Figure 9.12 Availability Overview: Make-to-Order Scenario

The transfer of requirements to production planning happens once MRP is executed for the concerned material or for the whole plant. After the MRP run, a planned order is automatically created and is visible in the stock/requirements list (Transaction MD04). In the example shown in Figure 9.13, planned order 6940 was created by the MRP to fulfill the requirement generated by order 378, item 10.

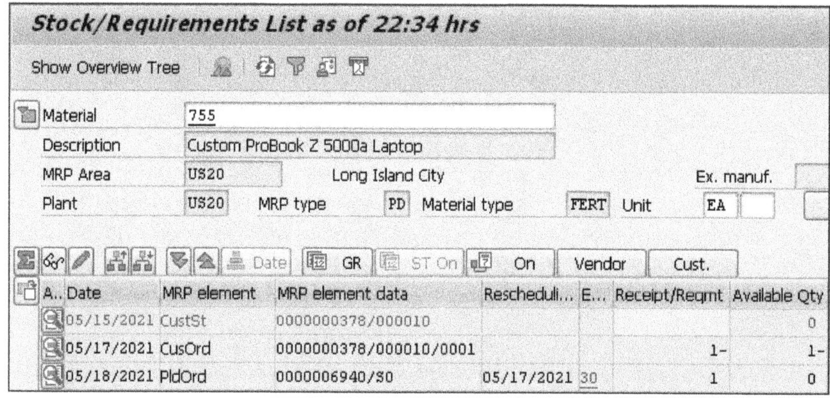

Figure 9.13 Stock/Requirements List: Make-to-Order Scenario

The planned order is assigned to the sales order item and has special procurement type E (in-house production), account assignment category E (Ind. cust. w. KD-CO), special stock E (orders on hand), and consumption E (sales order), as shown in Figure 9.14. These characteristics are all characteristics of a make-to-order planned order.

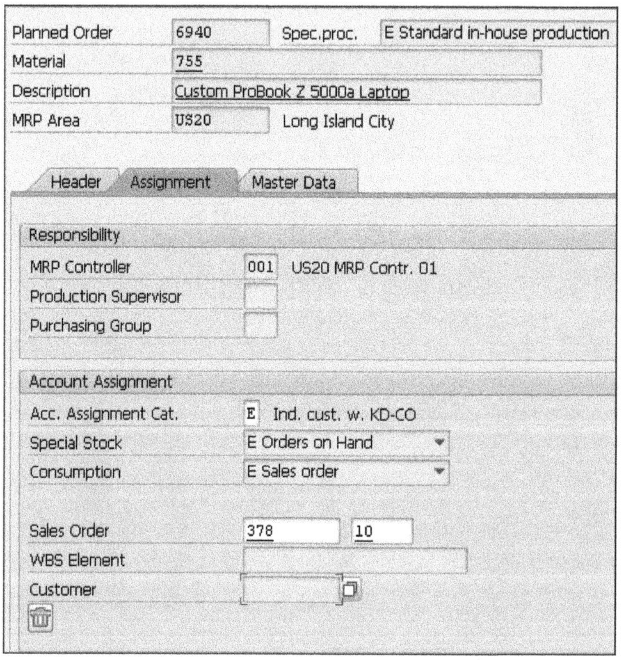

Planned Order 6940 Spec.proc. E Standard in-house production
Material 755
Description Custom ProBook Z 5000a Laptop
MRP Area US20 Long Island City

Header | Assignment | Master Data

Responsibility
MRP Controller 001 US20 MRP Contr. 01
Production Supervisor
Purchasing Group

Account Assignment
Acc. Assignment Cat. E Ind. cust. w. KD-CO
Special Stock E Orders on Hand
Consumption E Sales order
Sales Order 378 10
WBS Element
Customer

Figure 9.14 Planned Order: Make-to-Order Scenario

Next, the planned order is converted into a production order so the manufacturing process can start. Once the production order is completed, the finished goods are recorded into sales order-specific stock (special stock indicator E). An outbound delivery must be created based on the information in the schedule lines of the sales order item. Finished products can only be delivered to the customer through the assigned sales order. Therefore, to create a delivery, the availability check against the sales order stock must be successful. Finally, goods issues are posted with reference to the outbound delivery, and the billing document is generated and sent to the customer.

Warning

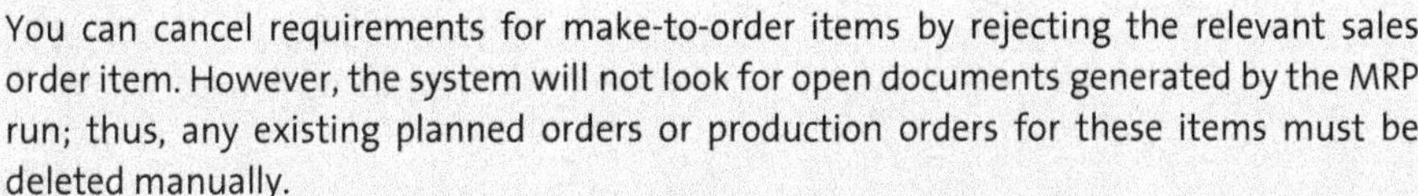

You can cancel requirements for make-to-order items by rejecting the relevant sales order item. However, the system will not look for open documents generated by the MRP run; thus, any existing planned orders or production orders for these items must be deleted manually.

9.7.2 Third-Party Order Processing

The third-party order process is a common scenario in the business world that enables your company to pass an order along to a third-party vendor who then ships the goods on your behalf directly to the customer. The vendor bills you based on the quantity delivered, and you bill the customer the same quantity.

Some key benefits of the third-party order process include the following:

- Reduces stocks and helps fulfill customer orders despite material shortages.
- Transfers customer requirements directly to external vendors, thus saving time and costs in activities like picking/packing and transportation planning.
- Invoices are sent to customers based on the quantities delivered by the vendor.
- A sales order may consist of a combination of third-party items and items shipped from the warehouse.

In this section, we'll discuss two different variants of the third-party order process available in SAP S/4HANA: the classic third-party process via sales item category TAS and the new third-party process with shipping notifications via sales item category CB1.

Note

You should not confuse third-party order processing with individual PO processing. The latter also forwards the order to a third-party vendor, but unlike the third-party order process, in the individual PO scenario, the vendor delivers the goods directly to you, and you deliver them to the customer yourself. In the standard system, the individual PO process is initiated by the item category TAB.

In both third-party order processes (TAS and CB1), a purchase requisition is automatically created by the standard third-party sales item. Then, the purchase requisition must be converted into a PO and sent to the vendor. If you expect a shipping notification from the vendor, we recommend using item category CB1 and then posting a statistical goods receipt when the shipping notification is received. After the statistical goods receipt is posted, the customer invoice can be generated based on the goods receipt quantities and issued to the customer. The billing document is created with reference to the sales order.

On the other hand, if the vendor does not provide a shipping notification, you can opt for item category TAS. You'll need to wait for the materials management invoice receipt to produce a customer invoice, however. The billing document is also created with reference to the sales order but, this time, based on the invoice receipt quantities. The diagram shown in Figure 9.15 illustrates the third-party process with shipping notification using item category CB1.

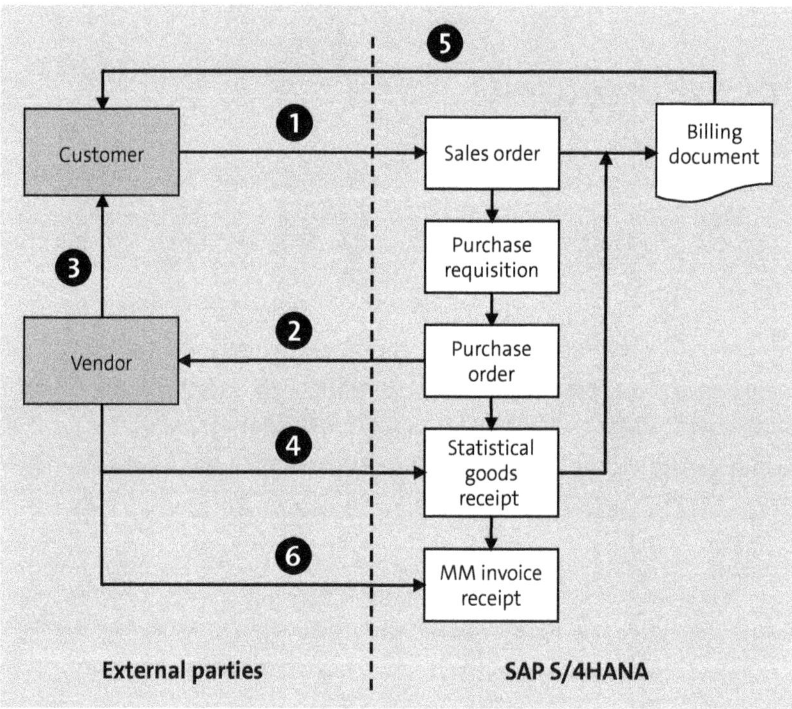

Figure 9.15 Third-Party Sales Process with Shipping Notification Overview (Item Category CB1)

The numbers shown in Figure 9.15 represent the sequence of events where the exchange of goods or information takes place between the company running SAP S/4HANA and the external parties. Let's briefly describe each of these numbered events:

❶ Customer PO

❷ PO to the vendor

❸ Goods delivered to the customer
❹ Vendor shipping notification
❺ Customer invoice
❻ Vendor invoice

To complement the previous diagram, a simple process flow for a third-party order, as displayed in the SAP Fiori factsheet app Sales Order, is shown in Figure 9.16.

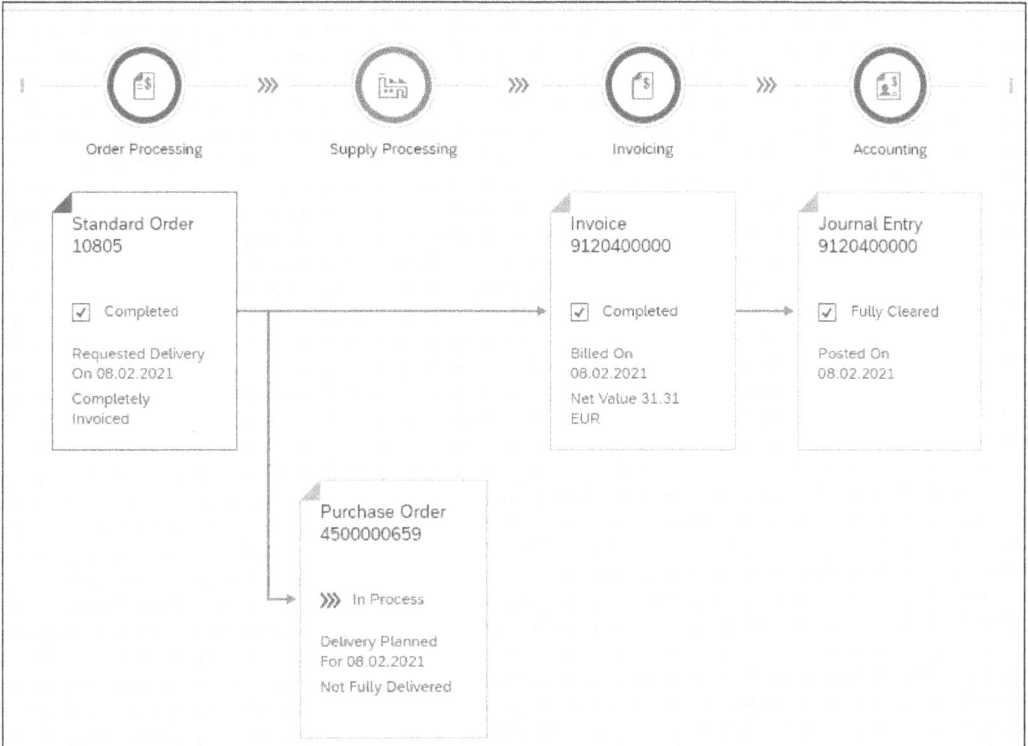

Figure 9.16 Process Flow in SAP Fiori: Third-Party Order Scenario

Note

When processing third-party orders, the ship-to party address can only be changed in the sales document. SAP S/4HANA does not allow you to change the delivery address in a third-party PO item. If you make any changes in the ship-to address in the sales order, the updated address will be automatically passed on to the purchase requisition and the PO.

As with the previous make-to-order production process, you must maintain some configuration and material master data settings before using any of the third-party order processes we've described.

For instance, if a particular material is always delivered from a vendor, enter "BANS" in the item category group field in the **Sales: sales org. data 2** view of the material master record if you want the system to automatically determine item category TAS. Alternatively, if you want item category CB1 (third-party with shipping notification) to be automatically determined, then you must specify item category group CBOR in the material master.

For materials usually delivered by your company, you must enter the item category group NORM in the material master. However, SAP S/4HANA allows you to manually change the item category during sales order processing to either TAS or CB1, provided these two item categories have been maintained in the item category assignment for item category group NORM and the relevant sales order type. By changing the item category to TAS or CB1 in the sales order, you can process stock materials as third-party items.

In the definition of both item categories TAS and CB1, the **Schedule Line Allowed** checkbox is selected since one purchase requisition item is created for each schedule line in the sales order item. From the schedule line data screen in the sales document, you can access the purchase requisition item information by double-clicking on the purchase requisition number under the **Procurement** tab.

The **Billing Relevance** field is another important field when it comes to third-party item categories. This field controls the basis for billing, as described in Chapter 4, Section 4.3. In the standard system, item category TAS is set up with billing relevance F, whereas item category CB1 is set up with billing relevance B, as shown in Table 9.3. Item category TAS is, therefore, not included in the billing due list until an invoice from the vendor is entered in the system against the PO item and is considered to be fully invoiced until the next vendor invoice is received. Whenever a new vendor invoice is received for a TAS item, a customer invoice is created for the quantity in the supplier invoice. In the case of item category CB1, the indicator has been set to B; thus, the billing due list takes the sales order item into account immediately and is considered to be fully billed once the invoice quantity matches the sales order item quantity.

The quantity billed is controlled by the **Billing Quantity** field in the copy control for billing (sales document to billing document) at the item level. For example, billing quantity indicator F has been given when copying item category TAS from sales document type OR (Standard Order) to billing document type F2 (Invoice), meaning the system calculates the quantity in the customer invoice based on the vendor's invoice receipt quantity. For item category CB1, the billing quantity indicator maintained in the copy control settings is E. As a result, when a sales order item with item category CB1 is billed, the system calculates the quantity in the customer invoice based on the goods receipt quantity for the corresponding PO item.

Table 9.3 compares some key elements of both third-party sales scenarios.

Sales Process	Item Category	Billing Relevance	Billing Quantity in Copy Control (Item)
Classic third-party order	TAS	F (Order-related billing. Status according to invoice quantity)	F (Invoice receipt quantity minus invoiced quantity)
Third-party order with shipping notification	CB1	B (Order-related billing. Status according to order quantity)	E (Goods receipt quantity minus invoiced quantity)

Table 9.3 Third-Party Processes with Item Category TAS versus Item Category CB1

9.8 Complaints Processing

In the business world, some transactions don't always go as planned. For example, after goods are delivered, a customer may complain that some boxes arrived damaged, or perhaps the quantity received was greater than the quantity ordered, and the customer wants to return the excess. Also, after reviewing the invoice, a customer may realize that the total net amount does not include a promotional discount. In these circumstances, an important task is for companies to record and solve any complaints received to build strong relationships with customers.

SAP S/4HANA supports the following processes in complaints processing:

- **Free-of-charge deliveries**
  Sales document type FD is used in the standard system to send samples or free products to your customers, so a customer complaint is not always the origin of a free-of-charge delivery. You must enter a value in the **Order Reason** field because this field is included in the incompleteness procedure assigned to order type FD. You can then create an outbound delivery document to ship the free-of-charge material to the customer. No billing is triggered in this scenario.
- **Free-of-charge subsequent deliveries**
  If a customer receives fewer goods than ordered or part of the shipment is damaged, you can use sales document type SD to send goods free of charge to your customer. A subsequent free-of-charge delivery is a sales document that must be created with reference to the original sales order, returns order, or contract. You can then create an outbound delivery document to ship the free-of- charge material to the customer. No billing is triggered in this scenario.

Note

Both sales documents types FD (Delivery Free-of-Charge) and SD (Free-of-Charge Subsequent Delivery) can use the same item category, that is, item category KLN (Free of Charge Item).

- **Returns**
A returns order is created in the system when a customer sends goods back, for example, if the wrong batch was delivered. In the standard system, sales document type RE is used to create customer returns. The **Check Credit Memo** billing block is assigned in customizing to order type RE to prevent a credit for returns from being created automatically by the billing batch job. A returns order can be created with reference to the original sales order, with reference to an invoice, or without reference. In all cases, a returns delivery is created in response to the returns order to post the goods to stock when they arrive in the warehouse. Once the goods are checked, the billing block is removed by an authorized business user, and the credit memo can be created. Alternatively, if the customer still wants the expected batch to be delivered, you can create a free-of-charge subsequent delivery order (type SD) with reference to the return and not issue a credit memo, as shown in Figure 9.17.

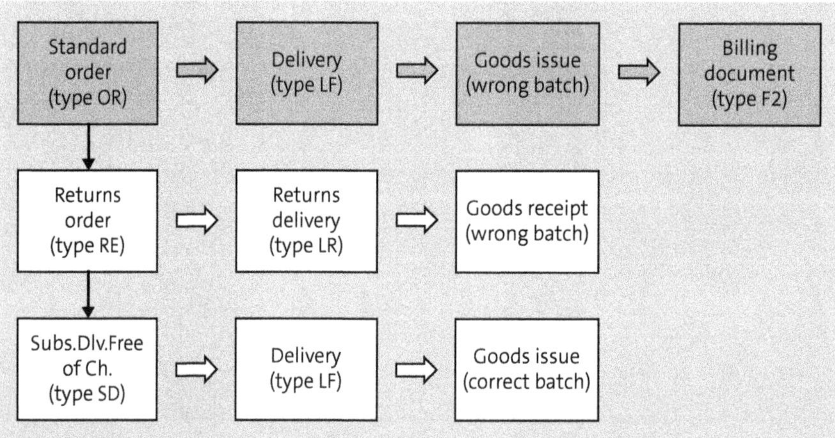

Figure 9.17 Returns Order with Free-of-Charge Subsequent Delivery

- **Credit memo requests**
If the customer receives damaged goods but the cost of returning them exceeds the value of the products, then you can ask the customer to scrap the damaged material and create a credit memo request to subsequently issue a credit note to compensate for the loss. To create a credit memo request, you can use sales document type CR. In the standard system, the **Check Credit Memo** billing block is assigned to sales document type CR. The billing block can be released by an authorized business user prior to the creation of the credit memo (billing type G2). Usually, credit memo requests are created with reference to a customer invoice, as shown in Figure 9.18, or with reference to a sales order. You can also create independent credit memo requests; however, we do not recommend this approach to ensure traceability. In this complaint process, no movement of material occurs.

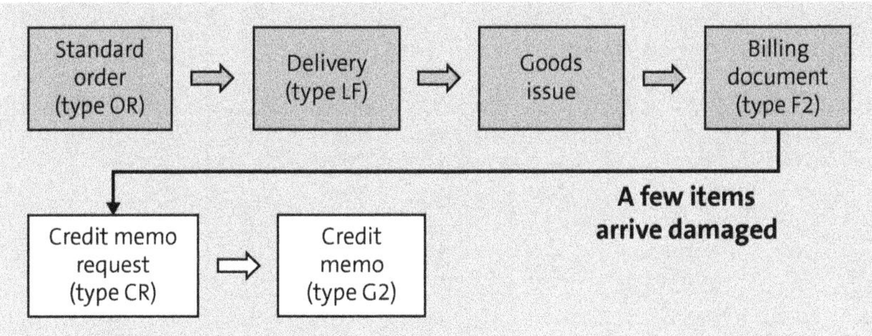

Figure 9.18 Credit Memo Request Process

- **Debit memo requests**
  If the price on the customer invoice is too low, you can then create a debit memo request to generate an invoice for that difference. In the standard system, the sales document type for debit memo requests is DR, and the **Check Debit Memo** billing block has been set in customizing to prevent debit memos from being triggered without manual supervision. Document type L2 is the billing type of the subsequent debit memo. As with the credit memo request, debit memo requests are typically created with reference to a customer invoice, as shown in Figure 9.19, or with reference to a sales order. You can also create independent debit memo requests if required. In this complaint process, no movement of material occurs.

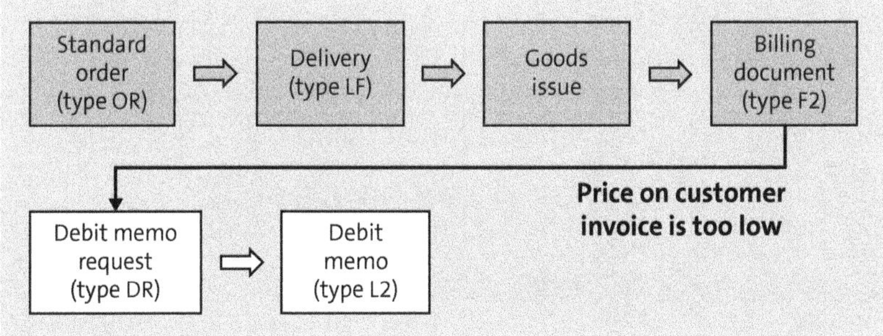

Figure 9.19 Debit Memo Request Process

- **Invoice correction requests**
  If the price charged to the customer was too high for one or more items in an invoice, besides creating a credit memo request, you can also use an invoice correction request to issue a credit to the customer. In the standard system, sales document type RK is used for invoice correction requests. As with a credit memo request, the **Check Credit Memo** billing block is automatically set when an invoice correction request is created. Once the sales document is reviewed and the billing block released, a credit memo type G2 can be generated. The same principle applies to complaints where the unit price in the invoice is correct, but the quantity is greater than the quantity ordered. Invoice correction

requests (type RK) can only be created with reference to an invoice, as shown in Figure 9.20. In this complaint process, no movement of material occurs.

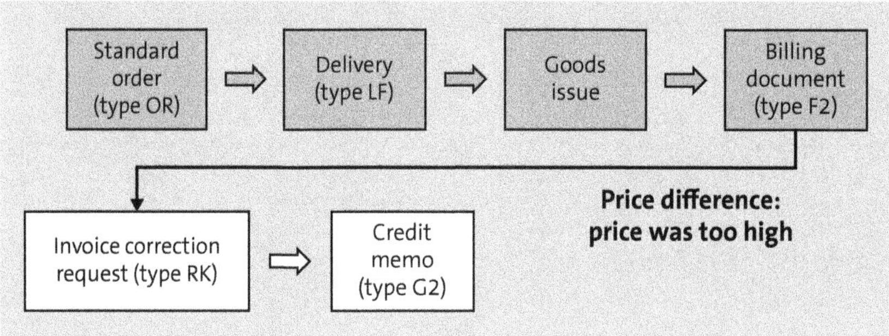

Figure 9.20 Invoice Correction Request Process

When generating an invoice correction request, two items are automatically created for each item copied from the invoice. The first item is always a credit item for the full amount and quantity in the source invoice item. The second item is a debit item where you must enter the correct amount or quantity that should have been invoiced. The difference between the credit and the debit item should appear at the header of the invoice correction request as a positive amount. In the example shown in Figure 9.21, a customer was charged $1,573.95 for 1 unit of material 252 but should have been charged $1,417.50. The correct amount has been entered in the debit item (item 20), and the total net value at the header is $156.45, which corresponds to the amount charged in excess. In this case, a credit memo (type G2) will be created for the difference amount of $156.45.

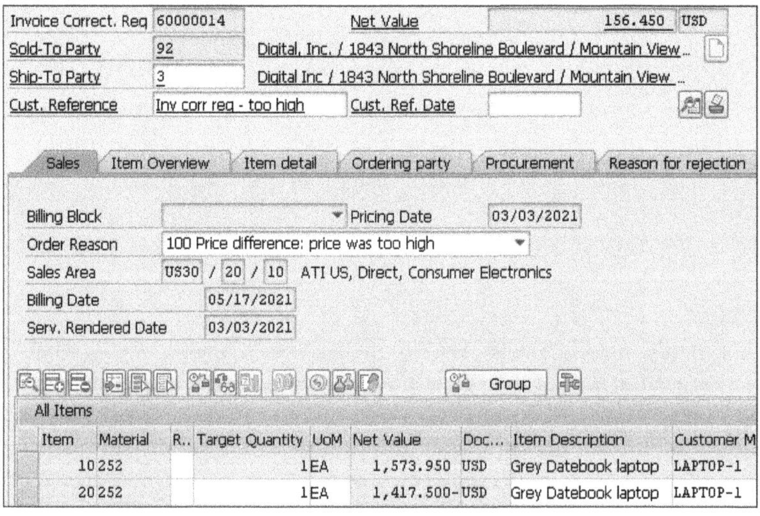

Figure 9.21 Invoice Correction Request in SAP S/4HANA

Most of the fields in the credit item are grayed out and cannot be edited. In a credit item, only the following data can be changed:

- Deletion
- Reason for rejection
- Billing block
- Account assignment

Note

If the difference between the credit and the debit item results in a negative amount, you may want the system to generate a debit memo from the invoice correction request and not a credit memo. This scenario arises when the amount charged in the invoice was too low. To resolve this issue, a new sales document type, for example, document type ZRK1 (Inv. Correct. Req. Debit Memo) must be set up with sales document category L (debit memo request) and order related billing type L2. As a result, before you create an invoice correction request, you must know if a credit memo or debit memo should be generated to solve the complaint and then select the sales document type accordingly—standard type RK for credit memos and custom type ZRK1 for debit memos.

9.9 Important Terminology

In this chapter, the following terminology was used:

- **Order-to-cash**
  Informally referred to as OTC or O2C, this end-to-end business process integrates all the sales and distribution activities required to transfer information from a sales transaction to financial accounting. The order-to-cash process in SAP S/4HANA consists of the following four main stages: pre-sales activities, order processing, shipping, and billing.
- **Sales order**
  A formal request from a customer to deliver a defined quantity of goods or provide a service within a specific time period. In SAP S/4HANA, a sales order can be created with reference to a quotation.
- **Inquiry**
  A formal, non-binding inquiry from a customer regarding products or services offered by your company.
- **Quotation**
  A legally binding offer valid within a certain timeframe that contains details about the conditions, including price, to deliver goods or provide a service to a customer. In SAP S/4HANA, a quotation can be created with reference to an inquiry.
- **Sales document flow**
  A function available from the sales, delivery, and billing document screens that enables a user to visualize all relevant preceding and subsequent documents in the chain of documents. The document flow can be displayed at the header level or the item level and is often used to help with problem resolution.

- **Credit memo request**
  A sales document used in complaints processing to request credit for a customer.
- **Debit memo request**
  A sales document used in complaints processing to request a debit memo for a customer.
- **Credit memo**
  A billing document that is created with reference to a credit memo request and used to credit the customer for the damaged products or to compensate for a miscalculated price on the invoice.
- **Debit memo**
  A billing document that is created with reference to a debit memo request and used to debit the customer for the disputed amount.
- **Correction invoice request**
  A sales document used in complaints processing to correct the quantity or price of one or more items in an invoice.
- **Make-to-order**
  A production process used to manufacture a product for a particular sales order; thus, finished goods are not usually kept in stock for long periods of time. Make-to-order stock can be easily identified by the special stock indicator E (Orders on Hand), which means the stock belongs to a specific sales order item.
- **Reference status**
  This status is updated when a sales document item is copied using the **Create with Reference** function. Depending on the quantity copied and the completion rule for the item, an item can have one of the following reference statuses: **Not Referenced**, **Partially Referenced**, or **Fully Referenced**.

9.10 Practice Questions

These practice questions will help you evaluate your understanding of the topics covered in this chapter. The questions shown are similar in nature to those found on the certification examination. Although none of these questions will be found on the exam itself, they will allow you to review your knowledge of the subject. Select the correct answers and then check the completeness of your answers in the next section. Remember that, on the exam, you must select all correct answers and only correct answers to receive credit for the question.

1. Which of the following sequence of documents represents a standard order-to-cash process in SAP S/4HANA?

☐ A. Inquiry → Quotation → Billing document → Sales order → Outbound delivery

☐ B. Quotation → Inquiry → Sales order → Outbound delivery

☐ C. Inquiry → Quotation → Sales order → Outbound delivery

☐ D. Sales order → Post goods issue → Outbound delivery → Billing document

2. Which of the following are key features of the SAP Fiori factsheet app Sales Quotations? (There are two correct answers.)

☐ A. Display schedule lines details for each of the items

☐ B. Navigate to the Customer – 360° View app

☐ C. Display relevant key facts of the preceding inquiry

☐ D. Display relevant key facts of the quotation

3. Which of the following statements are correct regarding alternative items in sales?

☐ A. Alternative items can only be used in inquiries.

☐ B. Alternative items can only be used in quotations.

☐ C. Alternative items can only be used in sales orders.

☐ D. Alternative items can be used in both inquiries and quotations.

4. In a sales order with three items (items 10, 20, and 30), payment terms CP00 have been copied from the customer master into the order header, as well as into all three items in the order. The user manually changes the payment terms for item 20 to R001 and saves the order. What would happen if another user subsequently changes the payment terms at the header level to TCPT?

☐ A. The payment terms are updated to TCPT on all items.

☐ B. The payment terms are updated to TCPT on items 10 and 30 only.

☐ C. The payment terms are updated to TCPT on item 20 only.

☐ D. The payment terms are changed at the header level, but none of the items are updated.

5. Which of the following data is proposed from the business partner master record of the ship-to party during sales order creation? (There are three correct answers.)

☐ A. Incoterms

☐ B. Tax

☐ C. Shipping conditions

☐ D. Ship-to address

☐ E. Delivering plant

6. Which of the following data is proposed from the business partner master record of the sold-to party during sales order creation? (There are two correct answers.)

☐ A. Delivering plant

☐ B. Incoterms

☐ C. Order reason

☐ D. Shipping conditions

7. Which master record has priority in terms of delivering plant determination in a sales order item?

☐ A. Material master record

☐ B. Business partner master record (sold-to)

☐ C. Business partner master record (ship-to)

☐ D. Customer-material information record

8. Which of the following fields are used during automatic shipping point determination in a sales order item? (There are three correct answers.)

☐ A. Shipping type

☐ B. Delivery priority

☐ C. Shipping condition

☐ D. Loading group

☐ E. Plant

9. Which of the following elements can influence the route determination for a sales document item? (There are two correct answers.)

☐ A. Material freight group

☐ B. Shipping conditions

☐ C. Unloading point

☐ D. Transportation group

10. True or False: If a rejection reason is selected for a sales document item, the reference status of the item is automatically set to **Fully Referenced**, and the item cannot be copied.

☐ A. True

☐ B. False

11. In the standard system, a quotation has been created for 5 units with reference to an inquiry. The total quantity in the inquiry item is 30, and no other subsequent documents exist. Which is the correct reference status value for the inquiry item?

- ☐ A. **Fully Referenced**
- ☐ B. **Not Referenced**
- ☐ C. **Partially Referenced**
- ☐ D. **Open**

12. True or False: In the standard cash sales process, both a delivery and a billing document are automatically generated when the sales order is posted, and a cash invoice is printed immediately from the billing document.

- ☐ A. True
- ☐ B. False

13. Which of the following are characteristics of a rush order transaction? (There are two correct answers.)

- ☐ A. The customer orders, picks up, and pays for the goods immediately.
- ☐ B. The system automatically creates a delivery when the order is saved.
- ☐ C. A customer invoice is automatically printed when the order is saved.
- ☐ D. The goods are picked up by the customer or delivered to the customer the same day the order is placed. Payment takes place later.

14. Which of the following statements describe the make-to-order sales process in SAP S/4HANA? (There are two correct answers.)

- ☐ A. You must use special item categories for controlling make-to-order production.
- ☐ B. Produced materials are received into consignment stock.
- ☐ C. The plant must be populated on the sales order item.
- ☐ D. The transfer of requirements to production planning is controlled by the strategy group in the material master record.

15. In a third-party order process, what customizing setting in SAP S/4HANA Sales controls whether the invoice receipt quantity or the quantity of goods received is relevant for billing?

- ☐ A. The copy control sales document to billing document at the schedule line level
- ☐ B. The copy control sales document to billing document at the item level

- ☐ C. The copy control delivery document to billing document at the item level
- ☐ D. The definition of the sales document item category

16. Which of the following documents used in complaints processing must be created with reference to an invoice?

- ☐ A. Returns order
- ☐ B. Free-of-charge subsequent delivery
- ☐ C. Credit memo request
- ☐ D. Invoice correction request

17. True or False: Although it is not recommended by SAP, you can create independent free-of-charge subsequent deliveries using standard sales document type SD.

- ☐ A. True
- ☐ B. False

9.11 Practice Question Answers and Explanations

1. Correct answer: **C**
   You cannot create a billing document before a sales order. Therefore, answer A is incorrect. You can create a quotation after an inquiry but not the other way around, which invalidates answer B. Post goods issue is always performed after the outbound delivery is created; thus, answer D is also incorrect. Only answer C remains as the correct answer, which is in line with the diagram shown earlier in Figure 9.1.
2. Correct answers: **B and D**
   The SAP Fiori factsheet app Sales Quotations does not display information at the schedule line level, nor does it show key facts of the preceding inquiry. The key facts displayed at the top of the screen always belong to the actual quotation selected. Therefore, answers A and C are both incorrect, leaving us with B and D as the only two correct answers to this question.
3. Correct answer: **D**
   You can offer your customer optional alternative materials by entering them directly below the item requested in both inquiries and quotations.
4. Correct answer: **B**
   The payment terms are an example of a general business data field. Thus, the value entered at the header is automatically copied to all items. However, if you manually change the payment terms for item 20 from CP00 to R001, the item will be included in the own business data item list, shown earlier in Figure 9.6.

As a consequence, if you then change the payment terms at the header level from CP00 to TCPT, item 20 won't be updated. Items 10 and 30 will be automatically updated as they don't show up in the own business data list. Answer B is, therefore, the only correct answer.

5. Correct answers: **B, D, and E**

   The data proposed from the ship-to party during sales order creation is the tax classification, the ship-to address, and the delivering plant.

6. Correct answers: **B and D**

   The data proposed from the sold-to party during sales order creation is the Incoterms and the shipping conditions.

7. Correct answer: **D**

   The sequence of master data objects used by the system to determine the delivering plant in a sales order item is:

   - 1. Customer-material information record
   - 2. Customer master record of the ship-to (assuming the material is maintained in the concerned plant)
   - 3. Material master record

   The customer-material information record always takes priority; thus, answer D is the correct answer to this question.

8. Correct answers: **C, D, and E**

   Shipping point determination is carried out for all relevant items based on the shipping condition at the header level, the loading group maintained in the material master record, and the plant in the sales order item.

9. Correct answers: **B and D**

   As shown earlier in Figure 9.7, the shipping condition in the header of the sales document and the transportation group maintained in the material master record are two fields used during route determination.

10. Correct answer: **B**

    False. Indeed, a rejected item cannot be copied, but the reference status of the item is not controlled by the rejection reason. The reference status is automatically set when the entire quantity has been copied or based on the completion rule defined for the item category.

11. Correct answer: **A**

    In the standard system, item categories for inquiries have been set up so the item reference status is set to **Fully Referenced** after the first reference and regardless of the quantity in the target sales document item. Therefore, answer A is the only correct answer to this question.

12. Correct answer: **B**

    False. In the standard cash sales process shown earlier in Figure 9.9, an outbound delivery is automatically generated when the sales order is posted. The billing document is created later, usually via a periodic batch job. Also, the cash sale invoice is an output generated from the sales order, not from the billing document.

13. Correct answers: **B and D**

    In a rush order scenario, the customer does not pay for the goods immediately. Both the billing document and the output are generated after the delivery of the goods depending on the payment terms negotiated with the customer. Billing happens usually through a periodic batch job. However, as soon as the sales order is saved, an outbound delivery is automatically created. Only answers B and D describe characteristics of the rush order process. This special sales process is illustrated in the diagram shown earlier in Figure 9.10.

14. Correct answers: **C and D**

    In SAP S/4HANA, the item category no longer takes priority when controlling the requirements type. The system first tries to find a requirements type using the strategy group in the material master, and if the strategy group has not been maintained, the system will use the MRP group. Thus, answer A is incorrect. In a make-to-order production scenario, produced materials are received in sales order-specific stock (special stock indicator E). Consignment stock (special stock indicator W) is not used in make-to-order production, making answer B a false statement. Therefore, the only two correct answers are answers C and D.

15. Correct answer: **B**

    The quantity used on the billing document item in a third-party order process (item categories such as TAS or CB1) is controlled by the **Billing Quantity** field in the copy control for *sales document to billing document* at the item level. For example, for item category CB1 the **Billing Quantity** indicator has been set to E; thus, the billing quantity is calculated as the goods receipt quantity minus the invoiced quantity. For item category TAS, however, this indicator is set to F meaning that the billing quantity, in this case, is calculated as the invoice receipt quantity minus the invoiced quantity.

16. Correct answer: **D**

    In the diagram shown earlier in Figure 9.20, an invoice correction request can only be created with reference to an invoice. This limitation exists because standard sales document type RK (Invoice Correction Request) has been defined with the value M in the **Reference mandatory** field, meaning any invoice correction request must be created with reference to a billing document.

17. Correct answer: **B**

    False. Standard sales document type SD (free-of-charge subsequent delivery) has been defined with value C in the **Reference mandatory** field, meaning any free-of-charge subsequent delivery can only be created with reference to a sales order. An example of a free-of-charge subsequent delivery created with reference to a returns order is represented in the diagram shown earlier in Figure 9.17.

9.12 Takeaway

You now understand the order-to-cash process in SAP S/4HANA from the inquiry to the journal entry and how the document flow allows you to visualize all relevant preceding and subsequent documents. You also learned how to use pre-sales documents, such as sales inquiries and sales quotations, to represent the business process prior to the creation of the sales order. Next, we covered the main sources of data in sales documents and explained the rules to determine key fields from both master records and customizing tables.

We then discussed the customizing parameters that control how the reference status of a sales document item is updated. We also saw how to use special sales orders, such as cash sales and rush orders. We reviewed two of the main special business processes in sales: the make-to-order production process and the third-party sales order process.

Finally, we listed the different tools available in SAP S/4HANA to manage and solve customer complaints.

9.13 Summary

In this chapter, we introduced you to the order-to-cash process and described important features in inquiries and quotations. The origin of data in sales documents was discussed as well as the rules followed by the system to update the reference status when subsequent documents are created. We looked at how you can use special sales orders and how to execute special business processes in sales. We ended this chapter by going through the complaints processes supported by SAP S/4HANA.

In the next chapter, you'll learn about how shipping transactions are processed in relation to sales order management and the customizing settings required to perform shipping processes.

Chapter 10
Shipping Process and Customizing

Techniques You'll Master

- Explain the structure of the outbound delivery
- Configure delivery documents
- Understand order combinations when creating outbound deliveries
- Set up automatic determination of relevant fields for outbound delivery creation
- Maintain settings for transportation and delivery scheduling
- Use goods issues for outbound deliveries

In this chapter, we'll focus on delivery management, a component of the Logistics Execution (LE) consisting of all the activities that must take place to deliver goods to customers. We'll explore the structure of the outbound delivery, discuss how to configure delivery types and delivery item categories, describe the prerequisites to combine items from different orders into one delivery, and explain the settings that control the determination of key fields during the creation of an outbound delivery. We'll also cover transportation and delivery scheduling as well as goods issue posting for individual and multiple outbound deliveries.

Real-World Scenario

You're an SAP S/4HANA Sales application consultant working on returns processing. Your customer requires the processing of the delivery due list for returns orders only. In this case, you would need to know how to configure the system to promote a returns-specific shipping point that is automatically determined for all returns sales document types. Using a different shipping point for returns and normal outgoing deliveries will allow creating and processing them separately, thus helping the warehouse manager prioritize and optimize customer order fulfillment functions.

An express shipping point is also a common need in companies operating a warehouse. This way, you can run the delivery due list more often to expedite urgent orders. For example, deliveries can be automatically created every hour for the express shipping point and once a day for the normal shipping point. The express shipping point can be proposed by the system based on the shipping condition or can be manually entered by the user in the sales order item if he or she knows the material has to be delivered with priority.

As a sales and distribution consultant, you should be able to define and assign new shipping points, configure shipping point determination, maintain picking/packing and loading times, and specify a factory calendar for determining dates.

10.1 Objectives of This Portion of the Test

The goal of this portion of the SAP S/4HANA Sales Certification exam is to test your understanding of the following topics:

- Explain the concept of the delivery document
- Define delivery types
- Define and assign delivery item categories
- Understand order combinations in shipping
- Configure shipping point determination
- Maintain route determination

- Maintain picking location determination
- Define scheduling by sales document type
- Define scheduling by shipping point
- Understand the concept of goods issue posting

Note

The Shipping Process and Customizing topic makes up 8%–12% of the total SAP S/4HANA Sales 2020 exam.

The content of this chapter is included in the Shipping topic of the SAP S/4HANA Sales 2020 Upskilling exam which also makes up 8%–12% of the total exam.

10.2 Outbound Deliveries

Once goods are sold and made available by the warehouse, the shipping component in SAP S/4HANA helps you to deliver the goods to your customer through the execution of the following tasks:

- Creation and execution of outbound deliveries
- Picking
- Packing
- Transportation
- Post goods issue

The outbound delivery, the central object used to manage all the shipping activities, can be created either manually by a user or automatically via periodic batch job using standard reports like the SAP Fiori app My Sales Order Due for Delivery. An outbound delivery can be created with reference to the following documents:

- Sales orders
- Stock transport orders
- Subcontract orders
- Projects

Note

You can also create an outbound delivery without any reference using Transaction VL01NO or, for SAP Fiori users, through the Create Outbound Delivery – Without Order Reference app.

Like a sales document, an outbound delivery contains a lot of data stored in both header and item fields. In an outbound delivery, you'll find information regarding the materials and quantities to be picked, picking deadlines, shipping dates, picking statuses, goods movement statuses, billing statuses, ship-to addresses, reference documents, etc.

The data stored in the delivery header is known as *general data* and is valid for the entire document. Some relevant header fields include the following:

- Shipping point
- Route
- Delivery date
- Picking deadline
- Planned goods issue date
- Actual goods issue date
- Gross and net weight of the entire delivery
- The volume of the entire delivery
- Sold-to and ship-to party numbers
- Delivery type
- Sales organization
- Departure and destination countries
- Overall statuses

Data at the item level only applies to the item concerned. Some item fields worth mentioning include the following:

- Material number
- Delivery quantity
- Pick quantity
- Sales unit of measure
- Delivery item category
- Plant
- Storage location for picking
- Batch number
- Gross and net weight of the item
- The volume of the item
- Reference document and item numbers
- Movement type
- Item statuses

Texts can be manually maintained or determined by the system, both at the header level and at the item level. Other information such as the shipping point, Incoterms, partners, or international trade data is only found in the header of the delivery. On the other hand, fields like the plant, storage location, or reference document are stored in the delivery item only.

Figure 10.1 shows an outbound delivery screen for a delivery document with four items. In this example, only three items are relevant for picking, and none of those three items have been picked yet, as indicated by the value **A** (not yet processed) in the highlighted column, the **Picking Status** column. Item 20 has picking status <blank>, which means the material is not relevant for picking.

Itm	Material	Description	Plnt	SLoc	R...	Deliv. Qty	Un	Picked Qty	Un	Batch	B	Picking	WM...
10	252	DateBook X Pro 1000a Laptop	US20	1000		2	EA		EA			A	
20	295	DateBook Laptop starter package	US20			1	EA	0	EA				
30	252	Grey Datebook laptop	US20	1000		1	UN		UN			A	
40	296	DateBook X Pro Power Cable	US20	1000		1	UN		UN			A	

Figure 10.1 Outbound Delivery Overview Screen

Before an outbound delivery is created with reference to a sales order, the system carries out several checks at the header level, the item level, and the schedule line level. If all the prerequisites are fulfilled, the delivery can be created.

At the header level, a sales order must pass the following checks:

- A delivery block is not set at the header level
- At least one item in the sales document is due for delivery

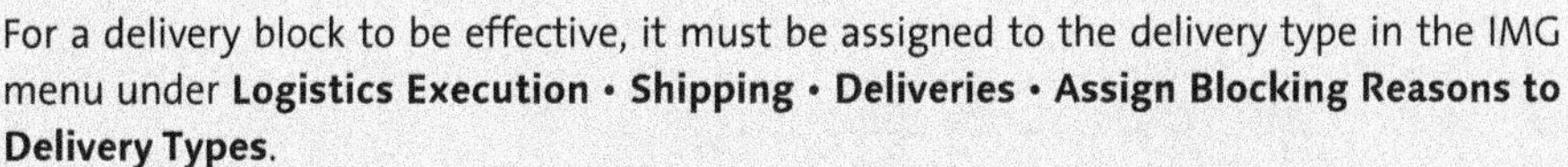

Warning

For a delivery block to be effective, it must be assigned to the delivery type in the IMG menu under **Logistics Execution • Shipping • Deliveries • Assign Blocking Reasons to Delivery Types**.

Also, the sales order item must fulfill the following conditions:

- The order quantity is greater than one.
- The item is complete; in other words, it must not show up on the incompletion log.
- The material cross-distribution-chain and distribution-chain specific statuses do not block the creation of the delivery.
- The quantity confirmed for the material is available on the selected date.

Finally, at the schedule line level, the following prerequisites must be fulfilled:

- The confirmed quantity must be greater than one.
- The selection date must be greater than or equal to the schedule line delivery date.
- The delivery block is not set for the schedule line.

When the outbound delivery document is created, the system updates the document flow, the delivered quantity in the sales order item, and the delivery status in the order at the header and item levels. The delivery status at the header level is calculated based on the delivery status of each item in the order. If the delivery status of all the items is **Not Delivered**, the entire order has the same status. Following the same logic, if all items have the delivery status **Fully Delivered**, the order also shows **Fully Delivered**. If you have both delivered items and items not yet delivered, then the entire order has the delivery status **Partially Delivered**, as shown in the next example.

Example

Let's say a customer that allows partial deliveries has placed an order for two materials, as shown in Table 10.1.

Item Number	Material	Order Quantity	Quantity Available
10	Mat1	30	30
20	Mat2	15	0

Table 10.1 Order to Be Delivered Partially

Since the material in item 20 is not available, an outbound delivery is created for item 10 only, which is delivered in full. After the delivery has been created, the delivery statuses in the order are updated as follows:

- Header delivery status: **Partially Delivered**
- Item 10 delivery status: **Fully Delivered**
- Item 20 delivery status: **Not Delivered**

10.3 Controlling Delivery Documents

Delivery types are used to control how a delivery is processed by the system. Each delivery type represents a specific business transaction in shipping. Table 10.2 shows some of the standard delivery types available in SAP S/4HANA.

Delivery Type	Description
LF	Outbound delivery
LO	Outbound delivery without reference
LR	Returns delivery
NL	Replenishment delivery
LB	Outbound delivery for subcontractor
EL	Inbound delivery
UL	Delivery for stock transfer

Table 10.2 Standard Delivery Types

Note

When a delivery is created with reference to a sales order, the system determines the delivery type from the default delivery type maintained in the shipping parameters of the sales document type. For example, in the standard version of the SAP S/4HANA system, delivery type LF is the default delivery type for sales document type OR, as described in Chapter 4.

In contrast to the sales document type we covered in Chapter 4, Section 4.2, a delivery type has fewer control settings. You can define delivery types in the IMG menu under **Logistics Execution • Shipping • Deliveries • Define Delivery Types**. Most settings used to configure delivery types are shown in Figure 10.2.

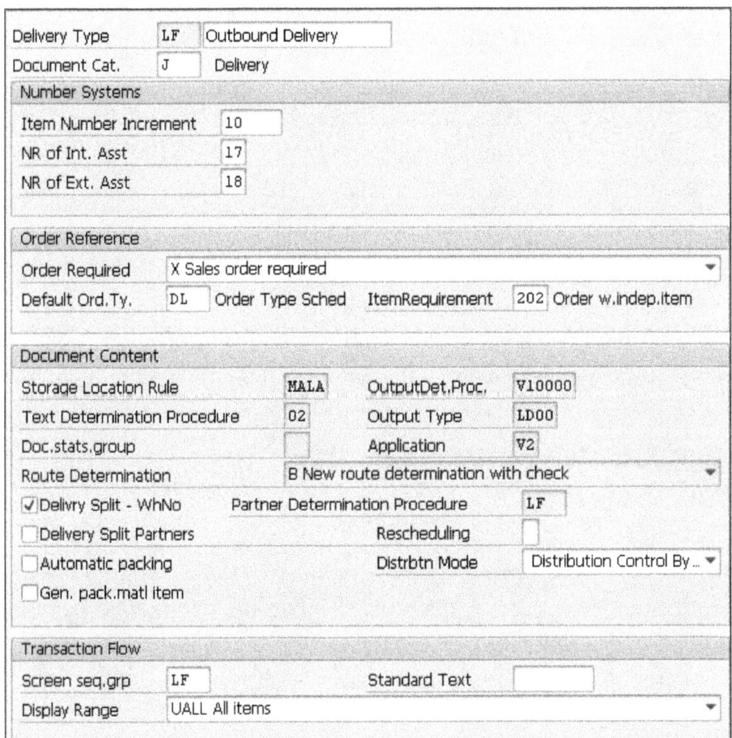

Figure 10.2 Defining a Delivery Type

The details of each and every element in this customizing activity are beyond the scope for the certification exam; however, you should know how the following settings affect the creation and processing of delivery documents:

- **General data**
  - **Document Cat.** (document category): As in the sales document type, this field is used to assign one of the predefined sales document categories to the delivery type. For instance, document category J is used for deliveries.

- **Order Required**: This field indicates whether the delivery can be created without reference to another document. If a preceding document is required, this field specifies which type of document. If you decide to allow the delivery to be created without a reference, you should maintain the default order type. Some of the values available in this field include the following.
  - <blank>: No preceding documents required
  - **B**: Purchase order required
  - **H**: Posting change with delivery
  - **L**: Delivery for subcontracting
  - **O**: Goods movement through inbound delivery / posting change / outbound delivery
  - **R**: Return delivery to the supplier
  - **X**: Sales order required
- **Default Ord.Ty.** (default order type for deliveries without order reference): This parameter must be maintained for delivery types that do not require a preceding document. In this field, you'll specify the order type that will be used by the system to get the control dataset, which is normally inherited from the sales document header into the delivery document in deliveries created with reference to a preceding document.
- **ItemRequirement** (requirement for item that does not refer to a sales order): If the delivery does not refer to a sales order, you can use this field to specify a requirement routine that applies to all items. If a requirement is not fulfilled, the delivery item cannot be processed further.

- **Document Content**
  - **Storage Location Rule**: This field controls how the system determines the picking location if this information is not copied from the preceding sales order. The commonly used rule MALA uses the combination of shipping point, plant, and storage condition.
  - **Partner Determination Procedure**: This field specifies which partner functions are allowed in the delivery and which are mandatory.
  - **Route Determination** (new route determination with or without a check): This field specifies whether a new route determination must be carried out during delivery creation. You can also configure the delivery type to check whether the alternative route entered manually matches the actual route defined in table TROAL (Routes: Allowed Actual Routes for Delivery) for the proposed route. The system will not check the route unless it is manually updated. The following values are available:
    - <blank>: No new route determination
    - **A**: New route determination without check
    - **B**: New route determination with check
    - **C**: New route determination for inbound deliveries

- **OutputDet.Proc.** (output determination procedure): Defines the output types allowed for the delivery type concerned and the sequence in which they are determined.
- **Delivry Split - WhNo** (delivery split according to warehouse number): This indicator controls whether items of different warehouse numbers are split up among two or more deliveries.
- **Delivery Split Partners**: If this checkbox is selected, two preceding document items can only be collected in one delivery if they are assigned to the same partner functions, same partner numbers, and the partners have the same address.

Note

As shown in Figure 10.2, some parameters in the delivery type screen are not editable. This limitation exists because specific activities in the IMG menu are used to maintain them. For example, to maintain the **Route Determination** field, you must select the IMG task under **Logistics Execution • Shipping • Basic Shipping Functions • Routes • Route Determination • Define New Route Determination by Delivery Type**.

Let's now discuss how to configure delivery item categories, which play a similar role to that of sales document item categories, in the sense that both control how the items are processed in their respective sales and distribution documents.

Table 10.3 shows some of the delivery item categories defined in the standard system. As with sales item categories, if the standard delivery item categories don't meet the shipping requirements of your customer, you can always define your own item categories in customizing.

Item Category	Description
TAN	Standard item (from standard order)
TANN	Free of charge item (from standard order)
TATX	Text item (from standard order)
KBN	Consignment fill-up
KEN	Consignment issue
KAN	Consignment pick-up
DLN	Standard item in a delivery without reference to an order

Table 10.3 Sample of Standard Delivery Item Categories

You can define item categories for deliveries in the IMG menu under **Logistics Execution • Shipping • Deliveries • Define Delivery Item Categories**.

Figure 10.3 shows the settings that define a delivery item category.

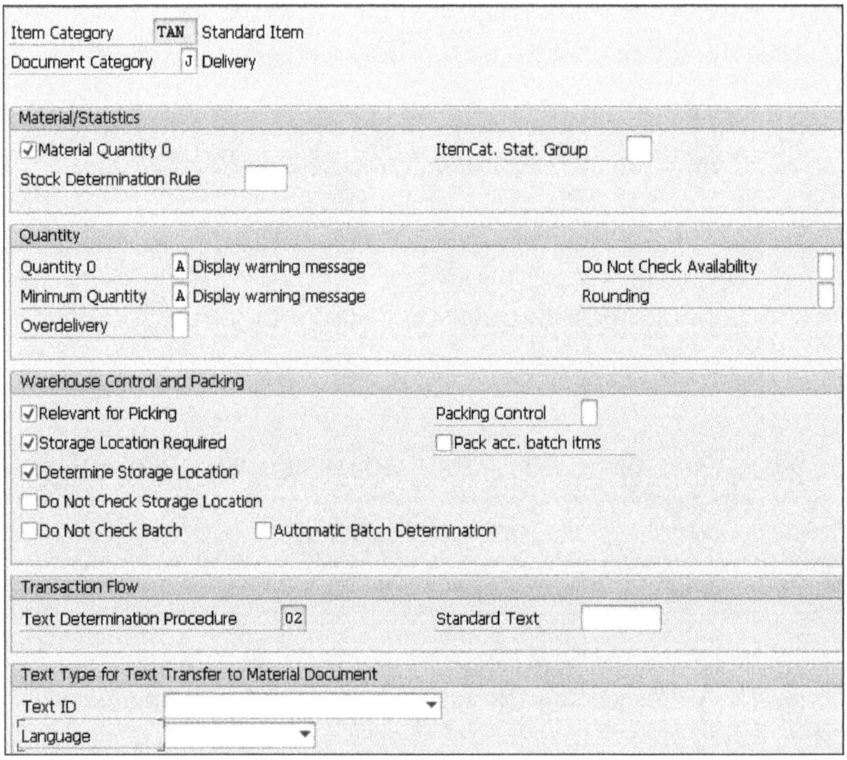

Item Category TAN Standard Item
Document Category J Delivery

Material/Statistics
Material Quantity 0 ItemCat. Stat. Group
Stock Determination Rule

Quantity
Quantity 0 A Display warning message Do Not Check Availability
Minimum Quantity A Display warning message Rounding
Overdelivery

Warehouse Control and Packing
Relevant for Picking Packing Control
Storage Location Required Pack acc. batch itms
Determine Storage Location
Do Not Check Storage Location
Do Not Check Batch Automatic Batch Determination

Transaction Flow
Text Determination Procedure 02 Standard Text

Text Type for Text Transfer to Material Document
Text ID
Language

Figure 10.3 Defining a Delivery Item Category

Again, we won't review every single function controlled by the delivery item category in this book, but you should familiarize yourself with the following parameters:

- **Material/Statistics**
  - **Material Quantity 0**: If this checkbox is selected, you don't need to specify a material quantity for the delivery item.
- **Quantity**
  - **Do Not Check Availability**: This indicator can be used to switch off the availability check for the delivery item with this item category. For example, the availability check is not required for return item categories.
  - **Quantity 0**: This field specifies how the system reacts to items with a quantity of zero. You can select one of the following options:
    - <blank>: No system response
    - **A**: Display warning message
    - **B**: Display error message when a delivery is created or changed
    - **C**: Display error message only when a delivery is created
  - **Minimum Quantity**: This indicator defines how the system reacts if the minimum quantity in the material master record or the customer-material information record is not met. You can choose from one of the following options:

    - <blank>: No system response
    - **A**: Display warning message
    - **B**: Cancel action and display error message
  - **Overdelivery**: This indicator determines the system response when the quantity delivered for an item exceeds either the quantity in the preceding sales order item or the overdelivery tolerance specified in the customer-material information record. The following three options are available:
    - <blank>: No system response
    - **A**: Display warning message
    - **B**: Cancel action and display error message
- **Warehouse Control and Packing**
  - **Relevant for Picking**: If selected, this checkbox activates the picking relevance for the outbound delivery item. For inbound deliveries, this option makes the item relevant for putaway. Picking should not be required for nonstock and service items.
  - **Storage Location Required**: If selected, a storage location must be entered for the item before the delivery can be fully processed.
  - **Determine Storage Location**: This checkbox tells the system to carry out automatic storage location determination in the delivery item.
  - **Packing Control**: This field indicates whether packing information should be entered for the delivery item before goods issue is posted. You can choose one of the following three options:
    - <blank>: Item can be packed.
    - **A**: Item must be packed.
    - **B**: Item cannot be packed.
  - **Automatic Batch Determination**: Select this checkbox to use automatic batch determination for batch-managed materials.
  - **Do Not Check Storage Location**: If selected, the system does not check whether the material entered in the delivery item has been maintained in the picking/putaway storage location.
  - **Do Not Check Batch**: If selected, the system does not check whether the batch entered exists in the database.

If an outbound delivery is created with reference to a sales order, the item category in the sales order is copied into the delivery item. For deliveries created without a preceding document or for independent items entered directly into the delivery, the item category is proposed through the rules defined in the item category determination for deliveries. These rules follow the same approach discussed in Chapter 4, Section 4.3, regarding item category determination in sales documents. As shown in Figure 10.4, the delivery item category is determined by the delivery type, in conjunction with the material item category group, the usage of the item, and the item category of the higher-level item. In this configuration table, the column highlighted in Figure 10.4 is the default item category proposed by the system, whereas all subsequent columns labeled as **MItC** contain the alternative *manual item categories* allowed in the delivery item.

Delivery item category determination

DlvT	ItCG	Usge	ItmC	ItmC	MItC	MItC	MItC	MItC	MItC
LF	NORM			TAN					
LF	NORM	C		TAN					
LF	NORM	CHSP		TAN					
LF	NORM	CHSP	KLN	KLN					
LF	NORM	CHSP	TANN	TANN					
LF	NORM	PACK		DLN	LX	DLP	KEN		
LF	NORM	PSEL	TAX	TAPS					

Figure 10.4 Defining the Item Category Determination in Deliveries

10.4 Order Combination in Delivery Processing

On the shipping screen of the customer master record, you can specify, at the sales area level, whether the customer requires complete deliveries. If the **Complete Delivery** checkbox is selected, the entire order must be delivered in a single outbound delivery. Otherwise, the order can be partially delivered in multiple deliveries. In Chapter 7, Section 7.2, we discussed the various delivery agreements available in the customer master record.

On the same screen, you'll also find the **Order Combination** checkbox, which indicates whether the customer allows several orders, or items from different orders, to be combined in a single delivery. The system takes this indicator into account when the delivery due list is executed and when additional orders or order items are manually added to an existing delivery. You can add additional items within Transaction VL02N or via the SAP Fiori app Manage Outbound Deliveries and then navigating to the menu option **Outbound Delivery • Deliver Sales Order**.

However, to combine multiple orders into a single delivery, besides having the **Order Combination** indicator selected in the order header, order items must fulfill the following criteria:

- Same shipping point
- Same ship-to partner number and ship-to address
- Same Incoterms
- Same sales organization
- Same route
- Same shipping conditions

Note

You can combine two or more items that have different goods issue dates, but the system will prompt you with a warning message. If you ignore this message, the system will use the earliest goods issue date of all the items as the goods issue date of the delivery.

Both the **Order Combination** and the **Complete Delivery** indicators are copied from the customer master record into the order header. If needed, you can change these selections in the sales order.

In summary, you have three options when creating outbound deliveries with reference, depending on your customer's requirements. These three different ways to create a delivery are shown in Figure 10.5.

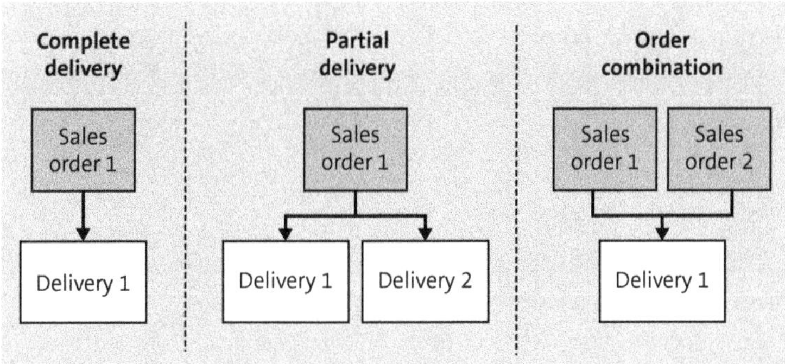

Figure 10.5 Different Ways to Create Outbound Deliveries in SAP S/4HANA

10.5 Adjusting Automatic Determination of Relevant Fields for Outbound Delivery Creation

An outbound delivery cannot be processed without a shipping point and route. A *shipping point* is the organizational structure responsible for processing and monitoring deliveries, whereas a *route* is used in transportation scheduling. Routes specify the means of transport, the transportation legs involved in the shipment, the transportation lead time, the carrier, etc.

Moreover, if the delivery item is relevant for picking, a picking location must be assigned. Therefore, the shipping point, route, and storage location for picking are all fields of special importance in the creation and processing of outbound deliveries. Fortunately, SAP S/4HANA provides mechanisms for the determination of these three fields. In this section, we'll explore how to configure the automatic determination of the shipping point, route, and picking location for outbound delivery creation.

10.5.1 Shipping Point Determination

As described in Chapter 2, a shipping point is part of the enterprise structure within Logistics Execution (LE). Shipping points are assigned to plants in the customizing path **Enterprise Structure • Assignment • Logistics Execution • Assign Shipping Point to Plant**. One plant can have several shipping points, but you can also assign a single shipping point to multiple plants.

You can configure the system to determine the shipping point when processing a sales document item. We suggest making sure the automatic shipping point determination has been set up correctly because the shipping point is a mandatory field during delivery creation. If the determination of the shipping point fails in the order, a user must enter the shipping point manually in the sales order item, so an outbound delivery can be created.

You maintain the shipping point determination in the IMG menu under **Logistics Execution • Shipping • Basic Shipping Functions • Shipping Point and Goods Receiving Point Determination**.

As shown in Figure 10.6, the determination of the shipping point in the sales document item is driven by the following three elements:

- The **Shipping Condition** in the sales order header, which is copied from the customer master record (**Shipping** tab for the relevant sales area data). The shipping condition represents the strategy used to deliver the goods to the customer. Some typical values are standard, pick up, and express.
- The **Loading Group** from the material master record (**Sales: General/Plant** view for the delivering plant in the sales item). A common loading group is assigned to materials that share the same loading requirements, for example, materials that use the same loading point or the same equipment during loading. Some typical values are crane, forklift, and manual.
- The delivering **Plant** in the sales order item. As described in Chapter 9, Section 9.4, a delivering plant can be copied from three master data objects: the material master, the customer master (of the ship-to party), or the customer-material information record.

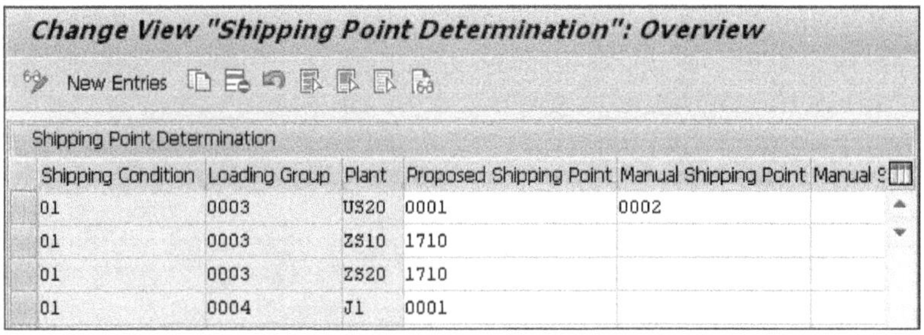

Change View "Shipping Point Determination": Overview

New Entries

Shipping Point Determination

Shipping Condition	Loading Group	Plant	Proposed Shipping Point	Manual Shipping Point	Manual S
01	0003	US20	0001	0002	
01	0003	ZS10	1710		
01	0003	ZS20	1710		
01	0004	J1	0001		

Figure 10.6 Shipping Point Determination

Besides the proposed shipping point, you can maintain up to 11 manual shipping points that can be entered by a user to override the default shipping point. For example, in the first row of the configuration table shown in Figure 10.6, the manual shipping point 0002 has been defined as an alternative shipping point to shipping point 0001, for the same combination shipping condition, loading group, and plant. If the default shipping point 0001 is out of operation, a business user can change the shipping point to shipping point 0002 before the delivery is created.

To avoid delays in the creation of deliveries, the shipping point has been added to the incompletion procedure assigned to the standard item categories that are relevant for delivery; for example, item category TAN. So, if the shipping point cannot be determined because, for example, the shipping condition is blank in the sales order header, the system adds the item to the incompletion log and informs the user that the document is incomplete due to missing mandatory data. The missing data must be entered as soon as possible to make the document complete. In this case, once the user populates the shipping condition, a new shipping point determination is carried out, and if successful, the sales document item is removed from the incompletion log.

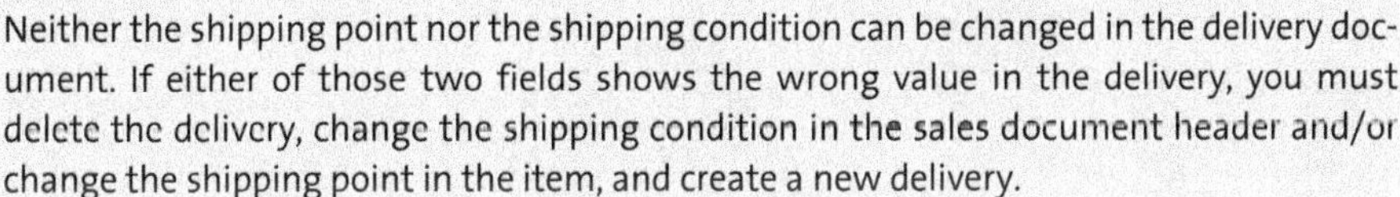

Warning

Neither the shipping point nor the shipping condition can be changed in the delivery document. If either of those two fields shows the wrong value in the delivery, you must delete the delivery, change the shipping condition in the sales document header and/or change the shipping point in the item, and create a new delivery.

10.5.2 Route Determination

Before you make settings for route determination, you must define routes in customizing. The route determines the itinerary and the mode of transport required to deliver the goods to the customer. Routes are useful for grouping sales order items according to shipping criteria. For example, you can carry out transportation planning and combine all sales items that require refrigerated trucks based on the transportation group maintained in the material master record.

To define routes in SAP S/4HANA, you must perform the following preliminary activities:

- Define the modes of transport (e.g., by road, by air, by post, etc.)
- Define the shipping types (e.g., truck, plane, ship, etc.)
- Define the transportation connection points (e.g., airport, railway station, border crossing, etc.)

Next, you can define the actual routes and route stages. Routes are maintained in the customizing path **Logistics Execution • Shipping • Basic Shipping Functions • Routes • Define Routes • Define Routes and Stages**. In this step, you'll define a

6-character alphanumeric key, enter a description, and maintain the relevant data for the following parameters of the route:

- Shipping type
- Shipping type of preliminary leg
- Shipping type of subsequent leg
- Forwarding agent
- Transit duration (or transit time) in calendar days
- Transit duration (or transit time) in hours/minutes
- Transportation lead time in calendar days
- Transportation lead time in hours/minutes
- Factory calendar of the forwarding agent
- Mode of transport

As a reference, Figure 10.7 shows the customizing screen where routes are defined. In this customizing activity, you can also define stages for each route, but this step is optional. On this screen, you must select a departure point from the list of transport connection points. Then, you'll select the destination point and move on to the next stage.

Routes

Route	Description	ST	PL	Un	ServcAgent	TransitDur	Trav. Dur.	TransLdTm.	Tr.Lead Ti...	Cal	Distance
TR0001	Truck/0days lead time/1day transit time	01				1.00				99	
TR0002	Truck/0days lead time/2days transit time	01				2.00				99	
TR0003	Truck/1day lead time /0days transit time	01						1.00		99	
TR0004	Truck/1day lead time /1day transit time	01				1.00		1.00		99	
TR0005	Truck/1day lead time /2days transit time	01				2.00		1.00		99	

Figure 10.7 Defining Routes and Stages

Once your routes have been defined, you can proceed to maintain the route determination. We've discussed route determination briefly in Chapter 9, Section 9.4, but in this context, we'll provide you with more details about the criteria for route proposal in both the sales order and the delivery.

In a sales order, route determination occurs based on the following elements:

- The **Country** and **Departure Zone** of the shipping point.
- The **Shipping Condition** in the sales order, which is usually maintained in the sold-to party and copied into the order.
- The **Transportation Group** of the material master record (**Sales: General/Plant** view for the delivering plant in the sales item). You can assign a common transportation group to materials that share the same transportation requirements. Some typical values are container, parcel, on pallets, and bulk product. We suggest including the transportation route as a mandatory field for the material master record to ensure the route is determined successfully.
- The **Country** and **Transportation Zone** of the ship-to party.

> **Note**
> If either the shipping condition or the ship-to party are changed in the sales order, the system prompts you to determine the route again for each item.

The diagram shown in Figure 10.8 illustrates how a route is determined for a sales order item.

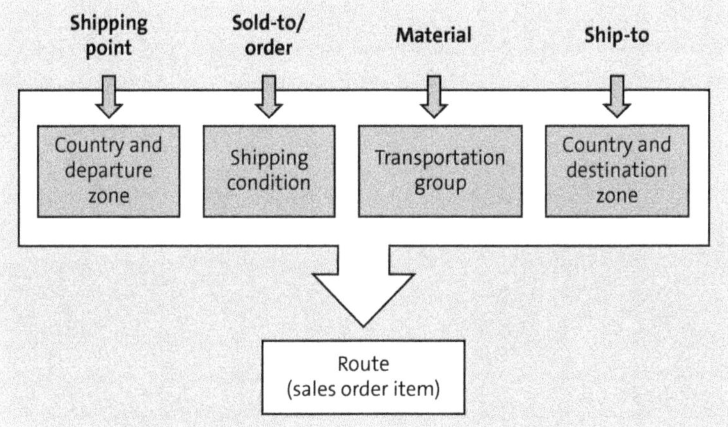

Figure 10.8 Route Determination in Sales Orders

If a route is found, when the sales order item is delivered, the system copies the route into the delivery document header. Since the route in the delivery is located at the header level, this route applies to all items in the delivery. However, unlike the shipping point or the shipping condition, a route can be manually changed in the delivery if required.

As described earlier in Section 10.3, the delivery type controls whether the route must be redetermined when an outbound delivery is created. This function is activated via the **Route Determination** parameter, shown earlier in Figure 10.2. In some cases, you may not know the actual weight of the goods until it's time to create the delivery, or you might just want to add the weight to the criteria for the proposal of the route to further refine the route determination. Therefore, performing another route determination at delivery creation makes sense so you can take the weight of the goods into account.

> **Example**
> Let's say a customer with shipping condition "express" places an order for 15 laptops to be delivered from your warehouse in San Francisco to an office in Los Angeles. Based on the shipping condition, transportation group, and departure and destination zones, the system determines a route to send the goods by air. However, shipments over 20 kilograms are too expensive to be delivered by air freight. So, for the shipping condition "express," you'll configure the system to choose a route by air only if the total weight is less than 20 kilograms; for heavier shipments, the system selects a route by road. In this case, each

box weighs 2.5 kilograms making the total weight 37.5 kilograms. As a result, the system switches from air freight to a route that has a transportation type by road in the delivery. You can thus use the weight group to redetermine the route for the delivery, now taking into consideration the total weight of the delivery.

A weight group is determined based on the total weight of the delivery. You can define weight groups in the following ways:

- Up to 1 kg
- Up to 20 kg
- Up to 100 kg
- Over 100 kg

The diagram shown in Figure 10.9 illustrates how a route is determined in a delivery while taking the weight group into account.

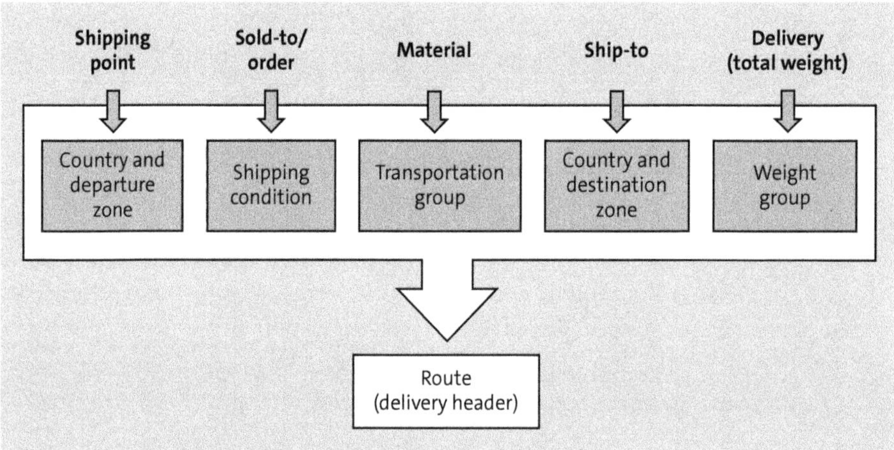

Figure 10.9 Route Determination in Deliveries

Finally, you can configure the system to check if the route found for the delivery is valid based on the route in the preceding order item (proposed route). This functionality is controlled by the value entered in the **Route Determination** field for the delivery type concerned. The route that the system determines during delivery processing is compared against the allowed actual routes maintained in the configuration table TROAL for a given proposed route. Users can still manually enter alternative routes in the sales document item.

You can define an alternative route (or routes) to the proposed route determined by the system in the IMG menu under **Logistics Execution • Shipping • Basic Shipping Functions • Routes • Route Determination • Define Allowed Actual Route by Proposed Route.**

10.5.3 Picking Location Determination

You've learned that the delivery item category controls whether or not the delivery item is relevant for picking. If the item is relevant for picking, a storage location is required so picking can be carried out. The goods are issued from the stock at the location entered in the delivery item.

If a storage location was specified in the preceding sales order item, this information is copied into the delivery item. Usually, the storage location is not part of the incompletion log of the sales item category, and the business users responsible for processing sales documents may encounter difficulty when selecting the correct picking location. Therefore, in most cases, the storage location field is left empty in the sales order.

To avoid delivery items that are relevant for picking without a picking location, SAP S/4HANA allows for the automatic determination of the picking location when a delivery is created.

The automatic determination of the picking location depends on the following factors:

- Whether the delivery item category is relevant for picking.
- Whether the automatic determination of storage locations has been activated for the delivery item category.
- Whether a picking rule has been assigned to the delivery type. This rule is visible in the **Rule for Determining the Storage Location for Picking** field in the definition of the delivery type.

You can assign picking rules to delivery types in the IMG menu path **Logistics Execution • Shipping • Picking • Determine Picking Location • Define Rules for Picking Location Determination**.

The standard system comes preconfigured with four picking rules:

- MALA: The most common picking rule across all delivery types and thus the only rule in scope for this certification.
- RETA: This rule is primarily used in the retail industry.
- MARE: A combination of rules MALA (preferred procedure) and RETA (backup procedure).
- MSRE: Designed for advanced returns scenarios.

The picking rule MALA is assigned to the standard delivery type LF (Outbound Delivery), shown earlier in Figure 10.2, and determines the correct picking location based on the following three elements:

- **Delivering Plant** of the delivery item.
- **Storage Condition** of the material to be delivered. The storage condition indicates how the goods should be stored in the warehouse. Some typical values could be ambient storage, cold storage, and deep-freeze storage. You'll find this

field in the material master record (in the **Plant data / stor. 1** view) and in the delivery on the item details screen for picking.

- **Shipping Point** of the delivery.

You can maintain the automatic determination of the picking location based on picking rule MALA in the IMG menu under **Logistics Execution • Shipping • Picking • Determine Picking Location • Assign Picking Locations**. Figure 10.10 shows the configuration table behind this customizing activity. In the first row, note that picking location **171A** is assigned to the combination of shipping point **NJ01**, delivering plant **1710**, and shipping condition **01**.

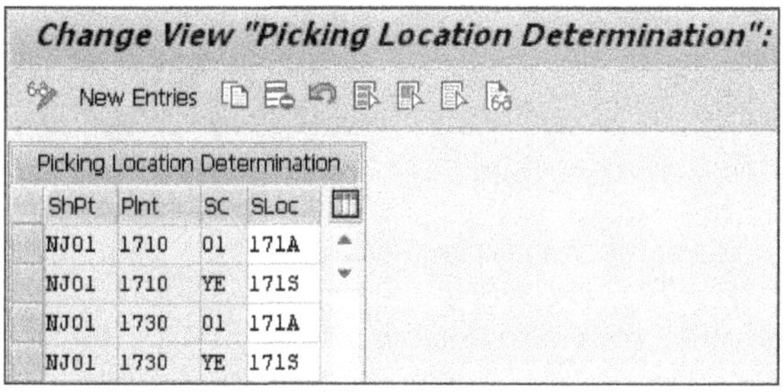

Change View "Picking Location Determination":

New Entries

Picking Location Determination

ShPt	Plnt	SC	SLoc
NJ01	1710	01	171A
NJ01	1710	YE	171S
NJ01	1730	01	171A
NJ01	1730	YE	171S

Figure 10.10 Picking Location Determination

You'll find the picking location in the outbound delivery next to the delivery plant on the picking overview screen, as shown earlier in Figure 10.1. The system allows you to change the picking location as long as the picking status is A (not yet picked). Once the picking status is updated to either partially or fully picked, the picking location cannot be changed. To change the picking location now, you must first undo the picking and then change the storage location.

10.6 Adjusting Delivery and Transportation Scheduling

Sometimes, a customer needs to receive goods on a particular date, and the user entering the order in the system specifies the requested delivery date on the schedule line of the sales order item. To ensure that goods arrive at the customer on the requested date, the system must determine if enough time is available to perform the associated shipping activities, such as picking, packing, and loading the materials; planning transportation; and conveying the goods to the customer.

As described in Chapter 7, Section 7.2, the system first carries out backward scheduling from the date of the requested delivery date to calculate the different deadlines in the shipping process, including the date when the material must be available in the warehouse. If the material availability date is in the past, the requested delivery date cannot be met. At this point, the system must perform forward scheduling to propose a realistic delivery date. All these calculations are only possible if you tell the system the times involved in performing the four main activities involved in delivery and transportation scheduling. Organized into a logical sequence of events, you'll maintain the following times:

- **Transportation lead time**
  The time, either in calendar days or in hours/minutes, required for planning a shipment when goods are to be delivered via a specific route. If working hours have been maintained in the referring shipping point, the system uses the transportation lead time in hours/minutes. Otherwise, the system uses the time in calendar days. Times in both calendar days and/or hours/minutes can be specified in the definition of the route, as shown earlier in Figure 10.7.
- **Pick/pack time**
  The time in days required to carry out the picking and packing activities. This time is determined using the shipping point, the route, and the weight group of the sales order item.
- **Loading time**
  The time in days required to load the goods in the sales order item. This time depends on the shipping point, the route, and the loading group of the material.
- **Transit time**
  Also referred to as the *transit duration*, the time in calendar days for the goods to arrive at the customer site via a specific route. This time is specified in the definition of the route, as shown earlier in Figure 10.7.

The times used in backward scheduling can be maintained by following the IMG menu path **Logistics Execution • Shipping • Basic Shipping Functions • Scheduling • Delivery Scheduling and Transportation Scheduling • Maintain Duration**. Figure 10.11 and Figure 10.12 show the customizing screens where you can enter loading times and pick/pack times, respectively.

Loading time

Shipping Point	Description of Shippi...	Route	Description of R...	Loading Group	Description of L...	L.time days	L.time hours
RF01	Shipping Point RF01	000001	Northern Route	0001	Crane	0.20	1:30
RF02	Shipping Point RF02	000001	Northern Route	0001	Crane	0.20	1:30
S100	Vehicle Shipping Point	000001	Northern Route	0001	Crane	0.20	1:30
S200	Parts Shipping Point	000001	Northern Route	0001	Crane	0.20	1:30

Figure 10.11 Maintaining Loading Times

Pick/pack time

Shipping pt	Route	WgtGr	P/P days	P/P hours	Description	Description	Description
RF01	000001	0010	0.50	3:30	Shipping Point RF01	Northern Route	Up to 10 kg
RF01	000001	0100	0.50	4:00	Shipping Point RF01	Northern Route	Up to 100 kg
RF02	000001	0010	0.50	3:30	Shipping Point RF02	Northern Route	Up to 10 kg
RF02	000001	0100	0.50	4:00	Shipping Point RF02	Northern Route	Up to 100 kg
S100	000001	0010	0.50	3:30	Vehicle Shipping Poi...	Northern Route	Up to 10 kg
S100	000001	0100	0.50	4:00	Vehicle Shipping Poi...	Northern Route	Up to 100 kg
S200	000001	0010	0.50	3:30	Parts Shipping Point	Northern Route	Up to 10 kg
S200	000001	0100	0.50	4:00	Parts Shipping Point	Northern Route	Up to 100 kg

Figure 10.12 Maintaining Pick/Pack Times

The system calculates the following the deadlines:

- **Delivery deadline**
  The requested delivery date, in other words, the date when the goods must arrive at the customer site. However, if the requested delivery date is not achievable, forward scheduling is carried out, and the delivery deadline becomes the earliest date on which the goods can be delivered.
- **Goods issue deadline**
  The date on which the goods should leave the warehouse to ensure that delivery takes place on the requested date.
- **Loading deadline**
  The date on which the goods must be on the loading bay ready to be loaded into the corresponding vehicle. Any vehicles used to ship the goods must be at the shipping location on the loading date.
- **Transportation scheduling deadline**
  The date on which the logistics department must start planning for the transportation of the goods to ensure the required vehicles are ready, and in the expected location, by the loading deadline.
- **Material availability deadline**
  The date on which the goods must be available in the warehouse so they can be picked and packed.

You'll find all these deadlines under the **Shipping** tab on the schedule line level, as shown in Figure 10.13.

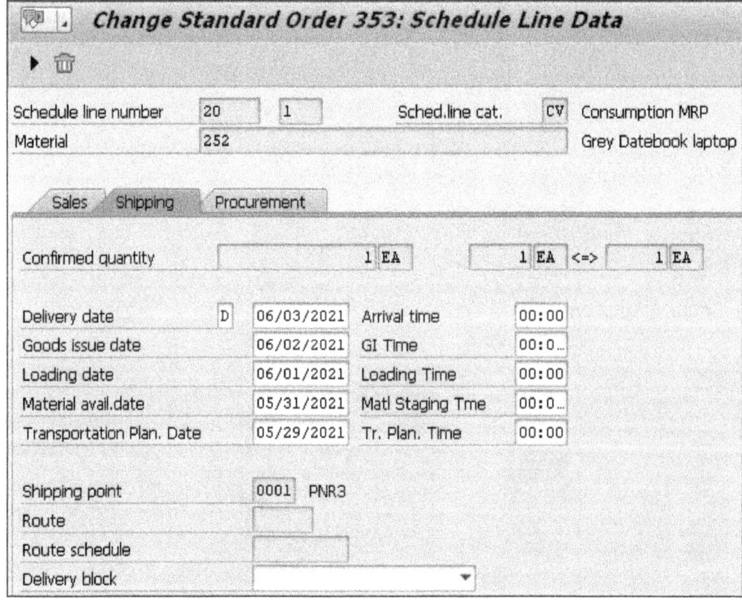

Figure 10.13 Schedule Line: Shipping Details

Assuming a requested delivery date has been entered in the sales order item, the system determines the deadlines via backward scheduling if the following criteria are met:

- For each sales document type, you've specified whether scheduling is to be carried out and, if so, which type.
- You've defined the indicators for scheduling for each shipping point.
- You've defined transportation lead times.
- You've defined pick/pack times.
- You've defined loading times.
- You've defined transit times.

SAP S/4HANA distinguishes between two types of scheduling depending on the times taken into account:

- Delivery scheduling
- Transportation scheduling

In backward scheduling, *delivery scheduling* uses the pick/pack time and the loading time to determine the material availability deadline and the loading deadline, respectively. On the other hand, *transportation scheduling* uses the transportation lead time and the transit time to determine the transportation scheduling deadline and the goods issue deadline, respectively. In customizing, you can define, for a given sales document type, whether delivery scheduling and/or transportation scheduling should be carried out. You can also indicate whether the system must only perform backward scheduling for all the items in a sales order with a certain sales document type.

Note

As described in Chapter 7, Section 7.3, backward scheduling is always carried out for both delivery and transportation scheduling. However, forward scheduling is only carried out if the material availability dates determined by backward scheduling fall in the past.

You can maintain the settings for delivery scheduling and transportation scheduling in the IMG menu under **Logistics Execution • Shipping • Basic Shipping Functions • Scheduling • Delivery Scheduling and Transportation Scheduling**. Figure 10.14 shows how you can configure automatic scheduling by sales document type.

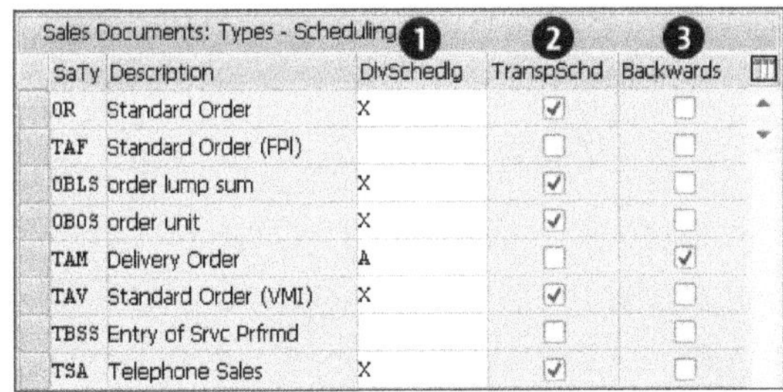

Sales Documents: Types - Scheduling ❶ ❷ ❸

SaTy	Description	DlvSchedlg	TranspSchd	Backwards
OR	Standard Order	X	☑	☐
TAF	Standard Order (FPI)		☐	☐
OBLS	order lump sum	X	☑	☐
OBOS	order unit	X	☑	☐
TAM	Delivery Order	A	☐	☑
TAV	Standard Order (VMI)	X	☑	☐
TBSS	Entry of Srvc Prfrmd		☐	☐
TSA	Telephone Sales	X	☑	☐

Figure 10.14 Defining Scheduling by Sales Document Type

In this context, you must take into consideration the following fields:

❶ **DlvSchedlg** (delivery scheduling): This field controls what type of delivery scheduling must be carried out for each sales document type. The following three options are available:

- <blank>: Delivery scheduling is off.
- **A**: Only the loading date is automatically calculated.
- **X**: Both the pick/pack date and the loading date are automatically calculated.

❷ **TranspSchd** (transportation scheduling): If this checkbox is selected, the system uses the transportation lead time and the transit time to determine the transportation planning date and the goods issue date.

❸ **Backwards**: If this checkbox is selected, the system only carries out backward scheduling for all the schedule lines in the sales document. You must be aware that, when this indicator is set, the system may determine a material availability date that lies in the past. Nonetheless, deactivating forward scheduling may help you identify backlogs in production and then take corrective actions to ensure your customers receive goods on time.

You can also define the type of delivery scheduling by shipping point. The settings shown in Figure 10.15 allow you to decide, for each shipping point, whether the pick/pack time and/or the loading time should be taken into account as well as how the system should determine both times should they be considered during delivery scheduling.

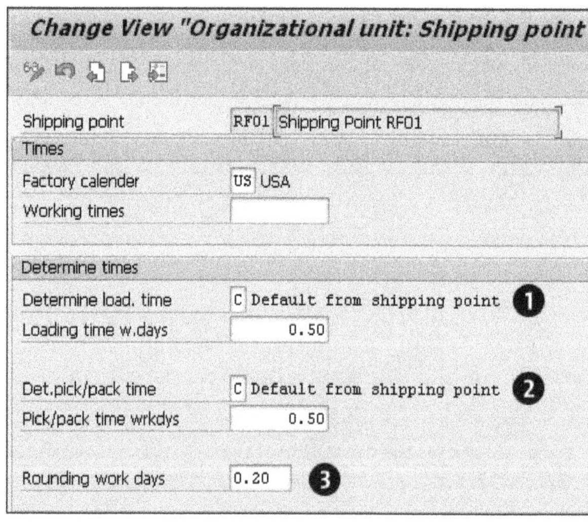

Figure 10.15 Defining Scheduling by Shipping Point

In this context, you must take into consideration the following fields:

❶ **Determine load. time**: This field dictates whether the loading time must be taken into account for the shipping point and, if so, the logic to determine the loading time. You must select one of the following four options:

  - <blank>: No loading time is determined.
  - **A**: A route-dependent loading time is determined based on the shipping point, the route, and the loading group. (The system reads entries in the customizing table shown earlier in Figure 10.11 where the route is populated.)
  - **B**: A route-independent loading time is determined based on the shipping point and the loading group. (The system reads entries in the customizing table shown earlier in Figure 10.11 where the route is empty.)
  - **C**: Only the loading time defined in the shipping point is used.

❷ **Det. pick/pack time**: This field indicates whether the pick/pack time must be taken into account for the shipping point and, if so, the logic to determine it. You must select one of the following four options:
  - <blank>: No pick/pack time is determined.
  - **A**: A route-dependent pick/pack time is determined based on the shipping point, the route, and the weight group. (The system reads entries in the customizing table shown earlier in Figure 10.12 where the route is populated.)
  - **B**: A route-independent pick/pack time is determined based on the shipping point and the weight group. (The system reads entries in the customizing table shown earlier in Figure 10.12 where the route is empty.)
  - **C**: Only the pick/pack time defined in the shipping point is used.

❸ **Rounding work days**: If you have not maintained any working times, you can use this field to specify a buffer time in days by shipping point, which will be automatically added to the loading time and the pick/pack time. You define this time as a safety margin to cater for possible delays in the pick/pack and loading activities.

> **Note**
> Both the transportation lead time and the transit time for transportation scheduling are always determined based on the route, as shown earlier in Figure 10.7.

10.7 Using Goods Issue for Outbound Deliveries

Let's now discuss the effects of the post goods issue in the SAP S/4HANA system. As soon as the goods leave the warehouse, a goods issue must be posted in the delivery to reflect the transfer of ownership. A goods issue posting is only possible for the entire delivery. The following are different ways to perform goods issues:

- **Goods issue posting for individual deliveries**
  This activity can be performed via Transaction VL02N or via the SAP Fiori apps Change Outbound Delivery or Manage Outbound Deliveries.
- **Goods issue in collective processing**
  You can use this option to post goods issues for multiple outbound deliveries. This activity can be performed in the foreground via Transaction VL06G or via

the SAP Fiori apps Manage Outbound Deliveries or Outbound Deliveries for Goods Issue. You can also schedule a background job to execute the program WS_MONITOR_OUTB_DEL_GDSI if you want the post goods issue to be performed automatically.

- **Goods issue for an entire shipment**
  This option is available if shipment documents are used as a basis for carrying out transportation planning. You can perform the post goods issue for a shipment document via Transaction VT02N or via the Change Shipment app.

However, to post goods issue for an outbound delivery, the following shipping activities must be completed:

- All the required data has been entered in the delivery. Depending on the settings in the material master record, you may need to enter batch or serial numbers as well.
- Picking has been completed for all the delivery items that are relevant for picking.
- All transfer orders have been confirmed for items picked using the warehouse management system (WMS) or Lean WM.

If all prerequisites are met, you can proceed and post a goods issue for a delivery or a group of deliveries. The posting of the goods issue causes the system to carry out the following functions:

- The overall goods movement status of the delivery is set to **Complete**.
- The delivery status of the preceding sales order is updated.
- The warehouse or plant stock is decreased by the delivery quantity.
- The postings to the general ledger accounts are updated.
- Delivery requirements in materials planning are reduced by the delivery quantity.
- If serial numbers are assigned to the delivery item, the system status of the serial number is updated.
- The document flow is updated to show the material document number of the goods issue.
- The majority of fields in the outbound delivery are no longer editable to prevent inconsistencies between the delivery and the goods issue document.

Note

Once a goods issue is posted for an outbound delivery, the delivery is added to the *billing due list*, and the billing document can be created as long as the following criteria are met:

- The delivery is relevant for billing.
- The overall credit status of the delivery allows billing.
- A billing block is not set on the delivery (at either the item level or the header level).

If a goods issue has been posted by accident, you can cancel the goods issue posting for the delivery. However, a goods issue reversal is only possible if the corresponding reversal movement type has been defined in customizing. For example, standard movement type 602 (RE goods issue delivery reversal) is the reversal movement type for 601 (GD goods issue delivery). If you have defined a custom movement type for goods issues, we recommend you also maintain associated reversal movement types.

You can reverse a goods issue for a delivery document via Transaction VL09 or via the Reverse GI –Outbound Delivery app or Manage Outbound Deliveries app. But if a delivery has been partially or fully invoiced, a goods issue cannot be reversed unless the billing document is canceled first.

Note
As with goods issue postings, partial goods issue reversals are not allowed. You can only reverse the goods issue for the entire delivery document.

Once a goods issue has been reversed, the overall goods movement status of the delivery is reset to **Not Yet Started**, and a new material document is created for the reversal and added to the document flow.

10.8 Important Terminology

In this chapter, the following terminology was used:

- **Shipping condition**
  Specifies the shipping strategy for the delivery of the goods to the customer (e.g., standard or express). Shipping conditions are used to define the route determination as well as determine the shipping point.
- **Shipping point**
  The physical location from which the truck departs. For instance, a shipping point can be defined as a warehouse or a collection of loading ramps. Each outbound delivery is processed by one shipping point.
- **Route**
  The path the truck (or other conveyance) follows to get to the address of the ship-to party. The route contains information about the different transportation legs, the forwarding agent, the transit time, transportation lead time, travel duration, and distance.
- **Transportation zone**
  A regional zone defined according to the transportation requirements of your business. For example, you can define transportation zones to correspond to the postal codes of the country where your company operates. This field is maintained in the addresses of objects such as customers, shipping points, etc. and is used by the system to determine the route.

- **Transportation group**
  A field in the material master record maintained at the plant level used to group materials that share the same route and transportation requirements.
- **Loading group**
  Defines a group of materials that share the same loading requirements, for example, materials that require the same amount of time for loading or require the same loading bay.
- **Shipping type**
  The type of shipping used to transport the goods to the customer. Examples of common shipping types are truck, mail, rail, airplane, or sea.
- **Backward scheduling**
  The process of scheduling the business process from the customer's requested delivery date backward in order to meet the specified delivery date.
- **Forward scheduling**
  The process of scheduling the business process from the material availability date forward in order to calculate an achievable delivery date.
- **Storage conditions**
  Defines the storage conditions required for a certain material in a specific plant, for example, ambient storage, cold storage, etc.
- **Picking**
  Refers to the process of physically moving the goods from a storage location to the picking area so goods are ready for loading and shipping.
- **Outbound delivery**
  A document in SAP S/4HANA used to record the data generated during the shipping process and to support basic functions in logistics such as picking, packing, transportation, and goods issues. An outbound delivery can be created with or without reference to a preceding document and contains information at the header level and the item level.
- **Goods issue posting**
  Usually the last step in the shipping process. Once goods leave the warehouse, ownership is transferred to the carrier, and the goods issue posting must be carried out for the outbound delivery to reduce the warehouse stock according to the delivery quantity.

10.9 Practice Questions

These practice questions will help you evaluate your understanding of the topics covered in this chapter. The questions shown are similar in nature to those found on the certification examination. Although none of these questions will be found on the exam itself, they will allow you to review your knowledge of the subject. Select the correct answers and then check the completeness of your answers in the next section. Remember that, on the exam, you must select all correct answers and only correct answers to receive credit for the question.

1. Which of the following ways of creating outbound deliveries are supported in SAP S/4HANA? (There are three correct answers.)

☐ A. Without any reference

☐ B. With reference to a billing document

☐ C. With reference to a sales order

☐ D. With reference to a stock transport order

☐ E. With reference to a shipment document

2. True or False: Unlike a sales order, an outbound delivery does not contain any data at the schedule line level.

☐ A. True

☐ B. False

3. Which of the following fields are only found in the header of the delivery document? (There are two correct answers.)

☐ A. Plant

☐ B. Shipping point

☐ C. Ship-to party number

☐ D. Reference document number

4. You must ensure that an outbound delivery will determine a picking storage location. What controls this functionality?

☐ A. The delivery schedule line category

☐ B. The delivery type

☐ C. The delivery item category

☐ D. The material master record

5. Which of the following settings can be configured in the definition of a delivery item category? (There are two correct answers.)

☐ A. Whether the delivery item must meet the minimum quantity maintained in the material master or customer-material record

☐ B. Whether the item is relevant for billing

☐ C. Whether an availability check must be carried out for the item

☐ D. Whether partial deliveries are allowed

6. Which of the following elements can be used to influence the determination of the delivery item category? (There are three correct answers.)

- ☐ A. Shipping point
- ☐ B. Item usage
- ☐ C. Item category group
- ☐ D. Route
- ☐ E. Delivery type

7. Which of the following functions are controlled by the delivery type? (There are two correct answers.)

- ☐ A. The delivery is relevant for picking.
- ☐ B. A credit limit check must be performed.
- ☐ C. A sales order is required as a basis for delivery.
- ☐ D. The rule for determining the storage location for picking.

8. True or False: Item category determination is not carried out if the outbound delivery is created with reference to a sales order.

- ☐ A. True
- ☐ B. False

9. A customer requests that two orders be combined into a single delivery. Which of the following sales order fields must have the same value to fulfill the criteria for order combination? (There are three correct answers.)

- ☐ A. Storage location
- ☐ B. Route
- ☐ C. Incoterms
- ☐ D. Ship-to party
- ☐ E. Division

10. True or False: If needed, you can manually override both the route and/or the shipping point determined by the system in the delivery.

- ☐ A. True
- ☐ B. False

11. Which of the following are characteristics of the route determination functionality in SAP S/4HANA? (There are three correct answers.)

☐ A. The system prompts you to redetermine the route for each item if the ship-to party changes in the order.

☐ B. The system prompts you to redetermine the route for each item if the shipping condition changes in the order.

☐ C. The system prompts you to redetermine the route for the item if the quantity changes in the order.

☐ D. The delivery item category controls whether the route must be redetermined when an outbound delivery is created.

☐ E. Route determination can be carried out with or without weight group.

12. Which of the following factors is a prerequisite for the automatic determination of the picking location in the delivery item? (There are two correct answers.)

☐ A. A picking rule has been assigned to the delivery type.

☐ B. The automatic determination of storage location has been activated for the delivery type.

☐ C. The delivery item category must be relevant for picking.

☐ D. The storage location has been assigned to a warehouse number.

13. Which of the following elements can be used to determine the loading time in delivery scheduling? (There are three correct answers.)

☐ A. Transportation group

☐ B. Route

☐ C. Weight group

☐ D. Loading group

☐ E. Shipping point

14. During backward scheduling, which of the following times are always determined using the route? (There are two correct answers.)

☐ A. Transit time

☐ B. Total replenishment lead time

☐ C. Loading time

☐ D. Transportation lead time

15. Which of the following functions are carried out on the basis of the goods issue document? (There are two correct answers.)

☐ A. Overall status for picking is updated.

☐ B. The document flow is updated.

☐ C. The warehouse or plant stock is increased by the delivery quantity.

☐ D. Delivery requirements in materials planning are reduced by the delivery quantity.

10.10 Practice Question Answers and Explanations

1. Correct answers: **A, C, and D**

   In SAP S/4HANA, depending on how the delivery type is configured, you can create deliveries without any reference using the SAP GUI Transaction VL01NO or the SAP Fiori app Create Outbound Delivery – Without Order Reference. You can also create deliveries with reference to sales orders and stock transport orders. However, you cannot create deliveries with reference to billing documents, which is a subsequent document that is only generated after a goods issue is posted for the outbound delivery. Shipments documents are used in transportation planning, and you can assign deliveries to them, but both deliveries and shipment documents are created independently of one another.

2. Correct answer: **A**

   True. The structure of the delivery is divided into data at the header level and at the item level. No schedule lines exist in the delivery. However, schedule line information is available in the sales order and is used to create the delivery document.

3. Correct answers: **B and C**

   Each delivery is always created for a single shipping point and is delivered to a single address; thus, both the shipping point and the ship-to party are found at the header level. The delivering plant is maintained at the item level, as shown earlier in Figure 10.1, since you can assign the same shipping point to multiple plants. The preceding document is also stored at the item level since SAP S/4HANA allows you to combine several sales orders into a single delivery.

4. Correct answer: **C**

   The **Determine Storage Location** checkbox in the delivery item category, shown earlier in Figure 10.3, controls whether the system automatically determines a storage location for the delivery item.

5. Correct answers: **A and C**

   The **Billing Relevance** field controls whether an item is relevant for billing. However, this field is not found in the definition of the delivery item category but rather in the definition of the sales document item category. The **Complete Delivery** indicator is maintained in the customer master record and is copied into the header of the sales document. Therefore, answers B and D are both incorrect, thus leaving only answers A and C as the only two correct answers to this question.

6. Correct answers: **B, C, and E**

   As shown earlier in Figure 10.4, the delivery type, the material item category group, and the item usage are fields used in the determination of the delivery item category. The shipping point and the route are important fields in delivery processing and transportation scheduling but have nothing to do with item category determination.

7. Correct answers: **C and D**

   The **Order Required** field in the definition of the delivery type controls whether a preceding document is required and, if so, the type of preceding document. When configuring the delivery type, you can also use the **Storage Location Rule** field to specify how the system determines the picking location when a delivery is created if a storage location has not been maintained in the sales order item. These two fields are shown earlier in Figure 10.2. Picking relevance is defined at the delivery item level, and the credit management functions are activated for the sales document item category; thus, both A and B are incorrect answers.

8. Correct answer: **A**

   True. When an outbound delivery is created with reference to a sales order, the item category in the sales order item is automatically copied into the corresponding delivery item. No item category determination is carried out for the delivery item in this scenario.

9. Correct answers: **B, C, and D**

   The three fields that must have the same value in the sales document items to meet the criteria for order combination are the route, the Incoterms, and the ship-to party. These three fields are found in the delivery header, so inconsistent values in any of these fields would automatically result in a delivery split. Storage location and division are, however, part of the delivery item data.

10. Correct answer: **B**

    False. In a delivery document, you can manually override the route but not the shipping point. The shipping point is not an editable field in any delivery document.

11. Correct answers: **A, B, and E**

    The country and transportation zone of the ship-to party and the shipping condition are factors that influence the route determination, so if either the ship-to party or the shipping condition in the order changes, the system will ask you whether you want to redetermine routes and shipping for all items. Route determination can also be carried out with or without considering the weight group but depends on which document is used as the base to determine the route. If the route is determined in the sales order, the weight group is not taken into account (shown earlier in Figure 10.8). Whereas if the route is determined in the delivery, the weight group is part of the search criteria (shown earlier in Figure 10.9).

12. Correct answers: **A and C**

    To activate the automatic determination of the picking location in the delivery document, you must assign a picking rule (e.g., MALA) in the definition of the delivery type and configure the delivery item category as relevant for picking. Answer B is incorrect since the **Determine Storage Location** checkbox is not maintained for the delivery type but instead defined in the delivery item category. Answer D is also wrong since the assignment of a storage location to a warehouse number does not affect the determination of the picking location.

13. Correct answers: **B, D, and E**

    As shown earlier in Figure 10.11, the determination of the loading time in delivery scheduling can be influenced by the shipping point, the route, and the loading group. The transportation group and the weight group are two factors that are never part of the search criteria to determine the loading time.

14. Correct answers: **A and D**

    The total replenishment lead time is maintained in the material master record (**MRP 3** view) and is thus not influenced by the route, so answer B is incorrect. The loading time can be determined using the route (route-dependent loading time) or without the route (route-independent loading time) depending on the option selected in the **Determine Loading Time** field for the shipping point, as shown earlier in Figure 10.15. Therefore, answer C is not correct, thus leaving only answers A and D as the only two correct answers. Both the transportation lead time and the transit time are specified in the definition of the route, as shown earlier in Figure 10.7.

15. Correct answers: **B and D**

    The overall picking status is updated when the picking is carried out and not at post goods issue; thus, answer A is not correct. Once the goods issue is posted for a delivery, the plant stock is decreased by the delivery quantity. By definition, performing the post goods issue cannot increase the plant stock, which invalidates answer C. Posting a goods issue causes the system to update the document flow and reduce delivery requirements in materials planning, among other functions. Therefore, answers B and D are the only two correct answers to this question.

10.11 Takeaway

At this point, you should now understand the structure of the outbound delivery, the different options available to create outbound deliveries, and the conditions that must be fulfilled by a sales order to allow a delivery to be created with reference.

In addition, you learned about the functions in deliveries controlled by the delivery type and by item categories. You understand the logic that determines the item category in deliveries as well as the conditions that must be fulfilled to combine multiple orders into a single delivery. You understand how the system determines the key fields needed for delivery creation and processing, such as the shipping point, the route, and the picking location.

We provided a good overview of transportation and delivery scheduling in the SAP S/4HANA system and the settings that control how shipping deadlines are calculated. Finally, you learned how to perform goods issues for outbound deliveries.

10.12 Summary

Logistics execution is usually treated as an independent business function when in fact it is deeply embedded in the order-to-cash process and represents a critical piece of the sales and distribution puzzle in SAP S/4HANA. Thanks to this chapter, you're now equipped with key concepts about configuration steps and processes in shipping, which will allow you to interpret and implement your business requirements in the system.

In the next chapter, we'll review basic tasks relating to billing and the customization settings required to carry out billing processes.

Chapter 11
Billing Process and Customizing

Techniques You'll Master

- Explain the structure of the billing document
- Configure billing documents
- Create billing documents in different ways
- Explain billing document splits and combinations
- Use preliminary billing documents
- Understand billing plans
- Process down payments using milestone billing plans
- Use invoice lists
- Set up the account determination procedure

The focus of this chapter is to explain the sales billing processes and configurations in SAP S/4HANA. We'll discuss billing document types, manual and automatic creation of billing documents, invoice combinations and invoice splits, preliminary billing documents, billing plans and down payments for sales orders, invoice lists, and account determination.

Real-World Scenario

As an SAP S/4HANA Sales consultant, you must be familiar with the most important features in billing to ensure invoices are generated and sent to customers according to your company's requirements.

For example, let's say your client sells expensive and sophisticated medical equipment to hospitals. Only specialized engineers can repair this type of equipment in the event of a malfunction. Spare parts are usually expensive too, so customers are offered the option to buy a service maintenance contract that guarantees a 48-hour response time to reported faults, including the replacement of any system components covered under the contract, along with quarterly equipment health checks. You've been told by the director of service support that all the most comprehensive service contracts must be billed monthly on the first day of the month. For all other service contracts, the billing date should be the 15th of the month.

To support this business requirement, you must set up a billing plan that calculates the monthly amount, settlement periods, and billing dates as indicated by your client. Also, your client wants the billing documents to be generated automatically so users don't need to remember to execute any transaction in SAP S/4HANA to create the necessary invoices.

Making the relevant customizing settings for the billing plan and scheduling jobs to execute the billing due list in the background are two examples of billing-related skills that you should have in your toolkit.

11.1 Objectives of This Portion of the Test

The goal of this portion of the SAP S/4HANA Sales Certification exam is to test your understanding of the following topics:

- Explain the concept of the billing document
- Define billing document types
- Process the billing due list manually and automatically by scheduling jobs
- Explain the concepts of invoice combination and invoice split
- Create preliminary billing documents
- Understand and configure billing plans

- Process down payments for sales orders using milestone billing plans
- Manage invoice lists
- Configure account determination

Note

The Billing Process and Customizing topic makes up 8%–12% of the total SAP S/4HANA Sales 2020 exam.

The content of this chapter is included in the Billing topic of the SAP S/4HANA Sales 2020 Upskilling exam, which makes up less than 8% of the total exam.

11.2 Controlling Billing Documents

The billing process can be defined as the business transactions required to create and send invoices to customers in order to request payment for the goods or services rendered. In an SAP S/4HANA system, billing transactions are assigned to a specific sales area and company code. The latter plays an important role when transferring billing data to financial accounting.

After an invoice is sent to the customer, assuming no complaints are raised regarding the products or services provided, the business transaction can be considered complete from the sales perspective.

In the SAP world, the term *billing document* refers to the following categories of documents:

- Customer invoices
- Credit memos
- Debit memos
- Pro forma invoices
- Cancellation documents

Billing documents are always created with reference to a preceding document. In this regard, the billing process can be order-related (if the preceding document is a sales document) or delivery-related (if the preceding document is an outbound delivery). Although in some industries not a common practice, you can create billing documents with reference to preliminary billing documents and billing document requests. Finally, you can create cancellation documents based on an existing billing document. Table 11.1 shows an overview of some of the reference documents defined in the standard system.

Billing Document	Reference Document
Invoice	Sales document or delivery
Credit memo	Credit memo request (sales document)

Table 11.1 Reference Documents Used in Billing

Billing Document	Reference Document
Debit memo	Debit memo request (sales document)
Credit memo for returns	Credit memo request for returns (sales document)
Pro forma invoice	Sales document or delivery
Invoice cancellation	Invoice
Credit memo cancellation	Credit memo
Invoice list	Invoices, credit memos, or debit memos

Table 11.1 Reference Documents Used in Billing (Cont.)

Tip

In the event that your company wants to ensure the invoice is only created after the goods have been shipped from the warehouse, you should configure the billing relevance of the item category so that the invoice is generated with reference to an outbound delivery.

In contrast, if your company wants to receive the payment before the goods are sent out, then you would configure the corresponding item category so that the invoice is created with reference to a sales order. This approach would also be the case when invoicing services since deliveries are not usually created for intangible items.

Note that the setting that controls the basis for billing is the indicator for billing relevance, which can be found in the definition of the sales document item category, as discussed in Chapter 4, Section 4.3.

The billing document matches the delivery structure in the sense that it only contains data at the header and item levels. At this stage in the sales process, the goods have already been delivered, so the billing document does not need data on the schedule line level. Figure 11.1 shows a billing document with two items, as presented to the user by the Display Billing Documents app. This display is the classic representation of a billing document used by Transaction VF03 that many SAP users that have worked with in SAP ERP 6.0 and older versions.

Invoice 90000236 (F2) Display: Overview of Billing Items

Display Document Flow | Accounting | Pricing Conditions Header | Change | Services

Invoice 90000236 Net Value: 330.750 USD

Payer: 4 Digital Inc / 1843 North Shore Boulevard / Mountain View CA 94043

Billing Date: 05/28/2021

Item	Material	Item Description	Billed Quantity	SU	Net Value	Curr...	Tax Amount
20	341	1.8" Inch 128x160 Tft Full Color LCD	1	EA	91.350	USD	0.000
30	299	TFT LCD Colour Display, 4.3in, 480x272	2	EA	239.400	USD	0.000

Figure 11.1 Display Billing Documents – VF03 App

However, with SAP S/4HANA and the SAP Fiori interface, SAP has introduced factsheet apps that provide overviews with contextual information about a particular business document. Figure 11.2 shows the same customer invoice but, this time, as seen in the Billing Document factsheet app. This SAP Fiori app displays a simplified overview of the billing document, where all the attributes in the **General Information** section belong to the header data of the document.

Figure 11.2 SAP Fiori Factsheet Billing Document App

As with other SAP S/4HANA documents, in the billing document, header fields apply to the entire document whereas item fields are only valid for the corresponding item. Table 11.2 shows a list of some of the header and items fields you'll find in a billing document.

Header	Item
Billing date	Material number
Net value of the entire billing document	Billed quantity
Partner numbers (sold-to, payer, bill-to)	Net value of the item
Incoterms	Number of the reference document
Terms of payment	Item number of the reference document
Document currency	Plant

Table 11.2 Billing Documents: Header and Item Fields

Header	Item
Company code	Weight
Posting status	Volume
Sales organization	Profit center

Table 11.2 Billing Documents: Header and Item Fields (Cont.)

As discussed in previous chapters, the processing of sales and distribution documents is usually controlled by the document type. As with sales documents and deliveries, with billing documents, the billing document type also controls the behavior of the billing document. Thus, each billing type is used in a specific business transaction. Table 11.3 shows some standard billing document types available in the SAP S/4HANA system.

Document Type	Description
BV	Cash sales
F1	Sales order-related invoice
F2	Delivery-related invoice
F5	Pro forma invoice for sales order
F8	Pro forma invoice for delivery
G2	Credit memo
IV	Intercompany billing
L2	Debit memo
LR	Invoice list
RE	Credit for returns
S1	Invoice cancellation
S2	Credit memo cancellation

Table 11.3 Standard Billing Document Types

At the time of billing document creation, the system automatically proposes the billing type based on the default billing type assigned to the reference document. Still, you can manually change the proposed billing type to another billing type if needed.

If none of the available billing types meet your individual functional requirements, you can define your own billing types or change the standard billing types in the customizing task **Sales and Distribution • Billing • Billing Documents • Define Billing Types**. However, as previously mentioned for other customizing elements, we recommend copying the closest existing standard billing type, changing the billing type key (starting with "Z"), and then making changes to the new billing type, thus leaving the standard types unchanged. Figure 11.3 shows a portion of the screen used to configure billing document types.

Billing Type F2 Invoice Created by SAP

Number systems

No. Range Int. Asst 19 Item No. Increment

General control

SD Document Category	M	Invoice	☐ Posting Block
Transaction Group	7	Billing Documents	☑ Statistics
Billing category			
Document type			
Negative posting		No negative posting	
Branch/Head office		Customer=Payer/Branch=sold-to party	
Credit memo w/ValDat	☐	No	
Invoice List Type	LR	Invoice List	
Rebate settlement			☑ Rel.for rebate
Standard Text			

Cancellation

Cancell.billing type	S1	Cancel Invoice
Copying Requirements		
Reference Number		
Assignment Number		

Account assignment/pricing

Act Det.G/L Act KOFI00 Account determination

Figure 11.3 Defining Billing Types

The details of each and every element in this customizing activity are beyond the scope for the certification exam, but you should be familiar with how the following settings influence the billing process:

- **Number systems**
  - **No. Range Int. Asst** (number range, internal assignment): This field indicates the internal number range assigned to the billing type. You cannot assign an external number range for a billing document.
  - **Item No. Increment**: This field indicates the increment used by the system to generate item numbers when the item number is not copied from the originating document.
- **General control**
  - **SD Document Category**: This field allocates the billing type to a specific business transaction and must be in line with the value selected in the **Transaction Group** field. This single-digit alphanumeric key is also used in sales document types and delivery types. The following are common values used for billing document types:

    - **M**: Invoice
    - **N**: Invoice cancellation
    - **O**: Credit memo
    - **P**: Debit memo
    - **S**: Credit memo cancellation
    - **U**: Pro forma invoice
  - **Posting Block**: If this checkbox is selected, the system will only transfer the billing document to accounting once the document is manually released to accounting. You can release billing documents to accounting via the following transactions or apps:
    - Transaction VF02 (Change Billing Documents)
    - Transaction VFX3 (Release Billing Documents for Accounting)
    - SAP Fiori app Manage Billing Documents
    - SAP Fiori app Change Billing Document
    - SAP Fiori app List Blocked Billing Documents
  - **Transaction Group**: This field controls the transactions with which you can process this billing type in billing and for which billing types indexes are updated for reporting purposes. Two values used in this field for billing include the following:
    - **7**: Billing documents
    - **8**: Pro forma invoices
  - **Statistics**: If this checkbox is selected, the system stores data from the billing document for statistical analysis.
  - **Invoice List Type**: This field specifies the document type that can be used to generate invoice lists for the billing type concerned.
- **Cancellation**
  - **Cancell.billing type** (cancellation billing type): This setting controls the default billing type for the cancellation billing document.
  - **Copying Requirements**: In this field, you'll specify the copying routine containing the requirements that must be met when a cancellation billing document is created with reference to an existing invoice.
- **Account assignment/pricing**
  - **Act Det.G/L Act** (account determination procedure, general ledger account): In this field, you'll maintain the account determination procedure used by the system to propose the general ledger account to which amounts will be posted.
- **Output/partners/texts**
  - **Output determ.proc.** (output determination procedure): This field indicates the output types that are allowed in a billing document if you decide to use

table NAST-based output determination in billing. If you choose the new output management framework for SAP S/4HANA in billing, then you must leave this field blank.

- **Header partners**: This setting allows you to assign a partner determination procedure that contains the partner functions the system automatically proposes at the header level. For example, in a billing document, you may decide to make sold-to, bill-to, and payer mandatory partner functions but make the ship-to party optional.
- **Item partners**: This setting allows you to assign a partner determination procedure that contains the partner functions the system automatically proposes at the item level.
- **TextDetermProcedure** (text determination procedure): This field determines the text determination procedure that contains the text types that can be used at the header level.
- **Text determ.proc.itm** (text determination procedure item): This field determines the text determination procedure that contains the text types that can be used at the item level.

Note

Since all billing documents are created with reference, the item category in the preceding document is always copied into the billing document item, and no item category determination is carried out during billing. Thus, you won't find a customizing task to define or assign item categories for billing documents in the IMG menu.

11.3 Creating Billing Documents

As mentioned earlier, in SAP S/4HANA, you can have delivery-related billing as well as order-related billing. You can even create a single invoice that refers to a sales order and a delivery at the same time, for example, an invoice for both services and physical products. Figure 11.4, Figure 11.5, and Figure 11.6 show the three different scenarios involved in the creation of an invoice, depending on the type of reference document for the billing item.

Figure 11.4 shows two different ways to generate order-related invoices. The upper flowchart represents a sales order with two items that are not relevant for delivery and a billing document where each item is created with reference to the corresponding sales order item. The lower flowchart shows a sales order with two items that have been copied to an outbound delivery to process the shipment; however, the billing document for both items is again created with reference to the sales order. This option is useful if you need to create an invoice before the goods are shipped.

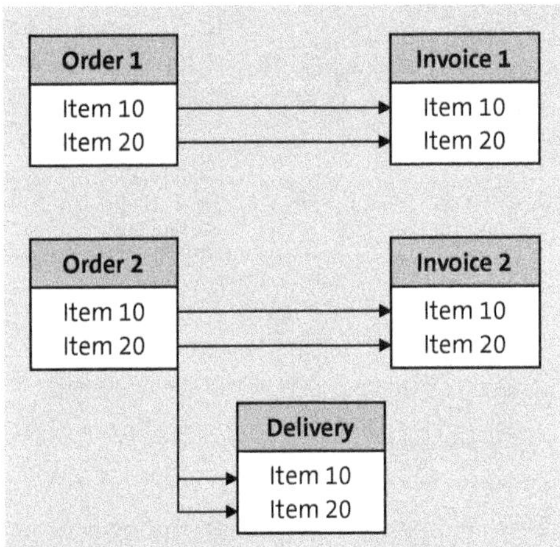

Figure 11.4 Order-Related Invoices

Figure 11.5 shows two possible document flows where the billing document is created with reference to an outbound delivery. The upper flowchart represents a sales order with two items that have been delivered, and a billing document where each item is created with reference to the corresponding delivery item. The lower flowchart shows a standalone delivery with two items and the subsequent billing document. You should use one of the two scenarios shown in Figure 11.5 if you want to make sure goods are shipped before an invoice is generated.

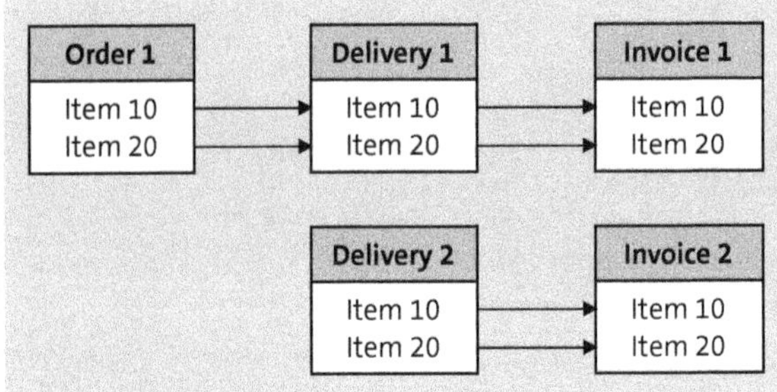

Figure 11.5 Delivery-Related Invoices

Figure 11.6 shows a special case of simultaneous order- and delivery-related billing. In this scenario, sales order item 10 is not relevant for delivery (service item) whereas item 20 must be shipped to your customer (physical product). Assuming the criteria for invoice split is not met, you can execute collective billing for both the sales order and the delivery to generate one billing document. As shown in this flowchart, invoice item 10 is created with reference to sales order item 10, and invoice item 20 is created with reference to the only item in the delivery, which is linked to sales order item 20.

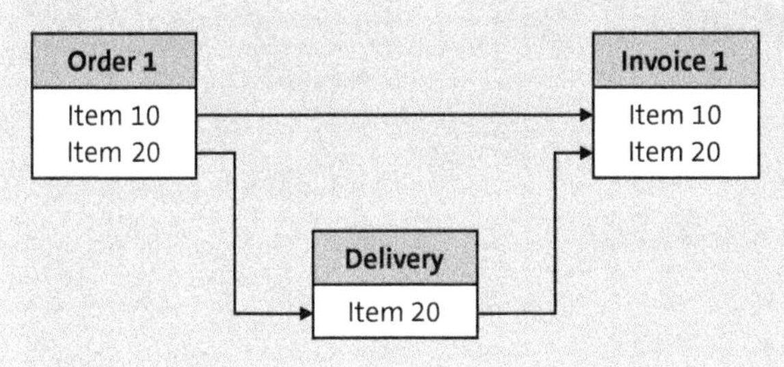

Figure 11.6 Order- and Delivery-Related Invoices

The creation of a credit memo or debit memo is a slightly different case to the invoice since you can either refer to a billing document (invoice) or a credit/debit memo request, as shown in Figure 11.7. The upper flowchart shows a credit (or debit) memo created with reference to item 10 of invoice 1, whereas the lower flowchart represents a scenario where a credit (or debit) memo request is used as a vehicle to generate the subsequent credit (or debit) memo for invoice 2 item 10.

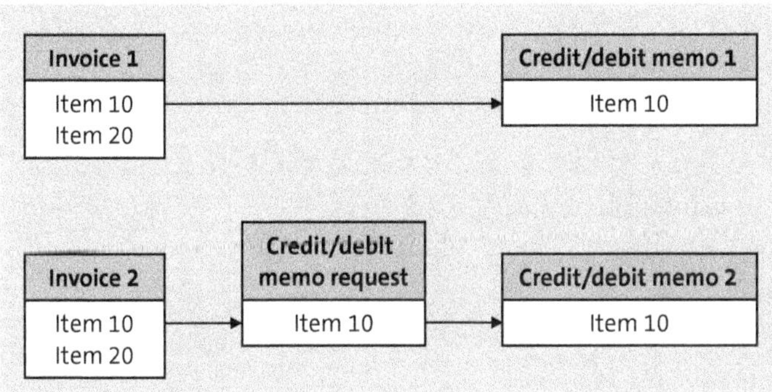

Figure 11.7 Creating a Credit Memo or Debit Memo

When creating a billing document, billing the entire reference document is not mandatory. By selecting the **Selection List** option in the Create Billing Documents – VF01 app, you can bill only for specific items or item quantities of the preceding document. For instance, you can decide to bill only 25 units out of the 100 units post goods issued on the delivery item. This approach is known as *partial billing*.

The quantity to be billed is defined by the **Billing Quantity** indicator in the copying control for billing documents at the item category level. This field dictates which quantity the system copies into the target billing document. We've already covered the values available in the **Billing Quantity** indicator in Chapter 6, Section 6.2. Table 11.4 shows some of the scenarios supported in the standard system.

Reference Document	Billing Type Example	Reference Quantity
Billing document based on order	Billing types G2 and L2	Order quantity minus invoiced quantity
	Pro forma invoice type F5	Order quantity
Billing document based on delivery	Billing types F1 and F2	Delivery quantity minus invoiced quantity
	Pro forma invoice type F8	Delivery quantity

Table 11.4 Billing Quantity versus Billing Type

Note

Other cases are defined in the standard system where the reference quantity is neither based on the order quantity nor on the delivery quantity. For example, in the third-party order process discussed in Chapter 9, (item category TAS), the quantity copied into the billing document is calculated as the incoming invoice quantity minus the quantity already billed.

In terms of the transactions or SAP Fiori apps available to create billing documents, SAP S/4HANA provides you with the following options:

- **Individually**
  You can create billing documents for an entire sales order, delivery, or billing document request. You can perform this function via the SAP GUI Transaction VF01 or the equivalent SAP Fiori app Create Billing Documents – VF01.
- **Manually in mass**
  You can process the billing due list manually via the SAP GUI Transaction VF04 or by using the following SAP Fiori apps:
  - Create Billing Documents – VF04
  - Create Billing Documents
  - Create Preliminary Billing Documents
- **Automatically in mass**
  This function allows you to process the billing due list in background mode as a scheduled batch job. For example, the system can process the billing due list every day at midnight. You can set up these jobs using Transaction VF06 or via the following SAP Fiori apps:
  - Schedule Billing Creation
  - Schedule Billing Creation for Preliminary Billing Documents

Let's now discuss the key features of, as well as some of the differences between, Transaction VF04 (Maintain Billing Due List) and its equivalent apps Create Billing Documents – VF04 and Create Billing Documents. Figure 11.8 shows the SAP Fiori tiles for both apps. At first glance, these two apps may look the same, and in fact, both can be used to create billing documents in mass, but a comparison of their characteristics exposes some differences.

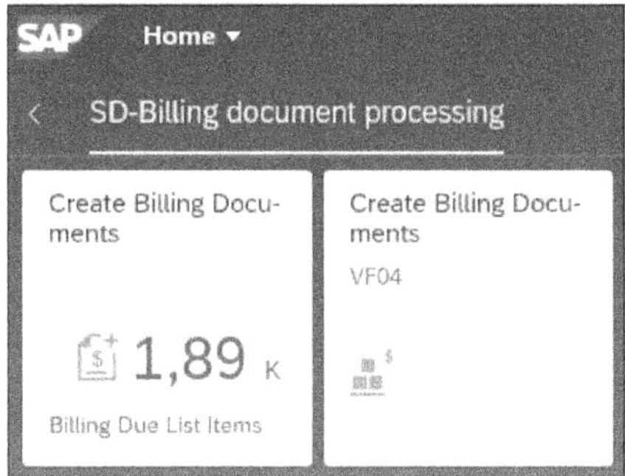

Figure 11.8 SAP Fiori Tiles to Create Billing Documents in Mass

Figure 11.9 shows the processing screen you'll be presented with after executing the Create Billing Documents app.

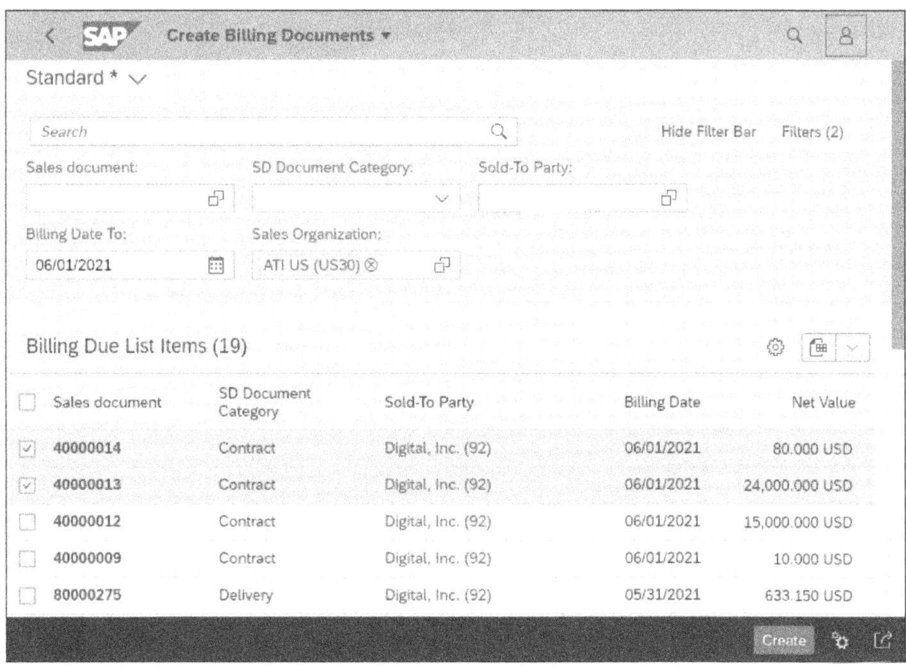

Figure 11.9 SAP Fiori App Create Billing Documents

The following are the main key features of the Create Billing Documents app:

- You can search for items in the billing due list using basic filters (e.g., document number, document category, and sold-to party) or using additional filters (e.g., sales organization, distribution channel, destination country, document currency, billing type, etc.).

- You can show and hide the filter bar displayed at the top of the screen shown in Figure 11.9.
- You can switch on/off the following billing settings by clicking on :
  - **Set billing date and type before billing**: If enabled, the system asks you to enter the billing date and billing type before creating billing documents.
  - **Create separate billing document for each item of billing due list**: If enabled, the system creates separate billing documents for each sales and distribution document marked on the billing due list.
  - **Automatically post billing documents**: If enabled, the system posts all billing documents created to accounting and triggers output.
  - **Display billing documents after creation**: If enabled, the system generates and displays temporary billing documents for the sales and distribution documents selected. Temporary billing documents have the status **Temporary** and use a specific number range for temporary documents with the prefix "S," as shown in Figure 11.10, where temporary number range S300000000 to S399999999 has been used. After reviewing the temporary billing document, you have the option of either saving and converting them into proper billing documents or discarding them, which returns you to the billing due list.

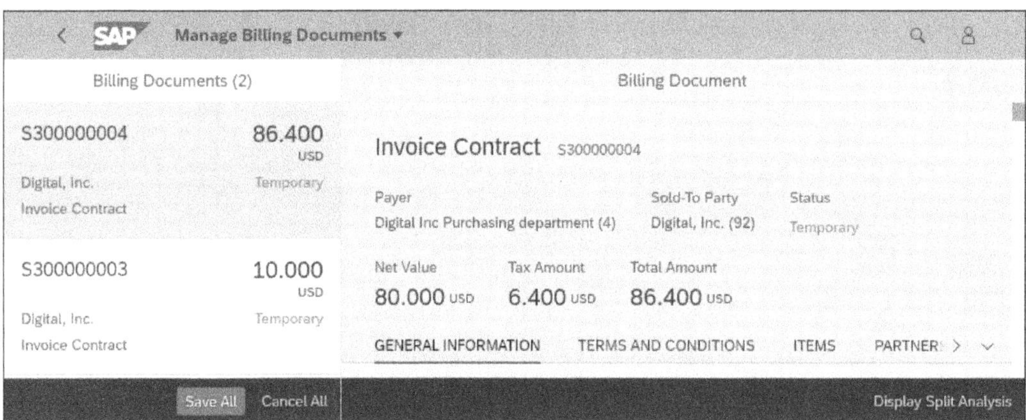

Figure 11.10 Temporary Billing Documents

Warning

The number range for temporary billing documents must be set up in customizing. Otherwise, you'll receive an error message if the billing setting **Display billing documents after creation** is enabled. For more information on how to set this number range up, please refer to SAP Note 2486592 for SAP S/4HANA 1909 or earlier or SAP Note 2922052 for SAP S/4HANA 2020.

- If two or more temporary billing documents are generated, you can carry out a split analysis by clicking on **Display Split Analysis** at the bottom-right corner of the screen.
- From the billing due list, you can navigate to the sales and distribution document's object page to display details of the document selected (e.g., sales order, delivery, contract, etc.).
- You can navigate to the customer's object page by clicking on the customer's name.
- You can export all items in the list to a spreadsheet file (*.xlsx).

On the other hand, if you choose to maintain the billing due list via Transaction VF04 or the SAP Fiori app Create Billing Documents – VF04, the system will present you with the selection screen shown in Figure 11.11.

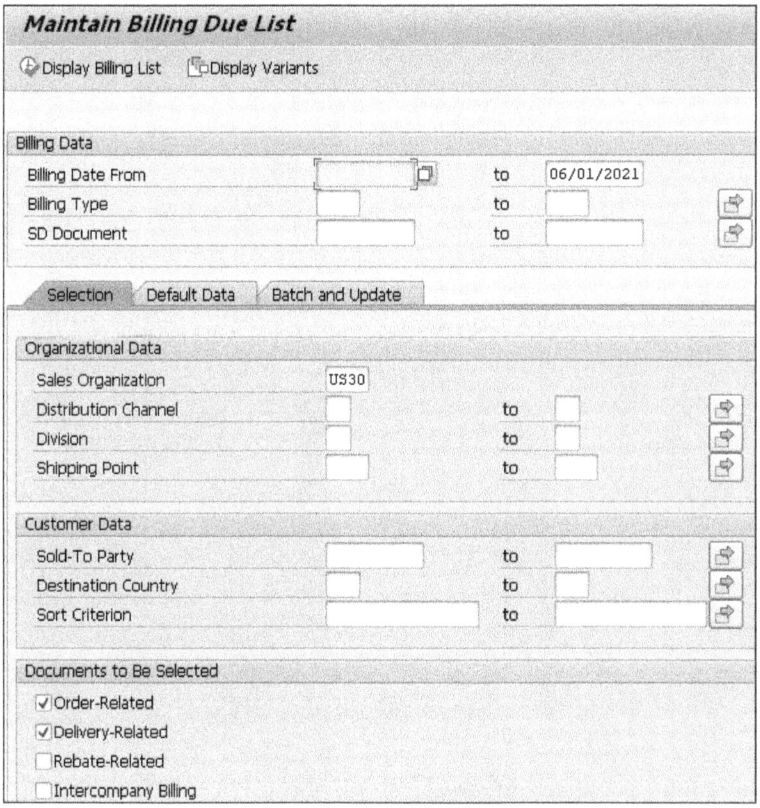

Maintain Billing Due List

Display Billing List   Display Variants

Billing Data

Billing Date From to 06/01/2021

Billing Type to

SD Document to

Selection | Default Data | Batch and Update

Organizational Data

Sales Organization US30

Distribution Channel to

Division to

Shipping Point to

Customer Data

Sold-To Party to

Destination Country to

Sort Criterion to

Documents to Be Selected

- [x] Order-Related
- [x] Delivery-Related
- [ ] Rebate-Related
- [ ] Intercompany Billing

Figure 11.11 Transaction VF04: Selection Screen

On this screen, you don't need to enter the individual documents to be billed; instead, you'll specify the selection criteria, and the system will find the documents for you. The sales organization, billing type, billing date from, and the billing relevance checkboxes at the bottom of the screen are fields typically used to restrict the list of documents to be processed. After executing the report, you'll see the list of sales and distribution documents to be billed, as shown in Figure 11.12.

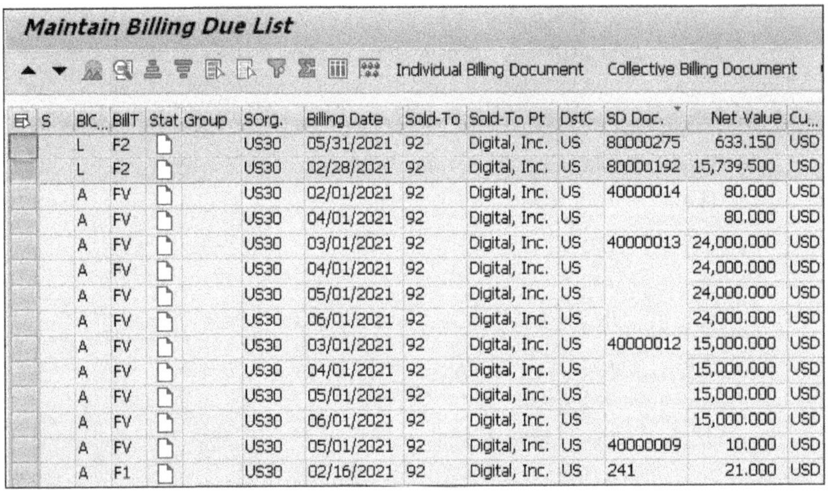

Maintain Billing Due List

Individual Billing Document Collective Billing Document

S	BlC...	BillT	Stat	Group	SOrg.	Billing Date	Sold-To	Sold-To Pt	DstC	SD Doc.	Net Value	Cu...
	L	F2			US30	05/31/2021	92	Digital, Inc.	US	80000275	633.150	USD
	L	F2			US30	02/28/2021	92	Digital, Inc.	US	80000192	15,739.500	USD
	A	FV			US30	02/01/2021	92	Digital, Inc.	US	40000014	80.000	USD
	A	FV			US30	04/01/2021	92	Digital, Inc.	US		80.000	USD
	A	FV			US30	03/01/2021	92	Digital, Inc.	US	40000013	24,000.000	USD
	A	FV			US30	04/01/2021	92	Digital, Inc.	US		24,000.000	USD
	A	FV			US30	05/01/2021	92	Digital, Inc.	US		24,000.000	USD
	A	FV			US30	06/01/2021	92	Digital, Inc.	US		24,000.000	USD
	A	FV			US30	03/01/2021	92	Digital, Inc.	US	40000012	15,000.000	USD
	A	FV			US30	04/01/2021	92	Digital, Inc.	US		15,000.000	USD
	A	FV			US30	05/01/2021	92	Digital, Inc.	US		15,000.000	USD
	A	FV			US30	06/01/2021	92	Digital, Inc.	US		15,000.000	USD
	A	FV			US30	05/01/2021	92	Digital, Inc.	US	40000009	10.000	USD
	A	F1			US30	02/16/2021	92	Digital, Inc.	US	241	21.000	USD

Figure 11.12 Transaction VF04: List of Documents to Be Billed

Now, you can select the documents you want to process (for example, deliveries 80000275 and 80000192 selected in Figure 11.12) and then choose one of the following five billing options available at the top of the screen:

- **Individual Billing Document**
  The system creates an individual billing document for each sales and distribution document selected in the billing due list. No combination is carried out.
- **Collective Billing Document**
  All sales and distribution documents selected are processed in the background, and the system tries to combine them as much as possible. Once all documents are processed, the system takes you back to the billing due list and updates the following columns:
  - **Selection indicator**: This column shows the value 1 if the document was selected and processed or × if the record was selected but not yet processed.
  - **Group**: A number is assigned to all sales and distribution documents processed in the same run. This number is known as the *number of collective run*.
  - **Processing status**: This column shows × if the record could not be processed and ✓ if the record was processed successfully.
- **Collective Billing Doc./Online**
  All sales and distribution documents selected are processed in the foreground, and the system tries to combine them as much as possible. The user is taken to Transaction VF01 (Create Billing Document).
- **Simulation**
  By clicking on the icon, you can simulate the **Collective Billing Doc./Online** billing option. A preview of each billing document is available, and you can display the log as well as the split analysis. No actual billing documents are saved in the database; thus, the status of the number range for billing is not updated.

When you go back to the billing due list, nothing has changed. All columns present the same values.

- **Save**
  If you click on the 🖫 icon, the system performs the **Collective Billing Document** billing option. However, in this case, once the documents are processed, you won't be taken back to the billing due list. The system displays the log screen where you can also display the billing documents created. Once you leave the log screen, you're taken back to the selection screen.

To summarize, the key features of Transaction VF04 and the equivalent SAP Fiori app Create Billing Documents – VF04 include the following:

- You can perform a split analysis to find out which fields caused the billing document split.
- You can produce a list of all the sales and distribution documents due for billing that meet the selection criteria.
- You can process the billing due list in simulation mode.
- You can generate individual billing documents for each record in the billing due list.
- You can merge multiple sales and distribution items into a single collective billing document, provided the billed reference documents have identical values for the fields defined as split relevant.
- You can sort and filter the billing due list.
- You can cancel all of the billing documents of a collective run. From the **Log of Collective Run** screen, you can navigate to **Documents**, select the collective processing number, and click on the **Reverse all** option.
- Unlike the Create Billing Documents app, Transaction VF04 generates a log with detailed information about the error messages returned by the system during the collective run.

Tip

You can access the log for a collective billing run at any time via Transaction V.21 (Log of Collective Run). On the selection screen, you can specify the number of a billing run, the processor of the billing due list, or the billing date. You also have the option of selecting a billing run, going to **Documents**, and reversing all billing documents of that run.

11.4 Analyzing Invoice Combinations and Invoice Splits

When the billing due list is processed for multiple billable reference documents, by default, the system tries to combine into a single billing document all sales and distribution documents with the same sold-to party, sales organization, billing date, and billing type. However, when processing multiple sales and distribution documents, the following four scenarios are all possible:

❶ The system bills one reference document into several billing documents (invoice split).

❷ The system combines several reference documents into a single collective billing document.

❸ The system creates an individual billing document for every reference document (one-to-one).

❹ The system bills multiple reference documents into several billing documents.

These four possible billing outcomes are shown in Figure 11.13.

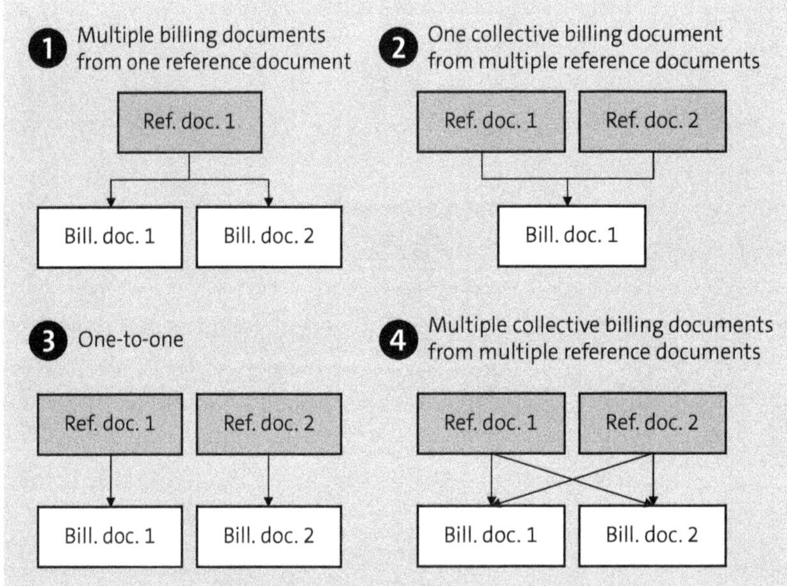

Figure 11.13 Billing Due List: Possible Outcomes

Which outcome is the result of a collective billing run for several billable reference documents depends on the following parameters:

- The billing setting **Create a separate billing document for each item of the billing due list** under the icon in the SAP Fiori app Create Billing Documents. If you enable this setting, then scenario 3 (one-to-one) will be the outcome of the billing run.
- The copy control settings for billing documents at the header level, which control how the **Assignment** and **Reference** fields are populated in the billing document. For example, when billing several deliveries, the system creates one invoice per delivery if you've selected option C (delivery number) in the copy control for either the reference number or the assignment number. This limitation exists because only one delivery number can be entered in each header field of the billing document.
- Any custom data transfer routines in copying control. For example, standard field VBRK/VBRP data transfer routine FV60C001 for delivery-related billing splits

the billing document if either the distribution channel (VBRK-VTWEB) or the division (VBRK-SPART) in the reference items or documents differ. If needed, you can create a copy of this data transfer routine and add more fields to the split criteria.

- Any discrepancies in split-relevant fields at the header or item level. The system always splits the billing document when fields that are copied to the newly created billing document header differ. This split occurs because the system has to create a new header for each of the different values.
- For the system to create a collective billing document, you must process two or more reference documents simultaneously. Processing a single reference document always results in one billing document.

Some standard fields, known as *standard split criteria*, will always cause an invoice split if they are not identical on the header or items of the reference documents. The standard split-relevant fields are shown in Table 11.5.

Field Descriptions	Technical Name	Partner Function (English)
Sales organization	VBRK-VKORG	
Distribution channel	VBRK-VTWEG	
Division	VBRK-SPART	
Sold-to party	VBRK-KUNAG	SP
Bill-to party	—	BP
Payer	VBRK-KUNRG	PY
Document currency	VBRK-WAERK	
Tax departure country	VBRK-LANDTX	
Tax destination country	VBRK-STCEG_L	
Payment terms	VBRK-ZTERM	
Incoterms (part 1)	VBRK-INCO1	
Incoterms (part 2)	VBRK-INCO2	
Incoterms version	VBRK-INCOV	
Incoterms location 2	VBRK-INCO3_L	

Table 11.5 Standard Fields That Always Cause a Billing Document Split

Warning

The *customer reference* field (field VBRK-BSTNK_VF) is a special case. This field will cause a billing document split only if populated, but VBRK-BSTNK_VF is only populated if you maintain the parameter **Reference Number** with value A in the copy control for billing documents at the header level.

In other words, if the **Reference Number** parameter is blank in the copy control for billing documents, you can combine two sales reference documents with two different customer reference numbers (field VBKD-BSTKD) in a single billing document because field VBRK-BSTNK_VF will be blank.

If any of the fields shown in Table 11.5 have a value, at the header or item level, that is not identical across the billed reference documents (or items), a billing document split will occur during collective billing. For technical reasons, you cannot clear any of these fields in the copy control to prevent a split, but other split-relevant standard fields can be cleared or overwritten in a custom data transfer routine to combine multiple sales and distribution documents into one billing document if required.

Note

Note that, since a sales organization belongs to just one company code, you cannot combine two sales and distribution documents with different company codes into a single billing document. Different company codes would always result in different sales organizations (each sales organization can only be assigned to one company code), and the sales organization is a field that always causes a billing document split, as shown in Table 11.5.

However, not all fields in the billing document header (table VBRK) cause a split. Table 11.6 shows the header fields that should not cause a billing document split.

Field Name	Technical Name
Number of the document condition	VBRK-KNUMV
Net value in document currency	VBRK-NETWR
Tax amount in document currency	VBRK-MWSBK
Billing document	VBRK-VBELN
Status for transfer to accounting	VBRK-RFBSK
Name of person who created the object	VBRK-ERNAM
Date of last change	VBRK-AEDAT
Date on which the record was created	VBRK-ERDAT
Entry time	VBRK-ERZET

Table 11.6 Standard Header Fields That Should Not Cause a Split

Example

Let's now take a look at few examples to better understand how the billing document split and convergence functionality works in SAP S/4HANA:

- **Example 1**
  You're billing two deliveries created with reference to two sales orders. The customer reference number (field VBKD-BSTKD) is different in each sales order. In copy control for the delivery document to the billing document at the header level, the Reference Number parameter has the value A (purchase order number). In this case, at least two billing documents are created, with the respective customer reference number copied to each header (field VBRK-BSTNK_VF).

- **Example 2**
  You're billing two deliveries created with reference to two sales orders. The customer reference number (VBKD-BSTKD) is different in each sales order. In copy control for the delivery document to the billing document at the header level, the reference number parameter is blank. In this case, assuming all the other split-relevant fields are identical, one billing document is created for both deliveries. The customer reference number field in the invoice header (field VBRK-BSTNK_VF) remains blank.
- **Example 3**:
  You're billing two deliveries created with reference to two sales orders. The payer is different in each sales order. In this case, at least two billing documents are created, with the respective payer copied to each billing document.
- **Example 4**
  You're billing one delivery created with reference to one sales order. Both documents (the sales order and the delivery) have three items. Two of the items in the sales order have the same payment terms, but the other item has different payment terms. In this case, at least two billing documents are created. One billing document will be created for the two items that have the same payment terms, and another billing document, for the item with different payment terms.

You can determine which fields caused (or will cause) a billing document split by performing a *split analysis*. This analysis shows you the split-relevant list of fields that have conflicting values, helping you understand why the system is unable to combine two or more reference documents into one collective billing document. You can carry out a split analysis in three ways:

- In Transaction VF04 or the Create Billing Documents – VF04 app: Before the billing documents are created, select the reference documents you want to combine in the billing due list and click on **Simulation**. Then, select the two simulated billing documents and click on **Split Analysis**. The list of differing header fields will be displayed, as shown in Figure 11.14.

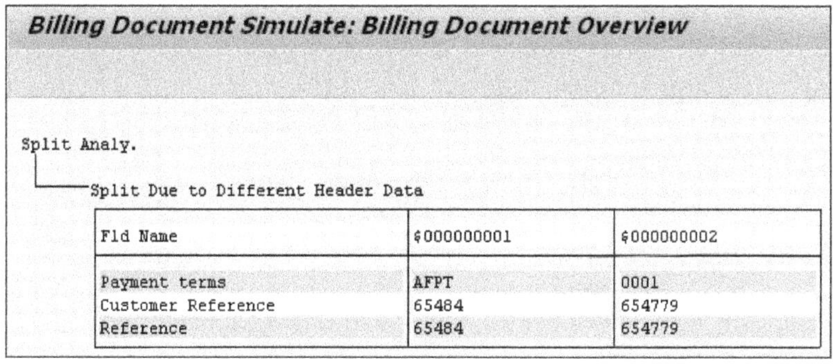
Billing Document Simulate: Billing Document Overview

Split Analy.

Split Due to Different Header Data

Fld Name	$000000001	$000000002
Payment terms	AFPT	0001
Customer Reference	65484	654779
Reference	65484	654779

Figure 11.14 Split Analysis in Transaction VF04

- Comparing two existing billing documents, using one of the following SAP Fiori apps:
  - **Manage Billing Documents app**
    Select the two billing documents you want to compare and click on **Display**

Split Analysis at the bottom of the screen. The fields causing the split are presented in a popup window, as shown in Figure 11.15.

Billing Document Comparison

	Split Field	90000282	90000281
Header	Payment terms	0001	AFPT
	Customer Reference	654779	65484
	Reference	654779	65484
	Assignment	0090000282	0090000281
	Payment reference	0090000282	0090000281

Close

Figure 11.15 Split Analysis in the Manage Billing Documents App

- **Display Billing Documents or Change Billing Documents apps**
  First, you'll enter the number of the first billing document that you want to analyze. Then, select the top menu option **More • Environment • Split Analysis** and enter the second billing document you want to compare. Again, the list of fields responsible for the split will be displayed in the same fashion as shown in Figure 11.15.

11.5 Using Preliminary Billing Documents

A preliminary billing document is nothing but a precursor for the final billing document and can be created with reference to a sales order, delivery, credit memo request, or debit memo request. This optional document shares the same structure as a billing document and its main purpose is to facilitate negotiations with business partners. Preliminary billing documents are commonly used in the professional services industry to prevent customer complaints, for example, due to inconsistencies in pricing.

Note

Preliminary billing documents are only available from SAP S/4HANA 2020 (software component S4CORE from release 105). To use preliminary billing documents, you must enable business function SD_BIL_PRELIMBILLINGDOC (Functionality for Preliminary Billing Documents) in the **Activate Business Functions** customizing activity (Transaction SFW5).

For further information on how to activate preliminary billing documents, please refer to SAP Note 2921988.

After the service has been provided, the customer can request a preliminary billing document to check that pricing, payment terms, etc. are in line with what was previously agreed. If any issues are found in the preliminary billing document, the customer will ask for an updated preliminary billing document. Once the customer confirms the preliminary billing document is accurate, a final invoice can be created and sent to the customer. The flowchart shown in Figure 11.16 provides a graphical overview of the process to bill services using preliminary billing documents.

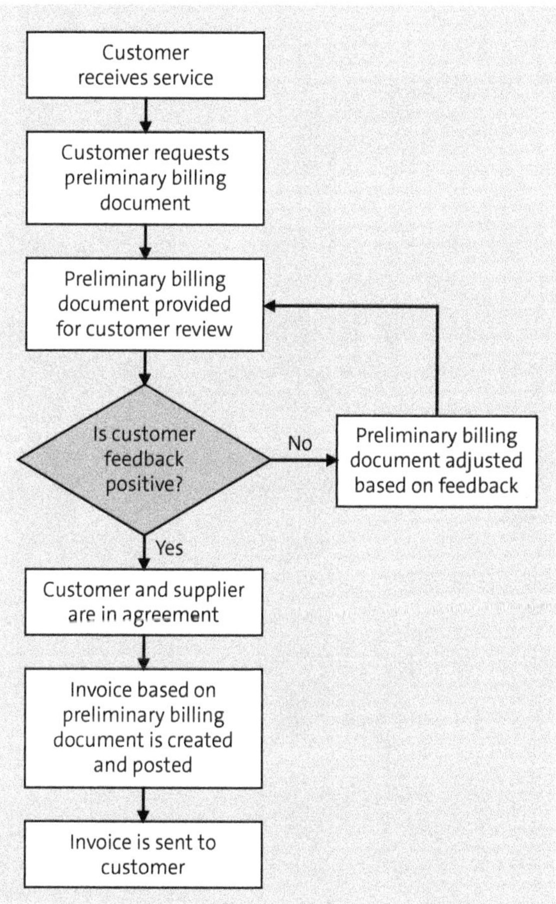

Figure 11.16 Preliminary Billing Documents in SAP S/4HANA

Some key features of preliminary billing documents include the following:

- Preliminary billing documents are optional.
- Preliminary billing documents are a predecessor to the final billing document.
- Preliminary billing documents are relevant for output.
- Preliminary billing documents can be rejected so they cannot be converted to billing documents.
- Preliminary billing documents can be finalized to prevent further changes.
- Preliminary billing documents can be changed as often as necessary.

- Preliminary billing documents are not relevant for posting to financial accounting.
- Preliminary billing documents have the same structure as standard billing documents.

Bear in mind that, once a preliminary billing document has been created for a sales and distribution document (e.g., a sales order), the billing status of the preceding document changes to **Completely Processed**. As a result, the preceding sales and distribution document is removed from the billing due list and cannot be billed directly. At this point, you have the following two options:

- You can create the final billing document with reference to the preliminary billing document.
- You can reject the preliminary billing document to reopen the preceding sales and distribution document for billing, so it reappears on the billing due list.

Like any other sales and distribution document, preliminary billing documents can have several statuses as they go through different phases. The four possible statuses a preliminary billing document can have are shown in Table 11.7.

Status	Description
In Progress	This status is automatically set when the preliminary billing document is created. You can change, finalize, reject, or convert a preliminary billing document into a final billing document.
Rejected	A preliminary billing document that has been rejected can no longer be converted into a billing document. The preceding document is reopened and can be changed or billed directly.
Finalized	A preliminary billing document with this status cannot be changed but can be converted into a billing document.
Completed	This status is set once the preliminary billing document is converted into a billing document. The preliminary billing document can no longer be changed. If the billing document is canceled, the preliminary billing document status is reset to **In Progress**.

Table 11.7 Preliminary Billing Document Statuses

Finally, Table 11.8 shows an overview of the SAP Fiori apps you can use to create and manage preliminary billing documents in SAP S/4HANA.

SAP Fiori App	Key Features
Create Preliminary Billing Documents	Search and filter the billing due list, maintain view settings, maintain billing settings, create preliminary billing documents, and display the details of any sales and distribution document

Table 11.8 SAP Fiori Apps Related to Preliminary Billing Documents

SAP Fiori App	Key Features
Change Preliminary Billing Documents	Add and remove attachments; add more items from documents in the billing due list; and change attributes such as prices, texts, payment terms, and billing date
Display Preliminary Billing Documents	Display header details, display item details, navigate to the document flow, and generate a PDF-based print preview
Manage Preliminary Billing Documents	Search and filter preliminary billing documents, create (final) billing documents based on preliminary billing documents, use the split analysis, reject preliminary billing documents, finalize preliminary billing documents, navigate to related apps, and maintain billing settings
Schedule Creation of Preliminary Billing Documents	Schedule the creation of preliminary billing documents; set up background jobs; specify selection criteria; copy and modify existing jobs; and specify default values such as billing type, billing date, and pricing date
Manage Preliminary Billing Document Workflows	Add a new workflow; configure, activate, or deactivate workflows; define the order in which to check the preconditions; and display/copy/delete an existing workflow

Table 11.8 SAP Fiori Apps Related to Preliminary Billing Documents (Cont.)

11.6 Billing Plans and Down Payment Processing

In this section, you'll learn about the two main types of billing plans that can be automatically determined for a sales order item: periodic billing and milestone billing. We'll also cover the function that allows down payment processing using milestone billing plans.

The billing plan of a given sales order item is determined from the item category and the billing relevancy I (Order-relevant billing - billing plan). The billing plan contains the dates when a specific amount or percentage rate should be invoiced. We'll analyze both periodic and milestone billing plans further in the corresponding subsections.

In customizing, you can define a new billing plan type or use the predefined billing plans for periodic and milestone billing. A billing plan type contains the basic control data for the billing plan and can be allocated to the sales document type if you want to use a header billing plan, which will be valid for all items that are assigned to the header billing plan. You can manually delete the assignment to the header billing plan for a specific order item and change some parameters, such as the billing rule or the horizon in a periodic billing plan. However, if you detach an item that has been partially billed from the header, you can no longer reattach it. In milestone billing, if an item is removed from the header, the system allows you to maintain dates and milestones in isolation from the header. But if this item is reassigned to the header billing plan, all manually set milestones will be deleted since the header data will take precedence again.

Billing plans are controlled by the item category. A billing plan type must be assigned to the item category with billing relevance I if you want the system to create a billing plan for the relevant order items.

Note

The billing plan type is automatically determined based on the item category and billing relevance I. This value is displayed in the sales document but cannot be changed.

The IMG menu path to configure billing plans is **Sales and Distribution • Billing • Billing Plan.**

11.6.1 Periodic Billing Plan

Periodic billing means the total item amount on the sales document is billed at periodic intervals defined by the billing dates in the billing plan until the end date is reached. A business use case for a periodic billing plan would be billing for a maintenance service contract.

In SAP S/4HANA, different overview screens exist for periodic and milestone billing plans. Figure 11.17 shows the overview screen for a periodic billing plan using the standard billing plan type 02. In this example, the amount of $10.00 is billed monthly on the first day of the next month, and the value **C** (invoiced) in the **BillSt** (billing status) column indicates that four invoices have already been issued for February, March, April, and May 2021.

Sales Document Item 10 Item category WVN Maint.Contract Item
Material 400000130 Laptop service agreement

Sales A | Sales B | Contract data | Shipping | Billing Document | Billing plan | Conditions | Account

Net value 110.000 USD

Billing plan

BillingPlanType	02 Periodic	50 Monthly on First of Mont...	InAdvance ☐	
Start date	02/01/2021	13 Contract Start (Contract hea...	Dates from	02/01/2021 11
End date	12/31/2021	14 Contract Ende (Contract hea...	Dates until	12/31/2021 09
Horizon	06/30/2022	52 Horizon 1 Year	Hdr ☐ Corr ☐	Cal-Id

Dates

Settlement	To	Billing date	Bill.value	Crcy	Block	BillSt	Man.	Cor	PayT	DCat
02/01/2021	02/28/2021	03/01/2021	10.000	USD		C	☐			01
03/01/2021	03/31/2021	04/01/2021	10.000	USD		C	☐			01
04/01/2021	04/30/2021	05/01/2021	10.000	USD		C	☐			01
05/01/2021	05/31/2021	06/01/2021	10.000	USD		C	☐			01
06/01/2021	06/30/2021	07/01/2021	10.000	USD		A	☐			01
07/01/2021	07/31/2021	08/01/2021	10.000	USD		A	☐			01
08/01/2021	08/31/2021	09/01/2021	10.000	USD		A	☐			01
09/01/2021	09/30/2021	10/01/2021	10.000	USD		A	☐			01
10/01/2021	10/31/2021	11/01/2021	10.000	USD		A	☐			01
11/01/2021	11/30/2021	12/01/2021	10.000	USD		A	☐			01
12/01/2021	12/31/2021	01/01/2022	10.000	USD		A	☐			01

Figure 11.17 Periodic Billing Plan Overview Screen

The system uses the start and end dates, the period (or billing rule), and the horizon to determine the billing dates in the billing plan. Let's discuss these fields in more detail:

- **Start date and end date**: The start and end date define the duration of the billing plan and are often copied from the start and end date of the service contract.
- **Periodic**: This field specifies the frequency with which the invoices must be created. Typical values are monthly, quarterly, and annually. The system uses the period (also known as **Rule for Origin of Next Billing/Invoice Date**) and the **InAdvance** checkbox to calculate the billing date for each settlement period in the periodic billing plan.
- **Horizon**: This field is used to determine the end date of the billing plan in the event no end date is entered for the contract.

You can also enable automatic credit memos by selecting the **Corr** (automatic creation of correction dates in billing plan) checkbox in the billing plan data. This option could be useful, for example, if a contract is canceled prior to the end date of the last settlement period that has already been billed. In this scenario, if the end date of the billing plan is changed to a date before the to-date of the last period billed, the system generates a credit memo for the period between the new billing plan end date (cancellation date) and the to-date of the period billed. The **Corr** indicator can be activated manually or defaulted via customizing.

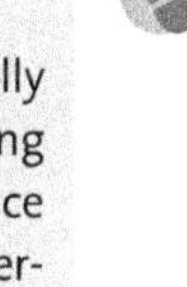

Example

Let's say you have a service contract for an old piece of equipment that is billed annually on January 1st. An invoice of $12,000 was sent to the customer for 2021 at the beginning of the year. The customer calls on May 16th asking to cancel the service contract since they would like to retire the old unit and buy a new one with improved features. The service manager accepts the requests and changes the end date of the billing plan from December 31st to May 16th, 2021. As a result of the end date change, part of the settlement period that was billed on January 1st is automatically credited, as shown in Table 11.9.

Billing plan start date = 01.01.2021

Billing plan end date = 05.16.2021

Corr = ✓

Settlement	To	Billing Date	Billing Value	Billing Type
01.01.2021	12.31.2021	01.01.2021	$12,000.00	FV
05.17.2021	12.31.2021	05.16.2021	–$7,483.87	G2

Table 11.9 Correction Dates in a Periodic Billing Plan

Finally, you should also know that, in periodic billing, the **Billing value** and **BillSt** columns are blank at the header level because pricing and status determination is only carried out at the item level in periodic billing.

11.6.2 Milestone Billing Plan

Milestone billing means the total item amount on the sales document is distributed over several billing dates based on milestones that mark the completion of specific stages of a project. When a milestone is reached, an invoice is sent to the customer for either a predefined amount or a percentage of the project cost. A good example would be a house improvement project to build an extension where the customer pays 2% of the cost upfront, an additional 8% when the design is completed and before execution, and 60% at the end of the execution, while the final 30% is paid 1 month after the build phase has been completed.

To enable the milestone billing plan function on a sales document item, you should assign the item category group 0005 (milestone billing) in the material master record, so the system determines the item category TAO (Milestone-Bill.Plan), which activates the milestone billing plan. Figure 11.18 shows the overview screen for a milestone billing plan using the standard billing plan type 01. In this example, the total amount for our home extension project is $50,000, which is distributed over six billing dates with the corresponding percentages to be billed at each stage, according to the milestones defined in the billing plan.

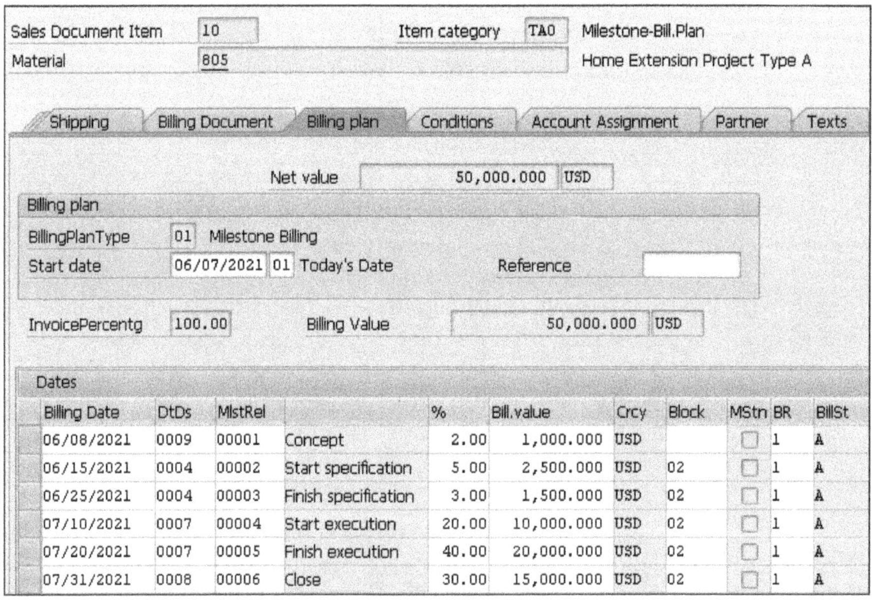

Figure 11.18 Milestone Billing Plan Overview Screen

You can connect each sales document item to a project in Project System. If a network plan has been opened in Project System, each schedule line in the sales document can be allocated to an event in the network, thus allowing billing dates in the billing plan to be updated with the actual dates of the milestones. Also, each billing date remains blocked for billing until the corresponding milestone is marked as completed in the Project System.

However, the functionality to update dates in the billing plan automatically is controlled by the **Fixed date** indicator in the definition of the default date category assigned to the billing plan type. The values available for the **Fixed date** field in the **Define and Assign Date Categories** customizing activity are shown in Figure 11.19.

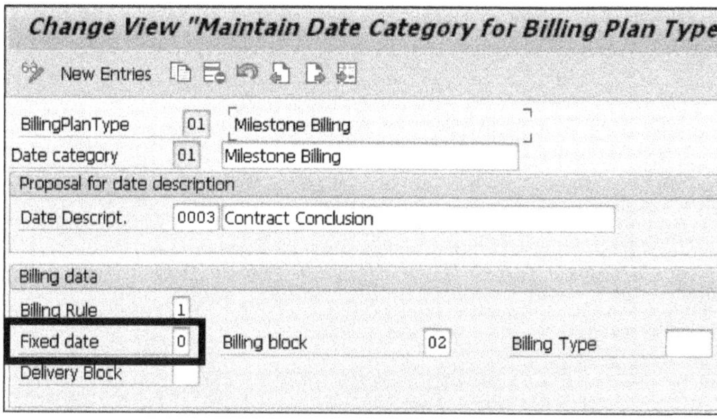

Figure 11.19 Maintaining a Date Category for the Billing Plan Type

The following options are available for the **Fixed date** field:

- <blank>: Assignment to milestones not possible.
- **0**: Fixed date—the date is not copied from the milestone.
- **1**: Planned/actual date from the milestone.
- **2**: Planned/actual date from the milestone if the date in the milestone is before the billing date in the billing plan.
- **3**: Planned/actual date from milestone if the date in the milestone is after the billing date in the billing plan.

If your company does not use Project System and no network is created, you can manually assign milestones to dates in the billing plan. In this case, you'll enter the value 0 in the **Fixed Date** field to fix the billing date and maintain dates directly in the billing plan.

11.6.3 Down Payment Processing Using Milestone Billing Plans

In this section, we'll show you how to use down payments in SAP S/4HANA. The down payments process allows customers to make payments upfront for expensive goods or services such as construction projects, machinery, and other kinds of equipment that are involved in the production of other products for sale.

Down payments can be made using milestone billing plans or using document conditions. For the purpose of this book, we'll focus on down payments using the milestone billing plan. Milestone billing plans can only be used for order-related billing.

Figure 11.20 shows an overview of the down payment process flow based on a milestone billing plan.

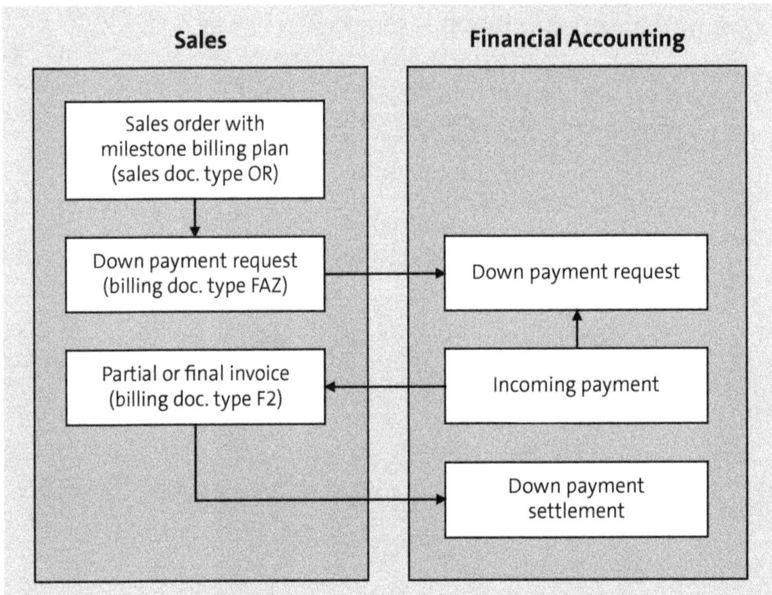

Figure 11.20 Standard Down Payment Process Using Milestone Billing Plans

This process can be summarized in the following four steps:

1. The down payment agreement is recorded in the sales order item by entering the deadlines in the billing plan. Each deadline is represented by the corresponding billing date. As mentioned earlier, the value of the down payment (fixed amount or percentage) depends on the billing rule (5 or 4) specified in the date category for the billing plan type.
2. Once a specific deadline is reached, the billing block must be removed, and the order, saved. At this point, a down payment request is created and sent to the customer. You can let the billing batch job generate the down payment request in the background or create it manually via the Create Billing Documents app or Transaction VF01. Once created, the down payment request is posted to financial accounting and can be found in the document flow of the sales order. Down payment requests can be canceled if needed with billing type FAS.
3. The customer makes a down payment, which is assigned to the down payment request.
4. The down payment is cleared when the first billing type F2 invoice is generated, which could be either a partial or final invoice. This step is achieved by adding the down payment underneath the corresponding billing items. The total cleared down payment amount cannot exceed the total amount of the respective billing item it is assigned to. This value is controlled by the calculation formula 48 assigned to pricing condition type AZWR. The total amount of the

partial or final invoice is posted to SAP S/4HANA Finance as a receivable. In the case of a partial invoice, depending on whether or not the down payment amount is greater than the total partial invoice amount, this invoice could either reduce or fully consume the down payments left to be cleared.

Example

Let's consider an example illustrating how a down payment for a given sales order item is represented as deadlines (billing dates) in the billing plan. Your customer places an order for an electric car on 01/01/2021, and a down payment is made for 40% of the total cost on the same day. A partial invoice is planned 5 months later, when the car is delivered on 06/01/2021. The final invoice will be sent to the customer 1 month after the partial invoice, that is, on 07/01/2021. Table 11.10 shows the main fields of the sales order item.

Item	Description	Quantity	Amount (USD)	Item Category
10	Roadster all-electric car	1	200,000	TAO

Table 11.10 Item Data

Table 11.11 shows the billing plan details of sales order item 10.

Billing Date	Description	%	Amount (USD)	Billing Type
01/01/2021	Down Payment	40%	80,000	FAZ
06/01/2021	Partial Invoice	30%	60,000	F2
07/01/2021	Final Invoice	70%	140,000	F2

Table 11.11 Billing Plan Data

In this example, the down payment request is the first billing document and is created on 01.01.2021 according to the billing plan. The total amount of the down payment request is $80,000, which represents 40% of the total amount ($200,000) of the sales order.

A partial invoice to clear part of the down payment made by the customer is issued on 06.01.2021. Since the partial invoice is for $60,000 (30% of the total amount) and the down payment made was for $80,000 (40% of the total amount), only $60,000 out of $80,000 is cleared in the partial invoice, as shown in Table 11.12.

Item	Description	Quantity	Amount (USD)	Item Category
10	Roadster all-electric car	1	60,000	TAO
11	Down Payment Clearing		60,000	TAO

Table 11.12 Partial Invoice for Billing Type F2 and Net Value $60,000

The down payment item is reduced from $80,000 to $60,000 to not exceed the partial invoice total amount. Because the partial invoice is now settled by clearing $60,000 from the down payment total of $80,000, the customer does not owe any additional money to settle the partial invoice.

As a result, the leftover down payment amount of $20,000 is carried forward and is proposed in the final invoice. Table 11.13 shows the two items in the final customer invoice.

Item	Description	Quantity	Amount (USD)	Item Category
10	Roadster all-electric car	1	140,000	TAO
11	Down Payment Clearing		20,000	TAO

Table 11.13 Final Invoice for Billing Type F2 and Net Value $140,000

The remaining down payment amount of $20,000 is cleared in the final invoice, which leaves $120,000 to be settled by the customer ($140,000 – $20,000 = $120,000).

Finally, to use down payments with milestone billing plans, you must maintain the following basic customizing settings:

- The value of a down payment request can be defined as either a fixed amount or as a percentage of the item value. This value is controlled by the **Billing Rule** field in the definition of the date category shown in Figure 11.19. A billing rule with values 4 or 5 will process the down payment on a percentage or fix amount basis, respectively.
- The billing type FAZ (down payment request) must be assigned in the definition of the date category, as shown in Figure 11.19.
- You must assign billing block 02 (complete confirmation missing) to billing type FAZ in the definition of the date category, as shown in Figure 11.19.
- Maintaining a deadline proposal for the milestone billing plan type where you'll specify a sequence of dates that are used as references for date determination.
- You must assign a pricing procedure to the sales document type that contains standard condition type AZWR. Also, you must maintain the **Print Type** field with the value X (printing at item level), pricing requirement 2 (item with pricing), and the calculation formula 48 (check down payments) for condition type AZWR.
- In the copy control settings for the down payment request, you must ensure the copying requirement routine 20 (Down pymnt req.item) is maintained at the item level (for example, where the reference sales document type is OR, the target billing document type is FAZ, and the item category is TAO).
- In the copy control settings for the final or partial invoice, you must ensure the copying requirement routine 23 (Item ord.ref. w/o AZ) is maintained at the item level (for example, where the reference sales document type is OR, the target billing document type is F2, and the item category is TAO).

Note

Other configuration steps in the area of SAP S/4HANA Finance are prerequisites for down payment processing, but we won't cover these steps in this book since these activities are beyond the scope for this certification. For more information regarding the customizing of down payments in both SAP S/4HANA Sales and SAP S/4HANA Finance, you can consult SAP Note 213526.

11.7 Using Invoice Lists

Invoice lists allow you to group together several billing documents that have a common payer and have already been posted to financial accounting. This functionality could be useful if your company sells goods or services to various customers that belong to a common buying group. In this scenario, the group payer receives and settles the invoice list containing all the billing documents designated to individual group members. Since the group payer must then collect payment from the individual group members, a common business practice is to offer a factoring discount to the group payer for the service provided.

Note

To use invoice lists, you must ensure a common payer is assigned to all customer master records to be included in the invoice list.

You can assign a factory calendar to the payer, which is used during the processing of invoice lists, so invoice lists are generated and sent to the payer at specified time intervals or on specific dates. This invoice list schedule can be maintained in the business partner master record, under the **Billing** tab of the **Sales and Distribution** view for the business partner role Customer (FLCU01).

An invoice list can contain both single and collective billing documents of the following three document categories:

- Invoice
- Debit memo
- Credit memo

The system can process these three document categories together during invoice list creation; however, credit memos will be grouped in a separate invoice list because a different type of invoice list is used for credit memos. The following two types of invoice lists are available in the SAP S/4HANA standard system:

- **LR**: This invoice list type is used for invoices and debit memos.
- **LG**: This invoice list type is used for credit memos.

Invoice list types are defined for billing types in customizing and then assigned to each billing type. For example, in the standard system, invoice type LR is assigned to standard billing document types F2 (delivery-related invoice) and L2 (debit memo), whereas invoice list type LG is assigned to billing document type G2 (credit memo). Figure 11.21 shows a portion of the screen with the parameters that define the invoice list type LR.

Billing Type	LR	Invoice List	Created by	SAP

Number systems

No. Range Int. Asst	19		Item No. Increment	

General control

SD Document Category	3	Invoice List	☐ Posting Block
Transaction Group	7	Billing Documents	☑ Statistics
Billing category			
Document type			
Negative posting		No negative posting	
Branch/Head office		Customer=Payer/Branch=sold-to party	
Credit memo w/ValDat	☐	No	
Invoice List Type			
Rebate settlement			☐ Rel.for rebate
Standard Text			

Cancellation

Cancell.billing type	LRS	Cancel Invoice List
Copying Requirements		
Reference Number		
Assignment Number		

Account assignment/pricing

Act Det.G/L Act	KOFI00	Account determination

Figure 11.21 Billing Type LR: Invoice List

To process invoice lists, the following prerequisites must be fulfilled:

- If the payer is entitled to a factoring discount, then pricing condition type RL00 must be assigned to the relevant pricing procedure. Also, in some countries like Germany, tax condition MW15 is required to reimburse the payer in advance for any tax liability due to the factoring discount.
- You must assign an invoice list type to each billing type that you want to add to an invoice list.
- Copying control **Billing Document to Billing Document** must be maintained. For example, in the standard copying control from F2 to LR, the copying requirements routine 16 (invoice list) has been maintained at the header level to check the accounting status of the preceding billing document is **Completely Processed**.
- A customer calendar must be defined and assigned to the payer in the **Billing** tab of the **Sales and Distribution** view if you want invoice lists to be processed at certain time intervals or on certain dates.

- If factoring discounts are used, you must maintain pricing condition records for condition type RL00 and, depending on the tax structure of the payer's country, also for tax condition MW15.
- To generate and output for the invoice list, you should maintain output condition records for output types LR00 (Invoice list) and RD01 (SingleInv.f.InvList).

You can maintain the settings for invoice lists by following the IMG menu path **Sales and Distribution • Billing • Billing Documents • Invoice Lists**.

In terms of the structure, an invoice list must be considered an individual billing document. Its structure is similar to that of a billing document and consists of a document header and document items. The header contains general information, such as the net value, taxes, factoring discount, billing date, payer, etc. Each item in the invoice list represents a single billing document, which can be either an invoice, a debit memo, or a credit memo. Figure 11.22 shows an invoice list containing four billing documents as seen in the Display Invoice Lists app. In this example, a factoring discount of 15% has been offered to the payer.

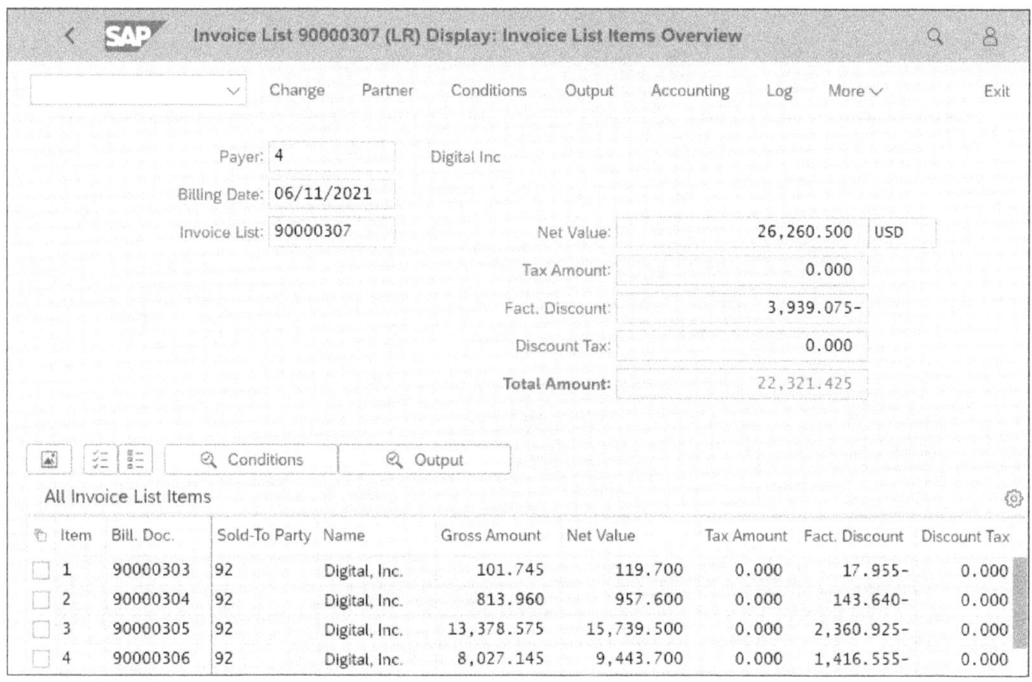

Figure 11.22 Display Invoice Lists App

As with any other billing document, a journal entry is created when an invoice list is released to accounting. In the case of invoice lists, the invoice list number is copied to the reference document number in the journal entry. This process allows invoice list numbers to be used when processing incoming payments.

Several transaction codes and SAP Fiori apps are available for managing invoice lists in SAP S/4HANA. Table 11.14 shows a list of the most relevant.

SAP Fiori App	Functionality	SAP GUI Transaction
Create Invoice Lists – VF21	Create invoice lists	VF21
Change Invoice Lists	Change header data for an invoice list	VF22
Display Invoice Lists	Detailed view of an invoice list	VF23
Create Invoice Lists – VF24	Create invoice lists in mass using the worklist	VF24
Invoice List	Simplified overview of key facts and details of the invoice list	
Manage Invoice Lists	Manage and cancel invoice lists	
Cancel Invoice Lists	Cancel invoice lists	VF26
List Invoice Lists	List of the invoice lists for a given payer	VF25

Table 11.14 Apps and Transactions Used to Process Invoice Lists

11.8 Setting Up Account Determination

In a standard order-to-cash scenario, the billing document can be considered the last sales and distribution document in the process. However, if the amount on the billing document is greater than zero, a subsequent document called a *journal entry* is required to transfer the individual amounts in the billing document to the corresponding general ledger accounts in SAP S/4HANA Finance. Once a billing document is released to accounting, the journal entry is generated and can be seen in the document flow, as shown in Figure 11.23.

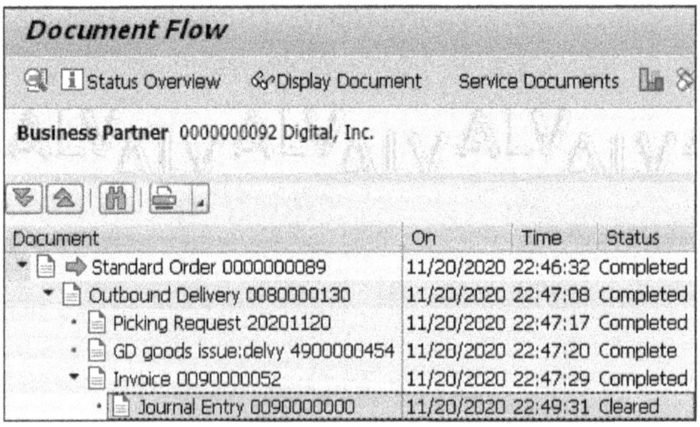

Document Flow

Status Overview   Display Document   Service Documents

Business Partner 0000000092 Digital, Inc.

Document	On	Time	Status
Standard Order 0000000089	11/20/2020	22:46:32	Completed
Outbound Delivery 0080000130	11/20/2020	22:47:08	Completed
Picking Request 20201120	11/20/2020	22:47:17	Completed
GD goods issue:delvy 4900000454	11/20/2020	22:47:20	Complete
Invoice 0090000052	11/20/2020	22:47:29	Completed
Journal Entry 0090000000	11/20/2020	22:49:31	Cleared

Figure 11.23 Journal Entry in the Sales Document Flow

Integration with SAP S/4HANA Finance is only possible if the account determination is carried out successfully. The system uses the condition technique to perform account determination, and the associated customizing settings are found via the menu path **Sales and Distribution • Basic Functions • Account Assignment/ Costing**. These settings are usually maintained by the financial accounting team.

However, as a sales and distribution consultant, you should be familiar with the following billing document-related parameters used by the system to determine the general ledger accounts for your pricing conditions and, eventually, to create the journal entry:

- Chart of accounts of the company code
- Sales organization
- Account assignment group of the payer: This field is available in the business partner master record, for business partner role FLCU01 (customer), under the **Billing** tab at the sales area level.
- Account assignment group of the material: This field is available in the material master record in in the **Sales: sales org. 2** view.
- Account key assigned to the pricing condition types in the pricing procedure. As a reference, some of the standard account keys include the following:
  - MWS: Sales tax
  - ERF: Freight revenues
  - ERL: Revenues
  - ERS: Sales deductions (e.g., discounts)
  - EVV: Cash settlement
- Other configurable billing document fields such as company code, item category, tax code, etc.

In a nutshell, account determination is the process of mapping general ledger accounts to account keys using the condition technique. Covering all the details behind the settings for account determination and how to set up the required condition records in customizing is beyond the scope of this book. Nevertheless, you should know that the account determination procedure can be defined via the following three customizing activities:

- **Revenue Account Determination**
  - Revenue accounts to which amounts are posted
  - Sales deduction accounts to which surcharges and discounts are posted
- **Reconciliation Account Determination**
- **Cash Account Determination**

If the account determination is carried out successfully, the data is automatically transferred to SAP S/4HANA Finance; however, if required, a posting block can be set at the billing document header to prevent the creation of the journal entry, as mentioned in Section 11.2, when we reviewed the configuration of the billing type. The posting block can be released manually for a single billing document via Transaction VF02 or the Change Billing Documents app or in mass via Transaction VFX3 or the List Blocked Billing Documents or Manage Billing Documents apps.

Finally, note that, in some cases, the system may not be able to determine the general ledger account for one or more account keys in the billing document. This error could be caused by either master data or configuration issues. The following are some reasons that can cause errors in account determination:

- The account assignment group in the customer master record of the payer has not been maintained or has an incorrect value.
- The account assignment group in the material master record has not been maintained or has an incorrect value.
- The customizing for account determination has not been set up correctly (e.g., incorrect access sequence, missing assignments of account keys or general ledger accounts, etc.).

Tip

To help you investigate the root cause of the errors in account determination, you can use the Display Billing Documents app or Transaction VF03 to simulate the account determination. Choose **More • Environment • Acc. Determ. Analysis • Revenue Accounts**. The analysis displays the determined general ledger accounts for the relevant pricing condition types.

11.9 Important Terminology

In this chapter, the following terminology was used:

- **Preliminary billing document**
  An optional intermediary document between the reference document (sales order, delivery, debit memo request, etc.) and the actual billing document (e.g., customer invoice). This document has the same structure as a billing document and is used for negotiation purposes.
- **Pro forma invoice**
  An invoice that looks like a customer invoice but does not need to be paid. Therefore, the billing document data in a pro forma invoice is not transferred to financial accounting, and no accounting document is generated. Pro forma invoices are required to clear customs when goods are exported and must contain information about the market value of the goods and other details as required by the responsible authorities in the country of destination.
- **Credit memo request**
  A sales document that is used in complaints processing to request a credit memo for a customer.
- **Credit memo**
  A billing document that is created with reference to an invoice or a credit memo request. This document reduces receivables in financial accounting.
- **Debit memo request**
  A sales document that is used in complaints processing to request a debit memo for a customer.

- **Debit memo**
  A billing document that is created with reference to an invoice or a debit memo request. This document increases receivables in financial accounting.
- **Invoice combination**
  The process of merging two or more sales and distribution documents into a single billing document. To create a collective billing document, the billable reference documents must have certain values in common.
- **Invoice split**
  The opposite of invoice combination. This term describes scenarios where the system creates several billing documents for each billed reference document. An invoice split can be caused by differing values in either standard or custom fields of billable reference documents. These fields are known as *split criteria*.
- **Invoice list**
  A document containing a list of billing documents with a common payer that have already been posted to financial accounting.
- **Billing plan**
  A form of invoice creation that specifies the dates when the given amount or percentage rate should be billed for a single item in a sales order. Two types of billing plans can be defined in SAP S/4HANA: periodic billing and milestone billing.
- **Milestone billing**
  A type of billing plan where the total item amount on the sales document is distributed over several billing dates based on milestones that mark the completion of specific stages of a project. When a milestone is reached, an invoice is sent to the customer for either a predefined amount or a percentage of the project cost.
- **Periodic billing**
  A type of billing plan where the total item amount on the sales document is billed periodically until a predefined end date is reached.
- **Down payment**
  A down payment is an initial upfront partial payment for the purchase of an expensive good or service. The payment represents a percentage of the full purchase price.
- **Billing date**
  The date on which the billing document must be created and posted to accounting. This date is usually part of the selection criteria when processing the billing due list.
- **General ledger account**
  General ledger accounts are used to keep track of its financial transactions and to prepare financial reports. General ledger accounts are classified into account categories, such as income, expense, assets, liabilities, and equity.

- **Account determination**
  The process whereby the system determines the general ledger accounts required to generate a journal entry.

11.10 Practice Questions

These practice questions will help you evaluate your understanding of the topics covered in this chapter. The questions shown are similar in nature to those found on the certification examination. Although none of these questions will be found on the exam itself, they will allow you to review your knowledge of the subject. Select the correct answers and then check the completeness of your answers in the next section. Remember that, on the exam, you must select all correct answers and only correct answers to receive credit for the question.

1. Which of the following are characteristics of the billing document in SAP S/4HANA? (There are two correct answers.)

☐ A. It contains data at the item level.

☐ B. It contains data at the schedule line level.

☐ C. It can be created without reference to a preceding document.

☐ D. The billing date is only found in the document's header.

2. True or False. You cannot assign an external number range to a billing document.

☐ A. True

☐ B. False

3. Which of the following customizing activities for billing documents are available in the IMG menu? (There are three correct answers.)

☐ A. Define billing item categories

☐ B. Define billing types

☐ C. Define item category determination in billing documents

☐ D. Assign invoice list type to each billing type

☐ E. Maintain copying control for billing documents

4. Which of the following parameters are available when configuring billing document types in SAP S/4HANA? (There are three correct answers.)

☐ A. Transaction group

☐ B. Incompleteness procedure

☐ C. Billing type for cancellation

☐ D. Posting block

☐ E. Billing block

5. True or False. At the time of billing document creation, the billing type is always defaulted based on the billing type associated with a specific sales document type and cannot be manually changed.

☐ A. True

☐ B. False

6. Which of the following are valid document flows when creating a billing document? (There are two correct answers.)

☐ A. Invoice → Delivery → Debit memo

☐ B. Credit memo request → Credit memo → Invoice

☐ C. Order → Invoice → Credit memo

☐ D. Delivery → Invoice

7. Which of the following are characteristics of the SAP GUI Transaction VF04 or the equivalent SAP Fiori app Create Billing Documents – VF04? (There are two correct answers.)

☐ A. You can display a detailed log for the collective run.

☐ B. You can process the billing due list as a simulation "trial run."

☐ C. You can switch on/off the billing setting **Automatically post billing documents**.

☐ D. It is not possible to perform a split analysis.

8. In terms of billing split or convergence, which of the following variables can influence the outcome of a collective billing run when processing multiple billable reference documents? (There are two correct answers.)

☐ A. Discrepancies in split-relevant fields at the header or item level

☐ B. A custom pricing requirement assigned to condition PPRO

☐ C. The billing setting **Automatically post billing documents** in the Create Billing Documents app

☐ D. A custom data transfer routine in the copying control for billing documents

9. True or False: During collective billing, different customer reference numbers (field VBKD-BSTKD) in two sales orders always cause a billing document split, despite all the other split-relevant fields having identical values.

☐ A. True

☐ B. False

10. Which standard fields always cause a billing document split when populated with differing values? (There are two correct answers.)

☐ A. Distribution channel

☐ B. Bill-to party

☐ C. Total net value

☐ D. Ship-to party

11. Which of the following statements regarding preliminary billing documents are true? (There are three correct answers.)

☐ A. They can be changed as often as necessary.

☐ B. To use them, you need to activate the corresponding business function in customizing.

☐ C. They are available from SAP S/4HANA 1909.

☐ D. They can be posted to financial accounting.

☐ E. A preliminary billing document is an optional predecessor to the final billing document.

12. You've been asked to set up the system so that invoices are created based on the completion of specific stages of a project. Which functionality supports this business requirement?

☐ A. Preliminary billing documents

☐ B. Milestone billing plans

☐ C. Periodic billing plans

☐ D. Invoice lists

13. True or False. SAP S/4HANA allows you to process down payments on the basis of either periodic or milestone billing plans in sales orders.

☐ A. True

☐ B. False

14. Which of the following are characteristics of an invoice list in the standard system? (There are two correct answers.)

☐ A. It can contain single or collective billing documents.

☐ B. A single invoice list can contain invoices, credit memos, and debit memos.

☐ C. It only contains billing documents with a common bill-to party.

☐ D. It only contains billing documents that have already been released to financial accounting.

15. Which of the following parameters are used by the system to perform account determination when a billing document is transferred to financial accounting? (There are three correct answers.)

- ☐ A. Account assignment group of the ship-to party
- ☐ B. Account key
- ☐ C. Sales organization
- ☐ D. Profit center in the material master record
- ☐ E. Account assignment group of the material

11.11 Practice Question Answers and Explanations

1. Correct answers: **A and D**
   All billing documents must always be created with reference to a preceding document, which can be either a sales document, delivery, or another billing document. Therefore, answer C is not correct. Also, the structure of a billing document matches the structure of the delivery document since it only contains data at the header and item level. No schedule lines exist in a billing document since the goods have been delivered and/or the services provided by the time the billing document is created. Thus, answer B is invalid, which leaves you with answers A and D as the only two correct answers to this question.
2. Correct answer: **A**
   True. As shown earlier in Figure 11.3, you can only define an internal number range for billing documents. External number ranges are not allowed to ensure the billing document numbers are unique and consecutive.
3. Correct answers: **B, D, and E**
   In SAP S/4HANA, all billing documents are created with reference, and the item category in the preceding document is always copied into the billing document item. As a result, no item category determination is carried out during billing; therefore, you won't find customizing tasks to define or assign item categories for billing documents in the IMG menu. Thus, both answers A and C are invalid, leaving you with answers B, D, and E as the only three correct answers to this question.
4. Correct answers: **A, C, and D**
   As shown earlier in Figure 11.3, for a given billing document type, you can maintain a transaction group, maintain a billing type for cancellation, and activate a posting block, among other parameters. These three settings are mentioned in answers A, C, and D, respectively. You cannot define an incompleteness procedure or a billing block for billing document types, so answers B and E are both incorrect.

5. Correct answer: **B**

   False. The system automatically proposes a billing type based on the default billing type assigned to the reference document. However, if needed, you can manually change the proposed billing type to another valid billing type. The billing document types allowed are defined in the copy control from sales documents and delivery documents to billing documents. For example, for a given delivery document type LF, you can create a customer invoice (billing type F2) or a pro forma invoice (billing type F8).

6. Correct answers: **C and D**

   Answer A is not correct because you cannot create a delivery with reference to an invoice; in fact, the opposite is true. Also, debit memos are created with reference to debit memo requests and not from a delivery document. Regarding answer B, you cannot create a customer invoice from a credit memo in the standard system. Therefore, answer B is also incorrect, which leaves you with answers C and D as the only two valid document flows out of the four options provided.

7. Correct answers: **A and B**

   In Transaction VF40 or the Create Billing Documents – VF04 app, you can generate multiple billing documents by processing the billing due list. In this transaction, you can display a detailed log for the collective run as well as process the billing due list in simulation mode. So, answers A and B are both correct. You can also perform a split analysis, as shown earlier in Figure 11.14 which invalidates answer D. Finally, answer C is also incorrect since the billing setting **Automatically post billing documents** is available in the Create Billing Documents app but not in Transaction VF04 or in the Create Billing Documents – VF04 app.

8. Correct answers: **A and D**

   Answer A is correct because any discrepancies in split-relevant fields at the header or item level will obviously cause an invoice split. We also mentioned that you can use the data transfer routine VBRK/VBRP in the copy control for billing documents to implement additional split criteria based on your customer requirements. For example, your company may want to split an invoice if the ship-to party in two or more reference documents differs. This scenario is not supported by the standard configuration, but it can be achieved via custom data transfer routines in the copying control. Answer D is, therefore, the second of the only two correct answers to this question.

9. Correct answer: **B**

   False. As mentioned in this chapter, the customer reference number is stored in field VBKD-BSTKD in the sales document and field VBRK-BSTNK_VF in the billing document. This field constitutes a special case and does not always cause a billing document split, which is the reason why it has not been included in the list of fields that form the standard split criteria shown earlier in Table 11.5. The customer reference number will cause a billing document split only if you maintain the **Reference Number** parameter with value A in the copy control for billing documents at the header level. But if the **Reference Number** parameter is blank in the copy control for billing documents, you can combine two sales documents with two different customer reference numbers (field VBKD-BSTKD) in one billing document since the field VBRK-BSTNK_VF will be blank.

10. Correct answers: **A and B**

    As shown earlier in Table 11.5, both the distribution channel and the bill-to party are part of the standard split criteria, which means that if the values between two or more reference documents differ, they always cause a billing document split. We strongly recommend you familiarize yourself with the list of fields that form the standard split criteria. Answers A and B are, therefore, the only two correct answers to this question.

11. Correct answers: **A, B, and E**

    Preliminary billing documents are a new function available SAP S/4HANA 2020 (software component S4CORE from release 105); thus, answer C is incorrect. Also, as discussed earlier, preliminary billing documents are optional documents that are not relevant for posting to financial accounting. Thus, answer D is invalid, which leaves you with answers A, B, and E as the only three correct answers to this question.

12. Correct answer: **B**

    The milestone billing plan is the functionality available in the standard system for managing the creation of billing documents based on milestones that mark the completion of specific stages of a project.

13. Correct answer: **B**

    False. In SAP S/4HANA, down payments can only be processed on the basis of either milestone billing plans or document conditions in sales orders. You cannot process down payments with periodic billing plans.

14. Correct answers: **A and D**

    In the standard system, invoice lists for credit memos use billing type LG, whereas invoice lists for invoices and debit memos use billing type LR. As a result, invoices, debit memos, and credit memos cannot be combined in the same invoice list since the invoice list types are different. Thus, answer B is invalid. Answer C is also incorrect since you can combine several billing documents with different bill-to parties in the same invoice list, as long as they all have the same payer. Therefore, answers A and D are the only two correct answers to this question.

15. Correct answers: **B, C, and E**

    The account key, sales organization, and account assignment group of the material are three of the billing document-related parameters used by the system to determine the correct general ledger accounts and create the journal entry. Thus, answers B, C, and E are all correct answers. The account assignment group of the payer is also used in account determination but not the account assignment group of the ship-to party. Finally, the profit center in the material master record is not a parameter used in account determination either. Therefore, answers A and D are both incorrect answers.

11.12 Takeaway

In this chapter, we started by reviewing the concept of the billing document in SAP S/4HANA and the most important parameters that define a billing document type. Next, we focused on the different ways you can create a billing document depending on the nature of the reference documents. We showed you how to process the billing due list both using the traditional transactions and the new SAP Fiori apps, as well as how to maintain the different settings and functions for generating billing documents and performing a split analysis.

We then went over the concept of preliminary billing documents, a new optional document commonly used in the professional services industry. We also explained the two types of billing plans available in the system and how to manage down payments using the milestone billing plan.

Finally, we provided an overview of the invoice list functionality and covered the basics of account determination in billing documents.

11.13 Summary

Billing is the process of formally requesting customer payments for products or services provided and, quite often, represents the final step of a sales transaction. This chapter described the use of billing functions in SAP S/4HANA and their integration with the financial accounting. With your new, improved understanding of the billing process, now you can explain how billing documents are structured and controlled. You've seen how to manage the billing due list both manually and automatically via background jobs. You now understand the concepts of invoice combination and invoice split, and you can perform a split analysis if needed.

You can distinguish between periodic and milestone billing plans, and you should be familiar with the use of down payments through milestone billing plans. You can also combine billing documents with a common payer via invoice lists and understand how account keys are mapped to general ledger accounts using the condition technique in order to transfer to SAP S/4HANA Finance the data required to create the journal entries.

In the next chapter, we'll go through the simplification list for SAP S/4HANA 2020 in the sales and distribution area.

Chapter 12
Simplifications

Techniques You'll Master

- Understand the simplification list for SAP S/4HANA 2020 in the sales and distribution area
- Understand how to set up and use business partners
- Understand sales and distribution outputs using Business Rules Framework plus (BRFplus)
- Understand the revised data model for sales tables

In this chapter, we'll explore the changes in functionality between SAP ERP 6.0 and SAP S/4HANA. You'll gain an understanding of the simplification list, which offers a high-level view of all these deltas. Once you understand how the simplification list is used, we'll drill down into more details about some of the key components that you'll need to master for the SAP S/4HANA Sales Upskilling certification exam.

Real-World Scenario

Let's say the company employing you is beginning the process of upgrading their SAP ERP 6.0 system to SAP S/4HANA 2020. One challenge they are facing is the concept of business partners and the approach they need to take to convert their existing customers and vendors. Your customer is also married to the SAP ERP 6.0 concepts of sales analytics through an external data warehouse, sales outputs via the table NAST approach, and the SAP ERP 6.0 view of sales data tables. Users will be heavily reliant upon your knowledge as an SAP S/4HANA consultant to inform them of what has changed and how they should approach the upgrade.

12.1 Objectives of This Portion of the Test

The goal of this portion of the SAP S/4HANA Sales Upskilling Certification exam is to test your understanding of the following topics:

- A sound overview of the changes in the sales and distribution area, which are contained in the simplification list
- Business partners and the conversion from the SAP ERP 6.0 data setup
- Tasks required as part of the conversion process from SAP ERP 6.0 to SAP S/4HANA
- The sales analytics model
- How outputs can be generated using BRFplus
- The simplification of the sales data tables

Note

The Simplification topic is not included in the SAP S/4HANA Sales exam but does comprise between 8% and 12% of the total SAP S/4HANA Sales Upskilling exam.

12.2 SAP S/4HANA Simplification List

Before we go into the details about some of the key deltas from SAP ERP 6.0 to SAP S/4HANA, let's first explore where these deltas can be viewed. Each version of SAP

S/4HANA has its own *simplification list*—a list of functions that have changed from the previous version, driving towards a more streamlined and simplified view of the system. The SAP S/4HANA 2020 simplification list can be found at the following link: *http://s-prs.co/v530800.*

The simplification list itself houses *simplification items*, which are areas of amendment to existing functionality in previous versions. Each simplification item has three components: the overall description, which describes the amendment; the recommendations for using the changes as well as the possible business impacts; and the SAP Notes that detail the changes.

As mentioned earlier, each version of SAP S/4HANA has its own simplification list, and therefore, getting a full picture of the changes from SAP ERP 6.0 to SAP S/4HANA can be difficult. However, this book is here to help! Throughout the book, many of the functionalities that have been explained, such as advanced available-to-promise (ATP), and many that are yet to be explained, refer to SAP S/4HANA functionality rather than to SAP ERP 6.0 functionality. Some key concepts, such as the business partner, will be explored in depth in this chapter, and many other concepts will be summarized—all in order to give you enough understanding to pass the certification exam on simplifications.

For completeness, the key deltas you should be aware of include the following:

- Business partner concept
- Settlement management instead of rebate processing
- BRFplus functionality for outputs
- Credit management amendments
- SAP S/4HANA for advanced ATP (detailed in Chapter 7)
- Removal of the SAP ERP SD foreign trade functionality from core SAP S/4HANA into SAP Global Trade Services (SAP GTS)
- Sales data model changes
- SAP S/4HANA embedded analytics (detailed in Chapter 1, Section 1.6)

In addition to these main changes, the simplification list also describes conversion activities when converting from SAP ERP 6.0 to SAP S/4HANA. These activities include how to run the custom code check for any transactions and programs that have been deprecated in SAP S/4HANA and how to optimize your conversion using parallel execution of conversion reports for sales objects.

Let's start on the main simplification issues by looking at the business partner concept in SAP S/4HANA.

12.3 Business Partners

The concept of business partners is not new to SAP S/4HANA. Business partners have been a part of SAP CRM for a long time.

However, what is new is that SAP S/4HANA mandates the use of the business partner, instead of using traditional methods of creating separate data objects for customers, vendors, and contacts. In SAP ERP 6.0, these objects would be created by such transactions as Transaction XD01, XK01, VD01, VAP1, etc. All these transactions have been deprecated in SAP S/4HANA in favor of the unified business partner maintenance transaction, Transaction BP. If you try to use these transactions, you'll be forwarded automatically to Transaction BP. For SAP Fiori users, the Maintain Business Partner app can be used.

Note

All the deprecated transactions that have been replaced by Transaction BP are Transactions FD01, FD02, FD03, FD05, FD06, FD32, FD33, FK01, FK02, FK03, FK05, FK06, FK08, MAP1, MAP2, MAP3, MK01, MK02, MK03, MK05, MK06, MK12, MK18, MK19, V-03, V-04, V-5, V-06, V-07, V-08, V-09, V-11, V+21, V+22, V+23, VAP1, VAP2, VAP3, VD01, VD02, VD03, VD05, VD06, XD01, XD02, XD03, XD05, XD06, XD07, XK01, XK02, XK03, XK05, XK06, and XK07.

Let's start with a quick definition of business partners in the context of SAP S/4HANA. A business partner can be defined as either a person, an organization, or a group of organizations that have a business relationship with your organization. Business partners are managed centrally within the system and assume specific roles for given business transactions.

Therefore, with this approach, one business partner can fulfill several roles: They can be a customer, vendor, contact person, sales rep, and more all at once—no need for separate master data objects. In other words, business partner general data, such as name, address, and contacts can be shared across multiple roles. Note, however, that business partners are not restricted to only one address—multiple address functionality is available.

The unified roles concept is a key advantage of the business partner in SAP S/4HANA. Importantly, for SAP, following the business partner approach makes sense because this development brings the core enterprise resource planning (ERP) system of SAP S/4HANA into line with all other SAP solutions (such as SAP CRM), offering a harmonized approach.

The separation of different types of business partner is performed by selecting a business partner role from the **Change in BP role** dropdown list, as shown in Figure 12.1.

The definition of roles can be carried out in the IMG by following the menu path **Cross Application Components SAP Business Partner • Business Partner • Define BP Roles**. Additionally, in this IMG path, useful help documentation can be found about the concept of business partner roles. Business partner roles define the role a business partner can play in a business process. Examples of these roles include:

- Customer
- Vendor
- Credit management (Section 12.6)
- Competitor

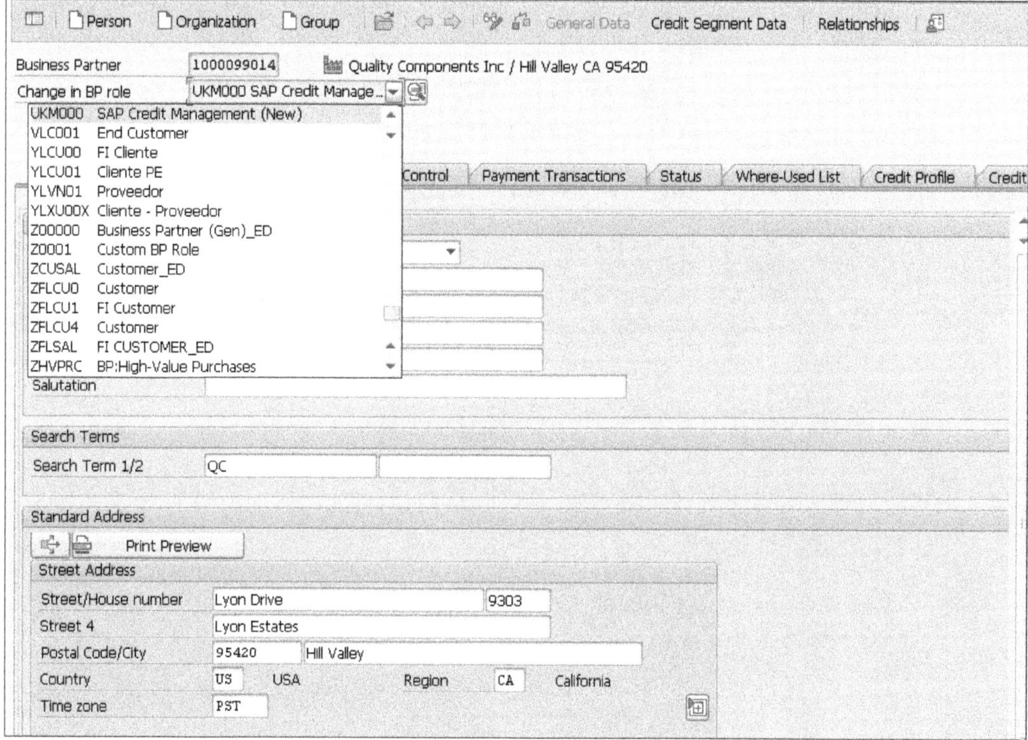

Figure 12.1 Business Partner Roles

Roles are freely definable within customizing.

The traditional SAP ERP 6.0 solution for customers and vendors meant that SAP would write out entries to the customer and vendor tables such as tables KNA1, KNVV, and LFA1. These tables have not disappeared in SAP S/4HANA, mainly because they are so deeply integrated into the sales and purchasing process. As such, note that these tables are still populated in the background by any creations or amendments made to business partners. Business partners themselves house the shared master data, such as name and address, in their own set of tables, which are prefixed BUT*. However, SAP S/4HANA will not automatically populate the old traditional tables—this step must be performed by your own configuration settings. You must define how the BUT* tables push data out to the existing customer and vendor tables. The definition of this approach is controlled via customer-vendor integration (CVI).

CVI is an important concept in SAP S/4HANA and underpins the business partner approach, including being central to how customers and vendors are converted from SAP ERP 6.0 systems into business partners in SAP S/4HANA—more on this topic later. Through CVI, your configuration tells SAP S/4HANA whether the customer and vendor tables must be populated for any business partner created. The IMG path for customizing CVI is **Cross Application Settings • Master Data Synchronization • Customer/Vendor Integration**. The individual paths to set up the link between business partners and customers and vendors is shown in Figure 12.2 and include the following options:

- **Set BP Role Category for Direction BP to Customer**
- **Set BP Role Category for Direction BP to Vendor**

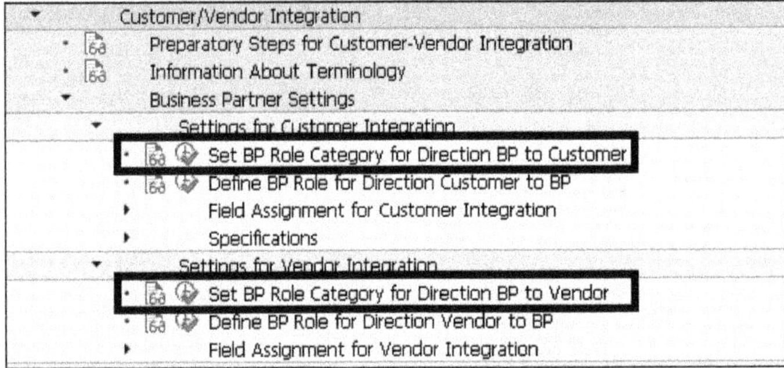

Figure 12.2 CVI: IMG

Once the CVI is set up, the table relationships are established, examples of which are shown in Figure 12.3.

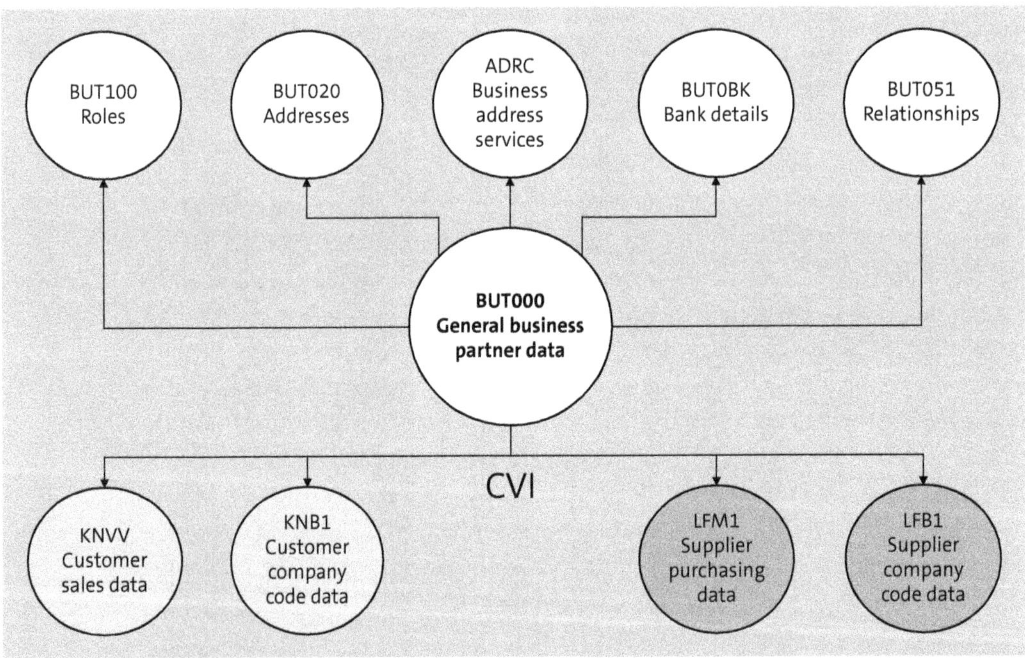

Figure 12.3 CVI: Table Relationships

As shown in Figure 12.3, the tables at the top of the diagram (prefixed BUT*) are business partner tables, populated as soon as a business partner is saved. In addition, address table ADRC is also populated at this time. Once the CVI configuration is established, saving a business partner in a customer or vendor role will also push data out to the tables at the bottom of the diagram, which are specific to customers or vendors.

An important point to note is how the CVI approach is used for conversions from SAP ERP 6.0 to SAP S/4HANA. As mentioned earlier, customers and vendors are set up as separate data objects in SAP ERP 6.0, and therefore, in order to begin the data conversion process, all customers and vendors must first be converted to business partners in the source SAP ERP 6.0 system. Thus, you'll need to set up your CVI configuration in your SAP ERP 6.0 system in reverse of the settings for your SAP S/4HANA system: **Set BP Role Category for Direction Customer to BP** in the IMG path shown earlier in Figure 12.2.

A defined process exists for carrying out this conversion, and its steps should be implemented in the correct order, in both your SAP ERP 6.0 system and your SAP S/4HANA system. These steps are as follows:

1. Implement CVI in your SAP ERP 6.0 system
2. Activate **Customer/vendor to BP** in the configuration settings in your SAP ERP 6.0 system
3. Convert existing customers/vendors to business partners in your SAP ERP 6.0 system
4. Continue using normal customer/vendor transactions such as Transaction XD01/XK01 in your SAP ERP 6.0 system (the CVI configuration will automatically convert any customers/vendors to business partners)
5. Monitor and correct any errors in the automated business partner creation (via Transaction MDS_PPO2)
6. Transfer business partners from SAP ERP 6.0 directly to SAP S/4HANA
7. Activate **BP to customer/vendor** in configuration settings in your SAP S/4HANA system

If these steps are followed in this order, then the conversion process should be seamless. Many of these steps provide little information for how to carry out the process, but our intention is to give you sufficient information for passing the certification exam, rather than a detailed step-by-step process, which would take up the whole book.

Note

Further detailed information can be found in *SAP S/4HANA Cookbook: Customer/Vendor Integration*, provided free by SAP: *http://s-prs.co/v530801.*

A user-friendly guide is also available as an E-Bite from SAP PRESS: *https://www.sap-press.com/4971/.*

With regard to the CVI process, note also that account groups are assigned to business partner groupings here. This step is important because the number range for the business partner is taken from the business partner grouping, whereas the number range for the customer is taken from the account group. You can assign number ranges in customizing by following the menu path **Cross-Application Components • Master Data Synchronization • Customer/Vendor Integration • Business Partner Groupings • Settings for Customer Integration • Field Assignment for Customer Integration • Assign Key • Define Number Assignment for Direction BP to Customer**. The **Same Numbers** checkbox can be activated to allocate the same number to both objects, as shown in Figure 12.4.

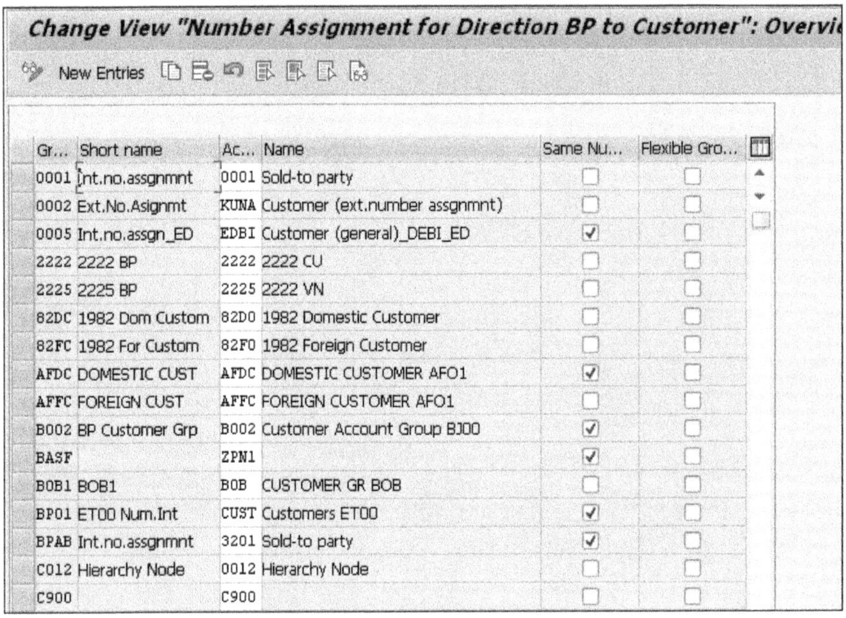

Change View "Number Assignment for Direction BP to Customer": Overvi

New Entries

Gr...	Short name	Ac...	Name	Same Nu...	Flexible Gro...
0001	Int.no.assgnmnt	0001	Sold-to party	☐	☐
0002	Ext.No.Asignmt	KUNA	Customer (ext.number assgnmnt)	☐	☐
0005	Int.no.assgn_ED	EDBI	Customer (general)_DEBI_ED	☑	☐
2222	2222 BP	2222	2222 CU	☐	☐
2225	2225 BP	2225	2222 VN	☐	☐
82DC	1982 Dom Custom	82D0	1982 Domestic Customer	☐	☐
82FC	1982 For Custom	82F0	1982 Foreign Customer	☐	☐
AFDC	DOMESTIC CUST	AFDC	DOMESTIC CUSTOMER AFO1	☑	☐
AFFC	FOREIGN CUST	AFFC	FOREIGN CUSTOMER AFO1	☐	☐
B002	BP Customer Grp	B002	Customer Account Group BJ00	☑	☐
BASF		ZPN1		☑	☐
BOB1	BOB1	BOB	CUSTOMER GR BOB	☐	☐
BP01	ET00 Num.Int	CUST	Customers ET00	☑	☐
BPAB	Int.no.assgnmnt	3201	Sold-to party	☑	☐
C012	Hierarchy Node	0012	Hierarchy Node	☐	☐
C900		C900		☐	☐

Figure 12.4 Business Partner Group to Account Group Assignment

12.4 Settlement Management

In SAP ERP 6.0, rebate processing was central to providing customers with rebates. This functionality has been overhauled in SAP S/4HANA, and the old Sales and Distribution (SD) and Materials Management (MM) rebates mechanism has been replaced with settlement management, which we discussed at a high level in Chapter 8, Section 8.5.

Numerous reasons justify the change from rebates to settlement management. System performance was a huge issue with the SAP ERP 6.0 rebates process, due to the large and unwieldy nature of the underlying index table (table VBOX). This table, which previously could have stored many millions of index records, has been retired in SAP S/4HANA since index tables are no longer needed because of the underlying power of the SAP HANA database (Section 12.8). All rebates are now processed through the condition contracts settlement process. This process allows rebates to given to customers and vendors alike, using the same process steps.

> **Note**
>
> Settlement management is mandated in SAP S/4HANA for rebate processing. The traditional method of rebate processing in SAP ERP 6.0 as now been retired, and the functionality is not available in SAP S/4HANA. Thus, any conversion to an SAP S/4HANA system will necessitate the condition contract settlement approach rather than SD rebates.

The creation of a condition contract is the first step in the process of condition contract settlement management, which can be accessed via Transaction WCOCO in SAP GUI. For SAP Fiori users, the Maintain Condition Contract app can also be used. In either case, you can create a condition contract for a customer or a vendor, dependent upon the condition contract type.

Once the condition contract is saved, all sales orders that relate to the condition contract will show the accrual condition in the **Conditions** tab. Sales orders created before the condition contract was created will not show this condition but are still relevant for settlement management.

Once a condition contract is active, it can be settled at any time via Transaction WB2R_SC, as shown in Figure 12.5, or the SAP Fiori app Settle Condition Contracts.

Settlement of Customer Condition Contracts

Settlement Control

Settlement Date		to	
Settlement Date Type		to	
Customer		to	
Settlement Execution Date		to	

Contract Selection

Condition Contract		to	
Contract Process Variant		to	
Condition Contract Type		to	
Condition Contract Category		to	
Company Code		to	
Purch. organization		to	
Purchasing Group		to	
Sales Organization		to	
Distribution Channel		to	
Division		to	
Status		to	

Default Data

Posting Date	03/06/2021
Document Date	03/06/2021
Activity Reason	

Figure 12.5 Settle Condition Contract Transaction W2BR_SC

Once executed, the credit memo will be automatically created and will be available in the condition contract in Transaction WCOCO.

The overall business volume for condition contracts can be tracked in Transaction WB2R_BUSVOL in SAP GUI or in the SAP Fiori app Display Business Volume.

12.5 Output Management through BRFplus

A significant change in the order-to-cash approach for SAP S/4HANA is the use of BRFplus for generating sales output documents. This change is firmly in the sales area, and therefore, you should expect this topic to appear in the SAP S/4HANA Sales Upskilling Certification exam. Thus, a good understanding of the functionality is required.

BRFplus itself is not a new concept and has been around in SAP for many years. However, its use in sales output determination is new to SAP S/4HANA. BRFplus can be used in many applications and is a useful tool for use in place of traditional ABAP coding to specify rules that apply to certain business scenarios.

The first thing to note is that BRFplus must be activated in your system before it can be used. This activation is carried out by your Basis administrator. For each application object, such as sales order or billing document, the relevant XML files must be imported into the system using Transaction BRFplus, from SAP Note 2248229. You can import the XML files for all application objects or simply import the ones you wish to use.

Once activated, you can access the BRFplus engine, which can be launched via the IMG menu path **Cross-Application Components • Output Control • Define Business Rules for Output Parameter Determination**, as shown in Figure 12.6.

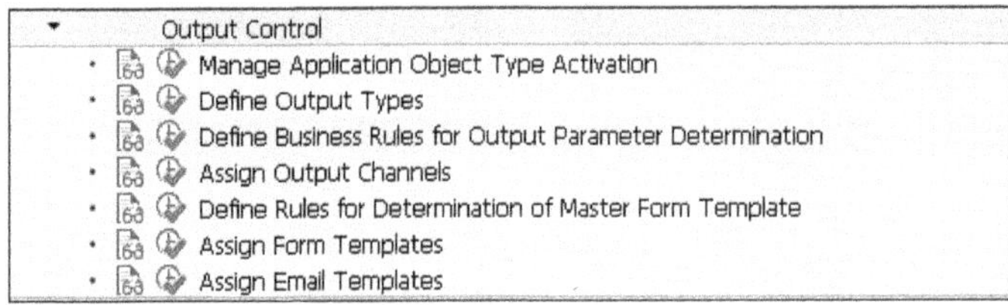

Figure 12.6 BRFplus: IMG Menu Path

Furthermore, although BRFplus outputs are recommended as the default approach for sales outputs, it is not mandatory. When embarking upon an SAP S/4HANA implementation, a key decision is whether to use BRFplus outputs or the more traditional NAST-based output method, as detailed in Chapter 6, Section 6.4. In the same IMG menu path (**Cross-Application Components • Output Control • Manage Application Object Type Activation**), the activation of BRFplus for outputs can be switched off.

This menu path takes you to a selection screen where you can switch on or off the activation status by object type, such as **Billing_document**, **Purchase_order**, **Sales_ document**, etc., as shown in Figure 12.7.

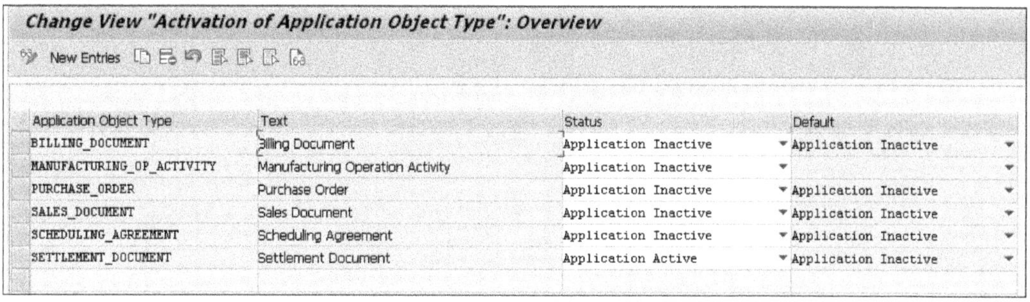

Change View "Activation of Application Object Type": Overview

New Entries

Application Object Type	Text	Status	Default
BILLING_DOCUMENT	Billing Document	Application Inactive	Application Inactive
MANUFACTURING_OP_ACTIVITY	Manufacturing Operation Activity	Application Inactive	
PURCHASE_ORDER	Purchase Order	Application Inactive	Application Inactive
SALES_DOCUMENT	Sales Document	Application Inactive	Application Inactive
SCHEDULING_AGREEMENT	Scheduling Agreement	Application Inactive	Application Inactive
SETTLEMENT_DOCUMENT	Settlement Document	Application Active	Application Inactive

Figure 12.7 BRFplus: IMG Deactivation by Object Type

A few limitations to using BRFplus include the following:

- The use of Intermediate Documents (IDocs), while supported by BRFplus, has limitations. You can only use IDocs in conjunction with BRFplus for the business applications that previously used table NAST.
- You cannot communicate with logical systems via BRFplus—only business partner-based communication is supported.
- Application Link Enabling (ALE) connections are not supported by BRFplus.

Assuming you are using BRFplus, let's walk through an overview of the concept, which will appear on the exam. Each step can be carried out in sequence from the menu options shown earlier in Figure 12.6.

Initially, an output type can be created, as with the NAST approach. Each business application, such as sales order or billing document, is represented by an application object type, and the output types should be assigned accordingly and given a descriptive text.

Now that the output type is defined, business rules must be set up for the output determination. By selecting the highlighted IMG menu path shown earlier in Figure 12.6 or using Transaction OPD, a web portal is opened in your browser. On this page, the application object can be selected from the **Show Rules For** dropdown list. You'll then see a set of available determination steps in the **Determination Step** dropdown list, such as the following:

- **Output Type**
- **Receiver**
- **Channel**
- **Printer Settings**
- **Email Recipient**
- **Email Settings**
- **Form Template**
- **Output Relevance**

Selecting each of these steps in turn opens the relevant *decision table* to display the available business rules. Figure 12.8 shows business rules for the **Billing Document** object and the **Output Type** decision table.

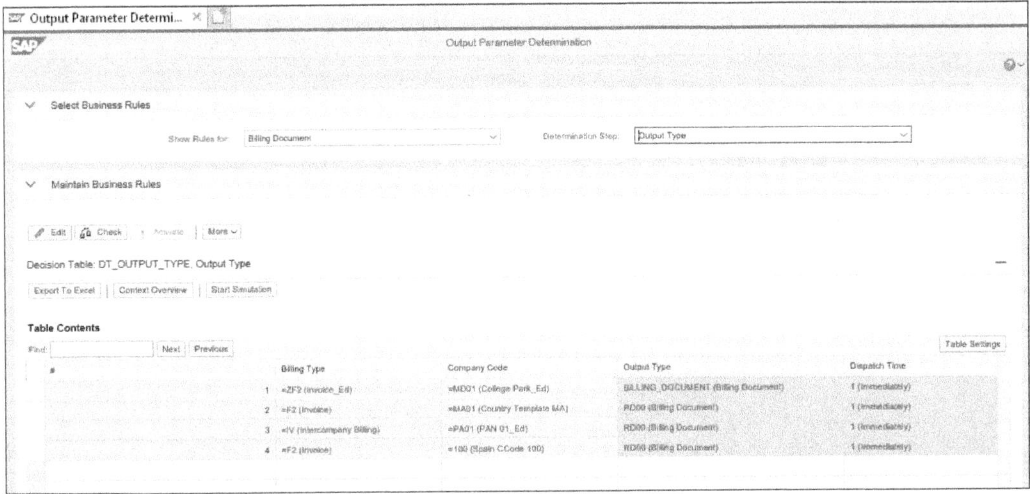

Figure 12.8 BRFplus: Business Rules

Each determination step allows you to define the required result for each setting. For example, the decision table related to the **Output Type** determination step will determine the required output type as a result. In this way, you can work your way through each determination step to define the results you require in each area. Don't forget that each step must be activated after you have completed it.

You'll need some understanding of decision tables to arm yourself with the appropriate knowledge to pass the certification exam in this area.

Decision tables are used in SAP S/4HANA to compare input values against predefined conditions in order to propose a result. The structure of decision tables is comprised of:

- One or more columns representing the combination of conditions against which the input values are compared
- One or more results columns if all conditions are satisfied

As with NAST condition tables, the BRFplus decision table is processed row by row in sequential order from top to bottom, reading the condition columns from left to right to achieve the result. Therefore, you must define your condition rows from most specific to least specific in order to achieve the correct result. Table 12.1 shows an example of a BRFplus decision table.

Country	Shipping Condition	Output Type
DE	01	Z101
US	01	Z102
<all>	01	Z103

Table 12.1 BRFplus: Example Decision Table

Once you've set up and activated all the decision steps, you can then return to your IMG menu path to update the further required settings

You can assign output channels via the menu path **Cross Application Components • Output Control • Assign Output Channels**. The allowed channels (i.e., electronic data interchange [EDI], email, IDoc, print, XML) are defined for a combination of the application object and the output type, as shown in Figure 12.9.

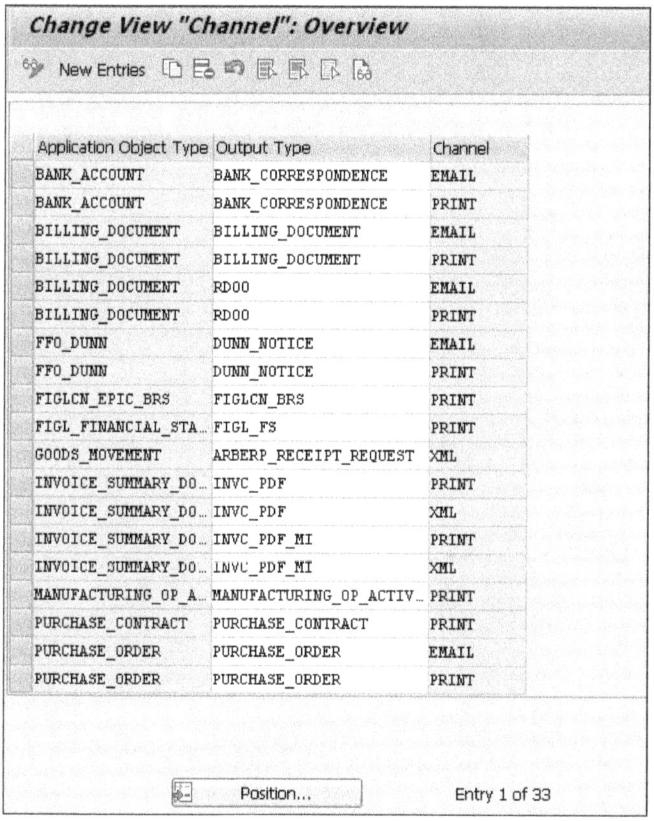

Change View "Channel": Overview

New Entries

Application Object Type	Output Type	Channel
BANK_ACCOUNT	BANK_CORRESPONDENCE	EMAIL
BANK_ACCOUNT	BANK_CORRESPONDENCE	PRINT
BILLING_DOCUMENT	BILLING_DOCUMENT	EMAIL
BILLING_DOCUMENT	BILLING_DOCUMENT	PRINT
BILLING_DOCUMENT	RD00	EMAIL
BILLING_DOCUMENT	RD00	PRINT
FFO_DUNN	DUNN_NOTICE	EMAIL
FFO_DUNN	DUNN_NOTICE	PRINT
FIGLCN_EPIC_BRS	FIGLCN_BRS	PRINT
FIGL_FINANCIAL_STA...	FIGL_FS	PRINT
GOODS_MOVEMENT	ARBERP_RECEIPT_REQUEST	XML
INVOICE_SUMMARY_DO...	INVC_PDF	PRINT
INVOICE_SUMMARY_DO...	INVC_PDF	XML
INVOICE_SUMMARY_DO...	INVC_PDF_MI	PRINT
INVOICE_SUMMARY_DO...	INVC_PDF_MI	XML
MANUFACTURING_OP_A...	MANUFACTURING_OP_ACTIV...	PRINT
PURCHASE_CONTRACT	PURCHASE_CONTRACT	PRINT
PURCHASE_ORDER	PURCHASE_ORDER	EMAIL
PURCHASE_ORDER	PURCHASE_ORDER	PRINT

Position... Entry 1 of 33

Figure 12.9 BRFplus: Defining Output Channels

Via the menu path **Cross Application Components • Output Control • Define Rules for Determination of Master Form Template**, you can specify additional rules to be applied in order to determine different form templates to be used, as shown in Figure 12.10. A *master form template* is a form that holds static data that is applicable to multiple forms, such as logos or footer data. The BRFplus rules as applied in the decision table will pass form deviation rule IDs to the table to check. If the entries match, then the rules are carried in **Number** order to define the templates to be used.

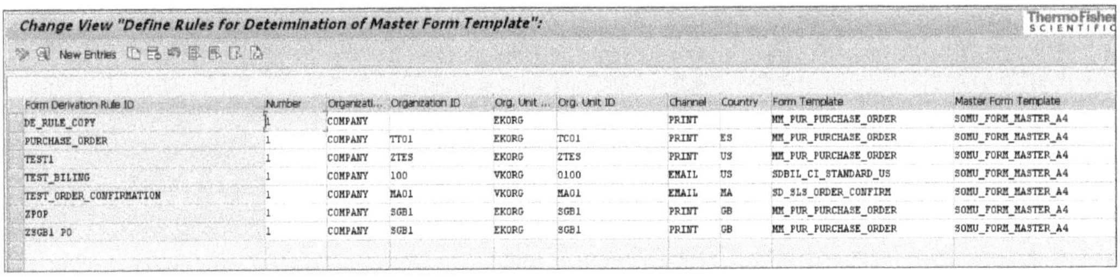

Change View "Define Rules for Determination of Master Form Template":

New Entries

Form Derivation Rule ID	Number	Organizati...	Organization ID	Org. Unit ...	Org. Unit ID	Channel	Country	Form Template	Master Form Template
DE_RULE_COPY	1	COMPANY		EKORG		PRINT		MM_PUR_PURCHASE_ORDER	SOMU_FORM_MASTER_A4
PURCHASE_ORDER	1	COMPANY	TT01	EKORG	TC01	PRINT	ES	MM_PUR_PURCHASE_ORDER	SOMU_FORM_MASTER_A4
TEST1	1	COMPANY	ZTES	EKORG	ZTES	PRINT	US	MM_PUR_PURCHASE_ORDER	SOMU_FORM_MASTER_A4
TEST_BILING	1	COMPANY	100	VKORG	0100	EMAIL	US	SDBIL_CI_STANDARD_US	SOMU_FORM_MASTER_A4
TEST_ORDER_CONFIRMATION	1	COMPANY	MA01	VKORG	MA01	EMAIL	MA	SD_SLS_ORDER_CONFIRM	SOMU_FORM_MASTER_A4
ZPOP	1	COMPANY	SGB1	EKORG	SGB1	PRINT	GB	MM_PUR_PURCHASE_ORDER	SOMU_FORM_MASTER_A4
ZSGB1 PO	1	COMPANY	SGB1	EKORG	SGB1	PRINT	GB	MM_PUR_PURCHASE_ORDER	SOMU_FORM_MASTER_A4

Figure 12.10 BRFplus: Defining Rules for Determination of Master Form Template

Via the menu path **Cross Application Components • Output Control • Assign Form Templates**, as shown in Figure 12.11, you can assign output forms to a combination of the **Application Object Type** and **Output Type** fields. This assignment can, of course, be overridden by a form deviation rule, as shown in Figure 12.10.

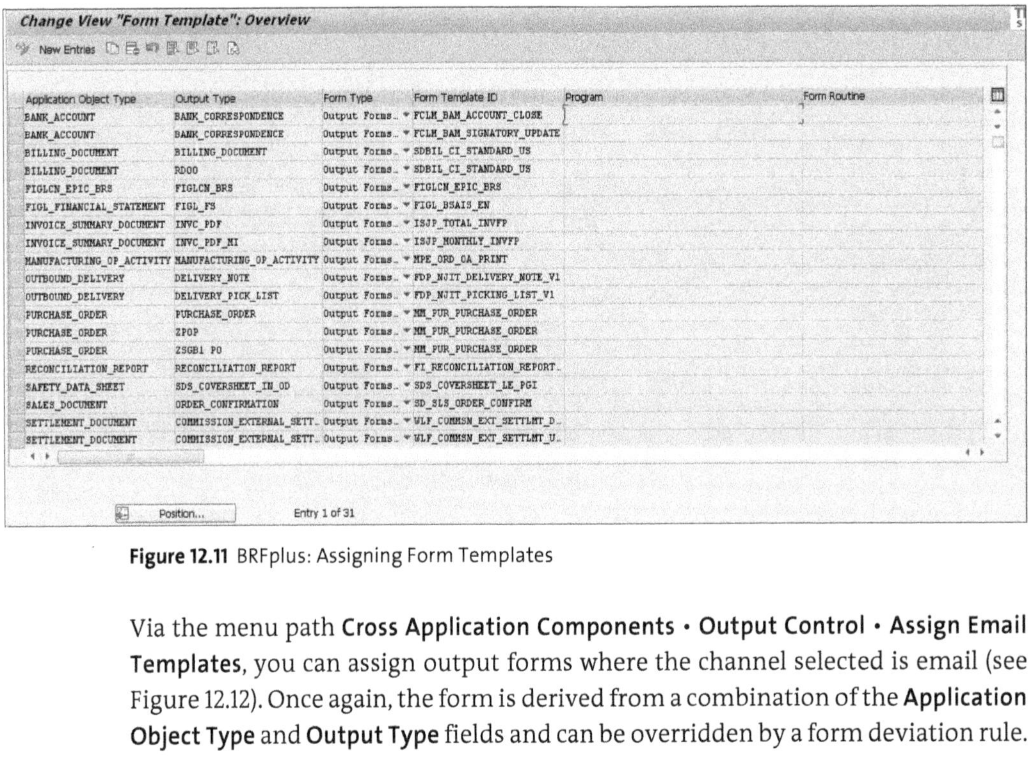

Change View "Form Template": Overview

New Entries

Application Object Type	Output Type	Form Type	Form Template ID	Program	Form Routine
BANK_ACCOUNT	BANK_CORRESPONDENCE	Output Forms...	FCLM_BAM_ACCOUNT_CLOSE		
BANK_ACCOUNT	BANK_CORRESPONDENCE	Output Forms...	FCLM_BAM_SIGNATORY_UPDATE		
BILLING_DOCUMENT	BILLING_DOCUMENT	Output Forms...	SDBIL_CI_STANDARD_US		
BILLING_DOCUMENT	RD00	Output Forms...	SDBIL_CI_STANDARD_US		
FIGLCN_EPIC_BRS	FIGLCN_BRS	Output Forms...	FIGLCN_EPIC_BRS		
FIGL_FINANCIAL_STATEMENT	FIGL_FS	Output Forms...	FIGL_BSAIS_EN		
INVOICE_SUMMARY_DOCUMENT	INVC_PDF	Output Forms...	ISJP_TOTAL_INVFP		
INVOICE_SUMMARY_DOCUMENT	INVC_PDF_MI	Output Forms...	ISJP_MONTHLY_INVFP		
MANUFACTURING_OP_ACTIVITY	MANUFACTURING_OP_ACTIVITY	Output Forms...	MPE_ORD_OA_PRINT		
OUTBOUND_DELIVERY	DELIVERY_NOTE	Output Forms...	FDP_MJIT_DELIVERY_NOTE_V1		
OUTBOUND_DELIVERY	DELIVERY_PICK_LIST	Output Forms...	FDP_MJIT_PICKING_LIST_V1		
PURCHASE_ORDER	PURCHASE_ORDER	Output Forms...	MM_PUR_PURCHASE_ORDER		
PURCHASE_ORDER	ZPOP	Output Forms...	MM_PUR_PURCHASE_ORDER		
PURCHASE_ORDER	ZSGB1 PO	Output Forms...	MM_PUR_PURCHASE_ORDER		
RECONCILIATION_REPORT	RECONCILIATION_REPORT	Output Forms...	FI_RECONCILIATION_REPORT...		
SAFETY_DATA_SHEET	SDS_COVERSHEET_IN_OD	Output Forms...	SDS_COVERSHEET_LE_PGI		
SALES_DOCUMENT	ORDER_CONFIRMATION	Output Forms...	SD_SLS_ORDER_CONFIRM		
SETTLEMENT_DOCUMENT	COMMISSION_EXTERNAL_SETT...	Output Forms...	WLF_COMMSN_EXT_SETTLMT_D...		
SETTLEMENT_DOCUMENT	COMMISSION_EXTERNAL_SETT...	Output Forms...	WLF_COMMSN_EXT_SETTLMT_U...		

Position... Entry 1 of 31

Figure 12.11 BRFplus: Assigning Form Templates

Via the menu path **Cross Application Components • Output Control • Assign Email Templates**, you can assign output forms where the channel selected is email (see Figure 12.12). Once again, the form is derived from a combination of the **Application Object Type** and **Output Type** fields and can be overridden by a form deviation rule.

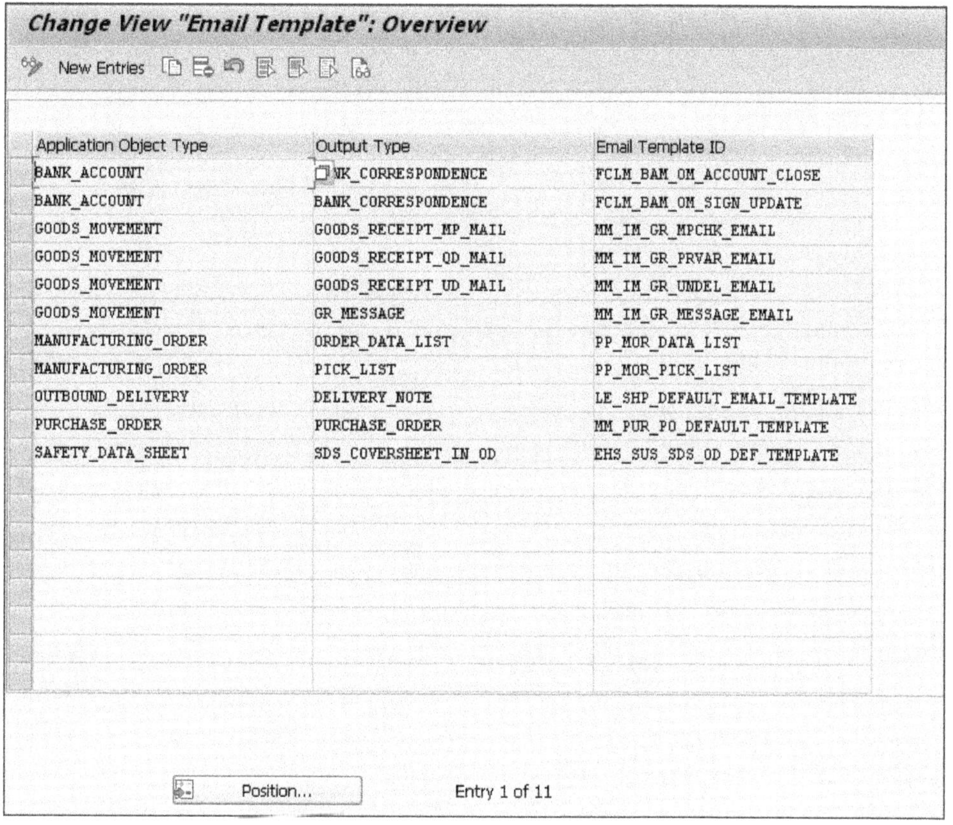

Application Object Type	Output Type	Email Template ID
BANK_ACCOUNT	NK_CORRESPONDENCE	FCLM_BAM_OM_ACCOUNT_CLOSE
BANK_ACCOUNT	BANK_CORRESPONDENCE	FCLM_BAM_OM_SIGN_UPDATE
GOODS_MOVEMENT	GOODS_RECEIPT_MP_MAIL	MM_IM_GR_MPCHK_EMAIL
GOODS_MOVEMENT	GOODS_RECEIPT_QD_MAIL	MM_IM_GR_PRVAR_EMAIL
GOODS_MOVEMENT	GOODS_RECEIPT_UD_MAIL	MM_IM_GR_UNDEL_EMAIL
GOODS_MOVEMENT	GR_MESSAGE	MM_IM_GR_MESSAGE_EMAIL
MANUFACTURING_ORDER	ORDER_DATA_LIST	PP_MOR_DATA_LIST
MANUFACTURING_ORDER	PICK_LIST	PP_MOR_PICK_LIST
OUTBOUND_DELIVERY	DELIVERY_NOTE	LE_SHP_DEFAULT_EMAIL_TEMPLATE
PURCHASE_ORDER	PURCHASE_ORDER	MM_PUR_PO_DEFAULT_TEMPLATE
SAFETY_DATA_SHEET	SDS_COVERSHEET_IN_OD	EHS_SUS_SDS_OD_DEF_TEMPLATE

Figure 12.12 BRFplus: Assigning Email Templates

12.6 Credit Management Amendments

In traditional SAP ERP 6.0 systems, credit management and credit limits are sales configuration items. However, in SAP S/4HANA, this configuration has been completely overhauled and is housed in the **Financial Supply Chain Management** area of the IMG. As such, the configuration of credit management should be carried out in conjunction with your finance analyst. The important takeaway is that the functionality for credit management is radically different in SAP S/4HANA. Since the customizing is carried out by a finance analyst, you only need to understand the concept of credit management in SAP S/4HANA and how it has changed from the SAP ERP 6.0 approach.

The change from SAP ERP 6.0 becomes evident when trying to use Transaction FD32 in the SAP ERP 6.0; you'll be presented with the error message shown in Figure 12.13.

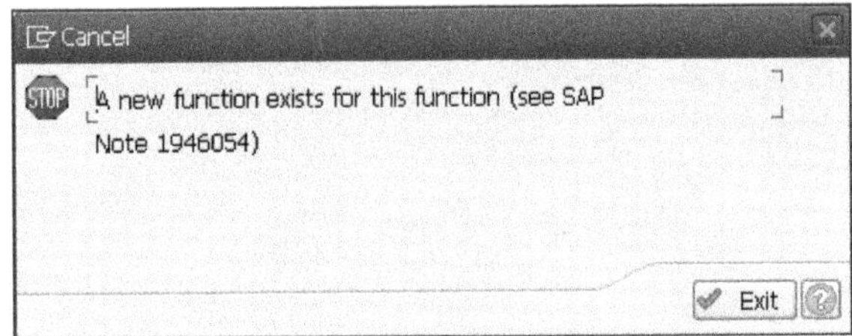

Figure 12.13 Credit Management Error in Transaction FD32

The advantages of the new SAP S/4HANA credit management approach can be summed up in the following ways:

- Higher degree of automation with automated calculation of credit scores
- Documented credit decisions
- Inclusion of the relationships between business partners in the calculation of credit limits
- Improved performance through the elimination of redundant data
- Standardized integration with external credit agencies

The additional options for the automatic calculation of credit limits and of risk class as well as internal scoring can be provided by credit rules engine, but these options require an additional license to be purchased. Furthermore, a standard cloud-based link to external credit agencies, such as Dun & Bradstreet, is provided out of the box but also requires additional licensing.

The activation of credit management can be performed, as with SAP ERP 6.0, for sales document types and item categories. These activations can be accessed via the following IMG paths:

- **Financial Supply Chain Management • Credit Management • Integration with Accounts Receivable Accounting and Sales and Distribution • Integration with Sales and Distribution • Determine Active Receivables Per Item Category**
- **Financial Supply Chain Management • Credit Management • Integration with Accounts Receivable Accounting and Sales and Distribution • Integration with Sales and Distribution • Assign Sales Documents and Delivery Documents**

All customer settings previously housed in the SAP ERP 6.0 Transaction FD32 are now housed in the business partner master data, by credit segment. A *credit segment* is an organizational unit used by the system to determine the credit limit for a customer. A credit segment is assigned to a credit control area, which itself is assigned to a sales area. Credit segment data can be viewed in Transaction BP when viewing the role **UKM000 SAP Credit Management**, as shown in Figure 12.14.

Within this view, the credit limit can be set. Note that the credit limit does not need to be a static figure. This limit can be defined manually in the business partner or can be calculated according to a defined rule in customizing. This rule is set by default according to the customizing setting maintained via the IMG menu path **Financial Supply Chain Management • Credit Management • Credit Risk Monitoring • Master Data • Create Rule for Scoring and Credit Limit Calculation**, as shown in Figure 12.15.

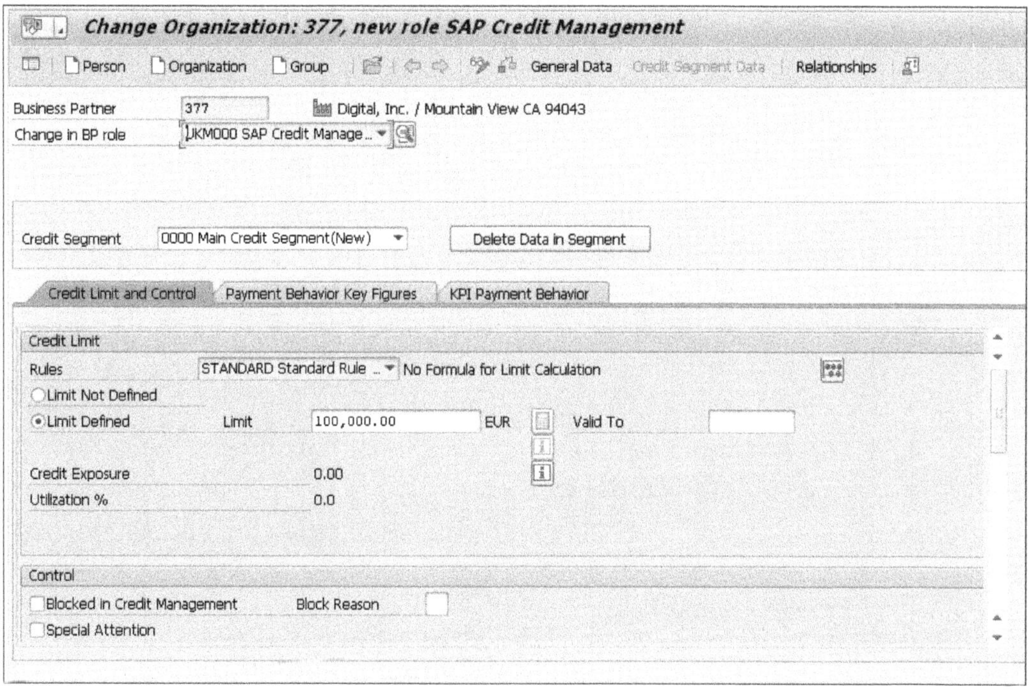

Figure 12.14 Credit Management Role in Business Partner

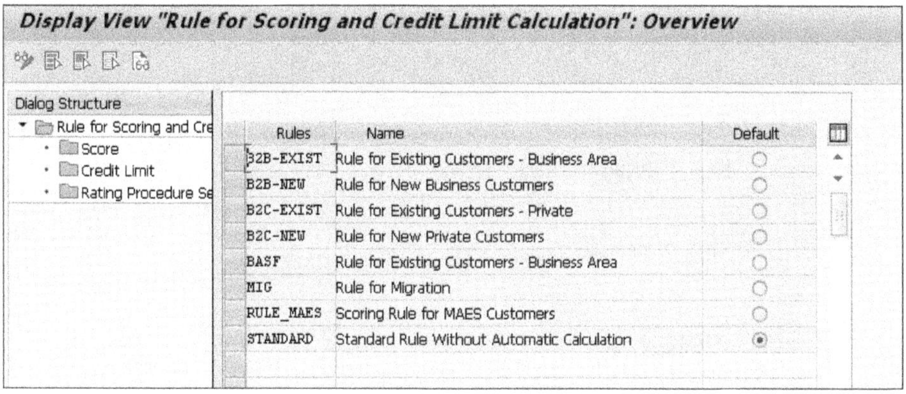

Figure 12.15 Rule for Scoring and Credit Limit Calculation

The credit limit rule is set in the **Credit Limit** folder by adding a predefined formula. The formula can be set in customizing too, via the same menu path but under **Define Formulas**.

12.7 Foreign Trade Amendments

Foreign trade in SAP S/4HANA has been moved out of the core sales functionality and is now located in the SAP GTS system. The traditional SAP ERP SD foreign trade functionality is no longer available for use—the functionality can only be found in SAP GTS and as such does not form a part of the SAP S/4HANA Sales suite of functions.

Nevertheless, SAP S/4HANA for international trade is available and includes some of the functionality associated with foreign trade and export services. For example, SAP S/4HANA for international trade allows you to perform the following activities:

- Set classifications for materials, such as commodity codes, intrastat service codes, and custom tariff numbers
- Control licenses for export purposes, including managing and releasing blocked documents
- Block sales documents for countries under embargo
- Manage intrastat declaration
- Integrate with SAP Watch List Screening

Note

Note that the maintenance of SAP S/4HANA for international trade is carried out exclusively in SAP Fiori, not in the backend SAP GUI.

The IMG path for SAP S/4HANA for international trade is **Governance, Risk and Compliance • International Trade**.

Let's deal with the sales-relevant elements of SAP S/4HANA for international trade one by one. SAP S/4HANA for international trade itself is a function in its own right in SAP S/4HANA, but the certification exam will require you to understand these changes as they relate to sales functions.

We won't cover the management of intrastat declarations since this task is a finance task rather than a sales task and will not be covered by the certification exam.

12.7.1 Set Classifications for Materials

Numbering schemes for commodity codes and intrastat service codes can be set up via the IMG menu path **Governance, Risk and Compliance • International Trade • Classification**.

Importantly, note that you cannot assign commodity codes to a material in the material master, as was the case in SAP ERP 6.0. Specific SAP Fiori apps are designed for this purpose. Similarly, for intrastat service codes, a suite of SAP Fiori apps available for updating these codes. These apps come in three flavors:

1. "Manage" type apps, such as Manage Commodity Codes and Manage Intrastat Service Codes. These apps allow you to create new codes with validity dates and other attributes such as units of measurement (UoMs) and languages.
2. "Classify" type apps, such as Classify Products – Commodity Codes and Classify Products – Intrastat Service Codes. These apps allow you to assign a specific material to a specific commodity code or intrastat service code.
3. "Reclassify" type apps, such as Reclassify Products – Commodity Codes and Reclassify Products – Intrastat Service Codes. These apps allow you to amend the assignment of commodity codes and intrastat service codes to materials.

12.7.2 Control Licenses and Embargos for Export Purposes

The control of the sale of goods under license was carried out in SAP ERP 6.0 in the Legal Control component. In SAP S/4HANA, this functionality has been made redundant and replaced by SAP S/4HANA for international trade. Note that the layout and operation of this functionality is quite different.

License control in SAP S/4HANA for international trade is a part of the trade compliance check functionality, and as such, the configuration must be activated, if required, by sales order type, delivery type, purchase order (PO) type, and company code. The customizing menu path is **Governance, Risk and Compliance • International Trade • General Settings • Trade Compliance**.

Additionally, via this IMG path, you can deactivate certain additional functions if the documents have been blocked by trade compliance (legal control, embargo, or SAP Watch List Screening). Figure 12.16 shows examples where, for legal control, documents that are blocked or have a decision pending will not be able to create purchase requisitions, outbound deliveries, goods issues, or outputs.

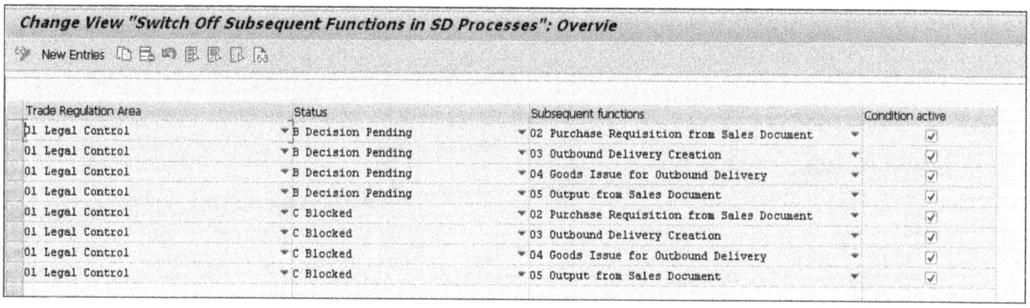

Change View "Switch Off Subsequent Functions in SD Processes": Overvie

New Entries

Trade Regulation Area	Status	Subsequent functions	Condition active
01 Legal Control	B Decision Pending	02 Purchase Requisition from Sales Document	☑
01 Legal Control	B Decision Pending	03 Outbound Delivery Creation	☑
01 Legal Control	B Decision Pending	04 Goods Issue for Outbound Delivery	☑
01 Legal Control	B Decision Pending	05 Output from Sales Document	☑
01 Legal Control	C Blocked	02 Purchase Requisition from Sales Document	☑
01 Legal Control	C Blocked	03 Outbound Delivery Creation	☑
01 Legal Control	C Blocked	04 Goods Issue for Outbound Delivery	☑
01 Legal Control	C Blocked	05 Output from Sales Document	☑

Figure 12.16 Trade Compliance Deactivation of Subsequent Functions

Once trade compliance is activated, you can continue to configure the setup of your licenses and embargos.

To set up legal control, you'll need to perform the following steps:

1. Create a partner role and assign the partner function you wish to use in legal control and embargos.

2. Create numbering schemes for commodity codes, intrastat service codes, and control classes. A control class is effectively an Export Control Classification Number (ECCN) for export control and has materials assigned to it. Numbering scheme content should then be created and assigned to the numbering scheme.
3. Create legal regulations, which can be a type of control required for specific product lines or could be an embargo on specific countries.
4. Create license types.

Once your customizing is completed, a suite of SAP Fiori apps related to legal control is available:

- Manage Licenses
- Classify Products – Legal Control
- Reclassify Products – Legal Control
- Manage Countries Under Embargo
- Resolve Blocked Documents – Trade Compliance

12.7.3 Integration with SAP Watch List Screening

SAP Watch List Screening is an external software as a service (SaaS) solution from SAP that is deployed in the cloud. This software offers real-time compliance checks for customers at the point of ordering. With a prebuilt integration with SAP S/4HANA, this solution can be switched on with an additional license. Once purchased, the SaaS solution can be activated in the IMG via **Governance, Risk and Compliance • International Trade • Integration with SAP Watch List Screening, • Activate Legal Regulations.**

12.8 Sales Data Model Changes

Due to the power of the SAP S/4HANA and its underlying SAP HANA database, significant changes have been made in the data model for sales and distribution, as is true for all modules, with the elimination of dozens of tables in favor of a cleaner, simpler end-to-end model.

Most notably for sales, index tables have been made redundant, eliminating the six document index tables VAKPA, VAPMA, VLKPA, VLPMA, VRKPA, and VRPMA as well as the rebate index table VBOX, which had caused many performance issues.

The status tables VBUK and VBUP have also been eliminated, and their contents moved into the standard sales header and item tables VBAK and VBAP.

Note

The status tables VBUK and VBUP still exist in SAP S/4HANA but are not populated by default. The reason they are retained is to ensure smooth conversion from SAP ERP 6.0 to

SAP S/4HANA. To enable the populating of tables VBUK and VBUP, SAP has provided function modules (such as SD_VBUK_READ_FROM_DOC for table VBUK), which can be used to populate these tables automatically.

The reasoning behind this change is simple: improved performance and smaller memory data footprint. Other tables have been retired too, relating to the following changes in the sales model in SAP S/4HANA:

- Export/import tables EIKP and EIPO, which are foreign trade tables, have been retired, along with the foreign trade functionality, into SAP GTS.
- Tables such as tables S066 and S067 have been retired since the Logistics Information System (LIS) is no longer supported by SAP S/4HANA. The embedded analytics functions and real-time reporting capabilities through core data services (CDS) views have made LIS redundant.
- The rebate table VBOX is no longer required since SD rebates have been replaced by settlement management in SAP S/4HANA.
- The output table NAST, while not redundant, is not essential to the SAP S/4HANA Sales landscape since the default output functionality is provided by BRFplus.

An additional important data model change is the extension of the **SD document category** field from 1 character to 4 characters—again adding greater flexibility.

The different overall data model between SAP ERP 6.0 Sales and Distribution (SD) and SAP S/4HANA Sales is shown in Figure 12.17, compared to the overall model shown in Figure 12.18.

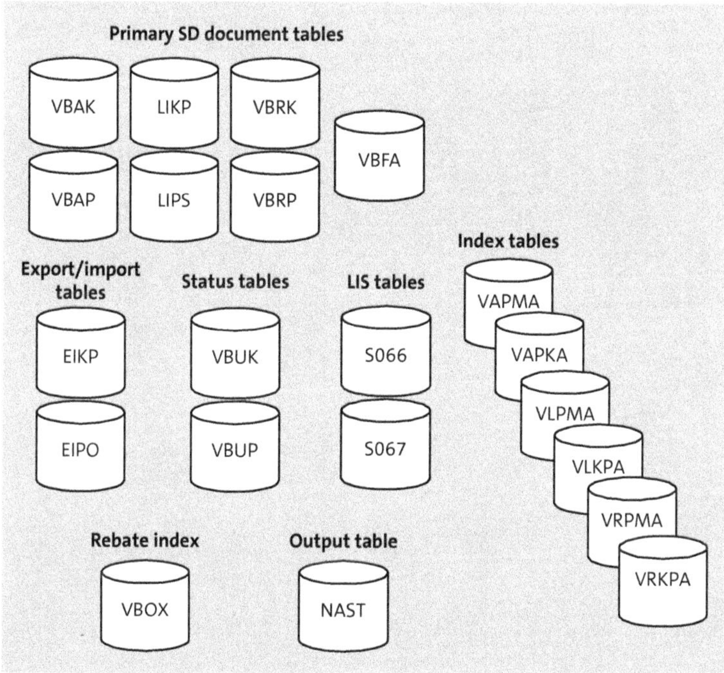

Figure 12.17 Sales and Distribution Data Model in SAP ERP 6.0

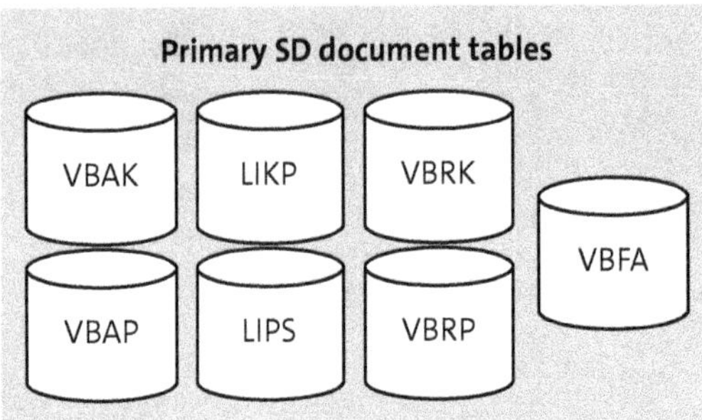

Figure 12.18 Sales Data Model in SAP S/4HANA

The new sales pricing table (table PRCD_ELEMENTS) houses pricing conditions and replaces the old cluster table KONV. As table PRCD_ELEMENTS is a transparent table, using this table in SELECT statements is much easier than with table KONV. Many fields in table PRCD_ELEMENTS have had their lengths extended, allowing for more flexible and powerful customizing in pricing. Figure 12.19 shows table KONV in SAP S/4HANA, with a description denoting that this table is obsolete and replaced by table PRCD_ELEMENTS, as shown in Figure 12.20. Both tables have highlighted fields showing the difference in length between the same fields of the two tables.

Dictionary: Display Table

Technical Settings Append Structure...

Transparent Table: KONV Active

Short Description: Conditions (Obsolete - replaced by PRCD_ELEMENTS)

Attributes | Delivery and Maintenance | Fields | Input Help/Check | Currency/Quantity Fields | Indexes

Search | Built-In Type | 1 / 70

Field	Key	Ini...	Data element	Data Type	Length	Deci...	Coordinate	Short Description	Group
MANDT	✓	✓	MANDT	CLNT	3	0	0	Client	
KNUMV	✓	✓	KNUMV	CHAR	10	0	0	Number of the Document Condition	
KPOSN	✓	✓	KPOSN	NUMC	6	0	0	Condition item number	
STUNR	✓	✓	STUNR	NUMC	3	0	0	Step Number	
ZAEHK	✓	✓	DZAEHK	NUMC	3	0	0	Condition Counter	
KAPPL			KAPPL	CHAR	2	0	0	Application	
KSCHL			KSCHA	CHAR	4	0	0	Condition type	
KDATU			KDATU	DATS	8	0	0	Condition pricing date	
KRECH			KRECH	CHAR	1	0	0	Calculation Type for Condition	
KAWRT			KAWRT	CURR	15	2	0	Condition Base Value	
KBETR			KBETR	CURR	11	2	0	Condition Amount or Percentage	
WAERS			WAERS	CUKY	5	0	0	Currency Key	
KKURS			KKURS	DEC	9	5	0	Condition exchange rate for conversion to local currency	
KPEIN			KPEIN	DEC	5	0	0	Condition Pricing Unit	
KMEIN			KVMEI	UNIT	3	0	0	Condition Unit in the Document	
KUMZA			KUMZA	DEC	5	0	0	Numerator for converting condition units to base units	
KUMNE			KUMNE	DEC	5	0	0	Denominator for converting condition units to base units	
KNTYP			KNTYP	CHAR	1	0	0	Condition Category (Examples: Tax, Freight, Price, Cost)	

Figure 12.19 Table KONV in SAP S/4HANA

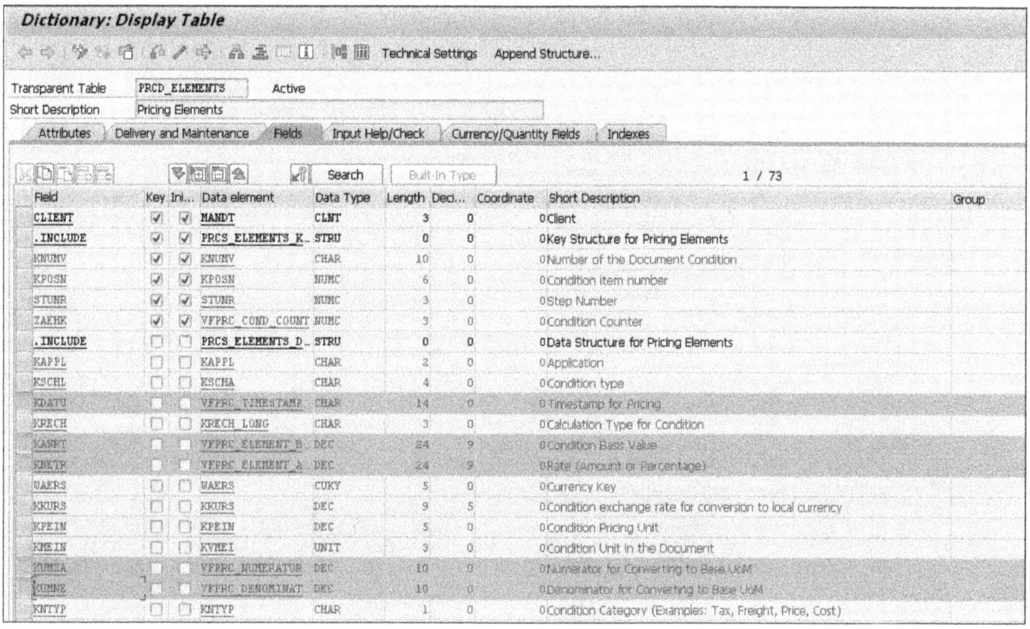

Dictionary: Display Table

Technical Settings Append Structure...

Transparent Table PRCD_ELEMENTS Active

Short Description Pricing Elements

Attributes | Delivery and Maintenance | Fields | Input Help/Check | Currency/Quantity Fields | Indexes

Search | Built-In Type | 1 / 73

Field	Key	Ini...	Data element	Data Type	Length	Deci...	Coordinate	Short Description	Group
CLIENT	☑	☑	MANDT	CLNT	3	0	0	Client	
.INCLUDE	☑	☑	PRCS_ELEMENTS_K…	STRU	0	0	0	Key Structure for Pricing Elements	
KNUMV	☑	☑	KNUMV	CHAR	10	0	0	Number of the Document Condition	
KPOSN	☑	☑	KPOSN	NUMC	6	0	0	Condition item number	
STUNR	☑	☑	STUNR	NUMC	3	0	0	Step Number	
ZAEHK	☑	☑	VFPRC_COND_COUNT	NUMC	3	0	0	Condition Counter	
.INCLUDE	☐	☐	PRCS_ELEMENTS_D…	STRU	0	0	0	Data Structure for Pricing Elements	
KAPPL	☐	☐	KAPPL	CHAR	2	0	0	Application	
KSCHL	☐	☐	KSCHA	CHAR	4	0	0	Condition type	
KDATU	☐	☐	VFPRC_TIMESTAMP	CHAR	14	0	0	Timestamp for Pricing	
KRECH	☐	☐	KRECH_LONG	CHAR	3	0	0	Calculation Type for Condition	
KAWRT	☐	☐	VFPRC_ELEMENT_B	DEC	24	9	0	Condition Basis Value	
KBETR	☐	☐	VFPRC_ELEMENT_A	DEC	24	9	0	Rate (Amount or Percentage)	
WAERS	☐	☐	WAERS	CUKY	5	0	0	Currency Key	
KKURS	☐	☐	KKURS	DEC	9	5	0	Condition exchange rate for conversion to local currency	
KPEIN	☐	☐	KPEIN	DEC	5	0	0	Condition Pricing Unit	
KMEIN	☐	☐	KVMEI	UNIT	3	0	0	Condition Unit in the Document	
KUMZA	☐	☐	VFPRC_NUMERATOR	DEC	10	0	0	Numerator for Converting to Base UoM	
KUMNE	☐	☐	VFPRC_DENOMINAT	DEC	10	0	0	Denominator for Converting to Base UoM	
KNTYP	☐	☐	KNTYP	CHAR	1	0	0	Condition Category (Examples: Tax, Freight, Price, Cost)	

Figure 12.20 Table PRCD_ELEMENTS in SAP S/4HANA

12.9 Important Terminology

In this chapter, the following terminology was used:

- **Business partners**
  In SAP S/4HANA, business partners are the default SAP S/4HANA approach to dealing with customers, vendors, and contacts. The use of business partners brings the SAP S/4HANA into line with other platforms in the SAP architecture, such as SAP CRM.
- **Business Rules Framework plus (BRFplus)**
  SAP ERP 6.0 SD output management using the NAST methodology has been deprecated in favor of using BRFplus by default to manage outputs. The NAST approach is still available in the system.
- **Condition contracts**
  Condition contracts are a part of settlement management. These condition contracts store all key information related to a specific agreement, such as customer data, prices, and validity dates.
- **Credit management**
  In SAP S/4HANA, credit management is housed within the **Financial Supply Chain Management** IMG menu path. These settings differ from traditional SAP ERP 6.0 credit management settings, allowing much more flexibility and functionality.

- **Customer-vendor integration (CVI)**
  Settings for business partner population of customer and vendor tables are controlled via CVI. Additionally, this approach allows for smooth conversion activities from SAP ERP 6.0 to SAP S/4HANA.
- **Decision table**
  A decision table is a list of business rules in a specific order that determine a given outcome within BRFplus. In SAP S/4HANA, this approach is the default approach for generating sales outputs.
- **Embargo control**
  Within SAP S/4HANA for international trade, you can embargo certain countries from a sales and distribution perspective. This activity can be carried out as part of your trade compliance approach.
- **Foreign trade**
  Foreign trade was a part of the SAP ERP 6.0 SD configuration and embedded into sales orders and deliveries. This functionality has been removed in SAP S/4HANA and replaced by SAP Global Trade Services (SAP GTS). Some of the functionality now appears within SAP S/4HANA for international trade.
- **SAP Global Trade Services (SAP GTS)**
  An external but integrated SAP application that supports import and export trade processes within the SAP S/4HANA system.
- **SAP HANA database**
  SAP's state-of-the-art, next-generation database, combining in-memory data with column-oriented storage. SAP S/4HANA can only use SAP HANA as its database.
- **SAP S/4HANA for international trade**
  SAP S/4HANA for international trade supports international trade compliance around commodity codes, export licenses, embargoes, intrastat declarations, and watch list screening.
- **Legal control**
  Legal control is the process by which you can restrict the ordering and shipping of goods to customers that hold export licenses.
- **Sales data model**
  The SAP S/4HANA data model for sales has changed drastically from SAP ERP 6.0. Tables have been deprecated, field definitions have been changed, and new tables have been created. These changes take advantage of the additional processing power of the SAP HANA database and simplify the overall underlying architecture of the sales process.
- **Settlement management**
  The settlement management functionality in SAP S/4HANA replaces the old SAP ERP 6.0 rebate functionality and is the powerful way to capture rebates and agreements with both customers and vendors. This functionality is much more flexible and operates using the advanced processing power of SAP HANA, resulting in a significant improvement in performance.

- **Simplification list**
  A document that details the improvements and additional capabilities in the most recent version of SAP S/4HANA.
- **SAP Watch List Screening**
  SAP Watch List Screening is an external cloud solution, subject to an additional license, which can carry out automated complex compliance checks for customers at ordering time.

12.10 Practice Questions

These practice questions will help you evaluate your understanding of the topics covered in this chapter. The questions shown are similar in nature to those found on the certification examination. Although none of these questions will be found on the exam itself, they will allow you to review your knowledge of the subject. Select the correct answers and then check the completeness of your answers in the next section. Remember that, on the exam, you must select all correct answers and only correct answers to receive credit for the question.

1. Which of the following is the purpose of the simplification list?

☐ A. To show a list of transactions that are no longer used

☐ B. To show a list of tasks needed for conversion from SAP ERP 6.0 to SAP S/4HANA

☐ C. To show a list of technical and business impacts that can be expected with an upgrade to that specific version of SAP S/4HANA

☐ D. To run a consistency check to run against your existing system to see what needs to change before upgrading to SAP S/4HANA.

2. Which of the following are simplification items that should be considered when moving to an SAP S/4HANA environment from an SAP ERP 6.0 environment? (There are three correct answers.)

☐ A. The SAP ERP SD foreign trade functionality has now been absorbed into the Materials Management (MM) module.

☐ B. SD rebates have been replaced by settlement management.

☐ C. Output management in SAP S/4HANA Sales is performed, by default, via BRFplus instead of the previous NAST-based approach.

☐ D. The ATP check has been enriched in SAP S/4HANA to include SAP S/4HANA for advanced ATP capabilities.

☐ E. The standard sales tables VBAK and VBAP do not need to be used anymore in SAP S/4HANA.

3. Your client asks you why they should implement business partners in their SAP S/4HANA project. Which of the following statements would you mention? (There are three correct answers.)

☐ A. Business partners are mandatory in SAP S/4HANA, and customers can no longer be created in transactions such as Transaction XD01.

☐ B. The business partner approach means we no longer need to think about partner functions such as sold-to party, ship-to party, bill-to party, and payer.

☐ C. The unified architecture of business partners means one data object can serve multiple roles.

☐ D. The business partner approach means that SAP S/4HANA conforms to the same architecture as other SAP products in the landscape.

☐ E. Every SAP user in the system has their own business partner record, thus improving transparency and visibility.

4. Which of the following statements is true about traditional customer master transactions, such as Transactions XD01 and VD01, in SAP S/4HANA:

☐ A. Attempting to use the transactions will redirect you to Transaction BP.

☐ B. Transactions are still available but are not recommended for use.

☐ C. Transactions have disappeared entirely and will not be recognized if used.

5. The CVI functionality can be used to perform which of the following activities?

☐ A. Set up the system so that all business partner data is stored in the BUT* tables without the need to copy to tables KNA1 and KNVV

☐ B. Ensure that, once business partners are created, the data is filtered down to the customer and vendor tables

☐ C. Set up a connection between customers and vendors so that any data added to the vendor master data object can be selectively copied to the customer master data object

6. When converting from SAP ERP 6.0 to SAP S/4HANA, several steps are required in order to convert customers and vendors to business partners. Which of the following steps are required? (There are two correct answers.)

☐ A. Implement CVI in your SAP ERP 6.0 system

☐ B. Stop all users from creating customers using Transaction XD01 in your SAP ERP 6.0 system using security authorizations

☐ C. Convert existing customers and vendors to business partners in your SAP ERP 6.0 system

☐ D. Create a Legacy System Migration Workbench (LSMW) migration script to convert the customers and vendors in your SAP ERP 6.0 system to business partners in your SAP S/4HANA system

7. Settlement management in SAP S/4HANA replaces SD rebates. What does this mean for the sales data model?

☐ A. The SD rebate type field has been extended from 1 character to 3 characters.

☐ B. Table VBOX is now redundant.

☐ C. Table VBOX can still be used but must be manually activated.

8. The BRFplus approach to outputs does come with a few limitations. Which of these limitations are true? (There are two correct answers.)

☐ A. Only business partners can be used for communication purposes.

☐ B. IDocs cannot be used at all by BRFplus.

☐ C. ALE is not supported by BRFplus.

☐ D. BRFplus is only available for sales outputs, and therefore, outputs for POs cannot be used.

9. True or False: The BRFplus approach to SD output management is mandatory in SAP S/4HANA.

☐ A. True

☐ B. False

10. Which of the following are the attributes of BRFplus decision tables? (There are two correct answers.)

☐ A. Multiple condition columns are allowed but only one result column.

☐ B. Rows in the decision table are read sequentially from top to bottom.

☐ C. When setting up the decision table, each step must be activated before it can be used.

☐ D. Master data value updates in the decision tables in BRFplus must always be transported through the development system to the production landscape.

11. Which of the following attributes are true for the SAP S/4HANA credit management approach to customer credit limits? (There are three correct answers.)

☐ A. Credit limits can be calculated using a predefined and customizable formula.

☐ B. Credit limits for customers are maintained using Transaction FD32.

☐ C. Customer credit limits can be calculated dynamically or set as a static value.

☐ D. Customer credit limits can be activated by sales document type and/or by item category.

☐ E. By default, customer credit limits are set by the integration with the Credit Rules Engine.

12. Settlement management has replaced the old SD rebates functionality in SAP S/4HANA, and rebates are now more flexible due to some of these changes. Regarding the difference between sales and materials management condition contracts in SAP S/4HANA, which of the following statements is true?

- ☐ A. Sales and materials management rebates can now be defined using the same functionality.
- ☐ B. Materials management rebates must now be defined and maintained in SAP Ariba.
- ☐ C. Rebates are processed in SAP Fiori and cannot be maintained in SAP GUI.

13. Where is the data formerly housed in the SAP ERP SD foreign trade functionality now found?

- ☐ A. Smart business embedded analytics apps in SAP Fiori
- ☐ B. SAP GTS
- ☐ C. SAP Analytics Cloud

14. SAP S/4HANA for international trade provides which of the following functions as standard, with no need for any additional licensing? (There are three correct answers.)

- ☐ A. SAP Watch List Screening
- ☐ B. Material classifications, such as commodity codes and intrastat service codes
- ☐ C. License control and embargoes
- ☐ D. Standard airway bill forms
- ☐ E. Intrastat declarations

15. The sales data model in SAP S/4HANA has been simplified in which of the following ways? (There are two correct answers.)

- ☐ A. Index tables are no longer needed due to the advanced processing power of the SAP HANA database.
- ☐ B. Index tables are no longer in the sales area but have been moved into SAP BTP to optimize performance.
- ☐ C. The status tables VBUK and VBUP have been moved into tables VBAK and VBAP.
- ☐ D. The status tables VBUK and VBUP do not exist in SAP S/4HANA.

16. The new pricing table PRCD_ELEMENTS has replaced table KONV in SAP S/4HANA, thus providing greater flexibility which of the following ways? (There are two correct answers.)

- ☐ A. All pricing master data previously found in the "A" prefixed tables is now found in table PRCD_ELEMENTS.
- ☐ B. Table PRCD_ELEMENTS now allows pricing to be passed freely to the business warehouse without any additional coding required.
- ☐ C. Table PRCD_ELEMENTS is a transparent table rather than a cluster table.
- ☐ D. Field lengths for key fields in the table have been lengthened.

17. A conversion from SAP ERP 6.0 to SAP S/4HANA requires which of the following steps to be undertaken? (There are three correct answers.)

- ☐ A. During the implementation project, customers in the SAP ERP 6.0 system must be set up as business partners using Transaction BP.
- ☐ B. You must use BRFplus as the new output management approach in SAP S/4HANA.
- ☐ C. Any SD rebates must be created in SAP S/4HANA as condition contracts.
- ☐ D. Customers and vendors must be converted to business partners.
- ☐ E. Credit management functionality must be moved to the financial supply chain management model.

12.11 Practice Question Answers and Explanations

1. Correct answer: **C**
   The simplification list is designed to identify specific impacts when upgrading to that version of SAP S/4HANA. These impacts can be business or technical, both of which are highlighted within the list. The simplification list is not designed to give you advice on conversion activities or how to run consistency checks for the purposes of conversion, although these tasks are mentioned where appropriate. Furthermore, the simplification list will not show you the transactions that have been made obsolete in SAP S/4HANA. Information is available in the SAP library for these requirements but is not included in the simplification list.

2. Correct answers: **B, C, and D**
   Understanding simplification items is important. From the list given, the only correct answers are that SD rebates have been replaced by settlement management, output management is now handled by default by BRFplus, and the ATP check can now include advanced ATP. The answer options illustrate the need to read a question thoroughly before answering. For example, answer E mentions the standard tables VBAK and VBAP as not being used anymore—this statement

is correct if we're talking about tables VBUK and VBUP, but not tables VBAK and VBAP!

3. Correct answers: **A, C, and D**
The business partner methodology is the only way to create customers and vendors in SAP S/4HANA. This change brings SAP S/4HANA in line with other SAP applications such as SAP CRM, that already use business partners. Note that some of the functionality around customers remains the same—such as the concepts of partner functions like sold-to, ship-to, etc. Nevertheless, the use of business partners allows a single data object to take on multiple roles in the system. SAP user records are a completely separate data object in SAP S/4HANA.

4. Correct answer: **A**
Obsolete transactions related to customer and vendor master records will now redirect you to the unified Transaction BP via a dialog message. As such, the transactions still exist, but only as a pointer to the correct transaction.

5. Correct answer: **B**
CVI exists so that there is a permanent link between a business partner and the customer or vendor tables, such as tables KNA1 and LFA1. Once the SAP S/4HANA system is established, business partner data is saved in the BUT* tables, but if the business partner is extended to the role "customer" then the data will be copied to all the customer master tables, such as tables KNA1 and KNVV.

6. Correct answers: **A and C**
Conversion from SAP ERP 6.0 to SAP S/4HANA involves some preparation work in both the source and the target system. In your SAP ERP 6.0 system, you must activate CVI in the direction of customer to business partner and convert all your customers and vendors to business partners. Your users can continue to create customers in the normal way using Transaction XD01.When you're ready to execute the conversion scripts, your SAP ERP 6.0 system will have business partners already created and ready for conversion.

7. Correct answer: **B**
Within the SD rebates functionality, table VBOX was a real problem for performance, with hundreds of thousands (and in many cases, millions) of data records housed in the table. Due to the power of the SAP HANA database, such index tables are no longer required, and table VBOX is therefore completely redundant in SAP S/4HANA.

8. Correct answers: **A and C**
Although the BRFplus functionality is quite powerful and flexible, its scope is limited by a number of factors. One factor is the requirement that outputs can only be communicated with business partners (meaning logical systems cannot be used), and another is that ALE system-to-system connections are not supported by BRFplus. IDocs can be used by BRFplus within these limitations, and outputs for BRFplus can be used for any business scenarios that were previously supported by NAST, including POs.

9. Correct answer: **B**

   False. While it is true that the BRFplus approach to sales and distribution outputs is the default approach in SAP S/4HANA, this functionality can be switched off, meaning that sales and distribution outputs can revert back to the NAST methodology.

10. Correct answers: **B and C**

    BRFplus decision tables act a lot like condition tables for NAST-based outputs, but with a few differences. Similar to NAST condition tables, BRFplus decision tables are read sequentially from top to bottom when the system attempts to determine a result. When adding data in the decision table, each step must be activated before it can be used. The table is made up of multiple condition columns, but unlike condition tables, it is also made up of multiple results columns, allowing full flexibility. Updates of the master data in the decision tables are carried out in Transaction OPD and do not need to be transported.

11. Correct answers: **A, C, and D**

    Credit limits in SAP S/4HANA financial supply chain management can be managed in a number of different ways. The first is to set a static credit limit against the business partner master record. However, you can also direct the system to calculate the credit limit from a customizable formula. The determination of the method used is carried out in the business partner transaction, Transaction BP. (Transaction FD32 is obsolete.) As with SAP ERP 6.0, credit limits in SAP S/4HANA can be activated by sales document type and by item category in the IMG. Additional options are available for automatic calculation of credit limits, calculations of risk class, and internal scoring, which are available as part of the Credit Rules Engine, but this solution is an additional licensed option and not a standard feature.

12. Correct answer: **A**

    The beauty of settlement management in SAP S/4HANA is that it harmonizes both sales and materials management functionality in the same area. As a result, condition contracts can be set up in the core system of SAP S/4HANA for customers and vendors alike. SAP Fiori apps are available for creating condition contracts, as are SAP GUI transactions (Transaction WCOCO).

13. Correct answer: **B**

    The old concept of data in the SAP ERP SD foreign trade functionality is obsolete in SAP S/4HANA. While some of this data has been moved to SAP S/4HANA for international trade, the recommended approach for SAP ERP SD foreign trade functions is to use SAP GTS.

14. Correct answers: **B, C, and E**

    Trade compliance functions are available in the core SAP S/4HANA system by using SAP S/4HANA for international trade functions. These functions cover material classifications, such as the allocation of commodity codes and intrastat service codes; legal control through the definition and maintenance of export licenses; restrictions on exports to countries under embargo; and intrastat declarations. The functionality also includes denied party screening through SAP Watch List Screening, but this capability requires an additional license.

15. Correct answers: **A and C**

    The power of the SAP HANA database has allowed some essential streamlining of the overall data model in SAP S/4HANA Sales. The key tables that have been rendered obsolete are the index tables and the status tables. Index tables have been completely removed and are not required anywhere in the SAP S/4HANA landscape. For status tables, such as tables VBUK and VBUP, these tables still exist in the system, but the status fields have been replicated in tables VBAK and VBAP for ease of use. It is important to note that tables VBUK and VBUP are still available and can be populated by activating standard function modules in the system.

16. Correct answers: **C and D**

    Table KONV in SAP ERP 6.0 was a cluster table, which meant that it was always difficult to select from using joins. Furthermore, data was restricted due to field length in the table. SAP S/4HANA has addressed both of these concerns by creating table PRCD_ELEMENTS, which is a transparent table rather than a cluster table, increasing ease of use. Additionally, the field lengths for some key fields in the table have been extended, meaning additional functionality has been opened for use. All pricing tables prefixed with "A" remain the same as in SAP ERP 6.0.

17. Correct answers: **C, D, and E**

    The conversion process from SAP ERP 6.0 to SAP S/4HANA requires that the CVI process for business partners is set up in the source system and that conversion from customers and vendors takes place in the source system. However, the business practice of creating customers and vendors in the traditional Transactions XD01 and XK01 can continue in the source system until the target SAP S/4HANA data migration is completed. The output management option using BRFplus is optional, not mandatory, because the NAST approach is still available for use in SAP S/4HANA. Rebates and credit management must change, however, with rebates moving to settlement management because condition contract and credit management functions have moved to the financial supply chain management module.

12.12 Takeaway

In this chapter, you learned about the concept of the simplification list and how it can provide an overview of areas that have changed in the most recent release. You now understand that upgrading an SAP system requires an appreciation of the goal of the simplification list and an understanding of the items therein. You're familiar, from this chapter and others, with the simplification items for SAP S/4HANA Sales, which differentiate this solution from SAP ERP 6.0, in the following ways:

- Business partner concept
- Settlement management instead of rebate processing
- BRFplus functionality for outputs
- Credit management amendments
- SAP S/4HANA for advanced ATP
- Foreign trade amendments
- SAP S/4HANA embedded analytics

You can now explain the rationale behind these simplifications and also how they are customized and used within SAP S/4HANA.

You now understand the various options available for an SAP S/4HANA implementation. Clearly, a system conversion from SAP ERP 6.0 to SAP S/4HANA is the option that will bring with it the most historical data to the new system. In this chapter, we discussed certain steps that must be undertaken with regard to the simplification items to ensure a smooth and effective conversion from SAP ERP 6.0 to SAP S/4HANA. These steps can be summarized in the following way:

1. Carry out the conversion steps in both your legacy and target systems to convert your customer and vendors to business partners.
2. If SD rebates are active in your legacy SAP ERP 6.0 system, these rebates must be created as condition contracts in SAP S/4HANA.
3. Your credit management settings should be updated to the new financial supply chain management settings.
4. You should decide whether to use BRFplus or traditional NAST-based outputs.
5. Data from the SAP ERP SD foreign trade functionality should be rehoused in SAP GTS, with some trade compliance settings configured in SAP S/4HANA for international trade.
6. You may need to carry out some custom code changes to take into consideration the new sales data model.
7. The use of SAP S/4HANA Sales is made much more efficient by the power of the SAP HANA database, as described in Chapter 1, Section 1.3. You can now explain how SAP S/4HANA Sales benefits from this additional processing power, allowing the system to simplify its sales data model, making some tables obsolete, and extending the functionality of other tables. This simplification has not only made the overall model easier to understand; it has also improved the flexibility of the system.

12.13 Summary

This chapter has focused heavily on additional SAP S/4HANA concepts that are now available to use as part of the standard sales functionality. A good understanding of these concepts and how they fit into the overall sales implementation approach is critical to becoming a certified SAP S/4HANA Sales consultant. Although this section represents a small part of the overall exam coverage, the contents contained in this chapter are vital to your understanding—you need, for example, to know that SD rebates and the SAP ERP SD foreign trade functionality are no longer supported in SAP S/4HANA in order to be an accomplished and trusted advisor for your clients.

In the next chapter, we'll dive more deeply into the leading-edge technologies available to you with your SAP S/4HANA system. These technologies include the SAP Business Technology Platform (SAP BTP) suite of applications in addition to add-ons such as SAP CoPilot.

Chapter 13
Next-Generation Technologies

Techniques You'll Master

- Understand the components of the next-generation suite of technologies as well as the SAP S/4HANA approach to innovation
- Describe accurately the phases of the SAP Activate implementation methodology and how they are utilized for fast and accurate deployments
- Define the intelligent enterprise and what it means for SAP S/4HANA customers as part of SAP Business Technology Platform (SAP BTP)
- Understand RISE with SAP and how it can help accelerate SAP migrations to SAP S/4HANA
- Understand additional add-on technologies, such as SAP CoPilot and the advanced user experience (UX), and how they impact your SAP S/4HANA implementation

In this chapter, we'll explore the suite of next-generation technologies that are available to SAP customers as part of their SAP S/4HANA deployments. These technologies can be implemented using the SAP Activate methodology, which was introduced in Chapter 1, Section 1.5. We'll discuss the various phases included in the SAP Activate approach and how this methodology is structured in order to streamline implementation projects. We'll also look at the new technologies that were formerly grouped together under the "SAP Leonardo" banner (although that product name has been retired) to give you insight into how these technologies can be used within the framework of what SAP refers to as the *intelligent enterprise*.

Real-World Scenario

When clients upgrade from SAP ERP 6.0, they are often unaware of the additional technologies that become available to them as part of their new SAP S/4HANA system. As an SAP S/4HANA Sales consultant, your task is to ensure that not only are they aware of the technologies, but they also know the best way to implement them. Your client will want to know which technologies are appropriate for their business and how they fit into the intelligent enterprise.

13.1 Objectives of This Portion of the Test

The goal of this portion of the SAP S/4HANA Sales Upskilling Certification exam is to test your understanding of the following topics:

- The intelligent enterprise
- SAP S/4HANA's built-in capabilities for the following technologies:
  - Machine learning
  - Blockchain
  - Big data
  - Data intelligence
  - Internet of Things (IoT)
  - Analytics
- Implementation methodologies for making the best use of these capabilities
- Understand real-world examples used in SAP S/4HANA, such as various next-generation SAP Fiori apps
- Understand the available add-ons that utilize these next-generation technologies

Note

The Next-Generation Technologies topic is not included in the SAP S/4HANA Sales exam but does comprise 8%–12% of the total SAP S/4HANA Sales Upskilling exam.

13.2 The Intelligent Enterprise

The *intelligent enterprise* is a phrase used a lot in advanced enterprise resource planning (ERP) circles, but what does this phrase actually mean? We'll answer that question in the following section, first by providing you with an overview of the intelligent enterprise and then by diving into its components.

13.2.1 Intelligent Enterprise Overview

To understand this concept, let's start with understanding the traditional role ERP has played over the years. Within this role, an ERP system is a vehicle for business data and business transactions—essentially no more than a system of record. In this approach, often complex models were built around an ERP system within data warehouses, data lakes, and other external applications, in order to try to extract value from this data. These external applications were typically spreadsheets and additional databases. This approach is shown in Figure 13.1, where the additional value extraction is represented as Excel spreadsheets, SQL, and further databases.

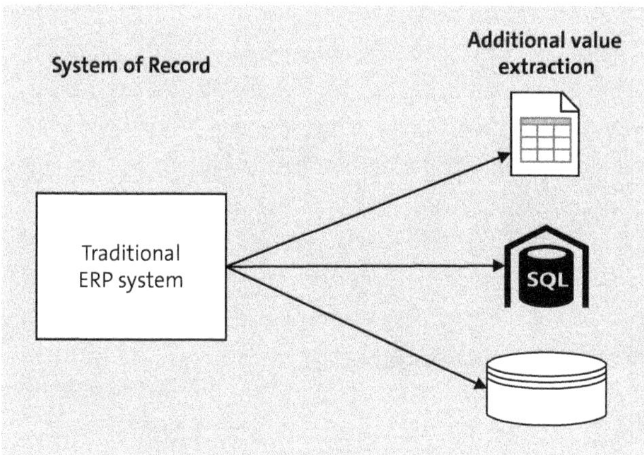

Figure 13.1 Traditional ERP Model

Compare this model with the intelligent enterprise model, where the full scope for additional value extraction is available in the same system. This important differential—with examples such as the SAP Fiori app Predicted Delivery Delay—offer next-generation technology capabilities in machine learning, embedded as standard within the SAP S/4HANA system without any need for additional interactions and without the need to build complex integrations to external applications. The intelligent enterprise model is shown in Figure 13.2. In this approach, real-time, live data is available within the intelligent enterprise and can be freely used by the system for intelligence capabilities such as embedded analytics, machine learning, blockchain, etc. This approach delivers complete visibility into the entire sales process from master data creation right through to cash collection, thus allowing real-time monitoring.

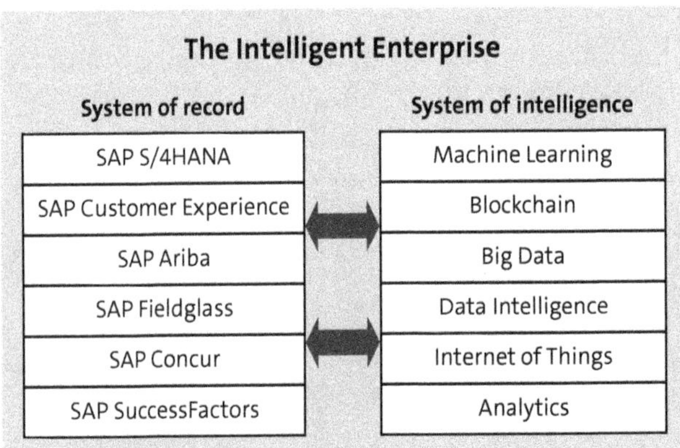

Figure 13.2 The Intelligent Enterprise

The technologies specified within the system of intelligence shown in Figure 13.2 are part of the suite of next-generation technologies, which are sometimes referred to as *intelligent technologies*, more recently branded as technologies within SAP BTP. Let's take some time to understand what is meant by this concept.

Intelligent technologies, as a concept, were introduced by SAP shortly after the advent of SAP S/4HANA and consist of a set of technologies as well as a methodology for enabling rapid digital transformation. At the time, these technologies were gathered under the now-retired SAP Leonardo product name, so you may find references to SAP Leonardo in some SAP literature. As far as the SAP S/4HANA Sales 2020 Upskilling Certification exam goes, these technologies will be listed as part of SAP BTP. SAP BTP encompasses not only the next-generation technologies dealt with in this chapter but also brings together these technologies alongside additional SAP capabilities, such as SAP BW/4HANA, SAP Analytics Cloud, and SAP Master Data Governance, as well as the SAP HANA database. SAP BTP should be viewed as an umbrella platform pulling together all the various cutting-edge threads of the SAP ecosystem.

As we focus on the next-generation technologies, note that the intelligent enterprise within SAP S/4HANA comprises of three components, as shown in Figure 13.3:

- Design thinking
- Next-generation technologies
- Implementation tools

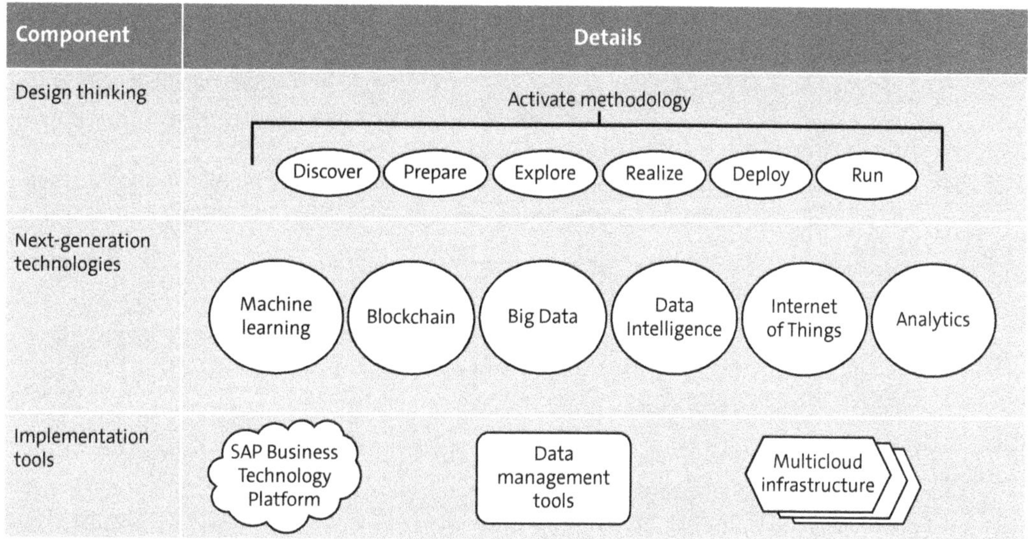

Figure 13.3 SAP Intelligent Enterprise Components

It's important to separate these components and to understand the differences between them.

13.2.2 Design Thinking

Design-led thinking is a component of the intelligent enterprise that focuses on the methodology for implementing and deploying SAP S/4HANA. Traditional SAP ERP 6.0 implementations relied on the Accelerated SAP (ASAP) methodology, which followed a waterfall approach to deployment, with distinct phases (blueprint, build, test, cutover, and go-live). The SAP Activate methodology replaces the ASAP method and is a more agile approach. We touched upon this topic in Chapter 1, Section 1.5, but let's conduct a deeper dive into the details now.

As shown in Figure 13.3, the SAP Activate methodology comprises six distinct phases. More detail about each of the phases is shown in Figure 13.4.

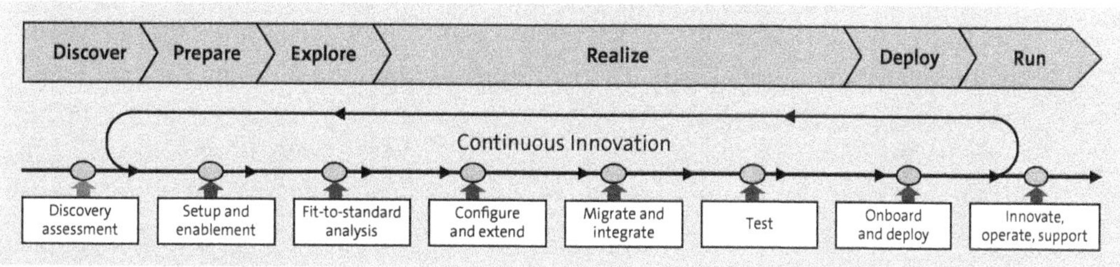

Figure 13.4 SAP Activate Phases

Three key pillars are involved in the SAP Activate methodology, all of which have been explored in Chapter 1:

- SAP Best Practices (Chapter 1, Section 1.4)
- Guided configuration
- SAP Activate methodology (Chapter 1, Section 1.5)

SAP Best Practices and guided configuration pillars can be grouped together as *industry accelerators*, allowing the rapid deployment of innovative solutions in weeks, for a fixed price for specific bundles.

The discover phase of SAP Activate can be considered a familiarization exercise. The project team will investigate the available SAP S/4HANA options and decide on the appropriate target architecture, for example, whether to deploy SAP S/4HANA Cloud or SAP S/4HANA. During this phase, trial editions of SAP S/4HANA can be utilized to help frame these decisions.

During the prepare phase, much of the project groundwork is undertaken. These steps include establishing the project executive sponsor, defining tracking and governance guidelines, and building up the project team. For an SAP S/4HANA project team to be successful, the following project roles and skillsets must be covered, at a minimum:

- Project manager
- SAP Basis and administration
- Security
- Finance
- Logistics
- Customer-vendor integration (CVI) expert
- Development
- Subject-matter experts from the business
- Testers
- SAP Fiori expert

The explore phase is where you'll start to see real tangible outcomes as solution definition begins to take shape. During this phase, the *fit-to-standard approach* is followed for business processes, whereby SAP Best Practices are explored with a view to fitting the best practice processes defined by SAP. The resulting fit-gap analysis provides a useful way for defining the scope of the project. Important additional tasks in this phase include the definition of a proof of concept, to be delivered to the business. Proof of concepts should be delivered as early as possible in the phase, so as to identify potential gaps in the process. As a result, early demonstrations are often provided as "low-fidelity" prototypes with only basic clickable content, which can be built in SAP-provided prototyping tools such as SAP Build. This prototyping tool allows you to mock up SAP Fiori screens according to business requirements to demonstrate the potential solution.

The realize phase is where the solution is built in SAP S/4HANA, based on the outcomes of the fit-to-standard and fit-gap exercises carried out in the explore phase.

The configuration is performed in short cycles, with regular feedback provided via solution walkthroughs with the business. This process is carried out in short, time-defined sprints. These sprints include predefined testing exercises, covering the following test types:

- Unit testing
- Functional testing
- Integration testing
- Regression testing
- Performance testing
- User-acceptance testing

The deploy phase is initiated when the build and test cycles are completed, and the project is entering final preparations for cutover and go-live. This phase is important in that the model for postproduction support is defined and agreed upon, the cutover plan is executed, and organizational readiness is established. During this phase, the SAP S/4HANA system goes live.

The run phase sees the solution established and stabilized and the handover to the standard production support model completed. A robust model for continuous improvement is established.

Each phase within the SAP Activate methodology has separate target outcomes, which should be met before progression to the next phase. These outcomes are shown in Table 13.1.

SAP Activate Phase	Target Outcomes
Discover	■ Business case definition ■ Target architecture decision
Prepare	■ Official project initiation ■ Project sponsorship established ■ Governance and tracking plan defined ■ Project plan definition ■ Project team members assigned
Explore	■ Fit-to-standard exercise ■ Fit-gap exercise ■ Master data load preparation completed ■ Proof of concepts delivered
Realize	■ Build and test signed off in defined sprints ■ Business validation ■ Data migration validation

Table 13.1 SAP Activate Phases and Target Outcomes

SAP Activate Phase	Target Outcomes
Deploy	▪ Postproduction support model established ▪ Cutover ▪ Go-live
Run	▪ Handover to production support ▪ Continuous monitoring established

Table 13.1 SAP Activate Phases and Target Outcomes (Cont.)

13.2.3 Implementation Tools and RISE with SAP

Before looking in detail at the technologies that make up the bulk of this chapter, let's first take a brief look at some of the tools used to implement the intelligent enterprise. To facilitate the accelerated implementation of SAP S/4HANA, we'll also explore the latest SAP program, announced in the early part of 2021: RISE with SAP.

SAP Business Technology Platform

As of early 2021, SAP decided to sunset the name *SAP Cloud Platform* in favor of *SAP Business Technology Platform (SAP BTP)*. Two of the most important offerings that run on SAP BTP are SAP Integration Suite and SAP Extension Suite, the names of which are a good indication of their capabilities.

SAP BTP is an open and extensible way to enable your digital transformation journey and is the foundational platform for the next-generation technologies described in this chapter. SAP BTP is provided by SAP as a platform as a service (PaaS) that allows you to develop your own capabilities in next-generation technologies using intuitive standard apps. Via your SAP BTP account, you can perform the following activities:

- Create simple and personalized UX apps via preconfigured templates.
- Integrate multiple applications with SAP S/4HANA. Imagine multiple, internet-enabled applications all connecting seamlessly with your ERP system via SAP BTP's ability to process vast volumes of real-time data concurrently.
- Real-time predictive analytics can be driven by applications and integrations defined in SAP BTP.

Data Management Tools

Data modeling tools such as SAP Data Intelligence offer comprehensive data management solutions, which can be pulled together from across your IT landscape. SAP Data Intelligence is the data orchestration layer within SAP BTP, and its goal is to bring together all your company's data into one place via the SAP Data Intelligence modeler. By bringing SAP data and non-SAP data together, as well as by providing standard integrations with other data management tools, the intent is to provide actionable business insights from across your entire IT landscape. In this way, SAP Data Intelligence can facilitate connection of IoT data as well as rapidly scalable machine learning.

Once your data is consolidated, this data can be transformed and augmented in order to create a searchable data catalog. Traditional extract, transform, and load (ETL) engines often operate by running regularly in batch mode. Even when ETL engines operate with real-time streaming, they cannot handle the huge volumes and constantly increasing number of systems that characterize a modern intelligent enterprise. SAP Data Intelligence is specifically built to handle big data in real time and at scale.

Multicloud Infrastructure

In the past, in the days of the SAP Cloud Platform, the platform was provided by the Neo environment exclusively. However, within the last year, SAP has moved to a more flexible approach, allowing the next-generation technologies available as part of the intelligent enterprise to be supported by any infrastructure provider you choose. In addition to the SAP-hosted cloud options, you can now also use hyperscalers like Microsoft Azure, Amazon Web Services (AWS), and Google Cloud Platform.

However, note that SAP BTP, Neo environment (note the updated name) will not be retired—no end-of-life process is planned. Instead, the environment is scheduled to be moved to a more modernized cloud infrastructure by the end of 2022.

RISE with SAP

The implementation of SAP S/4HANA can seem bewildering—a huge array of services and technologies are available as part of any implementation program. Where do you start?

RISE with SAP is the new business transformation program (business transformation as a service), aimed at making sense of all the components that make up the required technologies and processes required for a successful implementation project. The objective is to engage with SAP from an early stage, allowing you to transform your business at your own speed and on your own terms.

The following three steps are involved when using RISE with SAP:

1. **Business process redesign**
   SAP will analyze your business processes against industry benchmarks and with reference to the SAP Best Practices, which have been defined in conjunction with over 20,000 customers. The objective of this step is to tailor recommendations for your business, which may include automating business

processes through SAP Intelligent Robotic Process Automation (SAP Intelligent RPA).

2. **Technical migration to standard application landscape**
   As a result of these business process recommendations, modifications and customizations can be removed from the landscape in favor of a fit-to-standard approach. With this approach, the data layer can be harmonized, migration can be automated, and you can achieve near-zero downtime for your implementation.
3. **Cloud infrastructure**
   As a part of the RISE with SAP program, SAP offers maximum choice at a low cost. Any hyperscaler can be used to support the SAP S/4HANA deployment, with SAP defined as the single point of contact, regardless of who the hyperscaler is. In this way, RISE with SAP can commit to the following benefits:
   - 99.9% high availability
   - Up to 20% reduction in total cost of ownership (TCO)
   - 1800+ prebuilt integrations
   - 2100+ application programming interfaces (APIs) for open integration
   - 300+ artificial intelligence (AI) innovations
   - 1600+ apps available through the SAP App Center
   - 160+ standard connectors for third-party apps
   - Native embedding of Microsoft Teams in SAP S/4HANA

Additionally, organizations using RISE with SAP will receive free access to SAP Business Network.

Through these three structured and defined steps, the implementation of SAP S/4HANA using RISE with SAP can industrialize your project, ensuring the best use of the technologies available. Let's move on to explore those available technologies.

13.2.4 Intelligent Technologies

Now, we come to the bread and butter of this chapter—the next-generation technologies now available as a part of your SAP S/4HANA system. Note that these intelligent technologies are, in many cases, predefined and embedded in standard SAP applications such as SAP Fiori apps. As a result, these concepts are an important part of your understanding when implementing and recommending apps. Thus, we'll explore some examples later in this section.

First, however, let's define some of these next-generation technologies so we're comfortable with the concepts before proceeding.

Machine Learning

What is machine learning? In our context, machine learning is the ability of an application to teach itself to improve with each iteration and with each new data exposure. The objective of machine learning is to derive insights from large volumes of data to provide predictions about the future as well as lessons about the past.

Blockchain

A blockchain is a shared data ecosystem hosted on a distributed ledger with information exchanged via a peer-to-peer network. While this definition might be problematic, an example of blockchain technology can help for this context. The original and most famous use for a blockchain is the cryptocurrency Bitcoin. Each transaction carried out using Bitcoins is shared across a wide network of computers. The transaction is made up of blocks of data, with access to add or new block or to change only the previous block—thus, a chain of blocks or a "blockchain." Each computer has access to its own identical copy of the entire transactional record, and each addition to the record must be approved by the entire ledger. As a result, in practice, the transaction can be completed securely and without concern for hacking via the technology itself—no mediator or centralized organization is involved in these transactions.

Big Data

Modern organizational structures and technology feed off the lifeblood of data. What's new with big data are the three Vs of big data: volume, velocity, and variety. Imagine a sales transaction from start to finish. The data involved in this process can be complex and vast and may include the following kinds of data:

- Customer master data
- Lead transactional data
- Material master data
- Sales pricing master data
- Supply chain data, such as vendor, purchase price, delivery information, manufacturing lead times, bills of materials (BOMs), routings, etc.
- Inquiry transactional data
- Quotation transactional data
- Sales order transactional data
- Delivery transactional data, including warehouse data, manufacturing data, and transportation data
- Billing transactional data
- Accounting transactional data

Intelligent enterprises must have a strategy for dealing with this volume of data, which is expanding at a rapid pace across multiple disciplines. Furthermore, an additional challenge is connecting this data together, so the data collection can be viewed holistically rather than in functional silos.

Data Intelligence

The staggering amount of data collected by organizations requires well-defined and robust data analysis tools. Data intelligence is the analysis of all this big data into models for deriving insights and making business-driven decisions or, increasingly, for automating decisions.

Internet of Things

In our increasingly connected world, many devices are connected over the internet. These devices comprise IoT. A good example is a temperature control monitor placed in a temperature-sensitive parcel. This monitor would register the temperature inside the parcel and advise when and if the temperature reaches a point where the goods within become damaged or unusable.

Analytics

Analytics form the lifeblood of SAP S/4HANA. Real-time analytics, provided as an embedded functionality within the core SAP S/4HANA system, flows through every functional business process. In the past, analytics was often the final step in a business process—essentially not much more than reporting. However, with SAP S/4HANA, embedded analytics can be the starting point to launch a process, serve as a checkpoint at any stage, act as an early warning system, or complete a final reporting step. With this definition in mind, SAP has designed SAP Fiori apps that place analytics front and center of every process.

Now that we've explored the definitions of these crucial next-generation technologies, let's delve into their application in SAP S/4HANA with a focus on sales and distribution. Much of the certification exam will focus on specific examples within SAP S/4HANA applications, and therefore, we'll go through these examples as use cases.

13.3 Machine Learning: Predicted Delivery Delay App

SAP Fiori has a number of native machine learning apps that can be deployed in your SAP S/4HANA system to great effect. A useful example of this kind of capability is the SAP Fiori app Predicted Delivery Delay. This app can predict a potential delivery delay for a given specific customer or material, based on data input into the predictive model.

Once the SAP Fiori app is first opened, open sales document items are displayed in a graphical format, as shown in Figure 13.5.

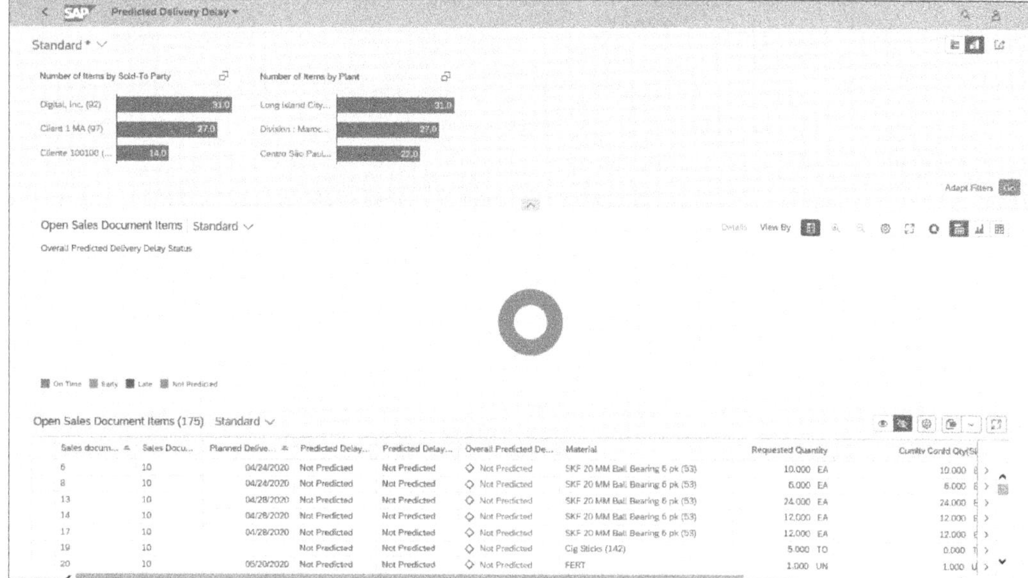

Figure 13.5 Predicted Delivery Delay in SAP Fiori

On this screen, the **Not Predicted** total encompasses all the open sales order items—this result is because the predictive model has not yet been established. The predictive model can be trained for this app using the SAP Fiori app Predictive Models. Prebuilt scenarios for predicted delivery delays are already available for activation. Once the predictive model is established, the statuses will resemble the screen shown in Figure 13.6.

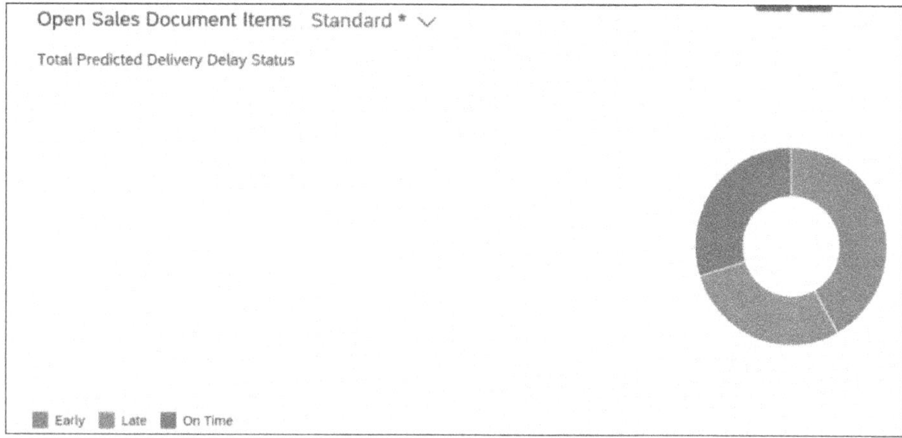

Figure 13.6 Predicted Delivery Delay Statuses

The algorithm behind the app uses previous data to build up a picture and predict whether sales order items will be late in two areas:

- Creation of the delivery
- Processing of the delivery

The previous data is used for the material and customer according to the rules set out in the predictive model. For example, if a sales order item has two deliveries against it, one of which has an actual goods movement date 5 days later than the planned goods issue date, and the other with an actual goods movement date 6 days later than the planned goods issue date, then the predicted processing delay will be 6 days (the highest of the two previous deliveries).

Table 13.2 shows how predicted delivery delays are determined based upon input data of the individual sales order items.

Sales Order Number	Predicted Delay (Days) in Delivery Creation	Predicted Delay (Days) in Delivery Processing	Overall Predicted Delivery Delay Status
1	1 (late)	–1 (early)	**Late**
2	–1 (early)	0 (on time)	**Early**
3	0	Not predicted	**On Time**
4	Not predicted	Not predicted	**Not Predicted**

Table 13.2 Predicted Delivery Delay Determination of Status

As usual with SAP Fiori apps, you can adapt the filters to suit your own required input parameters and save these parameters as a default if needed. This step is strongly recommended when using this app since the algorithm used to predict the delivery delay can cause potential performance issues if applied to all open sales order items in the system. At a minimum, filter requirements should involve a data range for the creation date and/or requested delivery date.

This app can show the predicted delivery delays for sales order items, with drill-down capabilities available to the sold-to party, the material master, and the sales document. By clicking on the predicted delivery delay status, you can drill down into items that are on time, early, late, or not predicted.

This app learns from previous delivery delays to inform you of potential future delays and thus allow you to take proactive action appropriately. This action could be to inform your customer about the delay as well as to intervene in the supply chain to rectify the situation.

13.4 Big Data: Sales Monitoring and Analytics

Across the entire landscape of functions for sales and distribution, a vast amount of data is available. A key challenge for users of modern ERP systems is how to make best use of that data to provide insights and value for your business. SAP S/4HANA provides a real helping hand with a whole suite of analytics apps in SAP Fiori, all of which harnesses the power of the SAP HANA database to make use of this big data.

All core business processes are covered by the sales monitoring and analytics processes, from inquiries right through to invoices. These monitoring apps are designed to help with the following activities:

- Gain a full understanding of your sales data in dashboard format
- Use key performance indicators (KPIs) and flexible drilldown capabilities to gain insights into sales and distribution processes
- Manage sales plans and sales performance monitoring
- Provide decision proposals based upon analytical insights
- Integrates seamlessly with other transactional processes, including with collaboration tools such as SAP CoPilot

Although not an exhaustive list, some key SAP Fiori apps that are a part of the suite of apps for sales monitoring and analytics include the following:

- **My Sales Overview app**
  This extremely powerful analytical app, shown in Figure 13.7, combines multiple other analytical apps into one dashboard with *cards* that display the data from the embedded apps. As is normal with SAP Fiori apps, adaptable filters are available, and each dataset has drilldown capabilities to open up more detailed views. Figure 13.7 shows further details.

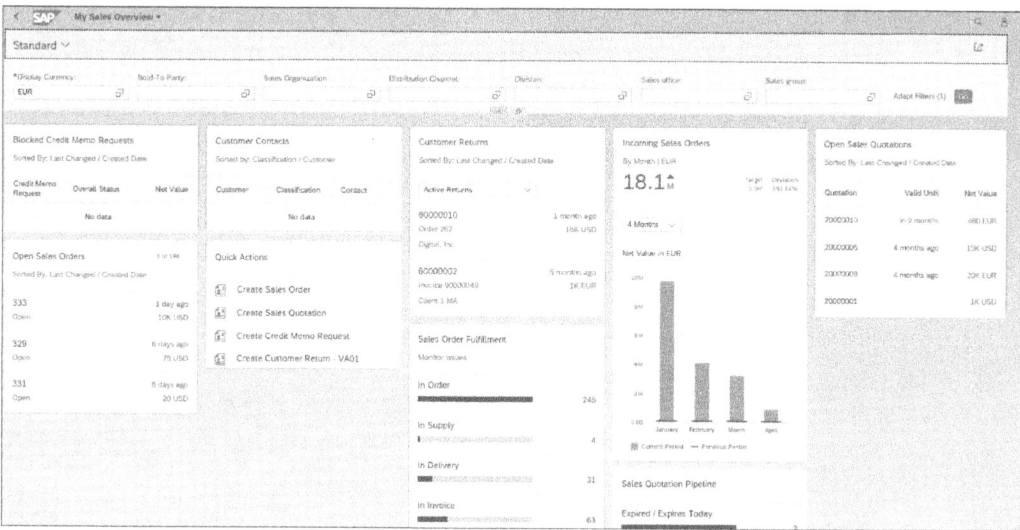

Figure 13.7 SAP Fiori My Sales Overview App

- **Customers Overview app**
  This app is available starting with SAP S/4HANA 2020 and is similar to the My Sales Overview app, in that it displays a list of cards within one app, but this time for a given customer.
- **Sales Management Overview app**
  This app is designed for executive-level analysis of the sales picture in your business. This view is provided by showing graphical overviews of data as cards, including incoming sales orders, blocked orders, and backorders. Again, full drilldown capabilities are included, as shown in Figure 13.8.

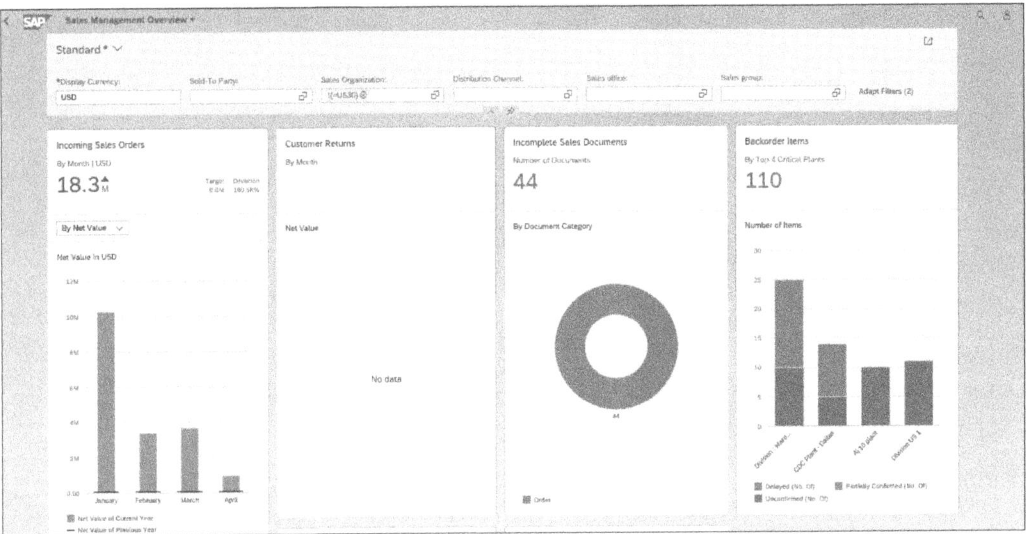

Figure 13.8 SAP Fiori Sales Management Overview App

- **Manage Sales Plans app**
  This app allows your sales teams to create sales plans for themselves whereby they can track performance over time. The creation of a sales plan could not be easier—just create your parameters in the app and then download the template to Microsoft Excel. Then, fill in the details of the plan and then upload the file back to the app, as shown in detail in Figure 13.9.

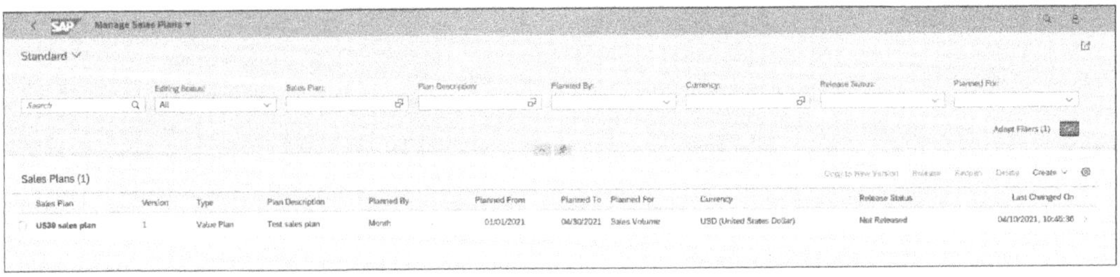

Figure 13.9 SAP Fiori Manage Sales Plans App

- **Sales Performance – Plan/Actual app**
  Once your sales plan is uploaded and released, this app can monitor the actual performance of your sales team versus the planned performance, in graphical format, as shown in Figure 13.10.

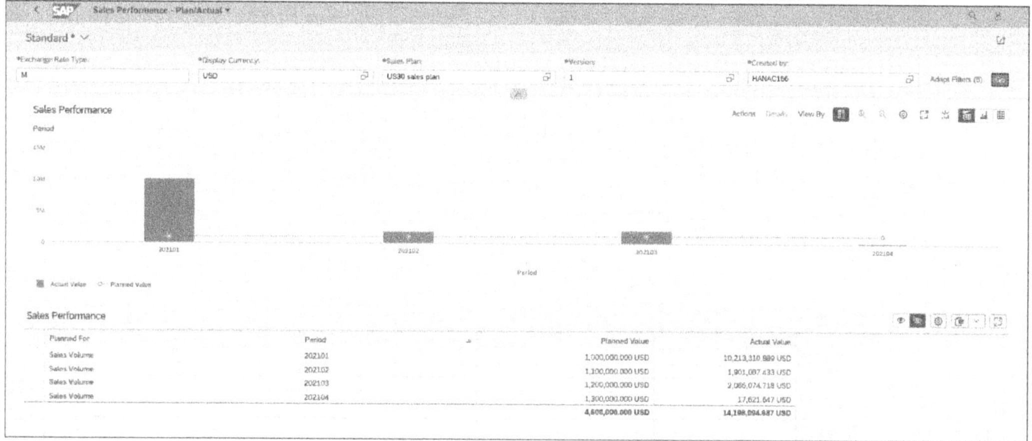

Figure 13.10 SAP Fiori Sales Performance App

- **Sales Quotations app**
  This app is a good example of an embedded analytics app using the power of the SAP HANA database to provide real-time analysis. The data is available as raw data or graphical data, with different types of graphs available. Additional drill-downs can be added, and the data in the app can be sliced and diced whichever way suits the user, as shown in Figure 13.11.

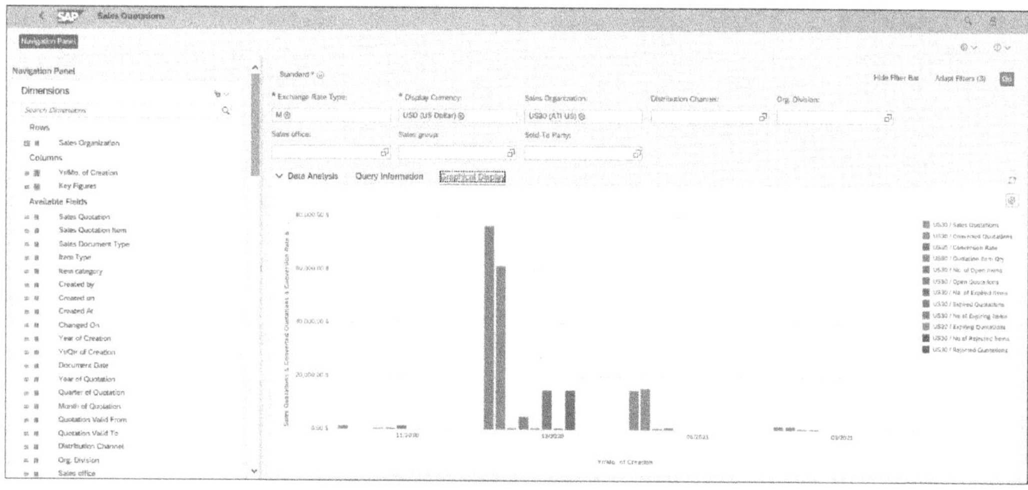

Figure 13.11 SAP Fiori Sales Quotations App

- **Incoming Sales Orders – Flexible Analysis app**
  This app is similar to the Sales Quotations app, with fully flexible analysis available for sales orders.
- **Sales Order Fulfillment – Analyze and Resolve Issues app**
  The intention of this app is to allow you to analyze and remove any impediments to the progression of a sales order through the entire process, as shown in Figure 13.12. This important SAP Fiori app utilizes SAP Smart Business technology (see Chapter 1, Section 1.7, for more details on SAP Smart Business). This app can be used to solve problems in the fulfillment process, with full drilldown capabilities into other analytical and factsheet apps, as well as the ability to show issues using different status colors.

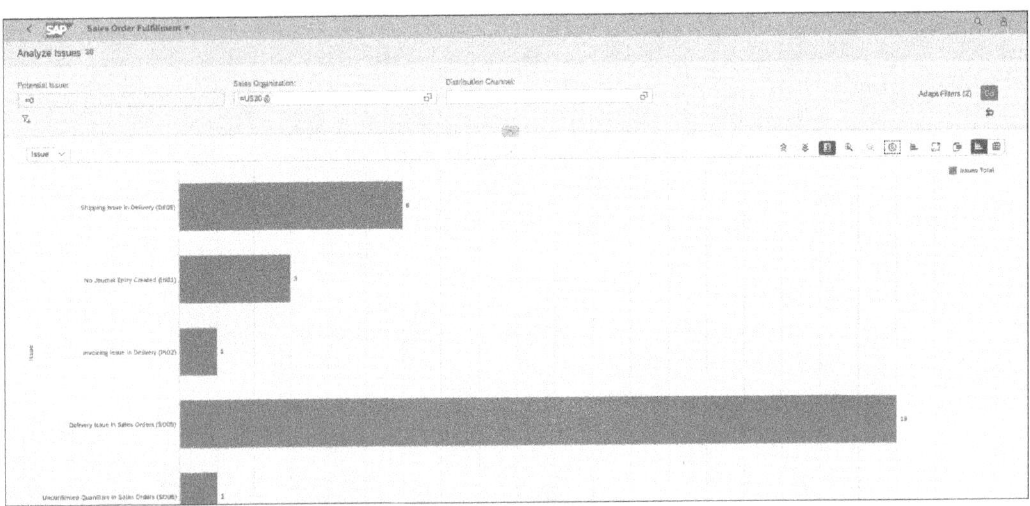

Figure 13.12 SAP Fiori Sales Order Fulfillment App

- **Track Sales Orders app**
  This app allows instant checking against the sales order that may be behind in terms of delivery creation, processing, invoice creation, or even payment. Fast actions are available for each sales order depending upon its status—for example, if a delivery block is set in the sales order, the fast action **Remove Delivery Block** will be available when that order is selected on the screen, as shown in detail in Figure 13.13.
- **Analyze Delivery Performance app**
  This app allows you to monitor the performance of your outbound logistics department. Again, filters and drilldowns are offered as standard, as shown in Figure 13.14.

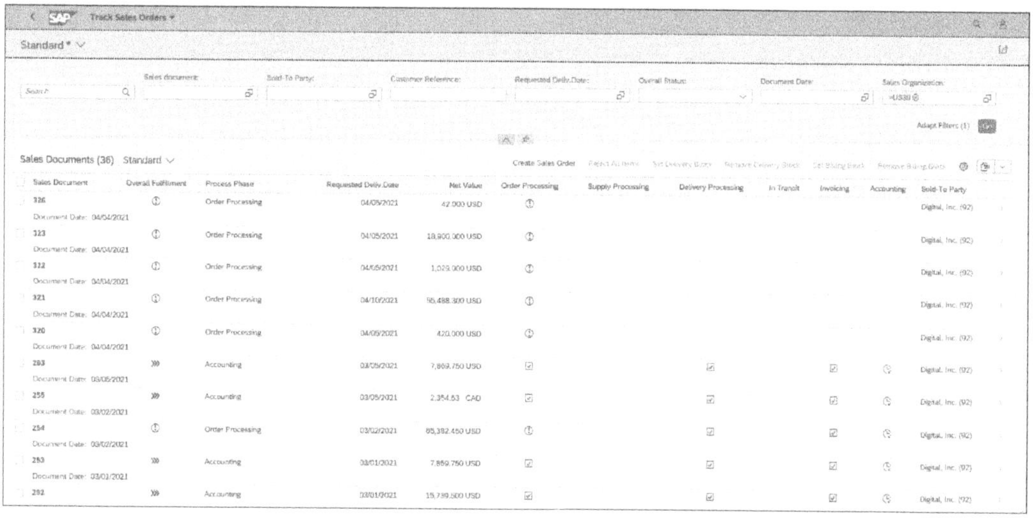

Figure 13.13 SAP Fiori Track Sales Orders App

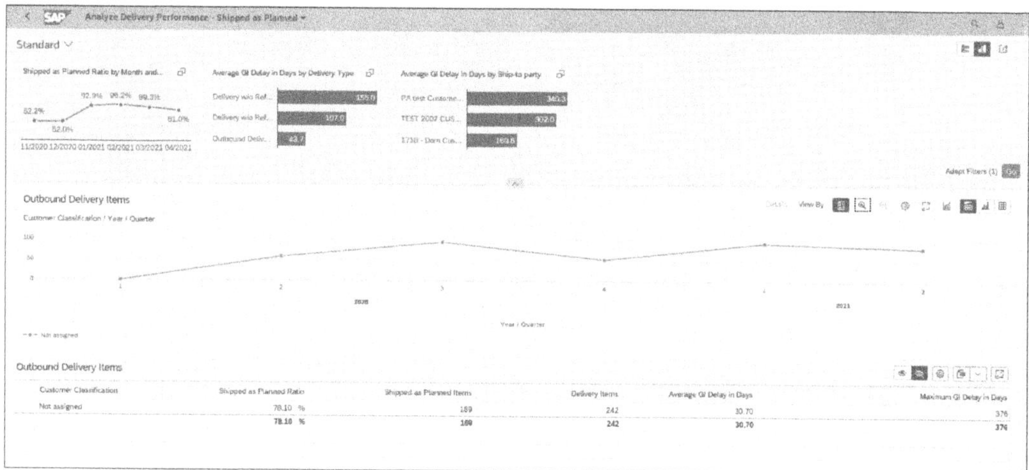

Figure 13.14 SAP Fiori Analyze Delivery Performance App

13.5 Intelligent Technology Add-Ons

As part of the SAP suite of next-generation technologies, you should be aware of many solutions available for use with SAP S/4HANA Sales. An overview of the following technologies is likely to be required for the certification exam:

- SAP CoPilot
- SAP Advanced Track and Trace for Pharmaceuticals
- SAP S/4HANA Cloud for customer payments
- SAP digital payments add-on

13.5.1 SAP CoPilot

SAP CoPilot is your digital assistant. Imagine you've logged on to your bank's website or your insurance company's website. Often, chat bots will be available, which can be used to answer questions in an automated fashion. SAP CoPilot is your entry point into this technology. This solution allows you to interact verbally or in written format, with SAP S/4HANA and display, amend, or create objects such as materials, sales order, invoices, etc.

In the same way that your bank's or insurance company's website uses chat bots to interact with its backend systems, SAP CoPilot can be used for the same purpose (i.e., interactions between external channels and platforms and SAP S/4HANA). SAP CoPilot can also be used to chat with other users, including sharing SAP documents and notes.

SAP CoPilot is available for SAP S/4HANA Cloud and SAP S/4HANA, but requires an SAP BTP account. This solution is available on multiple device types, as shown in Figure 13.15.

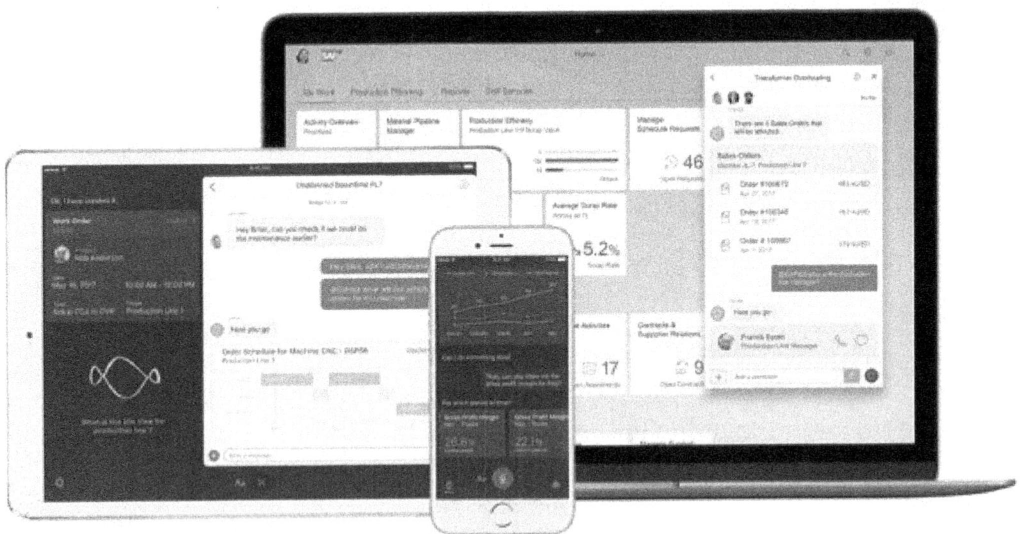

Figure 13.15 SAP CoPilot

SAP CoPilot is a bot integration hub, and as such, you can build your own bots throughout your company's entire technology landscape and connect them together via SAP CoPilot into a seamless digital assistant experience. Importantly, SAP CoPilot is an open and extensible application—in other words, new functionality can be added easily. To do this, SAP CoPilot uses the following technologies:

- Bots: Software applications used to conduct an online chat conversation via text or text-to-speech

- Agents: Assistance systems that use artificial intelligence to support you.
- APIs: Intermediary programs that enable two separate applications to talk to each other.

By using these capabilities, SAP CoPilot can be used with SAP and non-SAP applications.

13.5.2 SAP Advanced Track and Trace for Pharmaceuticals

SAP Advanced Track and Trace for Pharmaceuticals is a good example of SAP using blockchain technology to meet specific business goals. This technology add-on can verify the authenticity of products via serial numbers and batches. Each package in the production line is printed with a barcode and/or a QR code, which is passed through the SAP Advanced Track and Trace for Pharmaceuticals system into the blockchain. As a result, everyone in the supply chain (from the manufacturer, through the warehouse, shipper, stocker, and customer) can verify the authenticity of the package. This verification is performed by scanning the QR code or barcode on the package, which then interfaces with SAP Advanced Track and Trace for Pharmaceuticals and with the blockchain via SAP Integration Suite.

13.5.3 SAP S/4HANA Cloud for Customer Payments

SAP S/4HANA Cloud for customer payments is an online customer portal, where customers can perform the following activities:

- View the status of their account
- Review and download the billing information on their account, such as invoices, credit memos, etc.
- Make payments using credit cards and debit cards or online banking

SAP S/4HANA Cloud for customer payments is additional functionality which is available for SAP S/4HANA Cloud, but since this solution touches SAP S/4HANA Sales invoices, it is worth mentioning in this context.

13.5.4 SAP Digital Payments Add-On

The SAP digital payments add-on is part of the SAP BTP suite of products and allows you to connect SAP and non-SAP applications to external payment service providers (PSPs). By using the SAP Integration Suite, ERP systems such as SAP S/4HANA can be connected to these PSPs to facilitate secure, end-to-end processing of digital payments, including credit card payments, in real time. The payment is initiated in the backend system (either SAP or non-SAP) and is then passed through the SAP digital payments add-on to the relevant PSP in a secure fashion.

13.6 Important Terminology

In this chapter, the following terminology was used:

- **Intelligent technologies**
  Intelligent technologies is the umbrella term for defining a suite of technologies and innovation methodologies that make up SAP's intelligent enterprise. The term has been used more recently to replace the retired SAP Leonardo brand.
- **Intelligent enterprise**
  SAP's intelligent enterprise can be characterized as the ability of users to extract maximum value from their systems with live data, embedded analytics, and real-time monitoring, thus offering full visibility into core processes. The intelligent enterprise is achieved through interactions between next-generation technologies and innovative processes.
- **SAP Activate methodology**
  The SAP Activate methodology is the recommended approach for implementing SAP, along with agile guidelines. This methodology is designed to supersede the Accelerated SAP (ASAP) waterfall methodology.
- **Machine learning**
  Machine learning is the ability of an application to improve based upon previous data. The goal is to provide insight to action and make predictions.
- **Blockchain**
  Blockchain is a secure, tamper-proof distributed ledger, where data is passed throughout the ecosystem via a peer-to-peer network.
- **Big data**
  Big data is characterized by vast volumes of master and transactional data across various modules, which interact with each other.
- **Data intelligence**
  Data intelligence is the effective analysis of huge volumes of data into models for deriving insights and predictions.
- **Internet of Things (IoT)**
  IoT refers to the modern business landscape where multiple applications and devices are internet-enabled and can be connected together.
- **Analytics**
  This functionality is used for fast, real-time analytical model construction in SAP S/4HANA. This feature runs exclusively on the SAP HANA database and normally uses SAP Fiori as its frontend.
- **SAP CoPilot**
  SAP CoPilot is the digital assistant for the SAP ecosystem, providing chat bot functionality as well as connecting users.
- **SAP Business Technology Platform (SAP BTP)**
  SAP BTP is a platform as a service (PaaS) provided by SAP that can be used to integrate your next-generation technologies with each other and with external

applications. Additionally, SAP BTP can be used to extend your next-generation technology capabilities to provide a more effective UX.

- **Design thinking**
  Design thinking is the approach to innovation employed by SAP, including fast and repeated prototyping.
- **RISE with SAP**
  The new implementation program developed by SAP to industrialize and tailor your SAP S/4HANA implementation to your specific business goals.
- **SAP Advanced Track and Trace for Pharmaceuticals**
  A blockchain-enabled add-on technology that can be used to guarantee the authenticity of pharmaceutical drugs and supplies.
- **SAP S/4HANA Cloud for customer payments**
  SAP S/4HANA Cloud for customer payments is an online customer portal where customers can manage their financial settings.
- **SAP digital payments add-on**
  SAP digital payments add-on is an add-on technology that can be used to deliver secure payments to payment service providers (PSPs).

13.7 Practice Questions

These practice questions will help you evaluate your understanding of the topics covered in this chapter. The questions shown are similar in nature to those found on the certification examination. Although none of these questions will be found on the exam itself, they will allow you to review your knowledge of the subject. Select the correct answers and then check the completeness of your answers in the next section. Remember that, on the exam, you must select all correct answers and only correct answers to receive credit for the question.

1. What is meant by the term "SAP Business Technology Platform (SAP BTP)"?

☐ A. A suite of technologies used for enabling rapid digital transformation

☐ B. The SAP security solution employed within the SAP Fiori launchpad

☐ C. An SAP user forum for discussing next-generation technologies

2. Which of the following skills must be covered as part of an SAP S/4HANA implementation using the SAP Activate methodology? (There are three correct answers.)

☐ A. SAP Ariba expert

☐ B. SAP Basis and administration

☐ C. Functional specification expert

- ☐ D. Subject matter experts from the business
- ☐ E. CVI expert

3. The fit-to-standard approach is carried out in which phase of an SAP Activate implementation?

- ☐ A. Discover
- ☐ B. Prepare
- ☐ C. Explore

4. Which of the following are objectives of the run phase of an SAP Activate implementation? (There are two correct answers.)

- ☐ A. Cutover
- ☐ B. End-user training
- ☐ C. Handover to production support
- ☐ D. Continuous monitoring

5. SAP Integration Suite and SAP Extension Suite can be used for which of the following activities? (There are two correct answers.)

- ☐ A. Utilizing preconfigured templates to create UX apps
- ☐ B. Separating out transactional processes from analytical processes
- ☐ C. Analyzing whether to store SAP Fiori apps in the cloud or on-premise
- ☐ D. Integrating multiple external applications with your SAP S/4HANA system

6. Which of the following steps are involved with RISE with SAP for rapid tailored SAP S/4HANA implementations? (There are three correct answers.)

- ☐ A. Business process redesign
- ☐ B. Technical migration to standard application landscape
- ☐ C. Cloud infrastructure
- ☐ D. Fit-gap analysis
- ☐ E. Monitoring using SAP Solution Manager

7. Which intelligent enterprise technology is utilized in the Predicted Delivery Delay app?

- ☐ A. Blockchain
- ☐ B. Machine learning
- ☐ C. IoT

8. In the Predicted Delivery Delay app, what must be done before the app can be used?

☐ A. Standard function modules must be activated.

☐ B. The predictive model must be trained.

☐ C. Machine learning capabilities must be installed from sap.com.

9. Which of the following are true of the My Sales Overview app? (There are two correct answers.)

☐ A. Multiple cards are displayed with drilldowns into further apps.

☐ B. The app shows links to a suite of factsheet apps.

☐ C. Filters can be set to be applied automatically to the results in the app.

☐ D. The data in the app is refreshed overnight by batch jobs.

10. Which SAP Fiori app can be used to report against your sales team's performance?

☐ A. Sales Performance – Plan/Actual

☐ B. Monitor Sales Teams

☐ C. Sales Plans – Flexible Analysis

11. What is SAP CoPilot?

☐ A. A predictive tool to prompt you with tasks to carry out when you first log on

☐ B. A digital assistant used for conversational AI and as a chat bot

☐ C. An instant messaging tool

12. Which of the following SAP add-ons uses blockchain technology?

☐ A. SAP S/4HANA Cloud for customer payments

☐ B. SAP digital payments add-on

☐ C. SAP Advanced Track and Trace for Pharmaceuticals

13.8 Practice Question Answers and Explanations

1. Correct answer: **A**
   SAP BTP is the name employed by SAP to group together technologies used in innovation practices in SAP S/4HANA.
2. Correct answers: **B, D, and E**
   Several key skills are essential for an SAP Activate implementation. From the list of answers given, only the SAP Basis/administration, business subject matter experts, and the CVI expert are essential. In an SAP implementation with

SAP Ariba as the key supply chain capability, an SAP Ariba expert may be deemed essential, but this skill is not always required. Experts in all the functional areas would be need to be functional specification experts, so this expertise is not a requirement in its own right.

3. Correct answer: **C**

   The fit-to-standard approach is a key element of the explore phase of the SAP Activate methodology. During this phase, processes are examined to see whether they can be reengineered to fit to the standard SAP Best Practices model.

4. Correct answers: **C and D**

   The run phase is initiated after go-live, and as such, cutover activities and end-user training would already have been completed in previous phases. Key objectives of the run phase are handover to production support and continuous monitoring.

5. Correct answers: **A and D**

   SAP Integration Suite and SAP Extension Suite are the tools used for integrating with other applications as well as for extending existing apps and creating new apps. The creation of apps can be carried out using preexisting templates.

6. Correct answers: **A, B, and C**

   RISE with SAP is founded on the principles of analyzing and redesigning existing processes with a view to fitting to the standard best practice approach, carrying out a technical migration to a standard landscape, and finally allowing the flexibility to choose from a multicloud infrastructure.

7. Correct answer: **B**

   The Predicted Delivery Delay app uses machine learning technology to learn from previous deliveries in order to predict future delays.

8. Correct answer: **B**

   The Predicted Delivery Delay app is available for use out of the box, but the app must first be trained in order to be predictive. Predictive models are provided as standard.

9. Correct answers: **A and C**

   The My Sales Overview app displays key data from other analytical apps as cards in a dashboard format, with drilldowns into these additional apps. The data shown is real-time data (as with all embedded analytics in SAP S/4HANA), and filters can be applied automatically by users.

10. Correct answer: **A**

    The sales team performance is monitored against the sales plans you set up. This monitoring is performed via the Sales Performance – Plan/Actual app.

11. Correct answer: **B**

    SAP CoPilot is a digital assistant used as a chat bot as well as for conversational AI. Be careful with your answer to this question since SAP CoPilot can also be used as a tool to prompt you to carry out tasks when you log on and can also be used as an instant messaging tool with other users. However, these two options are not a good enough definition to explain what SAP CoPilot is. This question may seem unfair, but this style of question is common in SAP certification exams.

12. Correct answers: **C**
    SAP Advanced Track and Trace for Pharmaceuticals uses blockchain technology to ensure the authenticity of supplied products.

13.9 Takeaway

In this chapter, we introduced you to innovative, next-generation technologies and implementation technologies that together to make up the intelligent enterprise. You're now familiar with SAP BTP and can explain how next-generation technologies fit into that approach. The SAP Activate methodology is now clear from the content in this chapter, building upon the content explained in Chapter 1, Section 1.5.

You're familiar with the different types of technologies deployed within the intelligent enterprise and can also point to several examples used within the SAP Fiori UX.

Additionally, you can now identify next-generation add-ons and the underlying technologies they use.

13.10 Summary

A lot of the topics covered in this chapter are conceptual in nature. Gaining a good understanding of these capabilities is vital to becoming a truly rounded SAP S/4HANA consultant. We are confident that the technologies described within this chapter are likely to dominate the innovation approach of the SAP landscape in the years to come. As a result, the certification exam will test your understanding of the future of SAP S/4HANA Sales.

Do not be put off by the conceptual nature of these subjects and do not finish this chapter wondering how these technologies can be deployed. The path forward for SAP is to provide a framework for deployment for you. For instance, machine learning is already embedded in the Predicted Delivery Delay app; blockchain is embedded in SAP Advanced Track and Trace for Pharmaceuticals. What is key for you to understand is *how* these technologies work, so that these embedded apps and add-ons are not a mystery to you.

This chapter closes the overall certification guide. Together with your own research, education, and experience, this book represents the final piece of the puzzle for passing the SAP S/4HANA Sales 2020 Certification exam and the SAP

S/4HANA Sales 2020 Upskilling Certification exam. After reading through the practice questions in this book, if you now feel comfortable that you understand and can answer the questions correctly and confidently, then you're ready to test your newfound knowledge in a certification exam. We wish you the best of luck in your endeavors.

However, if you are successful in your certification efforts, please don't let this book gather dust on your bookshelf, never to be opened again. Hopefully, the content in this book can serve as a useful reference guide in your blossoming SAP consultancy career for years to come.

The Authors

Darío Franco is an experienced SAP S/4HANA business analyst who has more than 13 years of functional and technical SAP experience. His background is in sales and distribution, but he has practical knowledge across a wide range of functions such as order-to-cash, materials management, logistics execution, procure-to-pay, service management, warehouse management, ABAP programming, and EDI. His focus is on using SAP S/4HANA to create robust solutions that deliver real results to businesses.

Jon Simmonds is an experienced IT architect who has specialized in SAP systems for the past 19 years. His background is in sales and distribution, but he has practical knowledge across a wide range of functions such as order-to-cash, materials management, procure-to-pay, SAP Fiori, ABAP programming, and EDI. His focus is on enabling innovation and cost savings through SAP S/4HANA.

Index

A

B

D

J

K

L

M

N

O

Q

R

S

T

U

V

W

Y

- Set up your sales processes: sales order management, shipping and delivery, billing and invoicing, and more
- Configure your SAP S/4HANA system with step-by-step instructions
- Explore key sales reports and analytics

Christian van Helfteren

Configuring Sales in SAP S/4HANA

SAP S/4HANA Sales is here! Business partners, the material master, and critical sales workflows all require careful configuration—this guide has the expertise you need. Learn about key business processes for sales order management, billing and invoicing, available-to-promise, and more. From setup and configuration to your reporting options, this book has you covered!

766 pages, pub. 12/2019
E-Book: $79.99 | **Print:** $89.95 | **Bundle:** $99.99
www.sap-press.com/4907